Advances in Plant Transgenics: Methods and Applications

Ramalingam Sathishkumar
Sarma Rajeev Kumar
Jagadeesan Hema • Venkidasamy Baskar
Editors

Advances in Plant Transgenics: Methods and Applications

Springer

Editors
Ramalingam Sathishkumar
Plant Genetic Engineering Laboratory,
Department of Biotechnology
Bharathiar University
Coimbatore, Tamil Nadu, India

Jagadeesan Hema
Department of Biotechnology
PSG College of Technology
Coimbatore, Tamil Nadu, India

Sarma Rajeev Kumar
String Bio Private Ltd, IBAB Campus
Bangalore, India

Plant Genetic Engineering Laboratory,
Department of Biotechnology
Bharathiar University
Coimbatore, Tamil Nadu, India

Venkidasamy Baskar
Plant Genetic Engineering Laboratory,
Department of Biotechnology
Bharathiar University
Coimbatore, Tamil Nadu, India

ISBN 978-981-13-9623-6 ISBN 978-981-13-9624-3 (eBook)
https://doi.org/10.1007/978-981-13-9624-3

This Springer imprint is published by the registered company Springer Nature Singapore Pte Ltd.
The registered company address is: 152 Beach Road, #21-01/04 Gateway East, Singapore 189721, Singapore

Preface

The explosive human population around the globe leads to hunger and poverty, especially in developing countries. Moreover, the reduction in land area for crop cultivation due to urbanization and land degradation leads to starvation particularly in economically backward countries. In the resource-limited setting, malnutrition (deficiencies in micronutrients such as iron, zinc, and vitamin A) and food insecurity are the major risk factors that should be addressed. Therefore, it's important to enhance the food production with high nutritive value to resolve hunger and malnutrition. In addition, the climatic and various biotic factors, namely, bacteria, fungi, virus, insects, and herbivores, have severely affected the quality and quantity of crop production throughout the world. Bacterial and fungal pathogens decreased the crop yield by about 15%, and viruses reduce yields by 3%. These issues should be addressed with the advent of modern science, viz., plant genetic engineering. Transgenic plants are produced through the insertion of foreign gene or deletion of undesirable genes with the help of genetic engineering. The advancement in genetic engineering and plant molecular biology results in the development of a wide variety of transgenic plants with important agronomic traits, namely, biotic and abiotic stress tolerance in both the mono- and dicotyledonous plants. Despite being used in the Agri-sector, transgenic plants could also serve as a bioreactor for pharmaceutically important protein production.

Transgenic methods have advantages over conventional breeding methods in terms of addition, deletion, or modification of the gene or fine-tuning the gene of interest with reduced undesired changes: they permit interchange of genetic material across species and allow the augmentation of new genes into vegetative propagated crops such as potato, cassava, banana, etc. Moreover, it reduces the time required to release the product to market. Major areas have been used for the cultivation of transgenic soybean, tobacco, cotton, maize, and potato annually in different countries such as the USA, Canada, China, and Argentina. The adoption of transgenic plants in modern agriculture potentially decreased the usage of pesticides, enhanced crop yields, and improved farm profitability.

Moreover, modern genome editing tool offers more advantages than conventional transgenic approaches by permitting the changes to the endogenous plant's DNA including addition, deletion, and replacement of DNA of different lengths at targeted sites. In countries such as the USA, Brazil, and Argentina, the genome-engineered plants, which do not contain foreign DNA, will not be subjected to additional regulatory measures.

This book encompasses the most recent advances utilized for the production of transgenic plants and their potential applications in crop improvement and transgenic plants as bioreactors for the production of pharmaceutically important products as well. The volume is divided into three sections: Part I includes methods and technologies (related to transgenics), Part II contains the applications of genetic improvements of plants, and Part III explores the production of plant-made pharmaceuticals and other products.

Methods and Technologies

In order to develop a transgenic plant, the parameters include generation of novel expression vectors, improved genetic transformation methods, transgene integration, inheritance of transgenes, and screening of transgenics, which are required to be carefully considered to assure the success of the event. Therefore, this section deals with the plant tissue culture, cell culture, plastome engineering, and hairy root system for valuable material production. The latest information on the generation of new binary vectors for the plant genetic engineering and marker-free transgenic plant production are also discussed. In addition, the high-throughput screening methods for the transgenic plants have been described.

Application of Genetic Improvements

The conventional transgenic and genome editing methods were routinely employed in the development of transgenic plants with the desired agronomic traits such as high nutritive values and biotic and abiotic stress tolerance. This second section of the book deals with crop improvement such as horticulture crop production, generation of virus resistance, and stress tolerance of transgenic plants.

Production of Plant-Made Pharmaceuticals and Other Products

Transgenic plants to produce pharmaceutically useful products might represent the foundations of a new pharming industry. The low-cost production, rapid expandability, lack of human pathogens, and the capability to fold and assemble complex proteins properly that facilitates the plant system are preferable than animal- and microbial-based bioreactors. Hence, this section aims to provide the overview on the production of therapeutics in plant systems and expression strategies. Moreover, the key challenges associated with the products developed from transgenic plants were discussed. In addition, transcriptional engineering-based methods for the production of valuable products from microalgae are briefly discussed.

This book is written by experts mostly involved in plant genetic engineering research; hence, it is also essential for plant biotechnologists, agriculture scientists, postgraduate students, and plant biology researchers.

We would like to thank all our contributors who have made immense efforts to ensure the scientific quality of this book. We also thank our colleagues at Springer for their excellent and timely support.

Coimbatore, Tamil Nadu, India — Ramalingam Sathishkumar
Bangalore, Karnataka, India — Sarma Rajeev Kumar
Coimbatore, Tamil Nadu, India — Jagadeesan Hema
Coimbatore, Tamil Nadu, India — Venkidasamy Baskar

Contents

Editors and Contributors

About the Editors

Ramalingam Sathishkumar is currently a Professor of Biotechnology at Bharathiar University, India. He holds PhD from the School of Biological Sciences, Madurai Kamaraj University, India, and gained Post-doctoral training from the University of Hong Kong, Hong Kong. The thrust area of his research includes Plant Molecular Farming, Plant Metabolic Engineering, and DNA Barcoding. He has supervised 12 PhDs and mentored 4 Post-doctoral fellows till now. He has secured both national and international funding for his research. He has a US patent and filed an Indian patent. To his credit he has 70 peer-reviewed journal and 22 book chapter publications. Recently, he was elected as a subcommittee member on DNA barcoding by Indian Pharmacopeia Commission (IPC), Government of India. He is a consultant for many of the Indian herbal industries.

Sarma Rajeev Kumar is working as Scientist in String Bio Private Limited, India, after completing 5 years of post-doctoral research in CSIR-Central Institute of Medicinal and Aromatic Plants, Lucknow, and Research Centre, Bangalore. He received fellowships like Senior Research Associateship from the Council of Scientific and Industrial Research, Govt. of India, and Research Associateship from the Department of Biotechnology, Govt. of India. During his PhD tenure, he was awarded 1 year fellowship from the Foundation for Science and Technology, Portugal, to work with Prof. Birgit Arnholdt-Schmitt, University of Evora, Portugal. He has published 15 research papers in international peer-reviewed journals and authored 14 book chapters.

Jagadeesan Hema is currently an Associate Professor in Biotechnology at PSG College of Technology, India, and actively pursuing research in the area of environmental biotechnology, especially plant-microbe interactions in remediation. She holds a PhD in Plant Sciences from Madurai Kamaraj University, Madurai, India, and MSc. in Environmental Sciences from Jawaharlal Nehru University, New Delhi, India. She is the recipient of the Council of Scientific and Industrial Research (CSIR) Junior and Senior Research Fellowships for her doctoral research. She has 8 international journal articles and more than 11 conference publications. She has worked as Programme Officer (Scientist C) in "The Centre for Environment Education" (supported by the Ministry of Environment and Forests, Government of India, India).

Venkidasamy Baskar is working as a consultant in Texcity Biosciences Pvt Ltd., India. He completed his post-doctoral research in Bharathiar University, India, in the DST-SERB N-PDF scheme. He received fellowship from Konkuk University, S. Korea for doing PhD in Molecular Biotechnology. His area of expertise is genetic engineering of plants for improved traits such as stress tolerance, enhanced health and nutritive values. He obtained his Master's degree in Biotechnology from Bharathidasan University and Bachelor's degree in Microbiology from Madurai Kamaraj University. He has published 20 peer-reviewed publications in internationally reputed journals and 3 book chapters in international publishers.

Contributors

Pawan Kumar Agrawal Krishi Anusandhan Bhawan-I, ICAR, New Delhi, India

Furkan Ahmad Department of Natural Products, National Institute of Pharmaceutical Education and Research, SAS Nagar, Punjab, India

Sivalingam Anandhan ICAR-Directorate of Onion and Garlic Research, Rajgurunagar, Pune, Maharashtra, India

Paola Isabel Angulo-Bejarano Tecnologico de Monterrey, Centre of Bioengineering, Queretaro, Mexico

C. Appunu Division of Crop Improvement, Sugarcane Breeding Institute, Coimbatore, India

Srinivasan Balamurugan Key Laboratory of Eutrophication and Red Tide Prevention of Guangdong Higher Education Institutes, College of Life Science and Technology, Jinan University, Guangzhou, China

Venkidasamy Baskar Plant Genetic Engineering Laboratory, Department of Biotechnology, Bharathiar University, Coimbatore, Tamil Nadu, India

Gabriel Betanzos-Cabrera Tecnologico de Monterrey, Centre of Bioengineering, Queretaro, Mexico

R. K. B. Bharadwaj Plant Genetic Engineering Laboratory, Department of Biotechnology, Coimbatore, India

Nitin Bohra School of Genetics and Biotechnology, University of Trás-os-Montes and Alto Douro, UTAD, Vila Real, Portugal

Department of Biotechnology, National Institute of Technology, Warangal, Telangana, India

Hélia Cardoso ICAAM – Instituto de Ciências Agrárias e Ambientais Mediterrânicas, Instituto de Investigação e Formação Avançada, Universidade de Évora – Pólo da Mitra, Évora, Portugal

Irais Castillo-Maldonado Departamento de Bioquímica y Fitofarmacología, Centro de Investigación Biomédica, Facultad de Medicina, Universidad Autónoma de Coahuila, Torreón, Coahuila, Mexico

Soumya Chatterjee Defence Research Laboratory, DRDO, Tezpur, Assam, India

Marcos de Donato-Capote Tecnologico de Monterrey, Centre of Bioengineering, Queretaro, Mexico

Andreia Figueiredo Biosystems & Integrative Sciences Institute (BioISI), Faculdade de Ciências, Universidade de Lisboa, Campo Grande, Lisboa, Portugal

Juan Luis De la Fuente Jimenez Tecnologico de Monterrey, Centre of Bioengineering, Queretaro, Mexico

Dharmendra K. Gupta Ministry of Environment Forest and Climate Change (MoEFCC), Indira Paryavaran Bhavan, New Delhi, India

Santosh K. Gupta Defence Research Laboratory, DRDO, Tezpur, Assam, India

Ashirbad Guria Department of Plant Biotechnology, School of Biotechnology, Madurai Kamaraj University, Madurai, India

Jagadeesan Hema Department of Biotechnology, PSG College of Technology, Coimbatore, India

Gowtham Iyappan Plant Molecular Farming Laboratory, DRDO-BU Centre for Life Sciences, Bharathiar University, Coimbatore, India

Rajendran Jeyasri Department of Biotechnology, Science Campus, Alagappa University, Karaikudi, India

Kriti Juneja Department of Biotechnology, Indian Institute of Technology Roorkee, Roorkee, Uttarakhand, India

Subramanian Radhesh Krishnan Department of Biotechnology, Science Campus, Alagappa University, Karaikudi, India

Phytopharma Testing Laboratory, Herbal Division, T. Stanes & Company Ltd, Coimbatore, India

Sarma Rajeev Kumar Present Address: String Bio Private Ltd, IBAB Campus, Bangalore, India

Plant Genetic Engineering Laboratory, Department of Biotechnology, Bharathiar University, Coimbatore, Tamil Nadu, India

Sree Preethy Kuppuraj Plant Genetic Engineering Laboratory, Department of Biotechnology, Bharathiar University, Coimbatore, Tamil Nadu, India

Da-Wei Li Key Laboratory of Eutrophication and Red Tide Prevention of Guangdong Higher Education Institutes, College of Life Science and Technology, Jinan University, Guangzhou, China

Hong-Ye Li Key Laboratory of Eutrophication and Red Tide Prevention of Guangdong Higher Education Institutes, College of Life Science and Technology, Jinan University, Guangzhou, China

Jie-Sheng Liu Key Laboratory of Eutrophication and Red Tide Prevention of Guangdong Higher Education Institutes, College of Life Science and Technology, Jinan University, Guangzhou, China

Anindita Mitra Bankura Christian College, Bankura, West Bengal, India

Chakravarthi Mohan Molecular Biology Laboratory, Department of Genetics and Evolution, Federal University of São Carlos, São Carlos, SP, Brazil

Pandiyan Muthuramalingam Department of Biotechnology, Science Campus, Alagappa University, Karaikudi, India

Gauri Nerkar Division of Crop Improvement, Sugarcane Breeding Institute, Coimbatore, India

Rebecca Oziohu Omosimua Biotechnology Advanced Research Centre, Sheda Science and Technology complex, FCT-Abuja, Nigeria

Plant Genetic Engineering Laboratory, Department of Biotechnology, Bharathiar University, Coimbatore, Tamil Nadu, India

Shunmugiah Thevar Karutha Pandian Department of Biotechnology, Science Campus, Alagappa University, Karaikudi, India

Gopal Pandi Department of Plant Biotechnology, School of Biotechnology, Madurai Kamaraj University, Madurai, India

Sujay Paul Tecnologico de Monterrey, Centre of Bioengineering, Queretaro, Mexico

Augusto Peixe Departamento de Fitotecnia, ICAAM, Escola de Ciência e Tecnologia, Universidade de Évora – Pólo da Mitra, Évora, Portugal

Rita Pires Departamento de Fitotecnia, Escola de Ciência e Tecnologia, Universidade de Évora – Pólo da Mitra, Évora, Portugal

Pragadheesh VS Organic and Medicinal Chemistry Division, CSIR-Indian Institute of Chemical Biology, Kolkata, India

Francisco Roberto Quiroz-Figueroa Instituto Politécnico Nacional, Centro Interdisciplinario de Investigación para el Desarrollo Integral Regional Unidad Sinaloa (CIIDIR-IPN Unidad Sinaloa), Laboratorio de Fitomejoramiento Molecular, Guasave, Sinaloa, México

Bakshi Ram Division of Crop Improvement, Sugarcane Breeding Institute, Coimbatore, India

Manikandan Ramesh Department of Biotechnology, Science Campus, Alagappa University, Karaikudi, India

Eliel Ruiz-May Red de Estudios Moleculares Avanzados, Instituto de Ecología A. C., ClusterBioMimic®, Xalapa, Veracruz, Mexico

Shashank Sagar Saini Department of Biotechnology, Indian Institute of Technology Roorkee, Roorkee, Uttarakhand, India

Ramkumar Samynathan Plant Genetic Engineering Laboratory, Department of Biotechnology, Bharathiar University, Coimbatore, Tamil Nadu, India

Ramalingam Sathishkumar Plant Genetic Engineering Laboratory, Department of Biotechnology, Bharathiar University, Coimbatore, Tamil Nadu, India

Susana Serrazina Biosystems & Integrative Sciences Institute (BioISI), Faculdade de Ciências, Universidade de Lisboa, Lisboa, Portugal

Ashutosh Sharma Tecnologico de Monterrey, Centre of Bioengineering, Queretaro, Mexico

Lav Sharma CITAB – Centre for the Research and Technology of Agro-Environmental and Biological Sciences, University of Trás-os-Montes and Alto Douro, UTAD, Vila Real, Portugal

Priscila Yumi Tanaka Shibao Molecular Biology Laboratory, Department of Genetics and Evolution, Federal University of São Carlos, São Carlos, SP, Brazil

Flavio Henrique Silva Molecular Biology Laboratory, Department of Genetics and Evolution, Federal University of São Carlos, São Carlos, SP, Brazil

Rupesh Kumar Singh Centro de Química de Vila Real (CQ-VR), Universidade de Trás-os-Montes e Alto Douro, Vila Real, Portugal

Debabrata Sircar Department of Biotechnology, Indian Institute of Technology Roorkee, Roorkee, Uttarakhand, India

Jayanthi Soman Department of Biotechnology, PSG College of Technology, Coimbatore, India

Selvi Subramanian Department of Biotechnology, PSG College of Technology, Coimbatore, India

G. S. Suresha Division of Crop Improvement, Sugarcane Breeding Institute, Coimbatore, India

Varsha Tomar Department of Biotechnology, Indian Institute of Technology Roorkee, Roorkee, Uttarakhand, India

Juan Ignacio Valiente-Banuet Tecnologico de Monterrey, Centre of Bioengineering, Queretaro, Mexico

Xiang Wang Key Laboratory of Eutrophication and Red Tide Prevention of Guangdong Higher Education Institutes, College of Life Science and Technology, Jinan University, Guangzhou, China

Wei-Dong Yang Key Laboratory of Eutrophication and Red Tide Prevention of Guangdong Higher Education Institutes, College of Life Science and Technology, Jinan University, Guangzhou, China

Abbreviations

2,4-DPC	2,4-dichlorophenol
2-4D	2,4-dichlorophenoxyacetic acid
2-iP	2-isopentenyl adenine
A. rhizogenes	*Agrobacterium rhizogenes*
A. tumefaciens	*Agrobacterium tumefaciens*
ABA	abscisic acid
ADH	alcohol dehydrogenase
AGPAT	1-acyl-sn-glycerol-3-phosphate acyltransferase
ALS	acetolactate synthase
amiR	artificial microRNA
AMT	ammonium transporter
AP2	APETALA2
APX	ascorbate peroxidase
BA	benzyl adenine
BAP	benzylaminopurine
bar	phosphinothricin N-acetyltransferase
bFGF	basic fibroblast growth factor
bHLH	basic helix-loop-helix
BOD	biological oxygen demand
BR	brassinosteroids
bZIP	basic leucine zipper
CaMV35S	cauliflower mosaic 35S
CAT	catalase
CBM	constraint-based modeling
ChIP-qPCR	chromatin immunoprecipitation-qPCR
CK	cytokinins
CMV	cucumber mosaic virus
COD	chemical oxygen demand
CPMR	coat protein-mediated resistance
CRISPR	clustered regularly interspaced short palindrome repeats
crRNA	CRISPR RNA
CWR	crop wild relative
CWs	constructed wetlands
CYP450	cytochrome P450 monooxygenase

DDT	dichlorodiphenyltrichloroethane
DF-VSSF	downflow-VSSF
DGGE	denaturing gradient gel electrophoresis
DI-RNA	defective interfering RNA
DRB	double right border
DSB	double-stranded break
EENs	engineered endonucleases
ERF	ethylene response factor
FAPRI	Food and Agricultural Policy Research Institute
FISH	fluorescent in situ hybridization
FT-ICR-MS	Fourier-transform ion cyclotron resonance mass spectrometry
FTIR	Fourier-transform infrared spectroscopy
GA	gibberellins
GA_3	gibberellic acid
GBS	genotyping-by-sequencing
GC-MS	gas chromatography-mass spectrometry
GE	gene/genome editing
GEAC	Genetic Engineering Appraisal Committee
GFP	green fluorescent protein
GMO	genetically modified plants
GOGAT	glutamate-2-oxoglutarate aminotransferase
GPAT	glycerol-sn-3-phosphate acyltransferase
GR	glutathione reductase
gRNA	guide RNA
GS	glutamine synthetase
GUS	*β*-glucuronidase
GWA-GS	genome-wide association and genomic selection
HC-Pro	helper-component proteinase
HDGS	homology-dependent gene silencing
HPLC	high-performance liquid chromatography
HPLC-UV/DAD	HPLC with ultraviolet diode array detector
hpRNA	hairpin RNAs
HPT	homogentisate phytyltransferase
HPV	human papillomavirus
HR	homologous recombination
HSPs	heat-shock proteins
HSSF	horizontal subsurface
IAA	indole-3-acetic acid
IBA	indole-3-butyric acid
IL	interleukin
JA	jasmonic acid
KIN	kinetin
LC-MS	liquid chromatography-mass spectrometry
LEA	late embryogenesis abundant proteins

LSC	large single copy
MAS	marker-assisted selection
MAT	multi auto-transformation
MCS	multiple cloning site
MeJA	methyl jasmonate
miRNAs	microRNAs
MPMR	movement protein-mediated resistance
MYB	V-myb myeloblastosis viral oncogene homolog
N. benthamiana	*Nicotiana benthamiana*
NAA	naphthalene acetic acid
NAMR	nucleic acid-mediated resistance
NGS	next-generation sequencing
NHEJ	nonhomologous end-joining
NMR spectroscopy	nuclear magnetic resonance spectroscopy
nptII	neomycin phosphotransferase II
OE	overexpression
OPPP	oxidative pentose phosphate pathway
PA	polyamines
PAM	protospacer adjacent motif
PBP	bioelastic protein-based polymers
PEG	polyethylene glycol
PHB	polyhydroxybutyrate
pHBA	p-hydroxybenzoic acid
plastome	plant chloroplast genome
PMF	plant molecular farming
PMI	phosphomannose isomerase
PMPs	plant-made pharmaceuticals
PMVs	plant-made vaccines
PPV	plum pox virus
PRSV	papaya ringspot virus
PTGS	post-transcriptional gene silencing
PTM	post-translational modification
QTL	quantitative trait *locus*
RAM	root apical meristem
RBSs	reed bed systems
REs	restriction endonucleases
RGNs	RNA-guided nucleases
rhEPO	recombinant human erythropoietin
Ri	root-inducing plasmids
RISC	RNA-induced silencing complex
RNAi	RNA interference
RNPs	ribonucleoproteins
ROS	reactive oxygen species
RPMR	replicase protein-mediated resistance
RPS5A	ribosomal protein S5 A

RVD	repeat-variable diresidue
SAM	shoot apical meristem
satRNA	satellite RNA
SCMT	silicon carbide-mediated transformation
SE	somatic embryogenesis
sgRNA	single guide RNA
SH medium	Schenk and Hildebrandt medium
siRNA	small interfering RNA
SL	strigolactones
SLIC	sequence- and ligation-independent cloning
SOD	superoxide dismutase
SPME	solid-phase microextraction
SSC	small single-copy regions
STR	stirred tank reactors
TALENs	transcription activator-like effector nucleases
tasiRNAs	trans-acting small interfering RNAs
TC	tocopherol cyclase
TD	thermal desorption
T-DNA	transfer DNA
TDZ	thidiazuron
TF	transcription factor
TGM	targeted genome modification
TGS	transcriptional gene silencing
Ti	tumor inducing plasmids
TILLING	targeting-induced local lesions in genomes
TMV	tobacco mosaic virus
TOF MS	time of flight mass spectrometer
tracrRNA	trans-activating crRNA
TRBO	TMV RNA-based overexpression
TRFLP	terminal restriction fragment length polymorphism
TRSV	tobacco ringspot virus
TRV	tobacco rattle virus
UBQ	ubiquitin
UF-VSSF	upflow-VSSF
VIGS	virus-induced gene silencing
vir	virulence
VSSF	vertical subsurface
WMV	watermelon mosaic virus
WTPs	wastewater treatment facilities
ZFNs	zinc finger nucleases
ZYMV	zucchini yellow mosaic virus
γ-TMT	γ-tocopherol methyltransferase

Part I

Methods and Technologies

Plant Tissue Culture and DNA Delivery Methods

1

Jayanthi Soman, Jagadeesan Hema, and Selvi Subramanian

Abstract

Plant tissue culture techniques are used for in vitro culturing of plant parts (plant cells, tissues, organ) with the objective of plant improvement and commercial exploitation. Plant biotechnology is envisaged with the principles of plant tissue culture. Plant genomes respond differently to different culture techniques, conditions, explants, etc. Understanding the innate response conditions to culture techniques opens innumerable avenues for research and standardization. Till date the number of plant species considered recalcitrant is numerous. Plant tissue culture iterates the methods for clonal propagation, creating somaclonal variation and organ culture featuring monoploids, haploids, diploids, dihaploids, somatic embryogenesis, embryo rescue, cybrid developments, harnessing secondary metabolite production and plant transformation. Plant transformation results from targeted gene delivery into plant types for plant improvement and/or recombinant protein production. Transgenic plants are developed through gene delivery into single cell/tissue, regeneration into whole plants that are fertile and adapt to the normal plant propagation process. Delivery methods harness chemical methods, *Agrobacterium* mediated and particle bombardment, including in-planta transformation techniques. Each of the methods involves specificity with respect to transgene characters and host preferences along with target plant material. Techniques depend on specificity, precision, reliability and scope. Gene transfer techniques can greatly reduce the time taken in crop improvement against conventional breeding procedures and enlarge the array for desired plant improvement.

J. Soman · J. Hema · S. Subramanian (✉)
Department of Biotechnology, PSG College of Technology, Coimbatore, India
e-mail: selvi.bio@psgtech.ac.in

R. Sathishkumar et al. (eds.), *Advances in Plant Transgenics: Methods and Applications*, https://doi.org/10.1007/978-981-13-9624-3_1

Keywords
Plant tissue culture · Plant transformation · Transgenic plants · *Agrobacterium* · Bombardment · Plant improvement

1.1 Introduction

Improvement of crop plants by human being has taken a major leap after the invention of two major technologies, plant tissue culture and recombinant DNA technology. To date the art of growing plants and plant parts under in vitro conditions has evolved into mainstream botanical science. Tissue culture offers several diverse applications. Large-scale clonal multiplication of elite plant material, somaclonal variants generation and screening, secondary metabolite production, embryo rescue in interspecies/genus crosses and plant genetic transformation are some among the extensive list. The applications of plant tissue culture are categorized and given in Fig. 1.1.

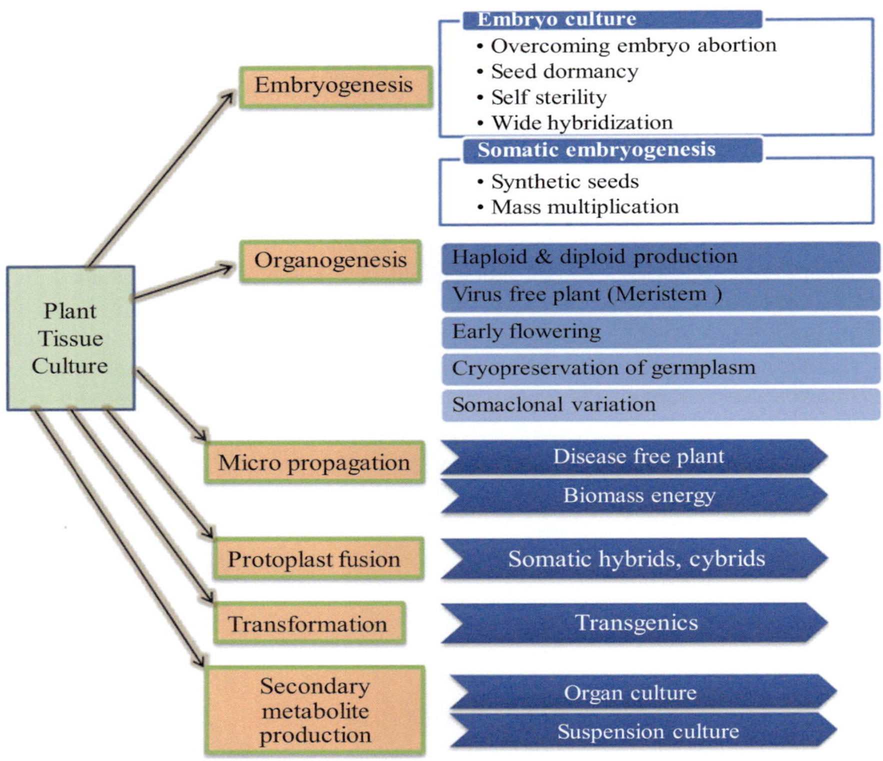

Fig. 1.1 Enumeration of plant tissue culture applications

1.2 Tissue Culture Methods

Habertland in 1902 established the mechanism of totipotency in in vitro-grown plant tissues. Skoog and Miller (1957) arrived at a relative concentration of auxin and cytokinin in order to induce organ regeneration from plant cells. Since then various explants, nutritional supplements and combinations of growth hormones have been achieved for many plants/crops/trees towards whole plant regeneration from single cells/plant tissues. Success in plant tissue culture relates to achievements in mass multiplication/secondary metabolite production/plant transformation and other methods to enhance the genetic potential of cultivated plants. The major methods put into use for plant improvements are discussed below.

1.2.1 Callus Culture

Callus refers to the undifferentiated cells produced with various explants (leaf, shoot, shoot tip, root, flower buds, anthers, axillary buds, etc.) and appropriate culture media. Calli when supplemented with growth hormones, maintained under specific day/night conditions, show ability to redifferentiate into whole plants or could be maintained as friable cells (Chavarriaga-Aguirre et al. Chavarriaga-Aguirre et al. 2016). Friable calli are extensively exploited under suspension cultures for secondary metabolite production (Valluri 2009; Perianez-Rodriguez et al. 2014; Nandagopal et al. 2018).

1.2.2 Shoot Tip Culture

Culture of meristem (0.05–0.1 mm) of the shoot tip along with the primordial and developing leaves is referred to as shoot tip culture. Mericloning or mersitemming is the culture of the meristem tissue into complete plants through plant tissue culture. The advantages include development of multiple shoots that can be propagated independently. The plant regeneration from meristems possesses advantages of multiplying a true-to-type plant exhibiting the same agronomic characteristics due to high genetic similarity (Ahmed and Anis 2014). Environmental influences in addition to the genotype on the explant at the time of retrieval for in vitro culture have a high impact on regeneration potential. Ahmed and Anis had eliminated the cumbersome two-step media requirement for shoot tip culture (shoot propagation and shoot elongation) adopting a combination of medium supplemented with different concentrations of BA, Kin and 2-iP (0.0–10 μM) for *Vitex trifolia* L, an important multipurpose plant for the Pacific traditional medicine. A similar result was also reported for *Bacopa monnieri* during subculturing up to three passages. The method is promising for easy and efficient micropropagation of genotypes that are economically important.

A combination of high cytokinin and low auxin favours multiple shoot proliferation in orchids. Adenine sulphate enhances shoot multiplication in many genotypes

(Naaz et al. 2014). A high percentage of shoot regeneration was achieved utilizing thidiazuron (0.5 mg L^{-1}) and indole-3-butyric acid (0.25 mg L^{-1}) with MS medium in an attempt to in vitro propagate the medicinal plant *Curculigo latifolia* (Babaei et al. 2014). Indole-3-butyric acid in combination with thidiazuron exhibited a synergistic effect on shoot regeneration. The basal medium was supplemented with different concentrations and combinations of thidiazuron (TDZ 0, 0.5, 1, 1.5 and 2 mg L^{-1}) and indole-3-butyric acid (IBA 0, 0.25 and 0.5 mg L^{-1}) for shoot regeneration and scalp induction. TDZ had proved effective in eliminating apical dominance and enhancing lateral bud development and resulted with adventitious shoot leading to axillary shoot emersion. The above combination paves way for micropropagating tough genotypes within a short period of time. Shoot organogenesis in *Curculigo orchioides*, *Pinus massoniana* and *Embelia ribes* proved promising with the above combination.

Cryopreservation of virus-free shoot tip cultures has been extensively exploited in potato (Kaczmarczyk et al. 2011). Cryogenic techniques like encapsulation/dehydration, encapsulation/vitrification and droplet vitrification have been used to cryostore meristem tissues (Feng et al. 2013). Pretreatment of shoot tips has been successful in maximizing regeneration ability of the cryopreserved shoot tips. Cryotherapy employs vitrification techniques through liquid nitrogen exposure where infected, vacuolated and differentiated cells could be eliminated by ultra-low temperature effects (−196 °C) allowing the meristamatic (cytoplasmic) cells to survive (Bettoni et al. 2016). However, the extremely small tissue size, low regeneration capacity and kind of viruses responsive to the treatment still remain a challenge that needs exploration.

1.2.3 Microspore Culture

Microspore culture has been the most promising method of plant tissue culture to incite direct embryogenesis or regeneration from callus since 1980 in egg plants. It is considered advantageous over the conventional anther culture due to the avoidance of intervening ploidy levels when derived from anther mother cells. Deriving double haploids from microspore-derived embryos provided promise to achieve maximum genetic uniformity, facilitating mapping of agronomically relevant traits and evolving new hybrids with homogeneity in minimal time periods. It also saves the cost of inbreeding depression in homozygous lines. The successful regeneration of whole plants from microspores has been increasing since then. Perera et al. (2014) were successful in reprogramming the microspores of *Manihot esculenta* towards androgenesis. NLN medium modified with either sodium ferrous ethylenediaminetetraacetic acid (NaFeEDTA) or CuSO4·5H2O induced sporophytic cell division in nonresponsive microspores. Haploid plants that can be multiplied vegetatively have the advantage of rapid development of completely homozygous lines towards efficient genotypic selection (Burbulis et al. 2011). Choice-type exclusion of chromosomes in hybrid embryos by use of male sterile plants achievable through androgenesis and inducing gynogenesis has been promising inclusion for microspore culture (Pretova et al. 2006). Evaluation of ploidy levels in regenerated calli/cells can be achieved through chromosome counts or flow cytometry (Ochatt 2008).

Double haploids have been developed in flax through optimization of donor plant growth conditions, pretreatment of appropriate explants and altering culture conditions such as temperature and media composition. New varieties of crop plants in wheat, rice, barley, eggplant, bell peppers, asparagus, etc. have been exploited through microspore culture. However differences among species and genotypes determine the rate of embrogenic responses for microspore-derived calli (Ferrie and Caswell 2011). Microspores can be isolated through mechanical crushing of surface-sterilized buds using a pestle and mortar, filtering of the slurry to spate cells of the anther walls from the microspores. Identifying of intact microspores under the microscope/collection through centrifugation was successful for few crops. More uniform and debris-free samples are possible through density gradients of maltose/sucrose/percoll (Maraschin et al. 2003). Microspores of barley and pepper responded well when extracted in liquid medium and allowed to dehisce (Supena et al. Supena et al. 2006). Separation of somatic cells from microspore is essential, for it may produce diploid calli and further challenge the search for haploid cells. Stress to donor plants or isolated microspores could induce embryogenesis and further whole plants. Conditions of pH, carrageenan oligosaccharides, heavy metals, 2-4D pretreatment, water stress, high humidity, nitrogen starvation, ethanol, c-irradiation, microtubule disruptive agents and electro stimulation are among the few novel means of stress conditions favouring embryogenesis of microspores. For some species spontaneous development of double haploids can generate homozygous DH plants. In the absence of spontaneous double haploidy, treatment with colchicine is promising to produce DH lines from in vitro-cultured microspores. Commercially exploited flowering plants like chrysanthemum that are highly heterozygous have benefitted through these microspore culture machineries (Wang et al. 2014). Addition of activated antioxidants to MS media and activated charcoal (aids reduce removal of phenolics from explants and hence tissue browning) has aided few genotypes for callus regeneration from microspore. Supplementation with substances like biotin, casein, coconut water, glutamine, inositol, PVP and silver nitrate has been widely reported beneficial for higher regeneration potential of microspores (Germana 2011).

1.2.4 Somatic Embryos

The ability of plant somatic cells to dedifferentiate to a totipotent embryonic stem cell is the process that has been exploited for the development of somatic embryos. The embryo from a somatic cell develops into a complete, fertile plant under appropriate culture conditions followed by hardening. Somatic embryogenesis has provided a great scope in woody plants and remains pivotal in clonal propagation, germplasm conservation, synthetic seed production and cryopreservation (Guan et al. 2016). The challenge lies in triggering the development of embryogenic cells from somatic cells. Study on molecular mechanisms involved in cell fate and regeneration of whole plants has been attempted for many plant species, owing to the advantages it presents through tissue culture. Somatic cells ideally pass through typical developmental stages including globular, torpedo and cotyledonary phases in dicot plants. In monocots the stages include globular, scutellar and coleoptilar. Conifers involve a distinguishing early and cotyledonary phases during the development of embryos from somatic cells (Quiroz-Figueroa et al. 2006).

There can be two ways of inducting somatic embryos, either directly or indirectly, avoiding the intermediate callus stage. This would shorten the developmental process and also avoid any variation created through somaclonal variation (Guan et al. 2016). Somatic embryos are advantageous with respect to utilization in synthetic seeds. Encapsulation methods comprising hydrated and desiccated forms are successful in creating synthetic seeds conducive for transport, storage and handling during sowing (Rai et al. 2009). Under circumstances these embryos can be cultured in vitro and taken to field as developed plantlets as well (Sharma et al. 2013). Vitrification of the developed embryos has also been proven successful in *Citrus* spp. (San José et al. 2015).

1.2.5 Embryo Rescue

Embryo rescue refers to the tissue culture technique that is capable of reviving any immature embryo into a whole plant. Excised embryos are cultured in vitro; the success of embryo rescue depends on the stage of maturity of the embryo at the time of excision, ploidy levels of the involved gametes and the optimum media composition along with culture conditions. Development of triploid plants in grapes was successfully cultured when embryos were excised 55 to 80 days after pollination, with a highest embryo emergence rate of 39.13% to 73.08%, respectively. Triploidy levels of the developed plantlets were confirmed through chromosome measurement of the regenerated tissues. The nuance required and the size of the seed handled bring out the challenge involved in culturing/rescuing embryos in vitro. A modified or an improved method of embryo rescue would encompass ovule culture as a promise to gain significant success in generating whole plants from compromised embryos as well. Ovule culture enables embryos to be cultured with minimal damage to the embryo. Ovules are removed before the time accessed for embryo abortion and cultured in vitro. Hybrids involving individuals with different ploidy levels are the means to create new germplasm. Interspecific hybrids involving *Sesamum indicum* and *Sesamum alatum Thonn* for phyllody resistance were achievable through culturing ovules (Rajeswari et al. 2010). The phytohormones and their combination were critical parameters for the germination of complete plantlets. Highest shoot regeneration frequency was observed for linseed variety 'Barbara' when MS media were supplemented with 2 mg L^{-1} BAP and 2 mg L^{-1} NAA while combination of 1 mgL^{-1} BAP and 2 mg L^{-1} IAA in an ovary-derived callus. The gynogenetic process of plantlet development is a critical phase for embryo rescue technique (Burbulis et al. 2011). The technique eliminates post-zygotic barriers limiting the use of interspecific hybridization to evolve elite, disease-resistant hybrids. Triploid breeding is a huge success with fruit crops owing to vigorous growth, thick cane, thick leaf, large fruit, seedlessness, high yield, adaptability and stress resistance. The technique is promising for shortening breeding cycles in many crops including early ripening stone fruits, triploid grapes (Yang et al. 2011). Also used for propagation are cryopreservation and germplasm exchange.

1.2.6 Cybrids

Cybrids envisage creation of novel combination of cytoplasm and nuclear genotypes (Guo et al. 2013). Creation of new alloplasmic combination of nuclear and organellar genomes is made possible through cybridization. The new recombinant leads to a unique genotype resulting in compatible variations in the plant. Changes with respect to phenotypes that are screened for improved characteristics are a breakthrough in plant improvement through cybridization. Development of citrus canker disease resistance has been achieved in few species (Ahmad et al. 2017, Omar et al. 2017). *Agrobacterium*-mediated leaf disc infiltration of *Phaseolus vulgaris*, the common bean that remains recalcitrant to tissue culture techniques, promises a method for improvement in field crops as well (Nanjareddy et al. 2016). Methods for protoplast isolation and in vitro conditions for regeneration have been prescribed for various crops (Nanjareddy et al. 2016, Bhalla and Singh 2008; Masani et al. 2014; Yu et al. 2017). Cybrid production combining the nucleus of one species and the cytoplasm of another is a valuable method for genetic improvement of various crops. This can be considered as a powerful non-GM technology with an utmost potential to bring in organellar genomes into crop improvement.

1.3 Plant Transformation Methods

Introducing DNA into plants is an important step in the creation of transgenic plants, as well as for transient expression of genes in plants. Plant transformation has been reported since 1980s. There is a wide variety of methods – physical, chemical as well as biological – which have been standardized for many economically important plants (Birch 1997). The efficiency of transformation varies with the plant at species and cultivar levels also. There is no universally efficient technique. The methods can also be categorized as direct and indirect based on whether the nucleic acid sequence is introduced directly or through another organism as in *Agrobacterium*-mediated method. The choice of the method depends on the efficiency of the method in delivering the genetic material into the respective plant cell, the explants' ability to regenerate and whether it is nuclear or plastid transformation. Totipotency of the plant cell ensures that the genetic element introduced into a single cell could be used to generate an entire plant with the transgene. Plastid transformation has the advantage of inheritable changes without the silencing effects observed with nuclear transformation, gene containment, higher expression levels, polycistronic expression, etc. (Dufourmantel et al. 2006). Many transformation methods have been described; of these *Agrobacterium*-based transformation, particle bombardment, electroporation and polyethylene glycol-based methods have been proven to produce transgenic plants.

1.3.1 *Agrobacterium*-Mediated Transformation

Agrobacterium tumefaciens (*Rhizobium radiobacter*) and *Agrobacterium rhizogenes* (*Rhizobium rhizogenes*) are Gram-negative soil bacteria known for their natural ability to transfer part of their genetic material to plants (Gelvin 2003; Hwang et al. 2017). These bacteria have Ti (tumour-inducing) and Ri (root-inducing) plasmids that are responsible for the crown gall and hairy root diseases in plants, mainly dicots (Otten et al. 2007). Two components of these plasmids are major contributors for the transfer of DNA to plant cells – the transfer DNA (T-DNA) and the virulence genes. T-DNA harbour the oncogenes coding for phytohormones responsible for the uncontrolled growth of the cells and those involved in the biosynthesis of amino acid sugar derivatives and are transferred to the host cell (Kiyokawa et al. 2009; Zupan and Zambryski 1995). The T-DNA is flanked by 25 base pair imperfect repeats – the left and right borders which are recognized by the vir gene products and are cleaved and transferred to the host DNA. Virulence genes (*virA*, *virB*, *virC*, *virD*, *virE*, *vir* and *virG*) are involved in the processing of the T-DNA and its transfer to the host cell (Hwang et al. 2017; Tzfira and Citovsky 2000; Zupan et al. 2000). The vir gene products transfer any genetic material in between the left and right borders. *Agrobacterium* can transfer DNA to wide host range including dicot and monocot plants, fungi and even human cells (Gelvin 2003). Although most practised methods involve wounding the tissue before infection, wounding per se is not required for induction of the process (Brencic et al. 2005).

This ability of *Agrobacterium* is manipulated and is one of the most widely used techniques for creating transgenic plants. This can be achieved by either co-integrative or binary vector strategies. Co-integrative strategy involves the integration of the transgene into the host *Agrobacterium*'s Ti plasmid (Van Haute et al. 1983; Zahm et al. 1984). In the binary vector strategy, the transgene is carried in a separate shuttle vector within the left and right border sequences along with suitable markers (de Framond et al. 1983) and is transformed into an *Agrobacterium* strain containing the disarmed Ti plasmid; it does not carry the oncogenes (phytohormone biosynthesis genes) but has the virulence genes. Such strains are called *vir helper* strains. The ability of the virulence gene products to recognize the border sequences in trans is exploited here. The co-integrative strategy is cumbersome but has the advantage of low copy number, while the binary vector was small and easy to manipulate (Gelvin 2003). Many binary vectors have been developed with multiple cloning sites, various types of plant as well as plant tissue-specific promoters and selectable markers and reporters within the T-DNA region (Hellens et al. 2000; Thole et al. 2007). Plants recalcitrant to *Agrobacterium*-mediated transformation can be subjected to various physical and chemical treatments including pH, temperature, light conditions, cell concentration and duration of co-cultivation, explant type and quality, preculture and hormone treatment, wounding to induce them to take up the DNA (Gafni et al. 1995; van Wordragen and Dons 1992).

1.3.2 Biolistic Method/Particle Bombardment Method

The use of small particles made up of tungsten, gold or any high-density metals coated with DNA when accelerated to very high speed and bombarded into live tissues can make the cells take the DNA up and integrate it into its genome (Iida et al. 1990; Klein et al. 1988, 1989; Sanford 1990). It is carried out using a gene gun or biolistic delivery system. Many plants that were recalcitrant to *Agrobacterium*-mediated gene transfer as well as a wide variety of plant tissues can be transformed by this method (Kikkert et al. 2004). Organellar genome, both chloroplast and mitochondria, can be transformed by this method. This can also cotransform both nuclear and organellar genome simultaneously (Elghabi et al. 2011). An expression cassette with the transgene and marker alone is sufficient for integration and expression (Fu et al. 2000; Vidal et al. 2006). The DNA to be delivered is coated onto the microprojectile in the presence of calcium chloride and spermidine, wherein spermidine is known to protect the DNA structure and improve binding (Klein et al. 1989). Protamines in place of spermidine have the added advantage of conferring protection against DNase activity (Sivamani et al. 2009). Polyethylene glycol and magnesium have been used effectively to transform wheat (Ismagul et al. 2018). Reducing the amount of DNA coated on the microprojectiles enhances single copy integrations (Lowe et al. 2009). Nanoparticles can also be used to deliver DNA using particle bombardment (Torney et al. 2007). Biolistic method is simple and can transform recalcitrant plants, especially monocots and the ability to transfer to chloroplast and mitochondria, and the main limitation is the requirement of specialized instrumentation and the possibility of introducing concatemers (Bates 1999).

1.3.3 Electroporation

Electroporation makes use of short high-voltage electrical pulses to a protoplast suspension and DNA in a electroporator cuvette and permeabilizing the membrane to take up the DNA (Bates 1999; Fromm et al. 1986; Zhang et al. 1998). Protoplasts are used, as the cell wall of the plant cell affects the movement of DNA into the cells. Only those plant species whose protoplasts can regenerate their cell walls are currently transformed using electroporation. The efficiency of the process is dependent upon the pulse length, type and duration (Rivera et al. 2012). It has been reported that use of PEG along with magnesium chloride enhances the transformation efficiency (Murakawa et al. 2008; Tyagi et al. 1989).

1.3.4 Polyethylene Glycol-Mediated Transformation

Polyethylene glycol-mediated DNA delivery into plant cells rely on the interactions between the DNA and cell membrane, as protoplast rather than cells is used in this technique. The protoplasts are mixed with DNA and incubated in the presence of PEG. This method although labour intensive has been found efficient in the transformation of various plants (Nugent et al. 2006).

1.3.5 Other Methods

Many novel methods have been developed and reported for transferring DNA into plant cells, but their feasibility and widespread applicability are yet to be proven. Some of these include silicon carbide-mediated transformation (SCMT) – wherein whiskers made of fine silicon carbide are mixed with plant cells and DNA and vortexed. These whiskers pierce cells and allow the entry of DNA (Kaeppler et al. 1992; Torney et al. 2007). Microinjection involves direct injection of the DNA into protoplasts by immobilization. It has been reported in plants which are difficult to transform by Agrobacterium mediated method and is known to reduce the developemnt of chimeric plants (Masani et al. 2014).

1.4 Recalcitrance in Plant Tissue Culture

Many of the above applications are constrained by the recalcitrant nature of several plant species and some specific genotypes in them. Inability of plant cells, tissues and organs to respond to tissue culture manipulations is recalcitrance. Tree species and plants with specialized secondary metabolite production often tend to be recalcitrant in nature. The major issue associated with recalcitrance is the interaction between genotype and both plant growth regulator type and concentration in in vitro culture conditions. Excess cytokinin production and preferential expression of certain transcription factors in developing embryo are considered as reasons for inability to regenerate (Ckurshumova and Berleth 2015).The choice and physiological state of explants and the genotype of the species play an important role in plant regeneration. The regeneration of shoots or roots must involve considerable genetic programming. As example sesame (*Sesamum indicum L.*) a recalcitrant crop for tissue culture and transformation is discussed in detail in this chapter.

1.4.1 Tissue Culture and Transformation in Sesame a Recalcitrant Plant

1.4.1.1 Somatic Embryogenesis and Indirect Organogenesis

In sesame shoot induction from callus was found to be the bottleneck (Taskin and Turgut 1997; Kim 2001; Kariyallappa et al. 2003). The first report on tissue culture in sesame was using shoot tip culture. Calli and organogenesis were induced from shoot tips with a single treatment of NAA and IAA. NAA was better for shoot differentiation, whereas IAA was better for root differentiation. Regeneration of plantlets by in vitro culture of apical meristems and hypocotyl segments was reported by Ram et al. (1990). They have successfully induced somatic embryos directly from the surface of the zygotic embryos of sesame in six cultivars, namely, Aceitera, Arawaca, Turen, Piritu, Maporal and Inama. The somatic embryo induction varied from 50 to 100%. Cotyledon explants of sesame cv. Muganli-57, cultured on MS medium supplemented with BAP (1, 2, 4 and 8 mg/l) and NAA (0.1 mg/l), formed callus with a frequency of 86–100%. Shoots were formed in the MS medium with BAP (4.0 or 8.0 mg/l) + NAA (0.1 mg/l) (Taskin and Turgut 1997). Xu et al. (1997) reported somatic embryogenesis in sesame *S. indicum L.* from cotyledon, root and subapical hypocotyl segments of 2–3 mm in length from 6 to 21 days old seedlings. Calli were induced on MS medium supplemented with 2, 4-D (2 mg/l). Embryos appeared on the callus surface when the callus was cultured on B5 medium + BAP (0.5 mg/l) +2, 4-D or NAA (0.5 mg/l). Conversion of embryos into plants was carried out on B5 medium supplemented with 0.5% activated charcoal and 3% sucrose with or without 3% mannitol and subsequently transferred to half strength MS medium containing Zeatin (0.1 mg/l) + GA_3 (1 mg/l) + $AgNO_3$ (10 mg/l). The optimal conditions for callus induction and shoot regeneration from hypocotyl and cotyledon of sesame *S. indicum* were reported by Younghee (2001). Shoot regeneration from hypocotyl (25.8%) and cotyledon (10%) was highest in the MS medium containing BAP (3.0 mg/l), NAA (0.5 mg/l), sucrose (30.0 g/l) and agar (8.0 g/l) with $AgNO_3$. Shoot regeneration from hypocotyl (20%) and cotyledon (8%) was promoted by $AgNO_3$ treatment at 15 μM/l. A new and rapid method was developed for callus induction by Lokesha et al. (2005); the surface-sterilized seeds were directly inoculated onto the MS media supplemented with the growth regulators like KIN, NAA and BAP, incubated in dark for 2–3 days and then transferred to continuous light. Callus was initiated in 5–7 days from hypocotyls, cotyledon leaves and roots concomitantly from every seedling. Primary calli were maintained by subculturing on NAA (1–3 mg/l), BAP (1–4 mg/l), $AgNO_3$ (5–25 μM) and TDZ (1–40 μM). Baskaran and Jayabalan (2006) studied the effect of various plant growth regulators on shoot proliferation from shoot tips and callus induction in hypocotyl and cotyledon explants of sesame *S. indicum* cv. VRI 1. MS medium supplemented with BAP (8.8–44.4 μM) and KIN (4.6 μM) increased shoot proliferation. Proliferated shoots were rooted in NAA (8.0 μM). Hypocotyl has shown better response for callus induction than cotyledon explants. Wadeyar et al. (2013) got the highest percentage of shoot regeneration in *S. indicum L.* from hypocotyls-derived callus on MS with TDZ 25 μM and 3 μM of IAA with 83.3% in variety DS-1 followed by 58% in variety E8 and 50% in variety TL with 53.2% of shoot regeneration and 1.5 mean number of shoots per callus. The shoot regeneration was also observed on MS with TDZ 25 μM and 2 μM of IAA with 75% in variety DS-1 followed by 58% in variety TL and 50% in variety E8 with 51.5% of mean shoot regeneration and 1.6 mean number of shoots per callus.

Hypocotyls from 14 days old in vitro-grown seedlings were used as explants for callus induction in indirect organogenesis which showed best response for callus induction (Muthulakshmi et al. 2013). Hypocotyl explants of two sesame varieties TMV 7 and SVPR 1 in MS medium supplemented with NAA (0.5 mg/l) + TDZ (1 mg/l) produced yellowish green friable calli (Fig. 1.2).

1.4.1.2 Adventitious Shoot Formation and Shoot Regeneration

Havgeppa and Rao (2013) induced direct somatic embryo induction without an intervening callus phase in (*S. indicum L.*) from 5 days old cotyledonary and hypocotyl explants. Maximum number of somatic embryos per explant was noted on MS medium supplemented with 3.0 mg/l 2, 4-D + 1.0 mg/l BAP. Conversion of somatic embryos into complete plantlets was achieved on MS medium supplemented with 1.0 mg/l BAP +0.5 mg/l ABA +5.0 mg/l $AgNO_3$.

Efficient regeneration protocols were developed using cotyledon and hypocotyl explants in sesame cv. Mtwara-2. They have studied the significant interaction between the hormone treatments and the macronutrients for shoot and root regeneration. MS medium supplemented with N_6 macronutrients (Chu medium) resulted in twice the shoot regeneration frequency. N_6 medium (Chu medium) with TDZ (20 μM) together with IAA (2.5 μM) was found to be the optimum growth regulator combination for shoot regeneration which gave a frequency of 63% and 4.4 shoots per regenerating explants (Were et al. 2006). Seo et al. (2007) established a high-frequency plant regeneration method via adventitious shoot formation in sesame using de-embryonated cotyledon as explants. Optimal medium for direct adventitious shoot formation was MS medium fortified with BAP (22.2 μM) + IAA (5.7 μM) + ABA (3.8 μM) + $AgNO_3$ (29.4 μM). Preculture of cotyledon explants on high sucrose concentration (6–9%) for 2 weeks and subsequent transfer to 3% sucrose enhanced the frequency of adventitious shoot formed. The shoots were transferred to MS medium containing NAA (2.7 μM) for rooting. Abdellatef et al. (2010) evaluated the in vitro regeneration capacity of sesame cultivars exposed to culture media containing ethylene inhibitors such as cobalt chloride and silver nitrate. Ethylene inhibitors had growth promoting

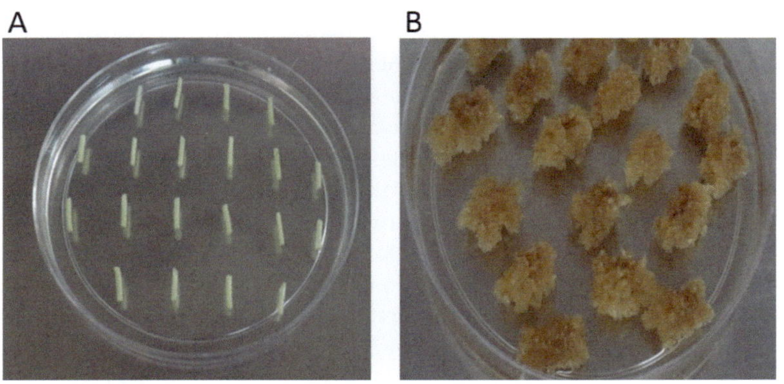

Fig. 1.2 Callus formation in sesame using hypocotyl explants. (**a**) Hypocotyl explants, (**b**) calli from hypocotyl on MS medium containing NAA (0.5 mg/l) + TDZ (1.0 mg/l)

effects due to the reduction in ethylene concentration followed by inhibition of ethylene action. MS medium enriched with BAP (1.0 mg/l) induced adventitious shoots in axenic seedling-derived shoot tips. Addition of ethylene inhibitors $AgNO_3$ (0.5–5.0 mg/l) enhanced the number of shoots as well as shoot length. Chattopadhyaya et al. (2010) established an efficient protocol for shoot regeneration from sesame internodes using the transverse thin cell layer (tTCL) culture method. A combination of BAP (2.0 mg/l) and NAA (0.5 mg/l) was found to be the best phytohormone combination for shoot-bud induction, with the maximum number of shoots obtained, when the tTCL sections were 0.5–1.0 mm thick and derived from 4 to 6 weeks old seedlings of sesame. Well-developed shoots were rooted on MS medium without phytohormones, and 80% of the regenerated plantlets were successfully established in soil.

De-embryonated cotyledons (Fig. 1.3a) were used as explants for direct organogenesis in sesame. Dissection of the embryonic axis of the cotyledon at the proximal end showed satisfactory adventitious shoot regeneration using MS medium containing BAP (6. mg/l), $AgNO_3$ (1 mg/l), IAA (1 mg/l), ABA (3 mg/l) and sucrose (3%). Proliferation was done on the same media, resulted in regeneration (Fig. 1.3b) of adventitious shoots (6–9 number) in TMV 7 and (4–7number) in SVPR 1 per cotyledon. Shoot elongation was observed in MS medium containing GA_3 (0.3 mg/l) after 3 weeks. The elongated shoots (Fig. 1.3c) were transferred to different concentrations of MS media and individual growth hormones like NAA, IBA and combination of growth hormones NAA, IBA and IAA with 1.5% and sucrose 3%. The MS media supplemented with hormone IBA (2 mg/l) showed more root growth in both the cv. TMV 7 and SVPR 1 than other hormones like NAA and IAA.

1.4.1.3 Agrobacterium-Mediated Transformation of Sesame

Although sesame having important economic value (*S. indicum L.*) and the recent availability of its genome sequence, a high-frequency transformation protocol is still not available because sesame is very recalcitrant for regeneration in vitro as it produces secondary metabolites. The only two existing *Agrobacterium*-mediated transformation protocols that are available have poor transformation efficiencies of less than 2%. Yadav et al. (2010) reported for the first time successful recovery of

Fig. 1.3 In vitro regeneration through direct organogenesis. (**a**) De-embryonated cotyledon of sesame (TMV7), (**b**) initiation of adventitious shoots, (**c**) elongated shoots

fertile transgenic plants of sesame with transformation frequency of 1.01%. From cotyledon explants inoculated with *A. tumefaciens* carrying a binary vector pCAMBIA2301 that contains a neomycin phosphotransferase gene (*npt*II) and a *β*-glucuronidase (GUS) gene (*uid*A) interrupted with an intron. Green shoots recovered from *A. tumefaciens*-infected explants on selection medium (MS) basal medium containing 25.0 μM benzyladenine (BA), 25.0 mg l-1 kanamycin and 400.0 mg/l cefotaxime were rooted on MS basal medium containing 2.0 μM indole-3-butyric acid and 5.0 mg/l kanamycin. The rooted shoots were established in soil and grown to maturity to collect seeds. The presence, integration and expression of transgenes in putative T_0 plants were confirmed by polymerase chain reaction (PCR), Southern blot hybridization and GUS histochemical assay, respectively.

Al-Shafeay et al. (2011) reported transformation of sesame cv. Sohag 1 by *Agrobacterium*. Different factors were considered in transformation experiments such as *Agrobacterium* optical density (0.1, 0.2, 0.5 or 1.0 O.D) and co-cultivation time (1, 2 or 3 days). In such conditions lower *Agrobacterium* density (0.1 and 0.2) resulted in a higher percentage of healthier explants with blue spots. *Agrobacterium* concentration at 0.1 O.D 600 with 2 days co-cultivation time was used. Several individuals were found to be positive in PCR screening experiments, and histochemical assay also confirmed the presence of faint blue colour in young leaves excised from greenhouse putatively transformed growing plants. Further analysis of putative individuals using RT-PCR indicated the presence of the 680 bp expected partial-GUS fragment. The transformation frequency was found to be 1.67%.

Bhattacharyya et al. (2015) established a protocol for efficient direct gene transfer through particle gun bombardment for sesame genotypes, viz. E8, Gulbarga local white and brown and RT273. Callus derived through direct seeding method was transformed with pABC plasmid carrying the *npt II* gene encoding kanamycin resistance and *GUS* driven by CaMV 35S promoter. Transformants were selected on MS supplemented with growth regulators NAA (0.5 mg/l), BAP (1.5 mg/l) and Kinetin (1.5 mg/l) and kanamycin sulphate 25 mg/l for first round of selection. The calli that proliferated on 25 mg/l kanamycin were transferred to medium containing 50 mg/l of kanamycin for second round of selection. The integration of transgenes in sesame plant genome was confirmed by PCR amplification and *Gus* histochemical assay. E8 and Gulbarga local white have shown positive response for kanamycin selection with an average transformation percentage of 12.5 and 7.5, respectively. In the case of E8, among all the treatment combinations, 7 days old calli with 4 hours prebombardment osmotic treatment and 9 cm target cell distance have given higher transformation percentage of 15 as against 10 in case of 15 days old calli with 4 hours osmotic treatment and 9 cm target cell distance. The integration of the gene was confirmed through PCR amplification with recovery of 800 bp band from transformed callus and development of blue colour with *GUS* histochemical assay.

Chowdhury et al. (2014) reported the most efficient gene transformation protocol into the *S. indicum L.* (cultivar VRI-1). They have used *Agrobacterium*-mediated gene transformation as a tool in which *A. tumefaciens* strain LBA4404 carrying the binary vector pCAMBIA2301 was used as a vector system which showed 42.66%

Fig. 1.4 *Agrobacterium*-mediated transformation of sesame. (**a**) Formation of adventitious shoots after co-cultivation with *Agrobacterium*, (**b**) elongated shoots, (**c**) root formation

Fig. 1.5 Transient GUS expression in putative transformants. (**a**) Leaf from control plant, (**b**) transformed leaf

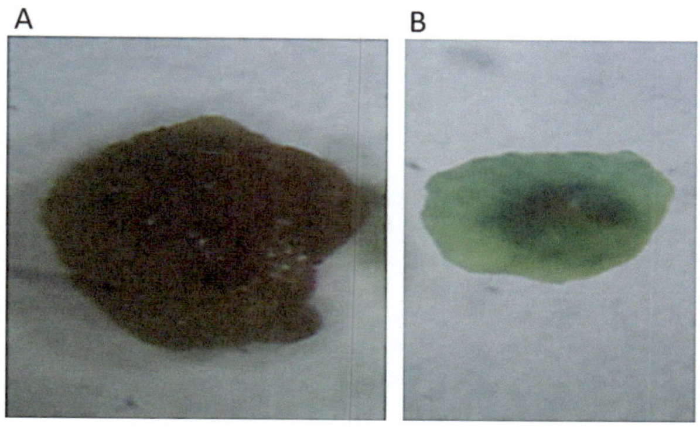

transformation efficiency. De-embryonated cotyledon was used as an explant which was firstly kept on pre-regeneration media for 6 days containing 9% sucrose. For the selection, the kanamycin concentration 50 mg/l in pre-regeneration, regeneration and selection medium helped to sort out nontransformed plants. Histochemical Gus assays have been used as a parameter to deduce the transformation frequency. Molecular analyses including PCR and RT-PCR were done to confirm the integration and expression of neomycin phosphotransferase II (nptII) gene in transgenic sesame lines. The analyses of T_0 and T_1 generation transgenic plants were done, and T_1 transgenic lines homozygous for the transgene were obtained.

Explants and medium which produced an efficient response in in vitro direct regeneration were used for *Agrobacterium*-mediated transformation in sesame. De-embryonated cotyledon explants were precultured for 4 days (Fig. 1.4a), and about 1 month old calli from hypocotyl were used in transformation experiments. An infection time of 15 min and co-cultivation period of 3 days were found to be optimum for de-embryonated cotyledon explants of *S. indicum*. The maximum number of explant survival on selection medium was achieved after 3 days of co-cultivation (Fig. 1.4b, c). Kanamycin at a concentration of 50 mg/l was used for preventing the multiplication and proliferation of untransformed shoots and to select putative transformants. Gus expression (Fig. 1.5) and PCR analysis were done to confirm the presence of gene of interest in putative transformants. The putatively transformed calli and co-cultivated explants of 8 to 10 weeks old were randomly selected for gus analysis. Presence of blue colour spots confirms the transient gene expression.

References

Abdellatef E, Ahmed MMM, Daffalla HM, Khalafalla MM (2010) Enhancement of adventitious shoot regeneration in sesame (*Sesamum indicum L.*) cultivar promo KY using ethylene inhibitors. J Phytol 2:61–67

Ahmad AO, Mayara M, Qibin Y, Fred G Jr, Christine D, James H, Jude W (2017) Production of three new grapefruit cybrids with potential for improved citrus canker resistance. In Vitro Cell Dev Biol Plant. https://doi.org/10.1007/s11627-017-9816-7

Ahmed R, Anis M (2014) Rapid in vitro propagation system through shoot tip cultures of *Vitex trifolia* L.– an important multipurpose plant of the Pacific traditional Medicine. Physiol Mol Biol Plants. https://doi.org/10.1007/s12298-014-0233-7

Al-Shafeay AF, Ibrahim AS, Nesiem MR, Tawfik MS (2011) Establishment of regeneration and transformation system in Egyptian sesame (*Sesamum indicum L.*) cvSohag 1. GM Crops 2:182–192

Babaei N, Ashikin N, Abdullah P, Saleh G, Abdullah LT (2014) An efficient in vitro plantlet regeneration from shoot tip cultures of *Curculigo latifolia*, a medicinal plant. Scient World J 2014:275028

Baskaran P, Jayabalan N (2006) *In vitro* mass propagation and diverse callus orientation on *Sesamum indicum* L. an important oil plant. J Agricult Technol 2:259–269

Bates GW (1999) Plant transformation via protoplast electroporation, in: plant cell culture protocols. Humana Press, New Jersey, pp 359–366. https://doi.org/10.1385/1-59259-583-9:359

Bettoni JC, Costa MD, Gardin JPP, Kretzschmar AA, Pathirana R (2016) Cryotherapy: a new technique to obtain grapevine plants free of viruses. Rev Bras Frutic 38(2)

Bhalla PL, Singh MB (2008) Agrobacterium-mediated transformation of Brassica napus and *Brassica oleracea*. Nature Proto 3:181–189

Bhattacharyya J, Chakraborthy A, Mitra J, Chakraborthy S, Pradhan S, Manna A, Sikdar N, Sen SK (2015) Genetic transformation of cultivated sesame (*Sesamum indicum* L. cv Rama) through particle bombardment using 5-day-old apical, meristematic tissues of germinating seedlings. Plant Cell Tissue Org Cult 123(3):455–466

Birch RG (1997) Plant Transformation: problems and strategies for practical application. Annu Rev Plant Physiol Plant Mol Biol 48:297–326. https://doi.org/10.1146/annurev.arplant.48.1.297

Brencic A, Angert ER, Winans SC (2005) Unwounded plants elicit agrobacterium vir gene induction and T-DNA transfer: transformed plant cells produce opines yet are tumour free. Mol Microbiol 57:1522–1531. https://doi.org/10.1111/j.1365-2958.2005.04763.x

Burbulis N, Blinstrubien˙e A, Kuprien˙e R (2011) Effect of genotype and medium composition on linseed (*Linum usitatissimum*) ovary culture. Biologia 66/3:465–469

Chattopadhyaya B, Banerjee J, Maiti MK (2010) Shoot induction and regeneration using internodal transverse thin cell layer culture in *Sesamumindicum* L. Plant Biotechnology Rep 4:173–178

Chavarriaga-Aguirre P, Brand A, Medina A, Prías M, Escobar R, Martinez J, Díaz P, López C, Roca WM, Tohme J (2016) The potential of using biotechnology to improve cassava: a review. In Vitro Cell Dev Biol Plant 52(5):461–478

Chowdhury S, Basu A, Kundu S (2014) A new high-frequency *Agrobacterium*- mediated transformation technique for *Sesamumindicum L.* using de-embryonated cotyledon as explants. Protoplasma:1–16

Ckurshumova W, Berleth T (2015) Overcoming recalcitrance-auxin response factor functions in plant regeneration. Plant Signal Behav *10*(7):e993293 de Framond AJ, Barton KA, Chilton MD (1983) Mini-Ti: a new vector strategy for plant genetic engineering. Biotechnol 5:262–269

Dufourmantel N, Tissot G, Garçon F, Pelissier B, Dubald M (2006) Stability of soybean recombinant plastome over six generations. Transgenic Res 15:305–311. https://doi.org/10.1007/s11248-005-5262-0

Elghabi Z, Ruf S, Bock R (2011) Biolistic co-transformation of the nuclear and plastid genomes: co-transformation of nuclei and plastids. Plant J 67:941–948. https://doi.org/10.1111/j.1365-313X.2011.04631.x

Feng C, Wang R, Li J, Wang B, Yin Z, Cui Z, Li B, Zhang ML, Wang Q (2013) Production of pathogen-free horticultural Cropsby cryotherapy of in vitro-grown shoot tips, Protocols for micropropagation of selected economically important horticultural crop series-methods in molecular biology, vol 994. Springer, New York, pp 463–482

Ferrie AMR, Caswell KL (2011) Isolated microspore culture techniques and recent progress for haploid and doubled haploid plant production. Plant Cell Tiss Organ Cult 104:301–309

Fromm ME, Taylor LP, Walbot V (1986) Stable transformation of maize after gene transfer by electroporation. Nature 319:791–793. https://doi.org/10.1038/319791a0

Fu X, Fu X, Duc LT, Fontana S, Bong BB, Tinjuangjun P, Sudhakar D, Twyman RM, Christou P, Kohli A (2000) Linear transgene constructs lacking vector backbone sequences generate low-copy-number transgenic plants with simple integration patterns. Transgenic Res 9:11–19. https://doi.org/10.1023/A:1008993730505

Gafni Y, Icht M, Rubinfeld B-Z (1995) Stimulation of agrobacterium tumefaciens virulence with indole-3-acetic acid. Lett Appl Microbiol 20:98–101

Gelvin SB (2003) Agrobacterium-mediated plant transformation: the biology behind the "gene-jockeying" tool. Microbiol Mol Biol Rev 67:16–37. https://doi.org/10.1128/MMBR.67.1.16-37.2003

Germana AM (2011) Anther culture for haploid and doubled haploid production. Plant Cell Tiss Organ Cult 104:283–300

Guan Y, Li S, Fan X, Su Z (2016) Application of somatic embryogenesis in woody plants. Front Plant Sci 7. https://doi.org/10.3389/fpls.2016.00938

Guo WW, Xiao XS, Deng XX (2013) Somatic cybrid production via protoplast fusion for citrus improvement. Sci Hort 163:20–26

Havgeppa H, Rao S (2013) Direct somatic embryogenesis in (*SesamumindicumL.*) Cv-e8 from cotyledon and hypocotyl explants. Intl J Appl Biol Pharm Technol 4(2):120–127

Hellens R, Mullineaux P, Klee H (2000) Technical focus: a guide to agrobacterium binary Ti vectors. Trends Plant Sci 5:446–451. https://doi.org/10.1016/S1360-1385(00)01740-4

Hwang H-H, Yu M, Lai E-M (2017) Agrobacterium -mediated plant transformation: biology and applications. Arab Book 15:e0186. https://doi.org/10.1199/tab.0186

Iida A, Seki M, Kamada M, Yamada Y, Morikawa H (1990) Gene delivery into cultured plant cells by DNA-coated gold particles accelerated by a pneumatic particle gun. Theor Appl Genet 80:813–816. https://doi.org/10.1007/BF00224198

Ismagul A, Yang N, Maltseva E, Iskakova G, Mazonka I, Skiba Y, Bi H, Eliby S, Jatayev S, Shavrukov Y, Borisjuk N, Langridge P (2018) A biolistic method for high-throughput production of transgenic wheat plants with single gene insertions. BMC Plant Biol 18:135. https://doi.org/10.1186/s12870-018-1326-1

Kaczmarczyk A, Rokka V-M, Keller JER (2011) Potato shoot tip cryopreservation. A review. Potato Res 54:45–79

Kaeppler HF, Somers DA, Rines HW, Cockburn AF (1992) Silicon carbide fiber-mediated stable transformation of plant cells. Theor Appl Genet 84–84:560–566. https://doi.org/10.1007/BF00224152

Kariyallappa DK, Lokesha R, Nair MK, Treertha Prasad D (2003) Breeding for alternaria blight resistant in sesame-classical V/s bio technological approaches. National seminar on advances in genetics and plant breeding. Impact of DNA revolution. University of Agricultural Science, Dharwad, pp 157–158

Kikkert JR, Vidal JR, Reisch BI. (2004) Stable transformation of plant cells by particle bombardment/biolistics. In: Methods in molecular biology. Transgenic plants: methods and protocols. Humana Press Inc, Totowa, pp 61–78

Kim YH (2001) Effects of BA,NAA,2-4,D and AgNO3 treatments on the callus induction and shoot regeneration from hypocotyls and cotyledon of sesame (*Sesamum indicum* L.). J Korean Soc Horticult Sci 42(1):70–74

Kiyokawa K, Yamamoto S, Sakuma K, Tanaka K, Moriguchi K, Suzuki K (2009) Construction of disarmed Ti plasmids transferable between Escherichia coli and agrobacterium species. Appl Env Microbiol 75:1845–1851. https://doi.org/10.1128/AEM.01856-08

Klein TM, Gradziel T, Fromm ME, Sanford JC (1988) Factors influencing gene delivery into Zea Mays cells by high-velocity microprojectiles. Nat Biotechnol 6:559–563. https://doi.org/10.1038/nbt0588-559

Klein TM, Kornstein L, Sanford JC, Fromm ME (1989) Genetic transformation of maize cells by particle bombardment. Plant Physiol 91:440–444. https://doi.org/10.1104/pp.91.1.440

Lokesha R, Shashidhara N, Janagoudar BS (2005) A new and rapid method of successful callusing in sesame (*Sesamumindicum* L.) through direct seeding. Plant Cell Biotechnol Mol Biol 8(1):85–88

Lowe BA, Shiva Prakash N, Way M, Mann MT, Spencer TM, Boddupalli RS (2009) Enhanced single copy integration events in corn via particle bombardment using low quantities of DNA. Transgenic Res 18:831–840. https://doi.org/10.1007/s11248-009-9265-0

Maraschin SF, Lamers GE, de Pater BS, Spaink HP, Wang M (2003) 14-3-3 isoforms and pattern formation during barley microspore embryogenesis. J Exp Bot 54:1033–1043. https://doi.org/10.1093/jxb/erg098

Masani MYA, Noll GA, Parveez GKA, Sambanthamurthi R, Prüfer D (2014) Efficient transformation of oil palm protoplasts by PEG-mediated transfection and DNA microinjection. PLoS One 9(5):e96831. https://doi.org/10.1371/journal.pone.0096831

Murakawa T, Kajiyama S, Fukui K (2008) Improvement of bioactive bead-mediated transformation by concomitant application of electroporation. Plant Biotechnol 25:387–390. https://doi.org/10.5511/plantbiotechnology.25.387

Muthulakshmi C, Pratik P, Kokiladevi E, Subramanian S (2013) Transgenic *Sesamum indicum* plants for alpha linolenic acid production. Amrita Bioquest, Amrita vishwa vidyapeetham

Naaz A, Shahzad A, Anis M (2014) Effect of adenine sulphate interaction on growth and development of shoot regeneration and inhibition of shoot tip necrosis under in vitro condition in adult Syzygium cumini L. – a multipurpose tree. Appl Biochem Biotechnol 173(1):90–102

Nandagopal K, Halder M, Dash B, Nayak S, Jha S (2018) Biotechnological approaches for production of anti-cancerous compounds resveratrol, podophyllotoxin and zerumbone. Curr Med Chem 25(36):4693–4717. https://doi.org/10.2174/0929867324666170404145656

Nanjareddy K, Arthikala MK, Blanco L, Arellano SE, Lara M (2016) Protoplast isolation, transient transformation of leaf mesophyll protoplasts and improved Agrobacterium-mediated leaf disc infiltration of *Phaseolus vulgaris*: tools for rapid gene expression analysis. BMC Biotechnol 16:53. https://doi.org/10.1186/s12896-016-0283-8

Nugent GD, Coyne S, Nguyen TT, Kavanagh TA, Dix PJ (2006) Nuclear and plastid transformation of Brassica oleracea var. botrytis (cauliflower) using PEG-mediated uptake of DNA into protoplasts. Plant Sci 170:135–142. https://doi.org/10.1016/j.plantsci.2005.08.020

Ochatt SJ (2008) Flow cytometry in plant breeding. Cytometry 73A:581–598

Omar AA, Murata M, Yu Q, Gmitter FG Jr, Chase CD, James HG, Grosser JW (2017) Production of three new grapefruit cybrids with potential for improve citrus canker resistance. In Vitro Cell Dev Biol Plant. https://doi.org/10.1007/s11627-017-9816-7

Otten L, Burr T, Szegedi E (2007) Agrobacterium: a disease-causing bacterium. In: Tzifra T, Citovsky V (eds) Agrobacterium: from biology to biotechnology. Springer, New York, pp 1–46

Perera PI, Ordoñez CA, Dedicova B, Ortega PE (2014, May 21) Reprogramming of cassava (*Manihot esculenta*) microspores towards sporophytic development. AoB Plants 6. pii: plu022. https://doi.org/10.1093/aobpla/plu022

Perianez-Rodriguez J, Manzano C, Moreno-Risueno MA (2014) Post-embryonic organogenesis and plant regeneration from tissues: two sides of the same coin? Front Plant Sci 5:219. https://doi.org/10.3389/fpls.2014.00219
Pretova A, Obert B, Bartosova Z (2006) Haploid formation inmaize, barley, flax and potato. Protoplasma 228(1–3):107–114
Quiroz-Figueroa FR, Rojas-Herrera R, Galaz-Avalos RM, Loyola-Vargas VM (2006) "Embryo production through somatic embriogénesis can be used to study", cell differentiation in plants. Plant Cell Tissue Organ Cult 86:285–301
Rai MK, Asthana P, Singh SK, Jaiswal VS, Jaiswal U (2009) The encapsulation technology in fruit plants – a review. Biotechnol Adv 27:671–679. https://doi.org/10.1016/j.biotechadv.2009.04.025
Rajeswari S, Thiruvengadam V, Ramaswamy NM (2010) Optimised embryo rescue method – Sesamum (production of interspecific hybrids between *Sesamum alatum* Thonn and *Sesamum indicum* L. through ovule culture and screening for phyllody disease resistance). S Afr J Bot 76:252–258
Ram R, Catlin D, Romero J, Cowley C (1990) Sesame: new approaches for crop improvement. In: Janick J, Simon JE (eds) Advances in new crops. Timber Press, Portland, pp 225–228
Rivera AL, Gómez-Lim M, Fernández F, Loske AM (2012) Physical methods for genetic plant transformation. Phys Life Rev 9:308–345. https://doi.org/10.1016/j.plrev.2012.06.002
Sanford JC (1990) Biolistic plant transformation. Physiol Plant 79:206–209. https://doi.org/10.1111/j.1399-3054.1990.tb05888.x
San José MDC, Corredoira E, Oliveira H, Santos C (2015) Cryopreservation of somatic embryos of *Alnus glutinosa* (L.) Gaertn. And confirmation of ploidy stability by flow cytometry. Plant Cell TissOrgCult 123:489–499. https://doi.org/10.1007/s11240-015-0853-9
Seo HY, Kim YJ, Park TI, Kim HS, Yun SJ, Park KH, Oh MK, Choi MY, Paik CH, Lee YS, Choi YE (2007) High frequency plant regeneration via adventitious shoot formation from deembryonated cotyledon explants of *SesamumindicumL.* In Vitro Cell Dev Biol 43:209–214
Sharma S, Shahzad A, da Silva JAT (2013) Synseed technology-A completesynthesis. Biotechnol Adv 31:186–207. https://doi.org/10.1016/j.biotechadv.2012.09.007
Sivamani E, DeLong RK, Qu R (2009) Protamine-mediated DNA coating remarkably improves bombardment transformation efficiency in plant cells. Plant Cell Rep 28:213–221. https://doi.org/10.1007/s00299-008-0636-4
Skoog F, Miller CO (1957) Chemical regulation of growth and organ formation in plant tissue cultured in vitro. Symp Soc Exp Biol 11:118–131
Supena EDJ, Suharsono S, Jacobsen E, Custers JBM (2006) Successful development of a shed-microspore culture protocol for doubled haploid production in Indonesian hot pepper (*Capsicum annuum* L.). Plant Cell Rep 25:1–10
Taskin KM, Turgut K (1997) *In vitro*regeneration of sesame. Turkish J Bot 21(1):15–18
Thole V, Worland B, Snape JW, Vain P (2007) The pCLEAN dual binary vector system for agrobacterium-mediated plant transformation. Plant Physiol 145:1211–1219. https://doi.org/10.1104/pp.107.108563
Torney F, Trewyn BG, Lin VS-Y, Wang K (2007) Mesoporous silica nanoparticles deliver DNA and chemicals into plants. Nat Nanotechnol 2:295–300. https://doi.org/10.1038/nnano.2007.108
Tyagi S, Spörlein B, Tyagi AK, Herrmann RG, Koop HU (1989) PEG- and electroporation-induced transformation in *Nicotiana tabacum*: influence of genotype on transformation frequencies. Theor Appl Genet 78:287–292. https://doi.org/10.1007/bf00288813
Tzfira T, Citovsky V (2000) From host recognition to T-DNA integration: the function of bacterial and plant genes in the agrobacterium – plant cell interaction. Mol Plant Pathol 1(4):201–212
Valluri JV (2009) Bioreactor production of secondary metabolites from cell cultures of periwinkle and sandalwood. Methods Mol Biol 547:325–335. https://doi.org/10.1007/978-1-60327-287-2_26
Van Haute E, Joos H, Maes M, Warren G, Van Montagu M, Schell J (1983) Intergeneric transfer and exchange recombination of restriction fragments cloned in pBR322: a novel strategy for the reversed genetics of the Ti plasmids of agrobacterium tumefaciens. EMBO J 2:411–417

van Wordragen MF, Dons HJM (1992) Agrobacterium tumefaciens-mediated transformation of recalcitrant crops. Plant Mol Biol Rep 10:12–36

Vidal JR, Kikkert JR, Donzelli BD, Wallace PG, Reisch BI (2006) Biolistic transformation of grapevine using minimal gene cassette technology. Plant Cell Rep 25:807–814. https://doi.org/10.1007/s00299-006-0132-7

Wadeyar BS, Lokesha R, Gayatree G, Sharanamma S (2013) Sesame shootregeneration-using different combinations of growth regulators. Mol Plant Breed 4(33):267–269

Wang H, Dong B, Jiang J, Fang W, Guan Z, Liao Y, Chen S, Chen F (2014) Characterization of in vitro haploid and doubled haploid *Chrysanthemum morifolium* plants via unfertilized ovule culture for phenotypical traits and DNA methylation pattern. Front Plant Sci 5:article 738

Were BA, Gudu S, Onkware AO, Carlsson AS, Welander M (2006) *Invitro* regeneration of sesame (*SesamumindicumL.*) from seedling cotyledon and hypocotyl explants. Plant Cell Tissue Organ Cult 85:235–239

Xu ZQ, Jia JF, Hu ZD (1997) Somatic embryogenesis in *Sesamum indicum* L.cv. Nigrum. J Plant Physiol 150:755–758

Yadav M, Chaudhary D, Sainger M, Jaiwal PK (2010) *Agrobacterium tumefaciens*-mediated genetic transformation of sesame (*Sesamumindicum L.*). Plant Cell Tissue Organ Cult 103:377–386

Yang C, Zhao T, Yu D, Gai J (2011) Mapping QTLs for Tissue Culture Response in Soybean (*Glycine max* (L.) Merr.). Mol Cells 32(4):337–342

Younghee K (2001) Effects of BA, NAA, 2, 4-D and $AgNO_3$ treatments on the callus induction and shoot regeneration from hypocotyl and cotyledon of sesame (*Sesamumindicum* L.). J Korean Soc Hort Sci 42:70–74

Yu G, Cheng Q, Xie Z, Xu B, Huang B, Zhao B (2017) An efficient protocol for perennial ryegrass mesophyll protoplast isolation and transformation, and its application on interaction study between LpNOL and LpNYC1. Plant Methods 13:46. https://doi.org/10.1186/s13007-017-0196-0

Zahm P, Hohmeyer C, Geider K (1984) Site-specific mutagenesis of the Ti plasmid by transformation of agrobacterium tumefaciens with mutagenized T-DNA fragments cloned in *E. coli* plasmids. MGG Mol Gen Genet 194:188–194. https://doi.org/10.1007/BF00383515

Zhang HM, Yang H, Rech EL, Golds TJ, Davis AS, Mulligan BJ, Cocking EC, Davey MR (1998) Transgenic rice plants produced by electroporation-mediated plasmid uptake into protoplasts. plant cell reports 7:379–384

Zupan JR, Zambryski P (1995) Transfer of T-DNA from agrobacterium to the plant cell. Plant Physiol 107:1041–1047

Zupan J, Muth TR, Draper O, Zambryski P (2000) The transfer of DNA from agrobacterium tumefaciens into plants: a feast of fundamental insights. Plant J 23:11–28. https://doi.org/10.1046/j.1365-313x.2000.00808.x

Cell Cultures and Hairy Roots as Platform for Production of High-Value Metabolites: Current Approaches, Limitations, and Future Prospects

2

Paola Isabel Angulo-Bejarano, Juan Luis De la Fuente Jimenez, Sujay Paul, Marcos de Donato-Capote, Irais Castillo-Maldonado, Gabriel Betanzos-Cabrera, Juan Ignacio Valiente-Banuet, and Ashutosh Sharma

Abstract

Current knowledge in the use of hairy root cell cultures is growing. The use of this technology in the production of secondary metabolites is also increasing with the general purpose of generating medicinal alternatives from ethnopharmaceutical origins. Strategies include either genetically modified hairy roots that overexpress genes that codify for key enzymes in complex metabolic pathways or the use of certain elicitors for the natural production of certain compounds in hairy roots. The importance of understanding basic root technology and the complex genetic regulation of root development helps in establishing well-designed strategies in order to obtain higher yields in terms of metabolites or recombinant protein production. Also, bioprocess factors such as the use of bioreactors, elicitors, nutrients, and phytohormones for the induction of compounds of economic importance are also growing. Thus, the aim of the present chapter is to summarize the most important aspects regarding the technology to generate these hairy roots and their main applications in novel fields such as phytoremediation.

P. I. Angulo-Bejarano · J. L. De la Fuente Jimenez · S. Paul · M. de Donato-Capote
G. Betanzos-Cabrera · J. I. Valiente-Banuet · A. Sharma (✉)
Tecnologico de Monterrey, Centre of Bioengineering, Queretaro, Mexico
e-mail: asharma@tec.mx

I. Castillo-Maldonado
Departamento de Bioquímica y Fitofarmacología, Centro de Investigación Biomédica, Facultad de Medicina Universidad Autónoma de Coahuila, Torreón, Coahuila, Mexico

R. Sathishkumar et al. (eds.), *Advances in Plant Transgenics: Methods and Applications*, https://doi.org/10.1007/978-981-13-9624-3_2

Keywords
Hairy root · Secondary metabolites · Recombinant protein · Bioreactors · Elicitors · Phytoremediation

2.1 Introduction

Plant biotechnology is tightly linked with modern human civilization development and can be traced back to the discovery of plant cell plasticity in the form of plant tissue culture dating back to 100 years which opened the door for new applications of plants such as plant tissue culture including plant micropropagation, direct and indirect organogenesis, and somatic embryogenesis, among others (Mauro et al. 2017). Further on, the advent of genetic transformation in the 1980s with the use of *Agrobacterium* strains increases the possibility of using plants for the expression of heterologous recombinant proteins (Chilton et al. 1982, Caplan et al. 1983; De Block et al. 1985). Furthermore, the capacity of certain strains of *Agrobacterium* to generate the hairy root syndrome revealed that this machinery could be used to induce the expression of either the production of hairy roots or to target the expression of heterologous proteins in root tissues (Chilton et al. 1982). This discovery made the production of secondary metabolites or newly produced proteins in hairy root cultures a new biotechnological tool. Nowadays, as the public searches for ethnopharmaceutical alternatives in plants, the use of this technology is facing new challenges such as the overexpression of natural occurring secondary metabolites in plants and its production in hairy root culture (Sharma et al. 2012, 2018).

2.2 Plant Cell Cultures

Plant tissue culture has long been a technology that has allowed the production of plants and several compounds present in them. The basis to use plant cell culture dates back to the beginning of the 1900s with the work of Gottlieb Haberlandt who was able to describe callus generation from adult plant cells and the consequent regeneration process (new plant formation). This capability of plants to regenerate themselves was named "totipotency" and has been applied for propagation of plants via cutting and grafting. In fact, this capacity was demonstrated for the first time in carrot cultures grown in vitro by Haberlandt in the 1950s (Ikeuchi et al. 2016; Eibl et al. 2018). The use of this technology thus centered in the generation of new plants to produce secondary metabolites with potential applications in the food, cosmetic, and pharmaceutical industries (Eibl et al. 2018). The plasticity of plant cells to regenerate is increased by the application of phytohormones also known as plant growth regulators. The change from a completely developed plant cell tissue into meristematic-like structures is caused by plant cell reprogramming of their cell cycles or by activation of undifferentiated plant somatic tissues (Ikeuchi et al. 2016).

The concentration ratio between auxins and cytokinins defines the cellular fate for plant cells: a high auxin to cytokinin ratio leads to root regeneration, while high ratio of cytokinins to auxins results in shoot regeneration (Skoog and Miller 1957). However, when both concentrations are almost in the same value, the formation of callus can be observed. Three main steps can be recognized during plant regeneration: first, explant cells respond to phytohormones and acquire organogenic competence, second quiescent cells reenter the cell cycle, and third, cellular fate is determined by the formation of newly formed primordia for each organ, this constitutes the key step during de novo organogenesis. Finally, during the last step in vitro morphogenesis is achieved (Zhao et al. 2008). Identity genes play a fundamental role during plant de novo organogenesis. Factors such as epigenetics also influence the expression of such genes altering the outcome of plant cells (Fransz and de Jong 2002).

Namely, organogenesis in plant cell cultures can be achieved by direct organogenesis, when the mother explant is exposed to varying concentrations of phytohormones (naturally occurring or via exogenous application) and produces new organs, for example, during micropropagation (Zhao et al. 2008). Indirect organogenesis includes the extra step of callus induction where the mother explant will experiment changes in its cell cycle programming leading to the proliferation of meristematic-like cells that generate calli with regenerative potential which in turn leads to the formation of de novo organs or complete plants (Angulo-Bejarano and Paredes-López 2011). Finally, another example of plant plasticity is somatic embryogenesis where plants are capable of generating embryos either from direct explants or via the callus formation step; these "synthetic embryos" will be able to regenerate new plants and a preferred strategy for plant regeneration after genetic transformation events since they imply that the genetic material in this type of embryo is homogenous and reduces the risk of chimeras during the regeneration of transformed plants (Angulo-Bejarano 2013).

2.3 Cell Suspension Cultures

Other applications for plant tissue culture include the use of cell cultures in liquid media, also known as cell suspension cultures. This type of cell cultures benefits from the production of calli in liquid media as well as the use of hairy roots for the liberation of compounds that have application in various biotechnological purposes (Eibl et al. 2018).

Plant tissue culture can be also utilized for the production of compounds derived from cell suspension extracts from dedifferentiated plant cells, calli, or hairy roots. General steps to produce dedifferentiated plant cells are explained in Fig. 2.1. In general terms this process consists in the induction, maintenance, and propagation of the callus culture in petri dishes followed with its transfer onto shake flasks or bioreactors to finally obtain the cell extracts. To achieve reliable results in terms of the compounds and its quality, it is necessary to start from healthy plant materials with proven regenerative capacity. However, compound production capacity is also affected by plant species, developmental stage, and the plant organ chosen for this process. As with all plant tissue culture procedures, choosing the correct

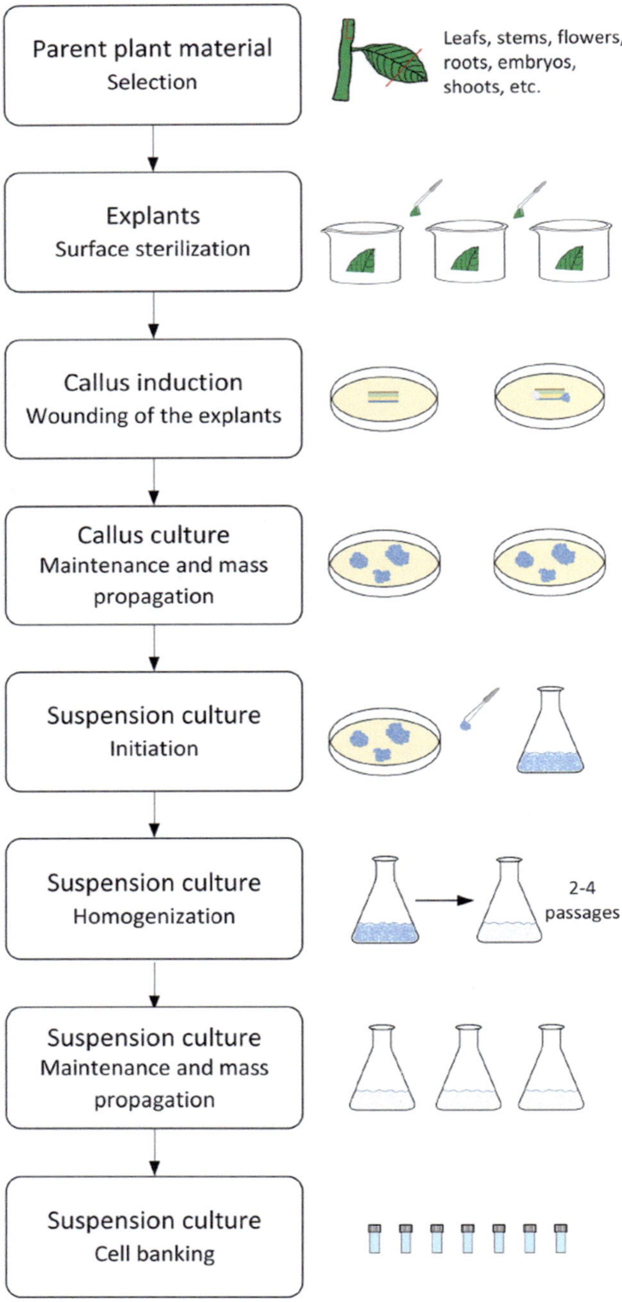

Fig. 2.1 Basic cell suspension culture steps (Eibl et al. 2018, used with permission)

phytohormone balance is vital, at first for callus induction (friable calli are preferred over compact calli, for their feasibility in liberating bioactive compounds into the liquid media) and second for bioactive compound production (Gutzeit and Ludwig-Müller 2014; Eibl et al. 2018).

Callus subcultures in new flask will be done once the calli formed start to grow in size, thus increasing its surface area. Factors that should be taken into consideration include the basal salts used for the growth and maintenance of calli, phytohormone concentration, as well as N, P, and sucrose levels (Murthy et al. 2014, Eibl et al. 2018). Cell suspension cultures are used for the production of important bioactive compounds, recombinant proteins, or secondary metabolites with potential applications in medicine. Examples of cell cultures used for medicinal purposes include the production of Taxol from *Taxus chinensis* and constitute one of the main procedures to obtain this compound (Onrubia et al. 2013). Cell cultures have also been applied for the production of anxiolytic compounds such as those present in *Galphimia glauca*, namely, galphimine B (Osuna et al. 1999). In general terms calli were obtained from hypocotyl explants and induced by the application of 2,4-dichlorophenoxyacetic acid (2,4-D). The compound was produced almost in the same magnitude as in uncultivated plants (Osuna et al. 1999).

The production of withanolides from *Withania somnifera* in cell suspension cultures has also been investigated (Sivanandhan et al. 2014). The application of several elicitors was also explored resulting in the expression of various types of withanolides. The highest values for total withanolides were found when utilizing shake flask cultures and bioreactors. Other strategies have included cell suspension cultures applying immobilized beads, such as in *Galphimia glauca* and the production of galphimine B (Osuna et al. 2008). When comparing the use of immobilized beads in bioreactors, the best result was obtained when applying agitation resulting in 1381 μg/L of the compound.

2.4 Cells Suspension vs Hairy Roots (Chromosome Number and Permeabilization)

In nature, different secondary products are hydrophobic and kept intracellularly either in the vacuole or cytosol and are insignificantly released into the culture medium. For increasing yields of wanted metabolites and decreasing costs of production, some methods to increase cell permeability to release the metabolites have been employed (Chandra and Chandra 2011). Among some agents that increase the cell permeability, we can find detergents, pH, sonication, oxygen stress, temperature, and calcium chelators, among others (Thimmaraju et al. 2003). Furthermore, suspension cell culture does not frequently release the secondary metabolites into the medium (Linden et al. 2001). Thus, hairy root culture could offer a way to pass this difficulty related with cell culture since there is a quick growth, less time, the capability to produce a variety of chemical compounds at the same time, and an easy maintenance (Chandra and Chandra 2011). For example, it was reported that application of Tween 20 in hairy roots of *Datura innoxia* led to the movement of large quantities of alkaloids from the cell toward the culture medium (Boitel-Conti et al. 1996). Another example was the achievement of betacyanin release from *Beta vulgaris* with oxygen starvation (Boitel-Conti et al. 1996). Products released into the medium can be retrieved by absorbents (Giri and Narasu 2000). Most of the works done that used differentiated cultures and no cell suspension cultures have been directed to hairy root transformation obtained from *A. rhizogenes*, which represent an organized continuous growth of plant tissue (Doran 1997). Another disadvantage in plant cell culture in many cases is a low metabolite production (Chandra and Chandra 2011), while the secondary product production is higher in hairy root cultures than in parent plants (Granicher et al. 1992). Moreover, biochemical and genetic stability along with efficient productivity of differentiated cultures gives plusses upon cell suspension cultures (Chandra and Chandra 2011).

2.5 Roots

2.5.1 Root Physiology

2.5.1.1 Root Architecture (Morphology, Topology, etc.)/Lateral Branches, Root Hairs

Root systems provide water and nutrient uptake for plant survival and health (Lynch, 1995). Hence, roots have high abilities for nutrient and water uptake and have symbiotic relation with mycorrhizal fungi and benefit from it. However, as time passes, the root gets older and loses pigmentation, death of cortical cells, and loses symbiotic relationships; less water and nutrient uptake are some aging symptoms (Wells and Eissenstat 2003). The root system configuration (root architecture) consists of the main root along with lateral roots and can differ between plant species and change because of soil compositions and availability of nutrients and water (Malamy 2005). The root architecture has three major processes: first, cell division in the meristem of the primary root which allows growth by adding new cells; second, the exploratory capacity increased by formation of lateral roots; and third, the formation of root hair for expansions in the total surface area (López-Bucio et al. 2003). In addition, this root system has five characters: topology (dissemination of the branches inside the system), diameter and length of internodes (links inside the system), and the two branching angles. Additionally, the root system has nodes (zones where new branches start) and internodes. All the nodes have the possibility to develop new branches which is determined by its positions on the root segment where it lies (primary, lateral, axis, etc.) (Fitter et al. 1991).

The root system is the result of a coordination between the action of biotic and abiotic factors and genetic endogenous programs (Malamy 2005). Moreover, plant growth and development are promoted by stem cells in the apical meristems (apical region of roots or shoots). This characteristic enables plants to adapt to environmental conditions with changes in morphology and organ development (Hodge et al. 2009). Furthermore, the root apical meristem (RAM) is the place where stem cells produce all underground organs and lateral roots (Sabatini et al. 2003). Some processes influence the root system architecture as the development of primary and adventitious roots; new lateral roots (branching) for nutrients perception; axial growth for new explorations in the soil; radial growth for more anchorage, strength, and storage capacity; and finally root decay and senescence when the uptake of resources is not effective (Hodge et al. 2009). An example is high orders of root that primarily serve in storage, transport, and as plant anchorage to the soil. These high-order roots are the point where new roots can develop (Wells and Eissenstat 2003). Another factor that influences roots is the age. Young plants with young roots have less complexity than older root systems (Iyer-Pascuzzi et al. 2010).

However, older roots represent the biggest possible surface area for nutrient and water uptake and from resources below the reach of young roots (Wells and Eissenstat 2003). Exploitation efficiency of the soil resources is performed by the architecture of the root system. Thus, high development of more branches in the main axis and long links in the root system is more efficient when the growth is limited due to low soil resources or is water deficient and has low nutrient supply, and root hair location, size, and number are valuable since they have a big impact in the absorptive surface parts of the root (Miller and Cramer 2004). Root hairs can be almost 70% of the total surface area of the root (López-Bucio et al. 2003). In addition, root branching pattern determines the root architecture. When nitrogen is in reduced quantities, the size of the root increases, thus, stimulating root growth (Miller and Cramer 2004). For example, thick roots have less surface area covered than fine roots and need more carbon for construction; however, fine roots may be costlier for preservation (Miller and Cramer 2004). Fine roots are usually short and highly branched (McCully 1999). These fine roots are white when they start growing and with age may become brown with no secondary thickening after some months of their production (Richards and Considine 1981). This color is a result of the lack of cortical and epidermal tissue (Wells and Eissenstat 2003).

On the other hand, development of plants and their secondary metabolite production is influenced by several factors such as phosphate and carbon concentration along with nitrogen source (Taya et al. 1994), temperature, light quality, pH of the medium, concentration of ions, and phytohormones (Yu et al. 2005). Also, several hormones such as ethylene, abscisic acid, brassinosteroids, and mainly auxins (Bao et al. 2004), as well as diverse nutrients like nitrate, iron, sulfate, and phosphate, control the formation of lateral roots (Jing and Strader 2019).

2.5.1.2 Soil Nutrients

The availability of soil nutrients is significant for plant growth, development, and productivity (Jing and Strader 2019). The most significant inorganic nutrient for plants is nitrogen (N) and in the greatest component of secondary metabolites, proteins, nucleic acids, and cofactors. It is also involved in growth, development, metabolism, modulating resource allocation in shoot-root, tuber and flower initiation, and senescence (Marschner 1995). Nitrogen is probably in charge of controlling the reactions of plants when there is a rising in the atmospheric carbon dioxide (CO_2) concentration ([CO_2]) (Aber et al. 1989). Therefore, increasing nitrogen disposal can enhance the growth of plants as a response to the rise of [CO_2] (Curtis and

Wang 1998). Plants can also modify their root system architecture in order to uptake more N with high CO_2 (Berntson et al. 1997). Therefore, Jackson et al. (1990) showed that physiological and architectural root responses could be the result of a trade-off among ambient [CO_2] and changes in nitrogen supply. Enzyme activity and gene expression are regulated by nitrogen and carbon status of the plant. The carbon accessible within the root is used for the delivery of reductant and for amino acid synthesis as carbon skeletons (Miller and Cramer 2004).

Moreover, plant tissues have nitrogen (N) as the principal mineral element that is mainly obtained by the roots from the soil. Nitrogen is found in the soil in different organic (peptides and proteins) and inorganic (ammonium (NH_4^+) and NO_3^-) forms. Inorganic nitrogen is used by microbes, and they immobilized it, causing the reduction of N making it available for plants if sufficient carbon is accessible to sustain the microbial biomass (Hodge et al. 2000). This influences the accessibility of nitrogen and its uptake by roots (Miller and Cramer 2004). However, ammonium (NH_4^+), nitrate (NO_3^-), and amino acids are the most abundant forms of nitrogen in the soil (Miller and Cramer 2004). After uptake of amino acids, they enter an amino compound pool in roots and afterward be assimilated directly into proteins, moved to the shoot, or deaminated in the root (Miller and Cramer 2004). Though NH_4^+ dominates the soil ratio solution, an increase in the accumulation of nitrogen can change NH_4^+ to NO_3^- (Aber et al. 1989). Plants need N during their growth which represents 2% of the total dry matter from the plant and is a constituent of several secondary metabolites, nucleic acids, proteins, and enzymes. However, nitrogen must adjust in order to contribute in the plants biochemistry (Miller and Cramer 2004). Thus, nitrogen enters the food chain mainly as NH_4^+ and NO_3^- (Miller and Cramer 2004). Nitrate acquired by roots is either transferred to the shoot, saved in vacuoles, or reduced throughout nitrate reductase enzymes to NO_2^-, and afterward ammonium can be oxidized in a method known as nitrification via NO_2^- to NO_3^- and included in a carbon skeleton in order to produce several amino acids such as glutamine and glutamate with the help of enzymes glutamine synthetase (GS) and glutamate-2-oxoglutarate aminotransferase (GOGAT) in the GS/GOGAT cycle (Miller and Cramer 2004). Nitrate induces nearly all the genes involved in de novo purine, pyrimidine, and deoxynucleotide synthesis, as well as nucleotide salvage, RNA synthesis and processing, and amino acid stimulation, but blocked assigned genes for nucleotide degradation, shikimate pathway (Scheible et al. 2004), and ammonium in soil is less accessible for root uptake (Lee and Rudge 1986). Some factors can influence negatively the process of nitrification as lack of water in the soil, low pH, low (5 °C) and high (40 °C) temperatures, anaerobic conditions (Lewis 1986), and NH_4^+ accumulation can arise (Britto and Kronzucker 2002). However, when ammonium is the predominant form of nitrogen in the soil, it can be toxic for certain species. This could be due to cation/anion (Chaillou and Lamaze 2001) and pH imbalance (Raven and Smith 1976). Another process that converts nitrates into nitrogen gases (NO, NO_2, N_2, N_2O) is called denitrification. This happens when the electron acceptor O_2 is switched with NO_3^- and the conditions allow it such as warm temperatures, available soil carbohydrates, high soil moisture, limited oxygen, and high nitrate concentration (Luo et al. 2000). Nitrogen is distributed in a heterogeneous way and needs the "nitrogen cycle" to transform from one form to another. This procedure involves soil microorganisms, as fungi and bacteria, which search for organic nitrogen molecules and through mineralization convert N into NH_4^+. Nonetheless, plants compete with microorganism for organic nitrogen, and this competition disables uptake from roots (Miller and Cramer 2004). Mineralization rates depend on temperature, aeration, and water content of the soil which in turn affects the microbial activity (Lewis 1986). Since mineralization is dependent of some factors, cold anaerobic soils control aerobic nitrification and nitrogen mineralization which produces soil rich in amino acids as the dominant variety of nitrogen. In contrast, warm aerobic soils have small amino acid quantities since mineralization occurs fast (Atkin 1996). Roots depend on the balance among mineralization, nitrification, and denitrification rates for uptaking nitrogen (Miller and Cramer 2004). Mostly, behind the root meristem occurs the most uptake of nitrate and ammonium (Miller and Cramer 2004). The net balance between passive efflux and active influx generates the total nitrate uptake, which is transported through the plasma membranes of cortical and epidermal cells from roots (Miller and Smith 1996). This transport needs energy from the cell no matter what total concentration of nitrate there is in the soil. The uptake is dependent of ATP which provides to the proton pumping ATPase that supports the gradient cross of two protons (H^+) throughout the plasma membrane down an electrochemical gradient (Miller and Smith 1996). Two different gene families have been recognized, *NRT1* and *NRT2*, for nitrate transportation (Williams and Miller 2001). Moreover, some members of both families are nitrate inducible, thus, suitable candidates for nitrate uptake from the soil. These members can be found in the root cortex and epidermis, including root hairs (Nazoa et al. 2003). However, some members of the families are expressed constitutively (Okamoto et al. 2003), especially root hairs, as *Arabidopsis AtNRT1.2* gene (Huang et al. 1999). On the other hand, ammonium transporter (AMT) gene family has also two different groups, *AMT1* and *AMT2* (Shelden et al. 2001). Some *AMT1* genes are located in root hairs (Ludewig et al. 2002). Several *AMTs* are constitutively expressed, but it depends on the availability of ammonium in the soil (von Wiren et al. 2000). White's (1996) studies revealed that ammonium ions can enter through potassium (K^+) channels into the cell. Furthermore, some genes specific from plants such as *ANR1*, *ABI4*, and *ABI5* have been associated with the response of lateral roots to nitrate (Zhang and Forde 2000; Signora et al. 2001) and genes needed for nitrate assimilation like *NII*, *NIA1*, and *NIA2* (Scheible et al. 2004). Also, nitrate is involved in enhancing the synthesis and exportation of cytokine toward the shoot (Sakakibara et al. 1998).

On the other hand, different nutrients that uptake from the soil can modify the root systems since the distribution is not homogenous. Potassium is vital for exudation processes, growth, movements, and osmoregulation (Szczerba et al. 2009), and boron increases the potassium influx and reduces the efflux by changing the plasma membrane permeability which causes a hyperpolarization of cell membranes of the roots (Cara et al. 2002). Also, aluminum increases citrate, malate, or succinate activity in roots (Andresen et al. 2011), and calcium at 100 μM concentration at least is required for integrity in root membrane (Tang et al. 2009). However, the absence of boron, zinc, and manganese modifies solute and ion flow under strong intensities of light causing a leakage of sucrose, amino acids, nitrate, and potassium since there is a loss in permeability of the membrane (Gherardi and Rengel 2004). Also, sulfate absence develops a branched root system and an increased density (López-Bucio et al. 2003). Another important micronutrient is phosphorus that is known to stimulate the lateral root growth (Williamson et al. 2001). However, phosphorus absence increases amino acid and organic acid exudation, promotes arbuscular mycorrhiza development, and decreases uptake of ions (Cl^-, Mg^{2+}, K^+, Ca^{2+}) and proton release (Tang et al. 2009). In addition, low phosphorus concentrations change root hairs by becoming longer and denser, thus, increasing biomass production (Ma et al. 2001). Similarly, zinc absence in plants makes them more efficient in iron mobilization but have a low concentration of vitamins (Zhang et al. 2006). Nevertheless, iron can become very limiting under specific soil conditions even if it is required for plant growth in small concentration (Guerinot and Yi 1994). Lack of iron increases the root hair elongation and formation, similarly as the absence of phosphorus, but this time the extra root hairs occupy positions that under normal conditions and iron availability are occupied by non-hair cells (Schmidt et al. 2000). Moreover, limitation of nitrogen in plants exhibits lateral root growth and anthocyanin accumulation in leaves but a reduction of chlorophyll (Scheible et al. 2004).

2.5.2 Root Metabolomics

2.5.2.1 Phytohormones and Root Metabolomics

Root growth and development have been widely linked with phytohormones (Saini et al. 2013). Among all the plant hormones, auxin plays a major role in controlling and regulating several stages of root development (Leyser 2006) and has been established as a master regulator (Jing and Strader 2019). Auxin transport and signaling are involved in the maintenance of apical dominance, adventitious root formation, lateral root initiation and development (Hodge et al. 2009), and signals of root gravitropism (Parry et al. 2001). It is known that auxin along with sucrose can influence lateral branch formation in roots (Tian and Reed 1999) and the number of lateral roots increased when a concentration of indole-3-acetic acid (IAA) is applied (Kerk et al. 2000). Auxin is produced in shoots and transported to roots in a polar way (Morris et al. 1969), and root cells are more sensitive than shoot cells to auxin concentrations (Thimann 1937).

On the other hand, IAA is the primary auxin in almost all of the plants and is in charge for several stages of root development and root system architecture (Overvoorde et al. 2010) as well as to be the first auxin to be applied in mediums for rooting (Cooper 1935). However, indole-3-butyric acid (IBA) is also an endogenous auxin in minor concentrations but promotes efficiently adventitious root development (Saini et al. 2013). Nevertheless, IBA is also used to enhance rooting and has been very efficient in many plant species. When comparing IAA and IBA, IBA had a greater ability to stimulate rooting due to its superior stability (Hartmann 1990). In addition, IAA showed more degradation in solid and liquid medium in light and dark periods than IBA (Epstein and Ludwig-Müller 1993). Small quantities of IAA can be transformed to IBA, and when the root system is extensive, the rate of conversion is higher. Therefore, IBA metabolic pathway could be analogous to IAA pathway via the tryptophan (indole + serine) but changing the serine for glutamate-γ-semialdehyde (Epstein and Ludwig-Müller 1993). Transport and distribution of auxin from young aerial tissues to the primary root through mature phloem trigger the cell polarity creating more root differentiation (Lewis et al. 2007). However, auxin can also be transported in shorter distances such as cell to cell auxin transport by specific auxin efflux and influx carriers (Vieten et al. 2007). Auxin transport gives optimum auxin concentration for development and growth of the root (Saini et al. 2013). Moreover, there are three major classes of auxin transporters: *like-aux1* (LAX) proteins are influx carriers (Swarup et al. 2001), pin formed (PIN) are efflux carriers (Blilou et al. 2005), and p-glycoproteins (PGPs) are involved in both efflux and influx of auxin but with a minor role (Geisler and Murphy 2006). The efflux transporter transports auxin out of the cell and its allocation in the root (Saini et al. 2013). Hence, auxin and the transporters are in charge of the root development. Auxin also interacts synergistically or antagonistically with other plant hormones such as cytokinins (CK), ethylene, gibberellins (GA), abscisic acid (ABA), jasmonic acid (JA), brassinosteroids (BR), polyamines (PA), and strigolactones (SL) causing cascade events controlling the root development and morphogenesis by integrating their signals in transport and distribution of auxin biosynthesis (Saini et al. 2013) (Fig. 2.2).

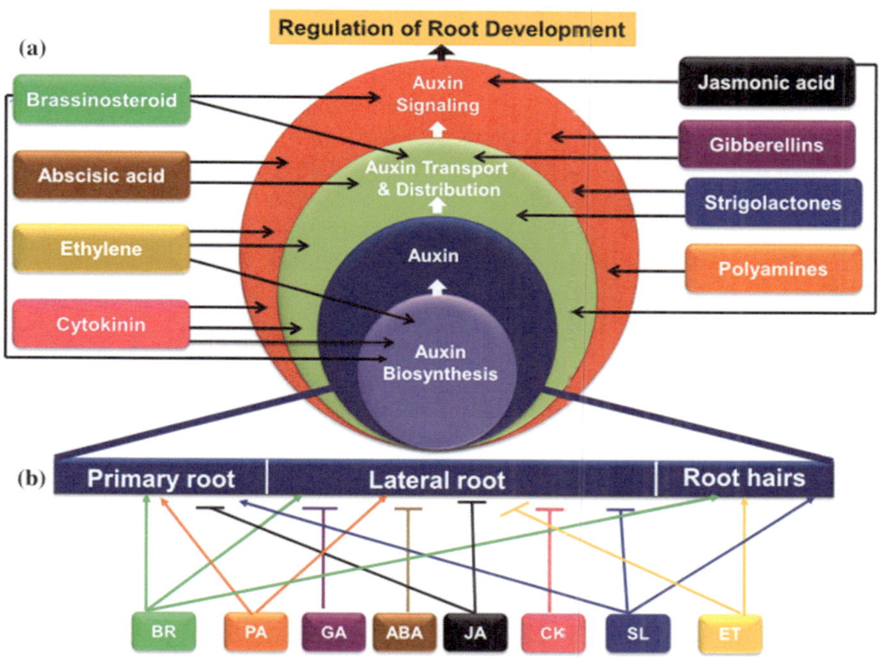

Fig. 2.2 Root and phytohormone-based signaling (Saini et al. 2013, used with permision)

Cytokinin is a phytohormone that has several roles and can inhibit root branching and growth (Riefler et al. 2006) since it suppresses auxin transport and signaling (Saini et al. 2013). However, this hormone can be perceived by several sensors of histidine kinases which are CRE1/AHK4 and helps in primary root elongation. It is mainly expressed in roots since its messenger RNA is localized mostly in the vascular cylinder and the pericycle of the root. The other two sensors are *AHK2* and *AHK3* (Higuchi et al. 2004). Mutations in this sensor do not have a strong plant phenotype change but alter the control of sulfate acquirement and physiological parameters in response of phosphate starvation (Franco-Zorilla et al. 2002). However, Riefler and co-workers (2006) showed that mutants *ahk2* and *ahk3* and in combination had a highly dense branched root system which guided to a stronger root improvement. These results compared with previous information indicate that root branching and primary root elongation are separate functions and these sensors are implicated in the regulation of these functions. Nevertheless, exogenous application of cytokinin stops the formation of lateral roots (Werner et al. 2003). Riefler and colleagues (2006) showed a quicker growth of primary root and a root branching increase with mutant cytokinin receptors. Cytokinin damages the expression of PIN genes in lateral root formation and stops the gradient required of auxin (Laplaze et al. 2007). Another hormone involved in the inhibition of cell elongation is ethylene by activating a local auxin response in the elongation zone of the root meristem (Hodge et al. 2009). Stepanova et al. (2005) showed that weak ethylene insensitive2/anthranilate synthase α1 (*WEI2/ASA1*) and wei7/anthranilate synthase β1 (*ASB1*) genes inhibited root growth by encoding rate-limiting enzymes of the tryptophan synthesis. Nevertheless, IAA in high concentrations stimulates ethylene production (Crozier et al. 2000), and ethylene in contrast reduces IAA content in roots (Saini et al. 2013). On the other hand, there are synergistic interactions in root gravitropism (Buer et al. 2006). Nevertheless, plants have to survive abiotic factors, and when drought affects the plant, the main root undergoes an elongation, and lateral roots are developed as a response to the abiotic stress. This response of roots is stimulated by abscisic acid (ABA, stress hormone). ABA has an impact in the final root architecture system since it has a function in lateral root meristem activity (De Smet et al. 2006). However, when there is no stress, ABA inhibits lateral root formation (Shkolnik and Bar-Zvi's 2010) study results showed the ABA inhibits lateral root formation. Another cross talk with auxin is with brassinosteroids, and several growth and development of roots are achieved (Choudhary et al. 2012). The gene *BRX* is strongly induced by auxin and is needed for optimal root growth (Mouchel et al. 2006). Moreover, BR can modify the efflux and influx auxin carrier cellular localization (Hacham et al. 2012). Additionally, auxin needs GA control for root growth and development (Fu and Harberd 2003) since GA degrades the repressor of GA insensitive (GAI) leading to root development and growth processes. GAI is highly expressed in root tips (Silverstone et al. 1997). In addition, Gou et al. (2010) propose that auxin-GA cross talk controls lateral root formation. Auxin also interacts with polyamines, and in low temperatures, they are involved in root architecture (Hennion and Martin-Tanguy 2000). The addition of PA to culture mediums with NAA and IBA strongly enhanced the root formation and growth (Saini et al. 2013) as well as hairy root growth and secondary metabolite production (Yang et al. 2010). Furthermore, jasmonic acid and auxin synergetic effects can control adventitious root initiation (Gutierrez et al. 2012) but inhibit primary and lateral root growth (Saini et al. 2013). Finally, SL is a new hormone that has been proposed to be in roots as the main site of biosynthesis and has been involved in primary, lateral, and root hair formation (Saini et al. 2013).

2.5.2.2 Anaerobic Metabolism

Under anaerobic or hypoxic conditions, the glycolytic intermediates of roots change in their levels, and an increase in glycolytic flux occurs, called the Pasteur effect (Davies 1980). This effect happens so that the cell can undergo the ATP utilization requirements even if the ATP from fermentation has lower efficiency. The enhancement of the glycolytic flux is complemented with the accumulation of several end products such as ethanol, alanine, lactate, and numerous amino acids and organic acids (Roberts et al. 1992). From these end products ethanol was the product in higher quantity, 50 times superior, in long-term hypoxic glycolysis compared to the other metabolites (Good and Muench 1993). Also, hypoxic conditions increased the content of pyruvate in root tissues as it was shown in *Hordeum vulgare* which increased its levels when compared with the aerobic control, maybe due to the Pasteur effect (Good and Muench 1993). Moreover, root tissues under anaerobic conditions exhibit an increase in different glycolytic enzymes such as alanine aminotransferase, lactate dehydrogenase, and alcohol dehydrogenase (Good and Crosby 1989). However, the concentration of aspartate tends to decrease (Good and Muench 1993).

2.5.3 Hairy Roots

2.5.3.1 Morphology

Plants are a valuable resource of secondary metabolites useful for mankind that applied in pharmaceuticals, flavors, fragrances, and food color, but the production of the plant for commercialization is too low. However, a higher concentration of metabolites generally occurs with low rates of growth and when cells are under stress (Chandra and Chandra 2011). The gram-negative soil bacterium (Guillon et al. 2006) *Agrobacterium rhizogenes* (natural genetic engineer) infection in plants leads to a phenotype of hairy roots as a consequence of a transference of a fraction from the plasmid DNA (T-DNA) into the genome of the host plant (Chilton et al. 1982). It stimulates the formation of several secondary functional roots with several root hairs termed "hairy roots" (Plasencia et al. 2016) that appear at the wounding site (Guillon et al. 2006). Also, organ cultures as *A. rhizogenes* look are more stable as a method for clonal propagation in vitro than callus or cell suspension cultures (David et al. 1984). The explants used for infection are generally young tissues of cotyledons, hypocotyl segments, petioles, young leaves, and sterile plantlets (Tomilov et al. 2007). The result of hairy root cultures is attractive since the root clones are long-term sterile and genetically stable, have absence of geotropism, and have higher growth rates even when compared with the cell suspension cultures with the fastest growing frequencies (Lorence et al. 2004). Transformed roots with

A. rhizogenes are well known to have an increase in growth rate and the ability to grow in media free of hormones and to develop with no plant growth regulators (Giri and Narasu 2000). In addition, the use of elicitors, cell permeabilization, entrapping released molecules in the medium, and precursor feeding are different ways to enhance the hairy root productivity (Guillon et al. 2006). Nevertheless, conditions such as bacterial strain and density, age and nature of explants, and infection protocol greatly influence the rate of transformation into hairy roots and their development (Bensaddek et al. 2018).

Moreover, it has been reported that hairy root cultures exhibit a stable secondary metabolite production (Hamill et al. 1987). These transformed roots can produce comparable amounts of secondary products with the mother plant. Furthermore, hairy root phytochemical production can be enriched by substrate feeding that intensifies the enzymatic machinery of plant cells, mostly enzyme that allowed oxidation, reduction, methylation, and glycosylation reactions (Li et al. 2003). However, the metabolic engineering combined with biosynthetic pathways could yield high-value secondary products (Chandra and Chandra 2011) by single gene overexpression (Hu and Du 2006). Hairy roots can have a high productivity of compounds, and the growth conditions can be easily controlled (Lorence et al. 2004). Another advantage of hairy roots is the production and accumulation of metabolites that are originally produced in aerial parts of the plant (leaves and shoots) (Chandra and Chandra 2011). An example is artemisinin that only accumulates on *Artemisia annua* aerial parts (Giri et al. 2000). All these features combined gives cultures of transformed roots a faster development than wild type, more quantity of lateral branching, a better biomass production, and a higher production of secondary metabolites on a free hormone medium (Chandra and Chandra 2011). An example is the two- to threefold higher content of stable alkaloid production in hairy roots than in non-transformed roots. Additionally, after 2 years of subculturing the hairy roots, the production of alkaloids was still qualitatively and quantitatively stable (Ciau-Uitz et al. 1994). Also, diverse genera of the solanaceous plants, such as *Datura*, *Duboisia*, *Scopolia*, *Atropa*, and *Hyoscyamus*, have been reported to produce tropane alkaloids in hairy root transformations (Tepfer 1990). Another advantage of hairy root cultures is the formation of xylem. Plasencia et al. (2016) reported a similar primary and secondary xylem formation between non-transformed and transformed roots at an equivalent distance from the apex. Moreover, in the same experiment, they showed that hairy root cultures can be used for identifying subcellular localization of proteins and the exploration of activity pattern promoters.

Furthermore, hairy root cultures offer enough amounts of material for different approaches such as biochemical, histochemical, and molecular biology (Plasencia et al. 2016) and can be used for a broad range of reasons, varying from studies of mycorrhization, nodulation, interactions of the plant with pathogens and/or nematodes, phytoremediation, rooting of recalcitrant species, and production of secondary metabolites (Georgiev et al. 2012). These approaches can be aided by hairy root formation since it allows a reduction in time for achieving functional analysis. A period of time after the infection, root fragments can be cultivated in vitro providing biological replicates and can restart growing (Plasencia et al. 2016). Additionally, this transformation allows quick analysis of transgenes in vivo, as, for example, the study of hairy roots from *Eucalyptus grandis* in fluorescent-tagged proteins (Plasencia et al. 2016). Nevertheless, each species and their individual clones have to be optimized separately since the hairy root culture may have different conditions of nutrient requirements (Hu and Du 2006).

2.5.3.2 Factors Influencing the Culture Medium

Development, greening, and secondary metabolite production of hairy root culture are controlled by temperature, light, pH, concentration of carbon source, ionic concentration, phytohormones, and oxygen stress (Chandra and Chandra 2011). The morphology of the roots changes with different effects of light such as light and dark. Hairy roots grown under light effects result in a greenish color and an increase of 400 times in fresh weight of hairy roots, and the opposed effect is obtained under dark effects as a pale brown color, and more importantly, after 28 days, the culture dies (Christen et al. 1992). Kino-oka et al. (2001) showed that light effects on hairy roots cultivated from *Ipomoea aquatica* had an enhancement in peroxide and superoxide dismutase activities and an increase in root growth. In addition, Jacob and Malpathak (2004) exhibit the influence of different temperatures in the greening, growth, and metabolite production of hairy roots of *Solanum khasianum*. The best temperature was 25 °C, while those cultures at 35 °C exhibit a reduced greening, growth, and metabolite production. Moreover, in the same experiment, the carbon metabolism was tested with different concentrations of sucrose. At 1% of sucrose, the metabolite production and greening of roots were enhanced, while development of hairy roots decreased. On contrary, at 5% of sucrose concentration, the metabolite was reduced, but the growth of hairy roots increased (Jacob and Malpathak 2004). This corroborates part of the results from Taya et al. (1989) stating that the best carbon source for a hairy root growth was sucrose. Furthermore, higher CO_2 concentrations showed a decrease in secondary metabolite production and inhibit hairy root growth in *Solanum khasianum* (Jacob and Malpathak 2004). Moreover, the ionic concentration has been studied by Christen et al. (1992), and their results on *Hyoscyamus albus* showed that the metal ion concentrations applied to hairy root culture medium such as Zn^{2+}, Ca^{2+}, or Fe^{2+} had little effect on growth. However, the addition of Cu^{2+} into the medium yielded a higher quantity of tropane alkaloids and improved root growth. Furthermore, high salt in mediums favors the formation of hairy roots in some plants, while mediums with low salt allow extreme bacterial multiplication in the medium, and hence the hairy roots need to be transferred to new mediums before incubation (Gamborg et al. 1976).

On the other hand, hormones such as auxins, ethylene, cytokinin, abscisic acid (ABA), and gibberellins (GA) have been studied in the development of hairy roots. Several studies indicate that plant hormones can change the morphology and development of roots (Weathers et al. 2005). One example is gibberellins (GA) which intensify the growth of numerous hairy root species, specially increasing lateral branching and root elongation, such as *Datura innoxia* (Ohkawa et al. 1989), *Artemisia annua* (Liu et al. 1997), and *Cichorium intybus* (Bais et al. 2001). Moreover, *A. annua* and *C. intybus* increase artemisinin and coumarin, respectively, in the same mediums with extra GA addition. Likewise, *A. annua* hairy root growth in medium stimulated with GA_3 reported an enhancement in artemisinin (Smith et al. 1997). Weathers et al. (2005) showed in their experiments that GA_3 had the highest values from several phytohormones (auxin, GA, ethylene, ABA, cytokinin) in producing length in the primary root, number of total lateral root measurements, and tip density. However, several studies displayed that a concentration of GA is more effective when it is low (Bais et al. 2001). Another hormone is ethylene which is recognized for its capacity to endorse senescence (Weathers et al. 2005). In previous studies, ethylene production in transformed roots demonstrates to be threefold higher, and the later inhibition of ethylene biosynthesis doubled their development (Biondi et al. 1997). Moreover, another hormone is abscisic acid (ABA), and some reviews reported stop of secondary metabolite production or accumulation but not the growth of hairy roots (Weathers et al. 2005). This was demonstrated by the results of Robbins et al. (1996) that showed a decrease in tannin content of *Lotus corniculatus* in ABA stimulated medium. Also, exogenous application of ABA did not affect the biomass of *Hyoscyamus muticus* (Vanhala et al. 1998). On the other hand, a decrease in hairy root culture biomass in *C. intybus* was observed with low levels of cytokinin and high levels of auxins, especially α-naphthaleneacetic acid (NAA) and indole-3-acetic acid (IAA) (Bais et al. 2001). Nevertheless, Rhodes et al. (1994) added auxins in cultures of transformed roots, and growth was stimulated. Shen et al. (1998) study results showed that hairy roots are 100- to 1000-fold more sensitive than normal roots to exogenous auxin. This reaction is probably to be species-specific, and this influences secondary metabolite production. The experiments from Weathers et al. (2005) showed that between hairy roots incubated with NAA and IAA at the same concentration, IAA inhibited more production of artemisinin in hairy roots than the synthetic auxin. In addition, Lorence et al. (2004) demonstrated that higher concentrations of auxin (NAA) increased the development of hairy roots but as well caused a dedifferentiation and stop in the production of camptothecin. In addition, 2,4-D also showed inhibition of the desired alkaloid production. Finally, cytokinin is principally known for its influence in shoot development in hairy root cultures but can also control secondary metabolite production and growth (Vanhala et al. 1998). High auxin to cytokinin levels in hairy root cultures of *C. intybus* presented a decrease in root development due to a quick disorganization and a reduction in the capacity of root cultures in producing secondary metabolites (Bais et al. 2001). However, complete hairy root growth is reduced when the phytohormone concentration increases (Bais et al. 2001).

2.5.3.3 Elicitors

Elicitors are microorganisms and chemical or physical factors that influence the physiological and morphological responses of plant cells cultured in vitro (Namdev 2007) that will trigger the defense mechanism of the cell and produce phytoalexins (Yoshikawa 1978). Also, it can be a substance applied in small concentrations to cells and start and/or improve the biosynthesis of desired compounds. To ensure survival and competitiveness of the plant, the biosynthesis of secondary metabolites is induced or enhanced by elicitation (Namdev 2007) as well as the secretion of secondary metabolites into the medium (Guillon et al. 2006). Bais et al. (2003) reported high amounts of secreted harmine and harmaline into the medium from *Oxalis tuberosa* after treated with elicitor *Phytophtora cinnamon*. Nevertheless, if cultures in early phase are elicited, the secondary metabolite formation is directly increased, but the biomass yield is lowered (Shilpa et al. 2010). Elicitation has been used in hairy root cultures as a significant approach for secondary metabolite production (Chandra and Chandra 2011), and this production is controlled by elicitors in different ways such as accumulation, vacuolar transit, and modulation of their biosynthesis or, on the contrary, with turnovers and degradation (Barz et al. 1990). Elicitors can be biotic (factors of pathogen or plant origin) or abiotic (physical, chemical, or mineral), and the combination of both can produce a synergetic effect and enhance the metabolite production more than the effect of a single elicitor (Chandra and Chandra 2011). One example is the synergetic effect of fungal elicitation and limitation of phosphate which produce an optimal quantity of secondary metabolites (Dunlop and Curtis 1991). Abiotic elicitors can include the use of selenium (Se), nickel sulfate ($NiSO_4$), and sodium chloride (NaCl) that have been implemented in mediums of hairy roots of *Panax ginseng* and showed an increase of saponins compared with the control (Jeong et al. 2006). Likewise, abiotic elicitors have been more exploited in recent time since the biotic counterpart is more expensive (Georgiev et al. 2007). However, Ge and Wu's (2005) studies prove that alternate addition of elicitors into the medium is more effective than synchronized addition of both elicitors. Since elicitation induces or increases the formation of secondary metabolites in hairy root cultures, it can be used for discovering new genes involved in biosynthesis of metabolites (Rischer et al. 2006). Furthermore, it can also be classified by on the basis like exogenous or endogenous elicitors (Namdev 2007). Some examples of endogenous elicitors are secondary plant messengers such as salicylic acid and methyl jasmonate (MeJA) which stimulates the secondary product production with no effects in the growth frequencies of cultures (Yaoya et al. 2004).

2.6 Genetically Modified Plants

2.6.1 Introduction

Crown gall disease and its study date back to more than 100 years ago, since then *Agrobacterium tumefaciens* was isolated and analyzed for its various effects in plants (Hwang et al. 2017). Even when the transfer of DNA from bacteria to plant cells is a natural occurring process, it was not until the beginning of the decade of 1980 that this genus was first described as a powerful genetic transformation agent (Chilton et al. 1982; Caplan et al. 1983; De Block et al. 1985). In fact in 1982, Chilton et al. described for the first time the capability of using *A. rhizogenes* for the introduction of T-DNA harbored in the Ri plasmid in the genome of plant cells (Chilton et al. 1982). Since then the number of articles describing its use for genetic transformation is large (Gelvin 2017).

The genetic transformation is done, namely, in strains harboring tumor-inducing (Ti) or root-inducing plasmids (Ri) which include *A. tumefaciens*, *A. vitis*, *A. rubi*, or *A. rhizogenes*. These particular types of plasmids codify for virulence genes named *vir* genes and can be expressed when plants release certain compounds, namely, from the phenolic family (Sarkar et al. 2018).

2.6.2 A. rhizogenes

Agrobacterium rhizogenes is a gram-negative bacterium found in the vicinity of plant roots (Mauro et al. 2017). Its identification dates back to 1930 (Ricker et al. 1930). Since then it was associated with the onset of hairy root syndrome affecting dicotyledoneous plants, with the trademark of exerting growth in main and lateral roots in the infection site (Mauro et al. 2017). In fact, several strains of *A. rhizogenes* are known to generate this phenomenon by transferring its "transfer DNA" or T-DNA with the help of root-inducing (Ri) plasmid into the plant cell (Tepfer 2016). This T-DNA is a fragment that is normally liberated from the Ri plasmid and enters the plant integrating into the plant cell genome. A set of repeated sequences that are conserved are found flanking the fragment borders (Mauro et al. 2017). The ultimate goal for the bacterium is to insert its DNA in the plant cell's roots allowing the production of opines which are excreted by these newly generated hairy roots into the surrounding environment which will be used by this microorganism as a food and energy source (Mauro et al. 2017). In biochemical terms, these molecules (opines) have low molecular weight and are produced due to the expression of a set of genes found near the *rol* genes. Popular opines generated by Ri plasmids include octopine, nopaline, and agropine, and their presence in such plasmids has been used as a classification system (Roy 2015).

Hairy root production relies in the expression of a set of genes that induce root formation and modulate the plant growth and differentiation (Ho et al. 2018). These oncogenes are known as *rol (**ro**oting **l**ocus) genes, with four of them being involved in this process*: *rol A*, *rol B*, *rol C*, and *rol D*. Since the main outcome of their expression is the formation of neoplastic material in the form of new roots, they were described as oncogenes. In addition to the presence of genes that affect phytohormone balance within the T-DNA region, the expression of *rol* genes in plants can also exert different effects in the auxin and cytokinin balance. In fact, although mainly conserved, they do show variations depending on the plant species, cultivar, or its physiological background (Mauro et al. 2017). In spite of being discovered a long time ago, the specific role for *rol* genes is currently understudied since clarity on their functions and synergism is lacking. Nevertheless the most accepted roles for these four genes are briefly described.

Rol A. The expression of *rolA* in genetically modified plants includes wrinkled leaves, a reduction in rooting capacity, small fruits sometimes lacking seeds, reduction in pollen viability, and dwarfism (as revised by Mauro et al. 2017 and Sarkar et al. 2018). However much of these effects are a result of the transgene expression level; in fact under normal *rol A* promoters, plants develop dwarfism (Schmülling et al. 1988), whereas when a 35S constitutive promoter was used, a significant reduction in overall plant growth was observed, increasing the negative impact of *rol A* gene in proper plant development (Dehio et al. 1993). Also, the effects can vary depending on the presence of plant growth regulators (auxins) and on the plant species or cultivar under analysis (Sarkar et al. 2018). Other studies have linked its effects with a reduction of the plants giberellin levels as well as with a reduction in the response of plants to these phytohormones (Bettini et al. 2016).

Rol B. It is associated with the main effects on hairy root induction since it interferes with phytohormone signaling pathways in transformed plants and plant cells. It alters the morphology of leaf and flower structures, allows for the formation of adventitious roots, changes root geotropism, and increases the response of roots to certain plant growth regulators (auxins) (Delbarre et al. 1994). It can also affect ROS (reactive oxygen species) signaling and repress apoptosis in transformed calli. Most of these effects have been associated with the activation of genes encoding transcription factors (MYB and BHLH) (Bulgakov et al. 2013). The *rolB* gene has a tyrosine phosphatase activity (Filippini et al. 1996) which can account for its interference with signal transduction pathways such as those governing response to auxin concentration changes in the environment either in nature or under controlled experimental conditions (Mauro et al. 2017).

Rol C. While *rol A* and *rol B* genes are more related with alterations in the auxin signalization pathway, the effects induced by the expression of *rol C* can be associated with an increase in cytokinin activity (Schmülling et al. 1988). Among the most common effects of *rolC* expression, a reduction in plant height, reduction in green color of leaves, and branching formation are found. As with the other members of this oncogenes present in *A. rhizogenes*, the effect of *rolC* expression varies according to plant species and cultivars (Sarkar et al. 2018).

2.6.3 Recombinant Protein Production

Among the different recombinant protein expression systems developed up until now days, several differences appear. In brief, some of them pose difficulties such as the inability to produce or secrete functional and highly complex proteins (bacterial systems), increase in contamination via toxic or viral molecules (mammalian cells), as well as their high cost in terms of process scaling up (mammalian cells), among others (Cardon et al. 2019). However, since the discovery of *Agrobacterium* strains that naturally transfer sections of their genome into plants, the advent of genetic transformation of plant cells (in vitro or complete plant systems) for the heterologous expression of recombinant proteins has offered a new window for the production of proteins with therapeutic potential at low cost, decreasing the risk for cross contamination with viral or toxic molecules such as with animal models (Horn et al. 2004). However, even when plants are considered an excellent model for these biotechnological purposes, the open-field production of such proteins possesses risk of cross contamination of genetically modified plant (GMO) pollen into the environment; strategies such as addressing the expression of recombinant proteins into organelles such as chloroplast or mitochondria have been described. Nevertheless, such strategies are time-consuming since the generation of completely transformed in vitro explants or completely transformed plants is painstaking and time-consuming; iterative cultures are needed to assure the homogeneity of recombinant protein expression avoiding the presence of chimeras (Yu et al. 2017). The use of the *A. rhizogenes* genetic transformation system to express and produce recombinant proteins in hairy roots has been a very useful strategy which has had broad acceptance in various plant models over the years (Gelvin 2017). The number of reports of this nature is increasing. Since its discovery in 1982, those reports have increased and have been revised over in the literature that we will focus in the main discoveries done in recent years.

The expression of the *hGL* gene which codifies for the human gastric lipase was expressed in hairy roots of *Brassica rapa* in liquid medium (Ekouna et al. 2017). Accordingly, the use of 2,4-D (2,4-dichlorophenoxiacetic acid, a synthetic auxin) helped to increase the levels of the recombinant protein 2.7 times when compared to untreated root cultures. The expression of chimerical peptides containing lactoferricin and lactoferrampin (LFchimera) both antimicrobial peptides present in lactoferrin of bovine origin hairy roots in *Nicotiana tabacum* was achieved. These types of antimicrobial peptides are promising candidates to be used as antibiotics (Chahardoli et al. 2018). Their expression and production were validated by means of conventional molecular biology techniques. The use of hairy roots generated 4.8 μg LF/g fresh weight. Other antimicrobial peptides expressed in hairy roots include the production of dermaseptin B1 (*DrsB1*) peptide in tobacco (Shams et al. 2019). In fact, the antimicrobial activity of the peptides demonstrated a positive effect against fungal and bacterial pathogens, in particular, against *Alternaria alternata* and *Pythium* spp. Furthermore, this approach can be used to generate genetically modified plants that produce plants resistant to diseases caused by these plant pathogenic fungi (Shams et al. 2019).

Naphatsamon et al. (2018) reported the expression of human β-glucocerebrosidase in hairy roots of *Nicotiana benthamiana*. These types of molecules are vital for the cure of certain diseases such as Gaucher disease which is characterized by an enzyme deficiency, namely, "glucocerebrosidase GCase." Traditional production of this complex molecule involves transgenic mammalian cells with the concomitant drawbacks associated with elevated prices and contamination risk. Thus, a strategy using transgenic roots is plausible. As a result, the production of this recombinant GCase in its full functional form generated an enzyme activity of 81.40 ± 17.99 units/mg total protein. The overall yield for the GCase was near to 1 μg/g of the root. The results demonstrated the feasibility of generating recombinant molecules of animal origin in plant hairy roots.

Other molecules of human origin expressed in hairy root systems include the production of recombinant human erythropoietin or "EPO" (rhEPO). In fact, this protein is normally produced in mammalian cells. Thus, Gurusamy et al. (2017) reported the expression of this protein in *Nicotiana tabacum* L. The expression vector designed allowed for a production of 66.75 ng/g of total soluble protein, and the authors indicated that the use of ER signal peptide was vital for this production in the hairy root system. An increase in the stabilization of the rhEPO produced in the hairy root system was observed after the addition of polyvinylpyrrolidone. As a result, a 5.6-fold increase of rhEPO was observed. Altogether these findings demonstrate that hairy root cultures are a powerful and innovative tool for the production of heterologous recombinant proteins with biotechnological and biopharmaceutical potential.

2.7 Applications

2.7.1 Phytochemical Production

2.7.1.1 Production of Secondary Metabolites

Roots accumulate several secondary metabolites with commercial interests; however, harvesting roots is a destructive procedure for the entire plant, and therefore, hairy root cultures especially of medicinal plants have gained attention (Guillon et al. 2006). Furthermore, most of the secondary metabolites from medicinal plants have a very difficult chemical synthesis (Namdev 2007). Hence, hairy root cultures offer several advantages for a constant production of beneficial secondary products such as an easy maintenance, a quick development, less time, and the capacity to synthesize a variety of chemical compounds. These cultures also have the ability to synthesize several different metabolites at the same time which is useful for cosmetics, food additives, and pharmaceuticals, thus proving to have commercial importance (Hu and Du 2006). Secondary metabolites include glycosides, flavonoids, tannins, resins, volatile oils, and alkaloids, among others (Namdev 2007). Thus, transformation of several medicinal plants with *A. rhizogenes* has been effective, and the production of secondary products with pharmaceutical aims had increased. Some significant alkaloids produced in cultures of hairy roots include plants such as *Atropa belladonna* L., *Catharanthus trichophyllus* L., and *Datura candida* L. (Sevon, 2002). In addition, ROOTec, a German company dedicated to hairy root culture in a broad scale, has succeeded in producing camptothecin and podophyllotoxin metabolites (Guillon et al. 2006) (http://www.rootec.com/en/home).

Hairy roots have also been used for the production of triterpenoid compounds of ethnopharmaceutical importance, as in the case of *Galphimia glauca* and the production of galphimine E, glaucacetalin A, and maslinic acid (Nader et al. 2006). The maximum root biomass was achieved at day 33, while the compound yield was found as the following: glaucacetalin A was 2.14 mg/L, and galphimine E and maslinic acid were found in concentrations of 0.11 and 0.43 mg/g, respectively (Nader et al. 2006). Hairy roots of valerian (*Valeriana officinalis*) have also been used to model the production of its main anxiolytic compound: valerenic acid, which belongs to the sesquiterpenoid family. Also the production of β-caryophyllene which has anti-inflammatory properties was also analyzed (Ricigliano et al. 2016). To evaluate the production of these compounds, a co-transformation strategy and the expression of several genes were used codifying for important enzymes in the sesquiterponoid pathway such as farnesyl pyrophosphate synthase (VoFPS) and valerenadiene synthase (VoVDS). The increase in the expression levels for these constructs was analyzed and led to suggest that there are rate-limiting steps in the production of valerian sesquiterpenoids such as the availability of farnesyl diphosphate and valerenadiene as well as the response to the use of jasmonic acid as an elicitor.

Ginseng (*Panax ginseng* C.A. Meyer) is a very popular plant mainly due to its medicinal properties and its link with Asian pharmacopeia which is very ancient. Hairy roots have been described as a mechanism for the recovery of important bioactive compounds in this plant. In general the use of hairy roots in this model has a lag period followed by an exponential growth phase where secondary metabolites start to accumulate. Accordingly to obtain higher yields, the use of bioreactors and elicitors is suggested (Jeong and Park 2006).

2.7.1.2 Productions of Compounds Not Located in Natural Roots

The metabolic pathway can be disturbed after transformation and could create new compounds that untransformed roots would not be available to produce (Hu and Du 2006) or by creating de novo metabolic pathways with the insertion of genes that code for associated enzymatic steps (Guillon et al. 2006). Nishikawa and Ishimaru (1997) study results exhibit an accumulation of glucoside conjugates of flavonoids in hairy roots of *Scutellaria baicalensis*, while the untransformed roots usually accumulate glucose conjugates. Also, a new compound from *Glycyrrhiza glabra* was detected with great antimicrobial activities called licoagrodione (Li et al. 1998). Thus, hairy roots can aid with the synthesis and isolation of new compounds.

2.8 Phytoremediation

Phytoremediation involves the use of plants to treat the increasing amount of environmental pollution worldwide. Even when there are various plant models used for this, one main problem is that plants have difficulty in metabolizing some organic compounds or in tolerating/accumulating toxic compounds such as heavy metals. Among the strategies used, hairy roots seem promising for the removal of certain inorganic and organic contaminants (Ibañez et al. 2016). In 2015, the use of hairy roots induced in *Sesuvium portulacastrum* (L.) for the removal of textile dyes was reported (Lokhande et al. 2015). This system allowed the removal of Reactive green 19A HE4DB as was proved by HPLC and FTIR analysis. The metabolites produced were considered nontoxic.

Also, hairy roots induced in *Ipomea carnea* J. were used in the decolorization of 25 different textile dyes and have been efficient against 15 of them (Jha et al. 2016). In fact, the use of these hairy roots was more efficient against Acid Red 114, which was decolorized in a 98%. The degradation of this compound was confirmed by HPLC and FTIR spectroscopy. GCMS analysis revealed that the compounds were degraded to form 4-aminobenzenesulfonic 2-methylaniline and 4-aminophenyl 4-ethyl benzene sulfonate. Cytotoxicity test revealed that the metabolites obtained are nontoxic. Altogether, this study demonstrated the potential use of plant hairy roots in the phytoremediation process of dyes derived from the textile industry (Jha et al. 2016).

The rich neoplastic root proliferation of hairy root cultures offers an advantage for phytoremediation, a process for elimination of pollutants resulted from industrial effluents (Guillon et al. 2006). Plant roots can absorb and gather heavy metals or change organic molecules that are toxic to safe forms (Suresh and Ravishankar 2004). Some examples are hairy roots of *Brassica juncea*, *Brassica napus*, *Helianthus annuus*, and *Cichorium intybus* that are able to detoxify the tetracycline, the pesticide dichorodiphenyltrichloroethane (DDT), as well as 2,4-dichlorophenol (2,4-DPC) and oxytetracycline from effluents of the industries (Suresh et al. 2005).

2.9 Limitations

2.9.1 Oxygen

One normal limitation of hairy root cultures is oxygen deficiency triggered by mass transfer conditions and poor mixing (Hu and Du 2006). Shiao et al. (2003) transformed root lines of *Arabidopsis thaliana* L. with two enzymes *Adh* and pyruvate decarboxylase to improve the low oxygen situations that disturb the development of hairy roots throughout fermentation. Their results showed a similar growth of hairy roots with low oxygen conditions and cultures with full aeration. Oxygen limitation also happens during the cultivation of transformed roots in bioreactors since the mass transfer is reduced and with a high tissue concentration of hairy roots many limitations can develop (Curtis 2000).

2.9.2 Reduction of Chromosomes Number in Subcultures

The chromosome number of hairy roots is identical to the number in wild-type plants. However, the percentage of normal chromosome number of hairy root cells after 4 months of subculture was decreased from 85.0% to 23.5%, and after 8 months only 4.1% of cells had a normal number of chromosomes (Xu and Jia 1996).

2.9.3 Alterations of Regenerated Plants Morphology

A shortcoming of regenerated plants from hairy roots culture is the morphology of several parts of the new plant. They exhibit reduced apical dominance, internode length, and leaf size with wrinkled leaves but also an abundant and plagiotropic root system and a rise in leaf explant's ability to differentiate roots in a free medium without phytohormones (Hamamoto et al. 1990). These irregularities in morphology are maybe due to somaclonal variation or insertion of external DNA, thus, originating genomic disturbances rather than T-DNA gene expression (Han et al. 1993). Moreover, untransformed plants show less mortality than transformed plants (Hu and Du 2006).

2.10 Bioreactors

Cultivation in bioreactors is the last step in the expansion of techniques for secondary metabolite production from in vitro plant systems (Bourgaud et al. 2001) since large-scale productions are limited due to environmental factors, low propagation rates, and restricted cultivation (Taya et al. 1989). Moreover, when a product has high value, it is reasonable to harvest roots, batch fermentation, and extract product, while in products with low value, it is preferred to determine a packed bed of roots to manage a continuous process in the reactor for large periods collecting from the effluent stream the desired product (Giri and Narasu 2000). New advances in bioreactor systems offer the opportunity to scale up the cultivation of transformed roots from small scale to large scale in industrial processes (Guillon et al. 2006). Nevertheless, several considerations such as chemical and physical parameter must be taken into account for a proper development on a bioreactor system (Giri and Narasu 2000). The major chemical factor for scale-up is the nutrient availability which can be maintained with rejuvenated medium. Furthermore, mass transfer is known as the availability of nutrients and water in any region of the network from hairy roots at different periods in the bioreactor (Giri and Narasu 2000). For an effective scale-up in bioreactors, the morphology, physiology, and elevated stress sensitivity of hairy roots have to be considered (Wysokinska and Chmiel 1997). On the other hand, there are several difficulties such as the excessive branching of hairy roots which can display a resistance to flow since they form an interlocked matrix (Giri and Narasu 2000). Also, one difficulty of scaling up is the delivery of nutrients in gas and liquid phases simultaneously. In liquid medium, the meristem growth of root cultures causes a root ball with young roots growing on the outside and inside older tissues' core (Giri and Narasu 2000). However, the main difficulty of hairy roots is the oxygen supply. With a deficient oxygen delivery toward the central mass of tissue, it arises in a pocket of dead tissue; hence, a continuous supply is needed in the bioreactor (Giri and Narasu 2000). At initial stages of the scale-up, the mediums have enough oxygen dissolved supporting the development of the inoculum. Afterward, mixing serves to supply dissolved oxygen and taking out the carbon dioxide. A unit of biomass uptake of oxygen in a unit of time is acknowledged as oxygen transfer coefficient (Giri and Narasu 2000). Figure 2.3 provides an example for the general steps in bioreactor process for hairy root induction.

Several categories of bioreactors have been used for transformed root cultivation (Curtis 2000).

2.10.1 Stirred Tank Reactors (STR)

This bioreactor has turbine blades or impeller that enables mass transfer. This type of bioreactor is not appropriate since the mechanical agitation leads to callus formation since the hairy roots are wounded by the agitation (Taya et al. 1989). Furthermore, cultivations of hairy roots done in stirred tank presented some inconveniences since the clumps of hairy roots growing stop the impeller rotation.

Fig. 2.3 General steps in hairy root production in bioreactor (Patra and Srivastava 2015, used with permission)

Nevertheless, polyurethane foam material is a proper support for hairy root cells, and the immobilization of cells gives a good cultivation result in airlift column bioreactors (Taya et al. 1989). In addition, modified STRs can be used with baffles and large impellers agitated at a very low speed or steel cages with the hairy roots inside the STR (Giri and Narasu 2000).

2.10.2 Submerged Bioreactors or Airlift

These bioreactors do not have impellers but are similar to STRs. They have aerators that serve with humidified air passing throughout a glass grid (Taya et al. 1989).

2.10.3 Bubble Column Reactor

These types of bioreactors are related to airlift bioreactors and create bubbles causing less stress; however, the bubbling rate has to be increased gradually as the hairy roots grow. It is convenient for organized structures like hairy roots (Buitelaar et al. 1991).

2.10.4 Turbine Blade Reactor

This is a bioreactor with airlift and STR combination. In this type of bioreactor, a stainless steel mesh separated the agitation from the cultivation space; thus, the hairy roots are not in interaction with the impeller, and the air is launched from the bottom and stir in the medium by an eight-blade impeller. These bioreactors are good for hairy root cultures (Kondo et al. 1989).

2.10.5 Gas-Sparged Reactor

These bioreactors are efficient for mixing and oxygenation since the humidified air is introduced from the bottom part throughout a glass sparger (Giri and Narasu 2000).

2.10.6 Spin Filter Reactor

Here the rotating filter mixes the cultures and in tandem lets fresh medium addition and old medium removal (Giri and Narasu 2000).

2.11 Current Needs

As stated previously, the use of hairy roots for the production of important secondary metabolites or heterologous recombinant proteins is growing. However, drawbacks still exist in terms of how to manipulate the genetic conditions for overexpressing complex metabolic pathways by different genetic transformation techniques. The advent of gene editing and the CRISPR-CAS9 technology and its application in the field of hairy roots is starting to show some results and constitutes a promising alternative to finally liberate hairy root potential in the production of important metabolites with applications in medicine (antibiotics or antimicrobial agents), alternative medicine, or phytoremediation of contaminated fields or water.

References

Aber JD, Nadelhoffer KJ, Steudler P, Melillo JM (1989) Nitrogen saturation in northern forest ecosystems. Bioscience 39:378–386

Andresen LC, Konestabo HS, Maraldo K, Homstrup M, Ambus P, Beier C, Michelsen A (2011) Organic matter flow in the food web at a temperate heath under multifactorial climate change. Rapid Commun Mass Spectrom 25:1485–1496

Angulo-Bejarano PI (2013) Desarrollo de estrategias para la utilización del nopal (Opuntia ficus-indica) como bioreactor vegetal. PhD Thesis. Centro de Investigación y de Estudios Avanzados del Instituto Politécnico Nacional Campus Guanajuato, Irapuato Guanajuato Mexico

Angulo-Bejarano PI, Paredes-López O (2011) Development of a regeneration protocol through indirect organogenesis in prickly pear cactus (*Opuntia ficus-indica* (L.) Mill). Sci Hortic 128(3):283–288

Atkin OK (1996) Reassessing the nitrogen relations of Arctic plants. A mini-review. Plant Cell Environ 19:695–704

Bais HP, Sudha G, George J, Ravishankar GA (2001) Influence of exogenous hormones on growth in secondary metabolite production in hairy root cultures of *Chicorium intybus* L. cv. Lucknow Local. In Vitro Cell Dev Biol Plant 37:293–299

Bais HP, Vepachedu R, Vivanco JM (2003) Root specific elicitation and exudation of fluorescent [beta]-carbolines in transformed root cultures of *Oxalis tuberosa*. Plant Physiol Biochem 41:345–353

Bao F, Shen J, Brady SR, Muday GK, Asami T, Yang Z (2004) Brassinosteroids interact with auxin to promote lateral root development in Arabidopsis. Plant Physiol 134:1624–1631

Barz W, Beimen A, Drae B, Jaques U, Otto C, Sue E, Upmeier B (1990) Turnover and storage of secondary products in cell cultures. In: Charlwood MJC, Rhodes BV, Rhodes MJC (eds) Secondary products from plant tissue culture. Clarendon, Oxford, pp 79–102

Bensaddek L, Villarreal ML, Fliniaux MA(2018) Induction and growth of hairy roots for the production of medicinal compounds. Electronic J Integrative Biosci, (1, Special Issue on Hairy Roots (A. Lorence and F. Medina-Bolivar, co-editors)), pp 2–9. ffhal-00602520f

Berntson GM, Wayne PM, Bazzaz FA (1997) Below ground architectural and mycorrhizal population level responses to elevated CO_2 in *Betula alleghaniensis*. Funct Ecol 11:684–695

Bettini PP, Santangelo E, Baraldi R, Rapparini F, Mosconi P, Crinò P, Mauro ML (2016) *Agrobacterium rhizogenes rolA* gene promotes tolerance to *Fusarium oxysporum* f. sp. lycopersici in transgenic tomato plants (*Solanum lycopersicum* L.). J Plant Biochem Biotechnol 25:225–233

Biondi S, Lenzi C, Baraldi R, Bagni N (1997) Hormonal effects on growth and morphology of normal and hairy roots of *Hyoscyamus muticus*. J Plant Growth Regul 16:159–167

Blilou I, Xu J, Wildwater M, Willemsen V, Paponov I, Friml J, Heidstra R, Aida M, Palme K, Scheres B (2005) The PIN auxin efflux facilitator network controls growth and patterning in Arabidopsis roots. Nature 433:39–44

Boitel-Conti M, Gontier E, Laberche JC, Ducrocq C, Sangwan-Norreel BS (1996) Inducer effect of Tween 20 permeabilization treatment used for release of stored tropane alkaloids in *Datura innoxia* Mill. hairy root cultures. Plant Cell Rep 16(3–4):241–244

Bourgaud F, Gravot A, Milesi S, Gontier E (2001) Production of plant secondary metabolites: a historical perspective. Plant Sci 161:839–851

Britto DT, Kronzucker HJ (2002) NH toxicity in higher plants: a critical review. J Plant Physiol 159:567–584

Buer CS, Sukumar P, Muday GK (2006) Ethylene modulates flavonoid accumulation and gravitropic responses in roots of *Arabidopsis thaliana*. Plant Physiol 140:1384–1396

Buitelaar RN, Langenhoff AAN, Heidstra R, Tramper J (1991) Growth and thiophene production by hairy root cultures of *Tagetus patula* in various two liquid phase bioreactors. Enzyme Microbiol Technol 13:487–494

Bulgakov V, Bulgakov P, Shkryl YN, Veremeichik GN, Gorpenchenko TY, Vereshchagina YV (2013) Recent Advances in the Understandingof Agrobacterium rhizogenes-DerivedGenes and Their Effects on StressResistance and Plant Metabolism. Adv Biochem Eng Biotechnol 134:1–22

Caplan A, Herrera-Estrella L, Inzé D, Van Haute E, Van Montagu M, Schell J, Zambryski P (1983) Introduction of genetic material into plant cells. Science 222(4625):815–821

Cara FA, Sánchez E, Ruiz JM, Romero L (2002) Is phenol oxidation responsible for the short-term effects of boron deficiency on plasma-membrane permeability and function in squash roots? Plant Physiol. Biochemist 40:853–858

Cardon F, Pallisse R, Bardor M, Caron A, Vanier J, Ele Ekouna JP, Guillet M (2019) *Brassica rapa* hairy root based expression system leads to the production of highly homogenous and reproducible profiles of recombinant human alpha-L-iduronidase. Plant Biotechnol J 17:505–516

Chahardoli M, Fazeli A, Ghabooli M (2018) Recombinant production of bovine Lactoferrin-derived antimicrobial peptide in tobacco hairy roots expression system. Plant Physiol Biochem 123:414–421

Chaillou S, Lamaze T (2001) Ammonical nutrition of plants. In: Morot-Gaudry J-F (ed) Nitrogen assimilation by plants. Science Publishers Inc, New Hampshire, pp 53–69

Chandra S, Chandra E (2011) Engineering secondary metabolite production in hairy roots. Phytochem Rev 10:371–395

Chilton MD, Tepfer DA, Petit A, David C, Casse-Delbart F, Tempe J (1982) *Agrobacterium rhizogenes* inserts T-DNA into the genomes of the host plant root cells. Nature 295:432–434

Choudhary SP, Yu JQ, Yamaguchi-Shinozaki K, Shinozaki K, Tran LS (2012) Benefits of brassinosteroid crosstalk. Trends Plant Sci 17:594–605

Christen P, Aoki T, Shimomura K (1992) Characteristics of growth and tropane alkaloid production in Hyoscyamus albus hairy roots transformed with *Agrobacterium rhizogenes* A4. Plant Cell Rep 11(12):597–600

Ciau-Uitz R, Miranda-Ham ML, Coello-Coello J, Chi B, Pacheco LM, Loyola-Vargas VM (1994) Indole alkaloid production by transformed and non-transformed root cultures of *Catharanthus roseus*. In Vitro Cell Dev Biol 30:84–88

Cooper WC (1935) Hormones in relation to root formation on stem cuttings. Plant Physiol 10:789–794

Crozier A, Kamiya Y, Bishop G, Yokota T (2000) Biosynthesis of hormones and elicitor molecules. In: Buchanan B, Guissem W, Jones R (eds) Biochemistry and molecular biology of plants. American Society of Plant Physiologists, Rockville, pp 850–929

Curtis WR (2000) Hairy roots, bioreactor growth. In: Spier RE (ed) Encyclopedia of cell technology, vol 2. Wiley, New York, pp 827–841

Curtis PS, Wang X (1998) A meta-analysis of elevated CO_2 effects on woody plant mass, form, and physiology. Oecologia 113:299–313

David C, Chilton MD, Tempe J (1984) Conservation of T-DNA in plants regenerated from 'hairy' root cultures. Biotech 2:73–76

Davies DD (1980) Anaerobic metabolism and the production of organic acids. In: Davies DD (ed) The biochemistry of plants, vol 2. Academic, New York, pp 581–611

De Block M, Schell J, Van Montagu M (1985) Chloroplast transformation by *Agrobacterium tumefaciens*. EMBO J 4:1367–1372

Dehio C, Grossmann K, Schell J, Schmülling T (1993) Phenotype and hormonal status of transgenic tobacco plants overexpressing the *rolA* gene of *Agrobacterium rhizogenes* T-DNA. Plant Mol Biol 23(6):1199–1210

Delbarre A, Muller P, Imhoff V, Barbier-Brygoo H, Maurel C, Leblanc N, Perrot-Rechenmann C, Guern J (1994) The *rolB* gene of *Agrobacterium rhizogenes* does not increase the auxin sensitivity of tobacco protoplasts by modifying the intracellular auxin concentration. Plant Physiol 105:563–569

De Smet I, Vanneste S, Inzé D, Beeckman T (2006) Lateral root initiation or the birth of a new meristem. Plant Mol Biol 60:871–887

Doran PM (ed) (1997) Hairy roots—culture and applications. Harwood Academic Publishers, Amsterdam, pp 1–31

Dunlop DS, Curtis WR (1991) Synergistic response of plant hairy-root cultures to phosphate limitation and fungal elicitation. Biotechnol Prog 7:434–438

Eibl R, Meier P, Stutz I, Schildberger D, Hühn T, Eibl D (2018) Plant cell culture technology in the cosmetics and food industries: current state and future trends. Appl Microbiol Biotechnol 102(20):8661–8675

Ekouna JPE, Boitel-Conti M, Lerouge P, Bardor M, Guerineau F (2017) Enhanced production of recombinant human gastric lipase in turnip hairy roots. Plant Cell Tissue Organ Cult (PCTOC) 131(3):601–610

Epstein E, Ludwig-Müller J (1993) Indole-3-butyric acid in plants: occurrence, synthesis, metabolism and transport. Physiol Plant 88:382–389

Filippini F, Rossi R, Marin O, Trovato M, Costantino P, Downey PM, Lo Schiavo F, Terzi M (1996) A plant oncogene as a phosphatase. Nature 379:499–500

Fitter AH, Stickland TR, Harvey ML, Wilson GW (1991) Architectural analysis of plant root system 1. Architectural correlates of exploitation efficiency. New Phytol 118:375–382

Franco-Zorilla JM, Martin AC, Solano R, Rubio V, Leyva A, Paz-Ares J (2002) Mutations at *CRE1* impair cytokinin-induced repression of phosphate starvation response in Arabidopsis. Plant J 32:353–360

Fransz PF, de Jong JH (2002) Chromatin dynamics in plants. Curr Opin Plant Biol 5:560–567

Fu X, Harberd NP (2003) Auxin promotes Arabidopsis root growth by modulating gibberellins response. Nature 421:740–743

Gamborg OL, Murashige T, Thorpe TA, Vasil IK (1976) Plant tissue culture media. In Vitro 7:473–478

Ge X, Wu J (2005) Induction and potentiation of diterpenoid tanshinone accumulation in *Salvia miltiorrhiza* hairy roots by β-aminobutyric acid. Appl Microbiol Biotechnol 68:183–188

Geisler M, Murphy AS (2006) The ABC of auxin transport: the role of p-glycoproteins in plant development. FEBS Lett 580:1094–1102

Gelvin SB (2017) Integration of *Agrobacterium* T-DNA into the plant genome. Annu Rev Genet 51:195–217

Georgiev MI, Agostini E, Ludwig-Müller J, Xu J (2012) Genetically transformed roots: from plant disease to biotechnological resource. Trends Biotechnol 30:528–537

Georgiev MI, Pavlov AI, Bley T (2007) Hairy root type plant *in vitro* systems as sources of bioactive substances. Appl Microbiol Biotechnol 74:1175

Gherardi JM, Rengel Z (2004) The effect of manganese supply on exudation of carboxylates by roots of lucerne (*Medicago sativa*). Plant Soil 260:271–282

Giri A, Narasu ML (2000) Transgenic hairy roots: recent trends and applications. Biotechnol Adv 18:1–22

Giri A, Ravindra ST, Dhingra V, Narasu ML (2000) Influence of different strains of *Agrobacterium rhizogenes* on induction of hairy roots and artemisinin production in *Artemisia annua*. Curr Sci 81:378–382

Good AG, Crosby WL (1989) Anaerobic induction of alanine aminotransferase in barley root tissue. Plant Physiol 90:1305–1309

Good AC, Muench DC (1993) Long-term anaerobic metabolism in root tissue. Plant Physiol 101:1163–1168

Gou J, Strauss SH, Tsai CJ, Kai F, Chen Y, Jiang X, Busov VB (2010) Gibberellins regulate lateral root formation in *Populus* through interactions with auxin and other hormones. Plant Cell 22:623–639

Granicher F, Christen P, Kapetandis I (1992) High yield production of valepotriates by hairy root cultures of *Valeriana officinals* L. var. sambucifolia Mikan. Plant Cell Rep 11:339–342

Guerinot ML, Yi Y (1994) Iron: nutritious, noxious and not readily available. Plant Physiol 104:815–820

Guillon S, Trémouillaux-Guiller J, Pati PK, Rideau M, Gantet P (2006) Hairy root research: recent scenario and exciting prospects. Curr Opin Plant Biol 9:341–346

Gurusamy PD, Schäfer H, Ramamoorthy S, Wink M (2017) Biologically active recombinant human erythropoietin expressed in hairy root cultures and regenerated plantlets of *Nicotiana tabacum* L. PLoS One 12(8):e0182367

Gutierrez L, Mongelard G, Floková K, Păcurar DI, Novák O, Staswick P, Kowalczyk M, Păcurar M, Demailly H, Geiss G, Bellini C (2012) Auxin controls Arabidopsis adventitious root initiation by regulating jasmonic acid homeostasis. Plant Cell 112(24):2515–2527

Gutzeit H, Ludwig-Müller J (2014) Plant natural products: synthesis, biological functions and practical applications. Wiley Blackwell, Weinheim

Hacham Y, Sela A, Friedlander L, Savaldi-Goldstein S (2012) *BRI1* activity in the root meristem involves post-transcriptional regulation of *PIN* auxin efflux carriers. Plant Signal Behav 7(1):68–70

Hamamoto H, Boulter ME, Shirsat AH, Croy EJ, Ellis JR (1990) Recovery of morphogenetically normal transgenic tobacco from hairy roots co-transformed with *Agrobacterium rhizogenes* and a binary vector plasmid. Plant Cell Rep 9:88–92

Hamill JD, Parr AJ, Rhodes MJC, Robins RJ, Walton NJ (1987) New routes to plant secondary products. Biotechnology 5:800–804

Han KH, Keathley DE, Davis JM, Gordon MP (1993) Regeneration of a transgenic woody legume (*Robinia pseudoacacia* L. black locust) and morphological alterations induced by *Agrobacterium rhizogenes*-mediated transformation. Plant Sci 88:149–157

Hartmann HT (1990) Anatomical and physiological basis of propagation by cuttings. In: Plant propagation, principles and practices. Prentice Hall Inc, pp 199–255

Hartman HT, Kester DE, Davies FT (1990) Plant propagation: principles and practices. Prentice-Hall of India, Englwood Cliffs. ISBN 0-13-681007-1

Hennion F, Martin-Tanguy J (2000) Amines of the subantarctic crucifer *Pringlea antiscorbutica* are responsive to temperature conditions. Physiol Plant 109:232–243

Higuchi M, Pischke MS, Mähönen AP, Miyawaki K, Hashimoto Y, Seki M, Helariutta Y, Sussman MR, Kakimoto T (2004) *In planta* functions of the Arabidopsis cytokinin receptor family. Proc Natl Acad Sci 101(23):8821–8826

Hodge A, Robinson D, Fitter AH (2000) Are microorganisms more effective than plants at competing for nitrogen? Trends Plant Sci 5:304–308

Hodge A, Berta G, Doussan C, Merchan F, Crespi M (2009) Plant root growth, architecture and function. Plant Soil 32:153–187

Horn ME, Woodward SL, Howard JA (2004) Plant molecular farming: systems and products. Plant Cell Rep 22:711–720

Hu ZB, Du M (2006) Hairy root and its application in plant genetic engineering. J Integr Plant Biol 48(2):121–127

Huang NC, Liu KH, Lo HJ, Tsay YF (1999) Cloning and functional characterization of an Arabidopsis nitrate transporter gene that encodes a constitutive component of low affinity uptake. Plant Cell 11:1381–1392

Hwang HH, Yu M, Lai EM (2017, Oct 20) *Agrobacterium*-mediated plant transformation: biology and applications. Arabidopsis Book 15:e0186

Ibañez S, Talano M, Ontañon O, Suman J, Medina MI, Macek T, Agostini E (2016) Transgenic plants and hairy roots: exploiting the potential of plant species to remediate contaminants. New Biotechnol 33:625–635

Ikeuchi M, Ogawa Y, Iwase A, Sugimoto K. (2016). Plant regeneration: cellular origins and molecular mechanisms. Development (2016) 143:1442-1451

Iyer-Pascuzzi A, Symonova O, Mileyko Y, Hao Y, Belcher H, Harer J, Weitz JS, Benfey PN (2010) Imaging and analysis platform for automatic phenotyping and trait ranking of plant root systems. Plant Physiol 152:1148–1157

Jackson RB, Manwaring JH, Caldwell MM (1990) Rapid physiological adjustment of roots to localized soil enrichment. Nature 344:58–60

Jacob A, Malpathak N (2004) Green hairy root cultures of *Solanum khasianum* Clarke—a new route to *in vitro* solasodine production. Curr Sci:1442–1447

Jeong CS, Chakrabarty D, Hahn EJ, Lee HL, Paek KY (2006) Effects of oxygen, carbon dioxide and ethylene on growth and bioactive compound production in bioreactor culture of ginseng adventitious roots. Biochem Eng J 27:252–263

Jeong GT, Park DH (2006) Enhanced secondary metabolite biosynthesis by elicitation in transformed plant root system: effect of abiotic elicitors. Appl Biochem Biotechnol 129–132:436–446

Jha P, Jobby R, Desai NS (2016) Remediation of textile azo dye acid red 114 by hairy roots of *Ipomoea carnea* Jacq. and assessment of degraded dye toxicity with human keratinocyte cell line. J Hazard Mater 311:158–167

Jing H, Strader LC (2019) Interplay of auxin and cytokinin in lateral root development. Int J Mol Sci 20(3):486

Kerk NM, Jiang K, Feldman LJ (2000) Auxin metabolism in the root apical meristem. Plant Physiol 122(3):925–932

Kino-oka M, Nagatome H, Taya M (2001) Characterization and application of plant hairy roots endowed with photosynthetic functions. Adv Biochem Eng Biotechnol 72:183–218

Kondo O, Honda H, Taya M, Kobayashi T (1989) Comparison of growth properties of carrot hairy root in various bioreactors. Appl Microbiol Biotechnol 32:291–294

Laplaze L, Benkova E, Casimiro I, Maes L, Vanneste S, Swarup R, Weijers D, Calvo V, Parizot B, Herrera-Rodriguez MB, Offringa R, Graham N, Doumas P, Friml J, Bogusz D, Beeckman T, Bennett M (2007) Cytokinins act directly on lateral root founder cells to inhibit root initiation. Plant Cell 19:3889–3900

Lee RB, Rudge KA (1986) Effects of nitrogen deficiency on the absorption of nitrate and ammonium by barley plants. Ann Bot 57:471–486

Lewis OAM (1986) Plants and nitrogen. Edward Arnold Publishers Ltd, London

Lewis DR, Miller ND, Splitt BL, Wu G, Spalding EP (2007) Separating the roles of acropetal and basipetal auxin transport on gravitropism with mutations in two Arabidopsis multidrug resistance-like ABC transporter genes. Plant Cell 19:1838–1850

Leyser O (2006) Dynamic integration of auxin transport and signalling. Curr Biol 16:R424–R433

Li W, Asada Y, Yoshikawa T (1998) Antimicrobial flavonoids from *Glycyrrhiza glabra* hairy root cultures. Planta Med 64:746–747

Li W, Koike K, Asada Y, Yoshikawa T, Nikaidoa T (2003) Biotransformation of low molecular-weight alcohols by *Coleus forskohlii* hairy root cultures. Carbohydr Res 338:729–731

Linden JC, Haigh JR, Mirjalili N, Phisaphalong M (2001) Gas concentration effects on secondary metabolite production by plant cell cultures. Adv Biochem Eng Biotechnol 72:27–62

Liu CZ, Wang YC, Ouyang F, Ye HC, Li GF (1997) Production of artemisinin by hairy root culture of *Artemisia annua* L. Biotechnol Lett 19:927–929

Lokhande VH, Kudale S, Nikalje G, Desai N, Suprasanna P (2015) Hairy root induction and phytoremediation of textile dye, Reactive green 19A-HE4BD, in a halophyte, *Sesuvium portulacastrum* (L.) L. Biotechnol Rep 8:56–63

Lopez-Bucio J, Cruz-Ramirez A, Herrera-Estrella L (2003) The role of nutrient availability in regulating root architecture. Curr Opin Plant Biol 6:280–287

Lorence A, Medina-Bolivar F, Nessler CL (2004) Camptothecin and 10-hydroxycamptothecin from *Camptotheca acuminate* hairy roots. Plant Cell Rep 22:437–441

Ludewig U, von Wirén N, Frommer WB (2002) Uniport of NH+4 by the root hair plasma membrane ammonium transporter *LeAMT1;1*. J Biol Chem 277:13548–13555

Luo J, Tillman RW, Ball PR (2000) Nitrogen loss through denitrification in a soil under pasture in New Zealand. Soil Biol Biochem 32:497–509

Lynch J (1995) Root architecture and plant productivity. Plant Physiol 109(1):7

Ma Z, Bielenberg DG, Brown KM, Lynch JP (2001) Regulation of root hair density by phosphorus availability in Arabidopsis thaliana. Plant Cell Environ 24:459–467

Malamy JE (2005) Intrinsic and environmental response pathways that regulate root system architecture. Plant Cell Environ 28:67–77

Marschner M (1995) Mineral nutrition of higher plants, 2nd edn. Academic, London

Mauro ML, Costantino P, Bettini PP (2017) The never ending story of *rol* genes: a century after. Plant Cell Tissue Organ Cult (PCTOC) 131(2):201–212

McCully M (1999) Roots in soil: unearthing the complexities of roots and their rhizospheres. Ann Rev Plant Physiol Plant Mol Biol 50:695–718

Miller AJ, Cramer MD (2004) Root nitrogen acquisition and assimilation. Plant Soil 274:1–36

Miller AJ, Smith SJ (1996) Nitrate transport and compartmentation. J Exp Bot 47:843–854

Morris DA, Briant RE, Thomson PG (1969) The transport and metabolism of 14C-indole acetic acid in intact pea seedlings. Planta 89:178–197

Mouchel CF, Osmont KS, Hardtke CS (2006) *BRX* mediates feedback between brassinosteroid and auxin signaling in root growth. Nature 443:458–461

Murthy HN, Lee EJ, Paek KY (2014) Production of secondary metabolites from cell and organ cultures: strategies and approaches for biomass improvement and metabolite accumulation. Plant Cell Tissue Organ Cult 118:1–16

Nader B, Cardoso-Taketa A, Pereda-Miranda R, Villareal MA (2006) Production of triterpenoids in liquid-cultivated hairy roots of *Galphimia glauca*. Planta Med 72:842–844

Namdev AG (2007) Plant cell elicitation for production of secondary metabolites: a review. Pharmacog Rev 1:69–79

Naphatsamon U, Ohashi T, Misaki R, Fujiyama K (2018) The Production of Human β-Glucocerebrosidase in *Nicotiana benthamiana* Root Culture. Int J Mol Sci 19(7):1972

Nazoa P, Vidmar JJ, Tranbarger TJ, Mouline K, Damiani I, Tillard P, Zhou D, Glass ADM, Touraine B (2003) Regulation of the nitrate transporter gene *ATNRT2.1* in *Arabidopsis thaliana*: responses to nitrate, amino acids and developmental stage. Plant Mol Biol 52:689–703

Nishikawa K, Ishimaru K (1997) Flavonoids in root cultures of *Scutellaria baicalensis*. J Plant Physiol 151:633–636

Ohkawa H, Kamada H, Sudo H, Harada H (1989) Effects of gibberellic acid on hairy root growth in the *Datura innoxia*. J Plant Physiol 134:633–636

Okamoto M, Vidmar JJ, Glass ADM (2003) Regulation of *NRT1* and *NRT2* gene families of *Arabidopsis thaliana*: responses to nitrate provision. Plant Cell Physiol 44:304–317

Onrubia M, Moyano E, Bonfill M, Cusidó RM, Goossens A, Palazón J (2013) Coronatine, a more powerful elicitor for inducing taxane biosynthesis in Taxus media cell cultures than methyl jasmonate. J Plant Physiol 170:211–219

Osuna L, Moyano E, Mangas S, Bonfill M, Cusidó RM, Piñol MT, Zamilpa A, Palazón J (2008) Immobilization of *Galphimia glauca* plant cell suspensions for the production of enhanced amounts of Galphimine-B. Planta Med 74(01):94–99

Osuna L, Pereda-Miranda R, Tortoriello J, Villarreal ML (1999) Production of the sedative triterpene galphimine B in *Galphimia glauca* tissue culture. Planta Med 65:149–152

Overvoorde P, Fukaki H, Beeckman T (2010) Auxin control of root development. Cold Spring Harb Perspect Biol 2(6)

Parry G, Delbarre A, Marchant A, Swarup R, Napier R, Perrot-Rechenmann C, Bennett MJ (2001) Novel auxin transport inhibitors phenocopy the auxin influx carrier mutation *aux1*. Plant J 25:399–406

Patra N, Srivastava A (2015) Artemisinin production by plant hairy root culturesin gas- and liquid-phase bioreactors. Plant Cell Rep 35(1):143–153

Plasencia A, Soler M, Dupas A, Ladouce N, Silva-Martins G, Martinez Y, Grima-Pettenati J (2016) Eucalyptus hairy roots, a fast, efficient and versatile tool to explore function and expression of genes involved in wood formation. Plant Biotechnol J 14(6):1381–1393

Raven J, Smith F (1976) Nitrogen assimilation and transport in vascular land plants in relation to intracellular pH regulation. New Phytol 76:415–431

Rhodes MJC, Parr AJ, Giuletti A, Aird ELH (1994) Influence of exogenous hormones on the growth and secondary metabolite formation in transformed root cultures. Plant Cell Tissue Organ Cult 38:143–151

Richards D, Considine JA (1981) Suberization and browning of grapevine roots. In: Brouwer R (ed) Structure and function of plant roots. Dr. W. Junk, London, pp 111–115

Ricigliano V, Kumar S, Kinison S, Brooks C, Nybo SE, Chappell J, Howarth DG (2016) Regulation of sesquiterpenoid metabolism in recombinant and elicited *Valeriana officinalis* hairy roots. Phytochemistry 125:43–53

Ricker AJ, Banfield WM, Wright WH, Keitt GW, Sagen HE (1930) Studies on infectious hairy root of nursery apple trees. J Agric Res 41:507–540

Riefler M, Novak O, Strnad M, Schmülling T (2006) Arabidopsis cytokinin receptor mutant reveal functions in shoot growth, leaf senescence, seed size, germination, root development, and cytokinin metabolism. Plant Cell 18:40–54

Rischer H, Orešič M, Seppänen-Laakso T, Katajamaa M, Lammertyn F, Ardiles-Diaz W, Goossens A (2006) Gene-to-metabolite networks for terpenoid indole alkaloid biosynthesis in *Catharanthus roseus* cells. Proc Natl Acad Sci 103(14):5614–5619

Robbins, MP, Evans TE, Morris P. (1996). The effect of plant growth regulators on growth, morphology and condensed tannin accumulation in transformed root cultures of Lotus corniculatus. Plant Cell Tiss. Organ Cult. 44:219–227; 1996

Roberts JKM, Hooks MA, Miallis AP, Edwards S, Webster C (1992) Contribution of malate and amino acid metabolism to cytoplasmic pH regulation in hypoxic root tips studied using nuclear magnetic resonance spectroscopy. Plant Physiol 98:480–487

Roy CS (2015) Gene transfer in higher plants for the development of genetically modified crops (GM crops). Int J Curr Adv Res 4:132–148

Sabatini S, Heidstra R, Wildwater M, Scheres B (2003) SCARECROW is involved in positioning the stem cell niche in the Arabidopsis root meristem. Genes Dev 17:354–358

Saini S, Sharma I, Kaur N, Pati PK (2013) Auxin: a master regulator in plant root development. Plant Cell Rep 32(6):741–757

Sakakibara H, Suzuki M, Takei M, Deji A, Taniguchi M, Sugiyama T (1998) A response-regulator homolog possibly involved in N signal transduction mediated by cytokinin in maize. Plant J 14:337–344

Sarkar S, Ghosh I, Roychowdhury D, Jha S (2018) The effects of *rol* genes of *Agrobacterium rhizogenes* on morphogenesis and secondary metabolite accumulation in medicinal plants. In: Biotechnological approaches for medicinal and aromatic plants. Springer, Singapore, pp 27–51

Scheible WR, Morcuende R, Czechowski T, Fritz C, Osuna D, Palacios-Rojas N, Schindelasch D, Thimm O, Udvardi MK, Stitt M (2004) Genome- Wide reprograming of primary and secondary metabolism, protein synthesis, cellular growth processes, and the regulatory infrastructure of Arabiposis in response to nitrogen. Plant Physiol 136:2483–2499

Schmidt W, Tittel J, Schikora A (2000) Role of hormones in the induction of iron deficiency responses in Arabidopsis roots. Plant Physiol 122:1109–1118

Schmülling T, Schell J, Spena A (1988) Single genes from *Agrobacterium rhizogenes* influence plant development. EMBO J 7:2621–2629

Shams MV, Nazarian-Firouzabadi F, Ismaili A, Shirzadian-Khorramabad R (2019) Production of a recombinant dermaseptin peptide in *Nicotiana tabacum* hairy roots with enhanced antimicrobial activity. Mol Biotechnol:1-12

Sharma A, Cardoso-Taketa A, Choi YH, Verpoorte R, Villarreal ML (2012) A comparison on the metabolicprofiling of the Mexican anxiolytic and sedative plant *Galphimia glauca* four years later. J Ethnopharmacol 141:964–974

Sharma A, Angulo-Bejarano PI, Madariaga-Navarrete A, Oza G Iqbal H, Cardoso-Taketa A, Luisa Villarreal M (2018) Multidisciplinary investigations on *Galphimia glauca:* a Mexican medicinal plant with pharmacological potential. Molecules 23(11) 2985

Shelden MC, Dong B, de Bruxelles GL, Trevaskis B, Whelan J, Ryan PR, Howitt SM, Udvardi MK (2001) Arabidopsis ammonium transporters, *AtAMT11* and *AtAMT1;2*, have different biochemical properties and functional roles. Plant Soil 231:151–160

Shen WH, Petit A, Guern J, Tempé J (1998) Hairy roots are more sensitive to auxin than normal roots. Proc Natl Acad Sci 85:3417–3421

Shiao TL, Ellis MH, Dolferus R, Dennis ES, Doran PM (2003) Overexpression of alcohol dehydrogenase or pyruvate decarboxylase improves growth of hairy roots at reduced oxygen concentrations. Biotechnol Bioeng 77:455–461

Shilpa K, Varun K, Lakshmi BS (2010) An alternate method of natural drug production: eliciting plant secondary metabolite using plant cell culture. J Plant Sci 5:222–247

Shkolnik-Inbar D, Bar-Zvi D (2010) *ABI4* mediates abscisic acid and cytokinin inhibition of lateral root formation by reducing polar auxin transport in Arabidopsis. Plant Cell 22:3560–3573
Signora L, De Smet I, Foyer CH, Zhang H (2001) ABA plays a central role in mediating the regulatory effects of nitrate on root branching in Arabidopsis. Plant J 28:655–662
Silverstone AL, Chang CW, Krol E, Sun TP (1997) Developmental regulation of the gibberellins biosynthetic gene *GA1I* in *Arabidopsis thaliana*. Plant J 12:9–19
Sivanandhan G, Selvaraj N, Ganapathi A, Manickavasagam M (2014) An efficient hairy root culture system for *Withania somnifera* (L.) Dunal. Afr J Biotechnol 13(43)
Skoog F, Miller CO (1957) Chemical regulation of growth and organ formation in plant tissue cultures in vitro. Symp Soc Experim Biol 11:118–131
Smith TC, Weathers PJ, Cheetham RD (1997) Effects of gibberellic acid on hairy root cultures of *Artemisia annua*: growth and artemisinin production. In Vitro Cell Dev Biol Plant 33:75–79
Stepanova AN, Hoyt JM, Hamilton AA, Alonso JM (2005) A link between ethylene and auxin uncovered by the characterization of two root-specific ethylene-insensitive mutants in Arabidopsis. Plant Cell 17:2230–2242
Suresh B, Ravishankar GA (2004) Phytoremediation – a novel and promising approach for environmental clean-up. Crit Rev Biotechnol 24:97–124
Suresh B, Sherkhane PD, Kale S, Eapen S, Ravishankar GA (2005) Uptake and degradation of DDT by hairy root cultures of *Cichorium intybus* and *Brassica juncea*. Chemosphere 61:1288–1292
Swarup R, Friml J, Marchant A, Ljung K, Sandberg G, Palme K, Bennett M (2001) Localization of the auxin permease *AUX1* suggests two functionally distinct hormone transport pathways operate in the Arabidopsis root apex. Genes Dev 15:2648–2653
Szczerba MW, Britto DT, Kronzucker HJ (2009) K+ transport in plants: Physiology and molecular biology. J Plant Physiol 166:447–466
Tang C, Han XZ, Qiao YF, Zheng SJ (2009) Phosphorus deficiency does not enhance proton release by roots of soybean [*Glycine max* (L.) Murr.]. Environ Exp Bot 67:228–234
Taya M, Yoyama A, Kondo O, Kobayashi T (1989) Growth characteristics of plant hairy roots and their cultures in bioreactors. J Chem Eng Japan 22:84–89
Taya M, Yakura K, Kino-Oka M, Tone S (1994) Influence of medium constituents on enhancement of pigment production by batch culture of red beet hairy roots. J Ferment Bioeng 77:215–217
Tepfer D (1990) Genetic transformation using *Agrobacterium rhizogenes*. Physiol Plantarum 79(1):140–146
Tepfer D (2016) DNA transfer to plants by *Agrobacterium rhizogenes*: a model for genetic communication between species and biospheres. In: Transgenesis and secondary metabolism. Springer, pp 1–41. ISBN 978-3-319-28669-3
Thimann KV (1937) On the nature of inhibitions caused by auxin. Am J Bot 24:407–412
Thimmaraju R, Bhagyalakshmi N, Narayan MS, Ravishankar GA (2003) Food-grade chemical and biological agents permeabilize red beet hairy roots, assisting the release of betalaines. Biotechnol Prog 19:1274–1282
Tian Q, Reed JW (1999) Control of auxin-regulated root development by the *Arabidopsis thaliana SHY2/IAA3* gene. Development 126:711–721
Tomilov A, Tomilova N, Yoder JI (2007) *Agrobacterium tumefaciens* and *Agrobacterium rhizogenes* transformed roots of the parasitic plant *Triphysaria versicolor* retain parasitic competence. Planta 225:1059–1071
Vanhala L, Eeva M, Lapinjoki S, Hiltunen R, Oksman-Caldentey K (1998) Effect of growth regulators on transformed root cultures of *Hyoscyamus muticus*. J Plant Physiol 153:475–481
Vieten A, Sauer M, Brewer PB, Friml J (2007) Molecular and cellular aspects of auxin transport-mediated development. Trends Plant Sci 12:160–168
von Wiren N, Lauter FR, Ninnemann O, Gillissen B, Walch-Liu P, Engels C, Jost W, Frommer WB (2000) Differential regulation of three functional ammonium transporter genes by nitrogen in root hairs and by light in leaves of tomato. Plant J 21:167–175
Weathers PJ, Bunk G, McCoy MC (2005) The effect of phytohormones on growth and artemisin production in *Artemisia annua* hairy roots. In Vitro Cell Dev Biol 41:47–53

Wells CE, Eissenstat DM (2003) Beyond the roots of young seedlings: the influence of age and order on fine root physiology. J Plant Growth Regul 21:324–334

Werner T, Motyka V, Laucou V, Smets R, Van Onckelen H, Schmülling T (2003) Cytokinin-deficient transgenic Arabidopsis plants show multiple developmental alterations indicating opposite functions of cytokinins in the regulation of shoot and root meristem activity. Plant Cell 15:2532–2550

White PJ (1996) The permeation of ammonium through a voltage- independent K+ channel in the plasma membrane of rye roots. J Membr Biol 152:89–99

Williams LE, Miller AJ (2001) Transporters responsible for the uptake and partitioning of nitrogenous solutes. Annu Rev Plant Physiol Plant Mol Biol 52:659–688

Williamson L, Ribrioux S, Fitter A, Leyser O (2001) Phosphate availability regulates root system architecture in Arabidopsis. Plant Physiol 2001(126):1–8

Wysokinska H, Chmiel A (1997) Transformed root cultures for biotechnology. Acta Biotechnol 17:131–159

Xu ZQ, Jia JF (1996) The reduction of chromosome number and the loss of regeneration ability during subculture of hairy root cultures of *Onobrychis viciaefolia* transformed by *Agrobacterium rhizogenes* A4. Plant Sci 120:107–112

Yang YK, Lee SY, Park WT, Park NI, Park SU (2010) Exogenous auxins and polyamines enhance growth and rosmarinic acid production in hairy root cultures of *Nepeta cataria* L. Plant Omics 3:190–193

Yaoya S, Kanho H, Mikami Y, Itani T, Umehara K, Kuroyanagi M (2004) Umbelliferone released from hairy root cultures of *Pharbitis nil* treated with copper sulfate and its subsequent glucosylation. Biosci Biotechnol Biochem 2004(68):1837–1841

Yoshikawa M (1978) Diverse modes of action of biotic and abiotic phytoalexin elicitors. Nature 275:546–547

Yu KW, Murthy HN, Hahn E-J, Paek KY (2005) Ginsenoside production by hairy root cultures of *Panax ginseng*: influence of temperature and light quality. Biochem Eng J 23:53–56

Yu Q, Lutz Kerry A, Maliga P (2017) Efficient Plastid Transformation in Arabidopsis. Plant Physiol 175:186–193

Zhang HM, Forde BG (2000) Regulation of Arabidopsis root development by nitrate availability. J Exp Bot 51:51–59

Zhang RQ, Tang CF, Wen SZ, Liu YG, Li KL (2006) Advances in research on genetically engineered plants for metal resistance. J Integr Plant Biol 48:1257–1265

Zhao XY, Su YH, Cheng ZJ, Zhang XS (2008) Cell fate switch during in vitro plant organogenesis. J Integr Plant Biol 50(7):816–824

Integrating the Bioinformatics and Omics Tools for Systems Analysis of Abiotic Stress Tolerance in *Oryza sativa* (L.)

3

Pandiyan Muthuramalingam, Rajendran Jeyasri, Subramanian Radhesh Krishnan, Shunmugiah Thevar Karutha Pandian, Ramalingam Sathishkumar, and Manikandan Ramesh

Abstract

Abiotic stress can inflict limitations on plant growth, developmental processes and also crop productivity. Here we have portrayed advances in omics tools in the view of conservative and contemporary approaches that could be used to unravel abiotic stress tolerance in rice. Under stressful conditions, plants can develop diverse molecular mechanisms to combat stress challenges, while it is not sufficient to protect them. Hence, speculation of this study is essential for understanding how plants react to adverse environmental conditions with the hope of enhancing the tolerance of plants to abiotic stress. It could be addressed by computational biology (bioinformatics); invigorated sequencing approaches in genomics have paved the way for various analytical applications. Focusing on the technological advances, multiple new omics such as the transcriptome, metabolome, hormonome, epigenome, proteome and phenome have emerged. An emphasis was given to systems approaches with respect to abiotic stress. In addition, the availability of rice whole genome information, advancement and development of omics studies has improved to address the identification of unique and combined abiotic stress responsive cellular metabolisms and this enables the interaction

P. Muthuramalingam · R. Jeyasri · S. T. K. Pandian · M. Ramesh (✉)
Department of Biotechnology, Science Campus, Alagappa University, Karaikudi, India

S. R. Krishnan
Department of Biotechnology, Science Campus, Alagappa University, Karaikudi, India

Phytopharma Testing Laboratory, Herbal Division, T. Stanes & Company Ltd, Coimbatore, India

R. Sathishkumar
Plant Genetic Engineering Laboratory, Department of Biotechnology, Bharathiar University, Coimbatore, Tamil Nadu, India

R. Sathishkumar et al. (eds.), *Advances in Plant Transgenics: Methods and Applications*, https://doi.org/10.1007/978-981-13-9624-3_3

between signalling pathways, molecular biological insights along with novel traits and their significance. Thus, this chapter provides the bioinformatics and systems biology aspects of abiotic stress responses by comparing it with the publically available omics and bioinformatics resources which could provide a base for detailed functional studies of stress tolerance in rice.

Keywords
Rice · Abiotic stress · Systems biology · Omics · Modelling · Genome · Transcriptome · Metabolome · Proteome · Hormonome · Phenome

3.1 Introduction

Rice is one of the widely used food crops, particularly in Asia. While the global rice yield has been gradually increasing from the Green Revolution in the 1960s, the ever-growing population and environmental changes is the major hurdle for its sustained production in the mere future. Besides, the number of biotic and abiotic stresses such as salinity, drought, flooding, metal, UV, high temperature, and cold significantly affects rice production. Hence, in order to increase crop yield and improve stress resistance, it is overbearing to analyse the possible biochemical and molecular adaptations of rice to multiple abiotic stresses.

A strong understanding on how the molecular and cellular morphology (phenotype) differs across multiple stress conditions still remains a major roadblock despite past decades of research. These kinds of circumstances exist in line with the publically available multiple high-throughput data such as genomics, transcriptomics, metabolomics, hormonomics, proteomics, phenomics and ionomics mainly owing to the lack of systematic analysis. In this regard, this chapter aims to delineate the integrative omics and systems approach to characterise and classify the rice molecular physiology under multiple stresses by integrative omics data mathematical network models.

3.2 Factors Affecting Rice Productivity

Moreover increasing food ultimatum and arable land shrinking, production of rice also faces major challenges by diverse environmental stressors, time and again simultaneously. Such stress factors can be divided into two major aspects: abiotic and biotic.

Abiotic stress factors include drought, salinity, cold, flooding, light, temperature, etc. Among these, drought stress can be considered as the predominant abiotic stressor, which can negatively affect crop yield and growth, even under supreme cultivation conditions (Hadiarto and Tran 2011). With water paucity, the stomatal conductance may close owing to the passive loss of turgor pressure in guard cells. Based on the available literature, uptake level of CO_2 is dramatically decreased and the high amount of light energy cannot fix any carbon is dissolute in the form of ROS (reactive oxygen species), such as peroxides and superoxides (de Carvalho 2008). These toxic compounds can alter or damage the DNA and proteins, and thus, perhaps leading to multiple lethal effects metabolism at cellular level. Opposite to water deficit, flooding stress is one of the major abiotic stress factors of crop yield and productivity as merely one third of the yield loss is directly exhibited to high amount of water or excess water (Blom and Voesenek 1996). Under flooding conditions, plants are logged and surrounded by water, which controls O_2 diffusion and in consequence highly hinders aerobic respiration. On the other hand, unlike several other food crops, rice is the most unique in its knack to survive O_2 stress via multiple adaptive mechanisms which include generation of fermentative energy (Perata and Alpi 1993; Guglielminetti et al. 1995). Thus, rice plant has been considered as a unique model plant for investigating flooding stress. Many of the rice varieties are salt-sensitive crops as salinity reduces plant growth and development stages specifically at the seedling stage (Negrao et al. 2011). During salt stress, the osmotic stress condition gradually develops and reduces the transpiration and expansion of leaf. Simultaneously to this dynamism, the defects in root Na^+ transporters can cause the senescence of leaves and/or death. Particularly, salinity gradually reduces photosynthesis which could affect plant growth and development. Light stress also affects plant developmental stages significantly: both wavelength (quality) and fluence (light intensity) impact plant growth and developmental stages in multiple aspects. As in water deficit conditions, high fluence also majorly inhibits photosynthesis and therefore crop yield and productivity (Osakabe and Osakabe 2012). Besides, wavelength affects diverse morpho-physiological processes in all plants, particularly in rice (Neff et al. 2000).

3.3 Rice Research in Past, Present and Future

Originally, rice research has been started almost 1000 years ago. Majority of the early research findings deals with rice cultivation techniques and breeding. These techniques include (1) screening and selection of rice varieties with appropriate characteristics as progenitors and storage for ensuing generations, (2) suitable land selection, (3) fortitude of optimal cultivation and suitable harvest times and (4) water management (Christou 1994). Rice-related scientific research activities started possibly during the European Rebirth, i.e. fourteenth to seventeenth century. During this period, plant anatomy was recognized with the innovation of microscopes, and the basic concepts about plant physiology such as release of O_2 and uptake of CO_2 and nutrient and water absorption through roots and plant sexuality was unveiled (Browne 2007). In the era of molecular biology research, though the exact origin cannot be pinpointed, majority of them might have started in the nineteenth century based on landmark sighting such as principles of fluence utilization in photosynthesis and structure of plant cells. This invention was used in the pre-twentieth century for diverse molecular characterization to investigate plant cells as individual entities. Multiple biochemical analyses were performed to mine the individual compounds acted in cellular growth. The details of metabolism, photosynthesis and transport were illuminated precisely and by mid of the twentieth century deals with molecular basis of plant reproduction and cellular growth was strongly established (Lack and Evans 2005). Later on, with the advent of recombinant DNA technology, plant and molecular biologists have used them in combinatorial aspects with molecular biology techniques to attain more insights into genes and their products in different plant cellular and molecular processes.

The eye-catching modern genomic era, the huge number of data sets and ever-increasing genome sequence of plants are now publicly available. *Arabidopsis* is the first completely sequenced plant genome (Arabidopsis Genome Initiative 2000). Among food crops, rice plant genome was the first sequenced in 2005 through IRGSP (IRGSP 2005). Rice is the most well-annotated C3 food crop, and its whole genome information availability has captured the special attention of plant scientists around the globe. Rice genome details and other facts are provided in Table 3.1.

Recently, the ever-increasing high-throughput omics technologies have paved the way for generating multiple omics data such as transcripome, proteome, metabolome, hormonome, ionome and phenome at genome level from rice plants grown

Table 3.1 Gramene rice genome information and FAOSTAT report

Genus/species and facts	*Oryza sativa*
Tribe	Oryzeae
Subfamily	Ehrhartoideae
Common name	Rice
World production	618,534,989 Mt
World harvest	153,783,818 Ha
Genome size	400–466 Mbp
Chromosome number	12
2n	24
Photosynthetic pathway	C3
Assembly	IRGSP-1.0, INSDC Assembly
Database version	94.7
Base pairs	375,049,285
Golden path length	375,049,285
Genebuild by	IRGSP
Genebuild method	Imported from RAP-DB
Coding genes	35,825
Non-coding genes	1017
Small non-coding genes	922
Long non-coding genes	95
Pseudogenes	9
Gene transcripts	43,404
FGENESH gene prediction	46,238
TE-related genes (MSU)	17,272
Short variants	28,179,246
Structural variants	1278

under various conditions. However, due to the availability of huge spectrum of cellular and molecular data, we are still far from characterizing the cellular and molecular responses of rice data which are highly under used. In this aspect, the systems biology field offers vast promise to integrate and incorporate every molecule of biological information which is available in public databases that can help understand the organism at the molecular physiological level, which could pave the way for next-level revolution in plant biological research.

3.4 Systems Biology Is a New Paradigm in Biological Research

Biological systems are highly complex, till, coordinated and organized at diverse levels, which deals with global range of activities. The past decades have made remarkable progress towards the molecular understanding of such complex systems, although in parts mainstream of the research in the twentieth century speculated on the chemistry of biological molecules and the elemental analysis of molecular machineries which produce and utilize them. The major development in such progressive way was made with mining of DNA, the kernel of all living systems, and its pivotal role in regulating the functioning of diverse levels of biological systems. From then, the ever-increasing sequencing and several high-throughput omics technologies have provided/generated the large amount of publically available data at diverse levels of the molecular hierarchy, i.e. genome, transcriptome, epigenome, proteome, metabolome and hormonome, and, thus, pave the way for diverse functional portfolios. Moreover, these omics approaches used to understand the complex network of the cellular organisms have enthused researchers to focus them in a different aspect by abstracting or indexing the biological processes as the models from mathematics using the wide experimental data. This era has been formally known as systems biology. The era of systems biology originally mimics biology, though studies on the living systems in provisions of mathematical representations. A few instances of systems biology usages include the development of yield in microbial cell factories for biotechnological industries, unveiling the cell-to-cell communications and molecular cellular evolution.

3.5 Developing the Parts and Generating the Components Data

The capability to represent the biological sections and their molecular crosstalks in the form of mathematical model and analyse them through suitable computational analysis techniques has been considered as a golden achievement of systems biology. In common, the mathematical models in systems biology involved two steps: (i) identification of systems and (ii) in silico and developmental analysis of mathematical models.

Systems analysis mainly involves the mining of the list of biological parts and their molecular crosstalks for delineating the biological process. As previously reported, biological processes and systems can operate and interact at diverse levels and be used to study the diverse genes, biological components and their molecular signalling. Correspondingly, a protein-protein interaction also is made at different levels. This could pave the way to study the biochemical reactions and their network, their interactions across and within genes and their products, i.e. metabolites, transcripts, proteins and signalling molecules.

After identification of system, the next stage is mathematical representation. As we mentioned before, since molecular biological activities are involved at diverse levels and many scales of space and time, the model type might be plural: it can be either (i) macroscopic or microscopic, (ii) stochastic or deterministic, (iii) continuous or discrete and (iv) temporal, steady state, spatial or spatio-temporal. The preference of modelling and simulation is particularly based on available data and the type of system being studied. The established models are reliable and accurate. So, it can be used further to decipher the novel hypotheses for developing new experimental horizons, leading to a replicate knowledge discovery process (Fig. 3.1).

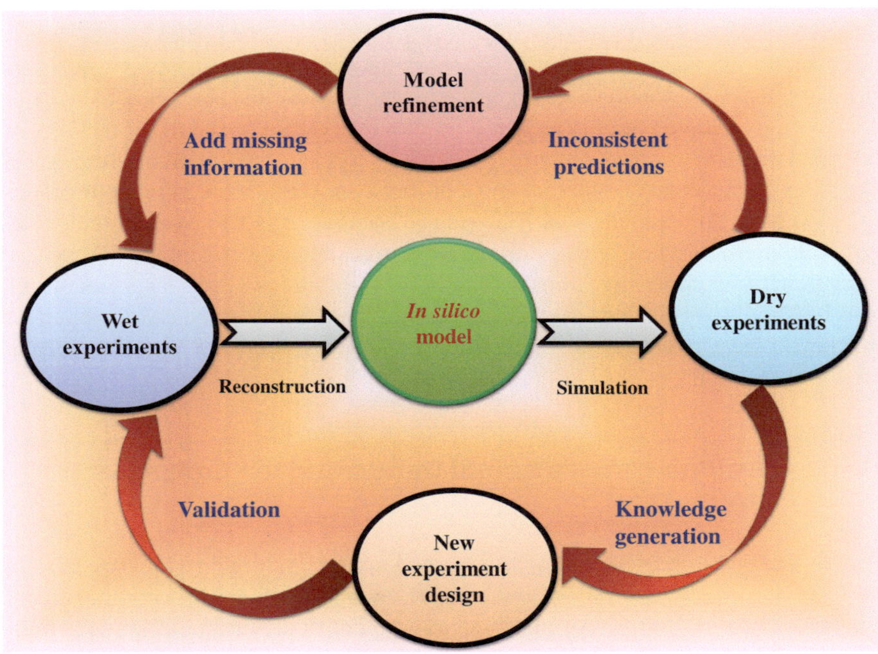

Fig. 3.1 Replicate model construction etiquette in systems biology (adopted from Kitano (2002))

3.6 Rice Omics

Publicly available rice plant and their whole genome sequences and the advent of omics based research, understanding the computational biology related tools has become highly needed to gain the novel molecular insights. Ever-increasing omics approaches such as genome, transcriptome, epigenome, proteome, metabolome, hormonome and interactome data sets will shed more light on several budding systematic and/or analytical applications. In addition, the advent of omics resources has allowed us to understand the molecular signalling, cellular systems containing properties of the individual species particularly on rice. Combined information from several omics-based resources is suitable to identify the molecular and cellular system insights and biological attributes that stimulate gene discovery and its annotated information (Table 3.2).

Table 3.2 Omics types and thematic properties

Omics	Thematic properties
Genomics	Nucleotide sequence, molecular marker, genome annotation, gene diversity, gene variation, transcription factors, gene family databases, SNP, marker-assisted selection (MAS)
Transcriptomics	Microarray, gene chip technology, Chip-seq, non-coding RNA, RNA fingerprinting, cDNA clones, ESTs, EST-SSRs
Epigenomics	Small RNAs (miRNA, siRNA), chromatin structure, methylC-seq, mRNA-seq, smallRNA-seq, DNA methylation, BS-seq, epigenetic map, histone modification, transposable elements, nucleosome-bound DNA modification
Proteomics	Protein peptide changes, subcellular localization, peptide engineering, interaction between protein and RNA, protein properties, modificome profiles, acetylome, methylome
Interactomics	Protein–protein interaction, cell cycle, Ca2+/ calmodulin-mediated signalling, auxin signalling, membrane protein–signalling protein interaction, biotic–abiotic stress interaction, protein–DNA interaction, protein–RNA interaction
Metabolomics	Metabolic map, metabolite profile, metabolic pathways, enzymes, systems analysis, metabolic measurements, metabolites, tissue- and organ-specific metabolites, metabolic modelling, metabolic simulation, metabolic network engineering
Hormonomics	Signalling molecules, auxin, cytokinin, gibberellins, ethylene, brassinosteroids, jasmonates, salicylic acid, strigolactone, peptide hormone, ABA regulatory network, transport and perception
Phenomics	Mutant lines, natural variations, species variation, phenotypic traits, phenotypic markers, structural traits, physiological traits, plant growth performance, integrated databases

3.7 Systems Biology Approaches Help to Understand Abiotic Stress Responses in Rice

Availability of rice genome sequences and transcriptome datasets, ever increasing omics-based research is a pivotal tool to understand the molecular signalling responses under abiotic stress conditions. This section will mainly focus on past research findings which have been used to analyse the abiotic stress responses in rice via systems biology approaches, particularly in omics data generation. With the ever-growing advancements in experimental techniques, the omics has been progressively applied to rice to unveil abiotic stress response modulators. Omics-based research studies on rice mainly speculating on understanding the molecular aspect of abiotic stress such as drought, salinity, flooding and combined abiotic stresses and their summary are given in Table 3.3. Publicly available repositories, databases and integrated omics tools (Table 3.4) were used to identify and annotate the unique and CAbS genes and their dynamism in rice.

3.8 View of Constraints-Based Modelling

The pivotal step in systems biology predominantly represents the biological processes and their significance in the mathematical form. Huge numbers of approaches have been discussed in the past for cellular modelling processes. In stipulations of kinetic modelling, metabolic modelling (Steuer et al. 2006), analysis of metabolic control (Reder 1988), cybernetic modelling (Kompala et al. 1984) and CBM (constraints-based modelling) (Price et al. 2004) are the four major approaches suggested. From these, approximately every metabolic network was investigated with the help of kinetic modelling framework. Usually, these kinds of models correspond to a particular component of metabolism by a set of formula where all equations may speculate the dynamic changes in involved metabolites concentration. The amount of depletion or formation of metabolite using an enzyme is generally corresponded by Hill equation or Michaelis–Menten kinetics. However, these models can deliver deep molecular process and their insights, which are practically not possible to model networks at large scale by using this method as kinetic variables may often, hurdle to measure, these things are not available for variety of biochemical reactions. On the other hand, CBM has a clear application over the kinetic models originally requires only the metabolic reaction information, low experimental data and flux capacity (Raman and Chandra 2009). Many genome-level metabolic networks have been reconstructed successfully for uncountable microbes and to a handful number of animals and plants using these approaches (Kim et al. 2012). Therefore, this CBM technique highly provides more insights into rice abiotic and cellular metabolism. The advantage and application of CBM methods and many of the software tools were available for employment of these analyses. CBM software applications herein can be grouped into any one of the 3 classes i.e. (1) stand alone, (2) toolbox based library and (3) web based, which depends upon the usability and software dependencies (Table 3.5).

Table 3.3 Usage of omics techniques to unravel abiotic stress responses in rice

Omics type	Type of abiotic stress	Summary	References
Transcriptome	Drought	EST-based analysis of drought stress-related genes in rice	Gorantla et al. (2007)
		Microarray analysis of salt and drought-induced genes in rice shoot, flag leaf and panicle	Zhou et al. (2007)
		Drought and high-salinity stresses, a comparative transcriptome analysis of rice under cold	Rabbani et al. (2003)
		Drought-sensitive rice varieties and transcriptome analysis of drought-tolerant	Degenkolbe et al. (2009)
		Drought-tolerant and lowland, microarray analysis of upland, drought-intolerant, rice varieties	Wang et al. (2007)
Proteome		Proteomic analysis of rice 2 to 6 days old under normal and drought stress conditions	Ali and Komatsu (2016)
		Proteomic analysis of 3-week-old rice seedlings grown under normal and drought stress conditions	Salekdeh et al. (2002)
Transcriptome	Salinity	Identification of salt stress-related genes through comparative microarray analysis	Dai Yin et al. (2005)
		Comparative transcriptome analysis of barley and rice during salt stress	Ueda et al. (2006)
		Salt tolerant and a wild type salt intolerant, comparison of transcriptional responses between a recombinant	Walia et al. (2005)
		drought and high-salinity stresses, a comparative transcriptome analysis of rice under cold	Rabbani et al. (2003)

(continued)

Table 3.3 (continued)

Omics type	Type of abiotic stress	Summary	References
		Flag leaf and panicle microarray analysis of salt and drought stress- induced genes in rice shoot	Zhou et al. (2007)
		Transcriptome analysis of rice plants experiencing salt stress between 15 min and 1 week from induction	Kawasaki et al. (2001)
Proteome		Comparative analysis of rice proteome between salt stress treated rice panicles and control	Dooki et al. (2006)
		Proteomic analysis of rice seedlings grown under salt stress and normal conditions	Kim et al. (2005)
		Proteomic analysis of long- and short-term salt-stress-responsive proteins in the rice leaf lamina	Parker et al. (2006)
		Proteomic analysis of 3-week-old rice seedlings grown under salt stress and normal conditions	Yan et al. (2005)
Metabolome		Metabolome analysis of rice leaf and root under salt stress conditions and control	Zuther et al. (2007)

(continued)

Table 3.3 (continued)

Omics type	Type of abiotic stress	Summary	References
Transcriptome	Flood	Transcriptome and metabolome analysis of rice seed germinated in anoxia and anoxia at 9 different time points	Narsai et al. (2009)
		ADH1 is one of the key enzymes involved in survival of rice under anoxia. Transcript profiling of alcohol dehydrogenase 1(ADH1)-deficient mutant	Takahashi et al. (2011)
		Microarray analysis of rice seed germinated in air and anoxia	Lasanthi-Kudahettige et al. (2007)
Proteome		Wheat during anoxic germination and comparative proteome and metabolome analysis of rice	Shingaki-Wells et al. (2011)
Metabolome		Metabolome and transcriptome analysis of rice seed in anoxia and anoxia at 9 different time points	Narsai et al. (2009)
		Wheat during anoxic germination and comparative proteome and metabolome analysis of rice	Shingaki-Wells et al. (2011)
		Anoxia and metabolome analysis of rice coleoptiles under air	Fan et al. (2003)
Transcriptome	Combined abiotic stress (Drought, salinity, metal, cold, submergence)	Improve the traits in stress-affected genes, gene identification in 48 developmental stages	Muthuramalingam et al. (2017)
Genome		Identification and analysis of rice tissues specific expression at various stress conditions from transcription factor	Muthuramalingam et al. (2018a)
Metabolome		Metabolome analysis of threonine metabolite in combined abiotic stress conditions	Muthuramalingam et al. (2018b)

(continued)

Table 3.4 Bioinformatics databases and tools for multiple omics approaches

Database name	Purpose	References
GRASSIUS	Comprehensive collection of Transcription factors and co-regulation transcription factors	Yilmaz et al. (2009)
Phytozome	Comparative genomics	Goodstein et al. (2012)
Gramene	Comparative functional genomics	Tello-Ruiz et al. (2016)
Plant-Reactome Gramene Pathways	Plant metabolic and regulatory pathway analysis	Tello-Ruiz et al. (2016)
Expression Atlas	Gene expression and biological conditions	Petryszak et al. (2016)
Gramene QTL Database	QTL prediction and analysis	Ni et al. (2009)
RiceNetDB	Genome-scale multiple-level network reconstruction and comprehensive rice genome annotated information	Liu et al. (2013)
RiceXPro	Repository of gene expression profiles derived from microarray analysis of tissues and organs	Sato et al. (2010)
RAP-DB	Comprehensive analysis of the genome structure and function of rice on the basis of the annotation	Sakai et al. (2012)
R-Oryzabase	Comprehensive major DNA sequences and literature-based database	Kurata and Yamazaki (2006)
MCDRP	Manually curated database of rice proteins	Raghuvanshi et al. (2016)
NCBI Ref Seq	Gene annotation	Pruitt et al. (2014)
Ensembl Plants	Genome sequenceGene annotation	Kersey et al. (2014)
Yale Plant Genomics	Epigenome analysis	Li et al. (2008), Wang et al. (2009), and He et al. (2010)
PRIN	Interactome analysis	Gu et al. (2011)
Rice SNP seek Database	Provides genotype, phenotype and variety information for rice	Alexandrov et al. (2014)
OryGeneDB data	Display sequence information	Droc et al. (2005)
Oryza sativa genome DB -PlantGDB	Displays high-quality spliced alignments for EST, cDNA, PUT and proteins	Duvick et al. (2008)
BGI rice Database	Sequence alignment, small RNA and transcriptome analysis, de novo genome assembly, gene structure and function annotation as well as evolutionary and comparative analysis	Zhao et al. (2003)

Table 3.5 Constraints-based modelling softwares and their applications

Type of the software application	Name	Tested version	URL	References
Stand-alone	SBRT	2.0.0	http://www.bioc.uzh.ch/wagner/software/SBRT/	Wright and Wagner (2008)
	BioOpt	–	http://129.16.106.142/tools.php?c=bioopt	Cvijovic et al. (2010)
	SurreyFBA	–	http://sysbio3.fhms.surrey.ac.uk/	Gevorgyan et al. (2011)
	MetaFluxNet	1.8	http://metauxnet.kaist.ac.kr/	Lee et al. (2003a, b)
	OptFlux	2.1	http://www.optux.org/	Rocha et al. (2010)
	GEMSiRV	–	http://sb.nhri.org.tw/GEMSiRV/en/	Liao et al. (2012)
	FASIMU	2.3.1	http://www.bioinformatics.org/fasimu/downloads	Hoppe et al. (2011)
Toolbox-based	SNAToolbox	1.0	http://bioinformatics.org/project/?groupid¼546	Urbanczik (2006)
	CellNetanalyzer/FluxAnalyzer	9.5	http://www.mpi-magdeburg.mpg.de/projects/cna/cna.html	Klamt et al. (2003) and Klamt et al. (2007)
	FBA-SimVis	–	http://fbasimvis.ipk-gatersleben.de/	Grafahrend-Belau et al. (2009)
	MetaFlux	–	http://www.biocyc.org/download.shtml	Latendresse et al. (2012)
	COBRAToolbox	2.0	http://opencobra.sourceforge.net/	Becker et al. (2007) and Schellenberger et al. (2011)

(continued)

Table 3.5 (continued)

Type of the software application	Name	Tested version	URL	References
Web-based	GSMN-TB	–	http://sysbio3.fhms.surrey.ac.uk/cgi-bin/fba/fbapy	Beste et al. (2007)
	Model SEED	–	http://seedviewer.theseed.org/seedviewer.cgi?page=ModelView	Henry et al. (2010)
	MicrobesFlux	–	http://tanglab.engineering.wustl.edu/static/MicrobesFlux.html	Feng et al. (2012)
	CycSim	1.0.0	http://www.biocyc.org/download.shtml	Le Fevre et al. (2009)
	WEbcoli	1.5	http://webcoli.org/	Jung et al. (2009)
	Acorn	–	http://sysbio3.fhms.surrey.ac.uk:8080/acorn/homepage.jsf	Sroka et al. (2011)
	FAME	–	http://f-a-m-e.org/	Boele et al. (2012)

3.9 Conclusion

Well-developed analytical platforms and constraints-based models are providing opportunities to address the omics scale outcomes, stimulating the researchers' knowledge and far standing basic questions in plant science. At the same time, we are mainly fronting diverse issues such as water, food, energy, security, climatic changes and global warming. Unravels the plant cellular and molecular functions, stress tolerance mechanisms that comes into view in particular plant species especially on rice is also significant to discover key players, modelling reconstruction, characterize the molecular network to enhance the agronomical traits in plants. An integration of broad-spectrum omics data sets and mathematical modelling, network reconstruction is then pivotal to improve the translational research to engineer plant systems.

3.10 Future Perspectives

The whole elucidation of cellular and molecular interaction underlying abiotic stress resistance, plant growth and developmental biology majorly needs overall investigation by omics-based systemic techniques, which act as the essential source for the application of various omics in plant biology. Among them, epigenomics, metabolomics, hormonomics, signalomics and phenomics have a pivotal role, because metabolites, signalling cassette and molecular and cellular elements are highly relevant to the plant physiological and morphological changes as compared with functional proteins and other biological molecules. Compared to the individual abiotic stressors, combined abiotic stressors severely affects the rice growth and productivity. Hence, future research in this thrust area will speculate the direction on the improvement of mathematical modelling and combination of omics especially epigenomics, metabolomics, hormonomics, signalomics and phenomics to analyse the effect of combined abiotic stresses on rice plant tissues will be more appropriate. These modern approaches will also provide the novel molecular and cellular insights about agronomical traits related to growth and development, amino acid biosynthesis, stress and hormone metabolisms. Therefore, integrated model and omics platforms enable accurate and realistic understanding about plant molecular systems physiology; these specific models will provide us global rice metabolism and guide crop improvement in an efficient way.

Acknowledgements The author Pandiyan Muthuramalingam (Rc.SO (P)/DBT-BIF/15207/2017 dated February 02, 2018) thanks the DBT-Bioinformatics Infrastructure Facility Scheme, New Delhi, India, for the financial support in the form of fellowship. The authors gratefully acknowledge the use of the Bioinformatics Infrastructure Facility, Alagappa University, funded by the Department of Biotechnology, Ministry of Science and Technology, Government of India grant (No.BT/BI/25/015/2012). The authors also thank RUSA 2.0 [F. 24-51/2014-U, Policy (TN Multi-Gen), Dept of Edn, GoI].

References

Alexandrov N, Tai S, Wang W, Mansueto L, Palis K, Fuentes R, Ulat V, Chebotarov D, Zhang G, Li Z, Mauleon R, Hamilton R, McNally K (2014) SNP-Seek database of SNPs derived from 3000 rice genomes. Nucleic Acids Res 43:D1023–D1027

Ali GM, Komatsu S (2016) Proteomic analysis of rice leaf sheath during drought stress. J Proteome Res 5:396–403

Becker SA, Feist AM, Mo ML, Hannum G, Palsson B, Herrgard MJ (2007) Quantitative prediction of cellular metabolism with constraint-based models: the COBRA Toolbox. Nat Protoc 2:727–738

Beste DJ, Hooper T, Stewart G, Bonde B, Avignone-Rossa C, Bushell ME, Wheeler P, Klamt S, Kierzek AM, McFadden J (2007) GSMN-TB: a web-based genome-scale network model of *Mycobacterium tuberculosis* metabolism. Genome Biol 8:R89

Blom C, Voesenek L (1996) Flooding: the survival strategies of plants. Trends Ecol Evolut 11:290–295

Boele J, Olivier BG, Teusink B (2012) FAME, the flux analysis and modeling environment. BMC Syst Biol 6:8

Browne J (2007) History of plant science. In: Roberts K (ed) Handbook of plant science, 2nd edn. Wiley, Chicester, pp 3–7

Christou P (1994) Rice as a global crop. In: Rice biotechnology and genetic engineering. Technomic Publishing Company, Inc, Lancaster, pp 1–38

Cvijovic M, Olivares-Hernandez R, Agren R, Dahr N, Vongsangnak W, Nookaew I, Patil KR, Nielsen J (2010) BioMet Toolbox: genome-wide analysis of metabolism. Nucleic Acids Res 38:W144–W149

Dai Yin C, Yong Hai L, Min S, Da L, Hong Xuan L (2005) Salt-responsive genes in rice revealed by cDNA microarray analysis. Cell Res 15:796–810

de Carvalho MHC (2008) Drought stress and reactive oxygen species. Plant Signaling Behaviour 3:156–165

Degenkolbe T, Do PT, Zuther E, Repsilber D, Walther D, Hincha DK, Kohl KI (2009) Expression profiling of rice cultivars differing in their tolerance to long-term drought stress. Plant Mol Biol 69:133–153

Dooki AD, Mayer-Posner FJ, Askari H, Aa Z, Salekdeh GH (2006) Proteomic responses of rice young panicles to salinity. Proteomics 6:6498–6507

Droc G, Ruiz M, Larmande P, Pereira A, Piffanelli P, Morel JB, Dievart A, Courtois B, Guiderdoni E, Périn C (2005) OryGenesDB: a database for rice reverse genetics. Nucleic Acids Res 34:D737

Duvick J, Fu A, Muppirala U, Sabharwal M, Wilkerso M, Lawrence C, Lushbough C, Brendel V (2008) PlantGDB: a resource for comparative plant genomics. Nucleic Acids Res 36:D959–D965

Fan TM, Lane A, Higashi R (2003) In vivo and in vitro metabolomic analysis of anaerobic rice coleoptiles revealed unexpected pathways. Russ J Plant Physiol 50:787–793

Feng X, Xu Y, Chen Y, Tang YJ (2012) MicrobesFlux: a web platform for drafting metabolic models from the KEGG database. BMC Syst Biol 6:94

Gevorgyan A, Bushell ME, Avignone-Rossa C, Kierzek AM (2011) SurreyFBA: a command line tool and graphics user interface for constraint-based modeling of genome scale metabolic reaction networks. Bioinformatics 27:433–434

Goodstein DM, Shu S, Howson R, Neupane R, Hayes RD, Fazo J, Mitros T, Dirks W, Hellsten U, Putnam N, Rokhsar DS (2012) Phytozome: a comparative platform for green plant genomics. Nucleic Acids Res 40:D1178–D1186

Gorantla M, Babu P, Lachagari VR, Reddy A, Wusirika R, Bennetzen JL, Reddy AR (2007) Identification of stress-responsive genes in an indica rice (*Oryza sativa* L.) using ESTs generated from drought-stressed seedlings. J Exp Bot 58:253–265

Grafahrend-Belau E, Klukas C, Junker BH, Schreiber F (2009) FBA-SimVis: interactive visualization of constraint-based metabolic models. Bioinformatics 25:2755–2757

Gu H, Zhu P, Jiao Y, Meng Y, Chen M (2011) PRIN: a predicted rice interactome network. BMC Bioinform 12:161

Guglielminetti L, Perata P, Alpi A (1995) Effect of anoxia on carbohydrate metabolism in rice seedlings. Plant Physiol 108:735–741

Hadiarto T, Tran LSP (2011) Progress studies of drought-responsive genes in rice. Plant Cell Reports 30:297–310

He G, Zhu X, Elling AA, Chen L, Wang X, Guo L (2010) Global epigenetic and transcriptional trends among two rice subspecies and their reciprocal hybrids. Plant Cell 22:17–33

Henry CS, DeJongh M, Best AA, Frybarger PM, Linsay B, Stevens RL (2010) High-throughput generation, optimization and analysis of genome-scale metabolic models. Nat Biotechnol 28:977–982

Hoppe A, Hoffmann S, Gerasch A, Gille C, Holzhutter HG (2011) FASIMU: flexible software for flux-balance computation series in large metabolic networks. BMC Bioinform 12:28

International Rice Genome Sequencing Project A (2005) The map-based sequence of the rice genome. Nature 436:793–800

Jung TS, Yeo HC, Reddy SG, Cho WS, Lee DY (2009) WEbcoli: an interactive and asynchronous web application for in silico design and analysis of genome-scale E: coli model. Bioinformatics 25:2850–2852

Kawasaki S, Borchert C, Deyholos M, Wang H, Brazille S, Kawai K, Galbraith D, Bohnert HJ (2001) Gene expression profiles during the initial phase of salt stress in rice. The Plant Cell Online 13:889–905

Kersey PJ, Allen JE, Christensen M, Davis P, Falin LJ, Grabmueller C, Hughes DS, Humphrey J, Kerhornou A, Khobova J, Langridge N, McDowall MD, Maheswari U, Maslen G, Nuhn M, Ong CK, Paulini M, Pedro H, Toneva I, Tuli MA, Walts B, Williams G, Wilson D, Youens-Clark K, Monaco MK, Stein J, Wei X, Ware D, Bolser DM, Howe KL, Kulesha E, Lawson D, Staines DM (2014) Ensembl genomes 2013: scaling up access to genome-wide data. Nucleic Acids Res 42:D546–D552

Kim DW, Rakwal R, Agrawal GK, Jung YH, Shibato J, Jwa NS, Iwahashi Y, Iwahashi H, Kim DH, Shim IS et al (2005) A hydroponic rice seedling culture model system for investigating proteome of salt stress in rice leaf. Electrophoresis 26:4521–4539

Kim TY, Sohn SB, Kim YB, Kim WJ, Lee SY (2012) Recent advances in reconstruction and applications of genome-scale metabolic models. Curr Opin Biotechnol 23:617–623

Kitano H (2002) Systems biology: a brief overview. Science 295:1662–1664

Klamt S, Stelling J, Ginkel M, Gilles ED (2003) FluxAnalyzer: exploring structure, pathways, and flux distributions in metabolic networks on interactive flux maps. Bioinformatics 19:261–269

Klamt S, Saez-Rodriguez J, Gilles ED (2007) Structural and functional analysis of cellular networks with CellNetAnalyzer. BMC Syst Biol 1:2

Kompala DS, Ramkrishna D, Tsao GT (1984) Cybernetic modeling of microbial growth on multiple substrates. Biotechnol Bioeng 26:1272–1281

Kurata N, Yamazaki Y (2006) Oryzabase. An integrated biological and genome information database for rice. Plant Physiol 140:12

Lack A, Evans D (2005) Understanding plants – methods in plant biology. In Plant biology. Taylor & Francis Group, New York, pp 5–16

Lasanthi-Kudahettige R, Magneschi L, Loreti E, Gonzali S, Licausi F, Novi G, Beretta O, Vitulli F, Alpi A, Perata P (2007) Transcript profiling of the anoxic rice coleoptile. Plant Physiol 144:218–231

Latendresse M, Krummenacker M, Trupp M, Karp PD (2012) Construction and completion of flux balance models from pathway databases. Bioinformatics 28:388–396

Le Fevre F, Smidtas S, Combe C, Durot M, d'Alche Buc F, Schachter V (2009) CycSim-an online tool for exploring and experimenting with genome-scale metabolic models. Bioinformatics 25:1987–1988

Lee DY, Yun H, Park S, Lee SY (2003a) MetaFluxNet: the management of metabolic reaction information and quantitative metabolic flux analysis. Bioinformatics 19:2144–2146

Lee SY, Lee DY, Hong SH, Kim TY, Yun H, Oh YG, Park S (2003b) MetaFluxNet, a program package for metabolic pathway construction and analysis, and its use in large-scale metabolic flux analysis of *Escherichia coli*. Genome Inform 14:23–33

Li X, Wang X, He K, Ma Y, Su N, He H (2008) High-resolution mapping of epigenetic modifications of the rice genome uncovers interplay between DNA methylation, histone methylation, and gene expression. Plant Cell 20:259–276

Liao YC, Tsai MH, Chen FC, Hsiung CA (2012) GEMSiRV: a software platform for Genome-scale metabolic model simulation, reconstruction and visualization. Bioinformatics 28:1752–1758

Liu L, Mei Q, Yu Z, Sun T, Zhang Z, Chen M (2013) An integrative bioinformatics framework for genome-scale multiple level network reconstruction of rice. J Integr Bioinform 10(2):94–102

Muthuramalingam P, Krishnan SR, Pothiraj R, Ramesh M (2017) Global transcriptome analysis of combined abiotic stress signaling genes unravels key players in *Oryza sativa* L.: an in silico approach. Front Plant Sci 8:759

Muthuramalingam P, Krishnan SR, Saravanan K, Mareeswaran N, Kumar R, Ramesh M (2018a) Genome-wide identification of major transcription factor superfamilies in rice identifies key candidates involved in abiotic stress dynamism. J Plant Biochem Biotechnol:1–18

Muthuramalingam P, Krishnan SR, Pandian S, Mareeswaran N, Aruni W, Pandian SK, Ramesh M (2018b) Global analysis of threonine metabolism genes unravel key players in rice to improve the abiotic stress tolerance. Sci Rep 8(1):9270

Narsai R, Howell KA, Carroll A, Ivanova A, Millar AH, Whelan J (2009) Defining core metabolic and transcriptomic responses to oxygen availability in rice embryos and young seedlings. Plant Physiol 151:306–322

Neff MM, Fankhauser C, Chory J (2000) Light: an indicator of time and place. Genes Develop 14:257–271

Negrao S, Courtois B, Ahmadi N, Abreu I, Saibo N, Oliveira M (2011) Recent updates on salinity stress in rice: from physiological to molecular responses. Crit Rev Plant Sci 30:329–377

Ni J, Pujar A, Youens-Clark K, Yap I, Jaiswal P, Tecle I, Tung C, Ren L, Spooner W, Wei X, Avraham S, Ware D, Stein L, McCouch S (2009) Gramene QTL database: development, content and applications. Database:Article ID bap005. https://doi.org/10.1093/database/bap005

Osakabe K, Osakabe Y (2012) Plant light stress. eLS

Parker R, Flowers TJ, Moore AL, Harpham NV (2006) An accurate and reproducible method for proteome profiling of the effects of salt stress in the rice leaf lamina. J Exp Bot 57:1109–1118

Perata P, Alpi A (1993) Plant responses to anaerobiosis. Plant Sci 93:1–17

Petryszak R, Keays M, Tang YA, Fonseca NA, Barrera E, Burdett T (2016) Expression Atlas update – an integrated database of gene and proteinexpression in humans, animals and plants. Nucleic Acids Res 44:746–752

Price ND, Reed JL, Palsson B (2004) Genome-scale models of microbial cells: evaluating the consequences of constraints. Nat Rev Microbiol 2:886–897

Pruitt KD, Brown GR, Hiatt SM, Thibaud-Nissen F, Astashyn A, Ermolaeva O, Farrell CM, Hart J, Landrum MJ, McGarvey KM, Murphy MR, O'Leary NA, Pujar S, Rajput B, Rangwala SH, Riddick LD, Shkeda A, Sun H, Tamez P, Tully RE, Wallin C, Webb D, Weber J, Wu W, Dicuccio M, Kitts P, Maglott DR, Murphy TD, Ostell JM (2014) RefSeq: an update on mammalian reference sequences. Nucleic Acids Res 42:D756–D763

Rabbani MA, Maruyama K, Abe H, Khan MA, Katsura K, Ito Y, Yoshiwara K, Seki M, Shinozaki K, Yamaguchi-Shinozaki K (2003) Monitoring expression profiles of rice genes under cold, drought, and high-salinity stresses and abscisic acid application using cDNA microarray and RNA gel-blot analyses. Plant Physiol 133:1755–1767

Raghuvanshi S, Gour P, Joseph SV (2016) Manually Curated Database of Rice Proteins (MCDRP), a database of digitized experimental data on rice. Current Plant Biol 8:26–30

Raman K, Chandra N (2009) Flux balance analysis of biological systems: applications and challenges. Brief Bioinform 10:435–449

Reder C (1988) Metabolic control theory: a structural approach. J Theor Biol 135:175–201

Rocha I, Maia P, Evangelista P, Vilaca P, Soares S, Pinto JP, Nielsen J, Patil KR, Ferreira EC, Rocha M (2010) OptFlux: an open-source software platform for in silico metabolic engineering. BMC Syst Biol 4:45

Sakai H, Shin Lee S, Tanaka T, Numa H, Kim J, Kawahara Y, Wakimoto H, Yang C, Iwamoto M, Abe T, Yamada Y, Muto A, Inokuchi H, Ikemura T, Matsumoto T, Sasak T, Itoh T (2012) Rice Annotation Project Database (RAP-DB): an integrative and interactive database for rice genomics. Plant Cell Physiol 54:1–11

Salekdeh G, Siopongco J, Wade LJ, Ghareyazie B, Bennett J et al (2002) Proteomic analysis of rice leaves during drought stress and recovery. Proteomics 2:1131–1145
Sato Y, Antonio BA, Namiki N, Takehisa H, Minami H, Kamatsuki K, Sugimoto K, Shimizu Y, Hirochika H, Nagamura Y (2010) RiceXPro: a platform for monitoring gene expression in japonica rice grown under natural field conditions. Nucleic Acids Res 39:D1141–D1148
Schellenberger J, Que R, Fleming RM, Thiele I, Orth JD, Feist AM, Zielinski DC, Bordbar A, Lewis NE, Rahmanian S et al (2011) Quantitative prediction of cellular metabolism with constraint-based models: the COBRA Toolbox v2. 0. Nat Protoc 6:1290–1307
Shingaki-Wells RN, Huang S, Taylor NL, Carroll AJ, ZhouW,Millar AH (2011) Differential molecular responses of rice and wheat coleoptiles to anoxia reveal novel metabolic adaptations in amino acid metabolism for tissue tolerance. Plant Physiol 156:1706–1724
Sroka J, Bieniasz-Krzywiec L, Gwozdz S, Leniowski D, Łącki J, Markowski M, Avignone-Rossa C, Bushell ME, McFadden J, Kierzek AM (2011) Acorn: A grid computing system for constraint based modeling and visualization of the genome scale metabolic reaction networks via a web interface. BMC Bioinform 12:196
Steuer R, Gross T, Selbig J, Blasius B (2006) Structural kinetic modeling of metabolic networks. Proc Natl Acad Sci 103:11868–11873
Takahashi H, Saika H, Matsumura H, Nagamura Y, Tsutsumi N, Nishizawa NK, Nakazono M (2011) Cell division and cell elongation in the coleoptile of rice alcohol dehydrogenase 1-deficient mutant are reduced under complete submergence. Ann Bot 108:253–261
Tello-Ruiz MK, Stein J, Wei S, Preece J, Olson A, Naithani S (2016) Gramene 2016: comparative plant genomics and pathway resources. Nucleic Acids Res 44:1133–1140
The Arabidopsis Initiative A (2000) Analysis of the genome sequence of the owering plant *Arabidopsis thaliana*. Nature 408:796–815
Ueda A, Kathiresan A, Bennett J, Takabe T (2006) Comparative transcriptome analyses of barley and rice under salt stress. Theor Appl Genet 112:1286–1294
Urbanczik R (2006) SNA-a toolbox for the stoichiometric analysis of metabolic networks. BMC Bioinform 7:129
Walia H, Wilson C, Condamine P, Liu X, Ismail AM, Zeng L, Wanamaker SI, Mandal J, Xu J, Cui X et al (2005) Comparative transcriptional profiling of two contrasting rice genotypes under salinity stress during the vegetative growth stage. Plant Physiol 139:822–835
Wang H, Zhang H, Gao F, Li J, Li Z (2007) Comparison of gene expression between upland and lowland rice cultivars under water stress using cDNA microarray. Theor Appl Genet 115:1109–1126
Wang H, Schauer N, Usadel B, Frasse P, Zouine M, Hernould M (2009) Regulatory features underlying pollination-dependent and -independent tomato fruit set revealed by transcript and primary metabolite profiling. Plant Cell 21:1428–1452
Wright J, Wagner A (2008) The systems biology research tool: evolvable open-source software. BMC Syst Biol 2:55
Yan S, Tang Z, Su W, Sun W (2005) Proteomic analysis of salt stress-responsive proteins in rice root. Proteomics 5:235–244
Yilmaz A, Nishiyama MY, Fuentes BG, Souza GM, Janies D, Gray J, Grotewold E (2009) GRASSIUS: a platform for comparative regulatory genomics across the grasses. Plant Physiol 149:171–180
Zhao W, Wang J, He X, Huang X, Jiao Y, Dai M, Wei S, Fu J, Chen Y, Ren X, Zhang Y, Ni P, Zhang J, Li S, Wang J, Wong G, Zhao H, Yu J, Yang H, Wang J (2003) BGI-RIS: an integrated information resource and comparative analysis workbench for rice genomics. Nucleic Acids Res 32:D377–D382
Zhou J, Wang X, Jiao Y, Qin Y, Liu X, He K, Chen C, Ma L, Wang J, Xiong L et al (2007) Global genome expression analysis of rice in response to drought and high salinity stresses in shoot, flag leaf, and panicle. Plant Mol Biol 63:591–608
Zuther E, Koehl K, Kopka J (2007) In Comparative metabolome analysis of the salt response in breeding cultivars of rice. Springer, pp 285–315

Green Biotechnology: A Brief Update on Plastid Genome Engineering

4

R. K. B. Bharadwaj, Sarma Rajeev Kumar, and Ramalingam Sathishkumar

Abstract

Plant genetic engineering has become an inevitable tool in the molecular breeding of crops. Significant progress has been made in the generation of novel plastid transformation vectors and optimized transformation protocols. There are several advantages of plastid genome engineering over conventional nuclear transformation. Some of the advantages include multigene engineering by expression of biosynthetic pathway genes as operons, extremely high-level expression of protein accumulation, lack of transgene silencing, etc. Transgene containment owing to maternal inheritance is another important advantage of plastid genome engineering. Chloroplast genome modification usually results in alteration of several thousand plastid genome copies in a cell. Several therapeutic proteins, edible vaccines, antimicrobial peptides, and industrially important enzymes have been successfully expressed in chloroplasts so far. Here, we critically recapitulate the latest developments in plastid genome engineering. Latest advancements in plastid genome sequencing are briefed. In addition, advancement of extending the toolbox for plastid engineering for selected applications in the area of molecular farming and production of industrially important enzyme is briefed.

R. K. B. Bharadwaj
Plant Genetic Engineering Laboratory, Department of Biotechnology, Coimbatore, India

S. R. Kumar
Present Address: String Bio Private Ltd, IBAB Campus, Bangalore, India

Plant Genetic Engineering Laboratory, Department of Biotechnology, Bharathiar University, Coimbatore, Tamil Nadu, India

R. Sathishkumar (✉)
Plant Genetic Engineering Laboratory, Department of Biotechnology, Bharathiar University, Coimbatore, Tamil Nadu, India
e-mail: rsathish@buc.edu.in

R. Sathishkumar et al. (eds.), *Advances in Plant Transgenics: Methods and Applications*, https://doi.org/10.1007/978-981-13-9624-3_4

Keywords
Enzymes · Lettuce · Molecular farming · Plastome · Tobacco · Vaccines

4.1 Introduction

The expected global population by 2050 will be approximately 9.1 billion (High-Level Expert Forum, FAO, October 2009; http://www.fao.org). To nourish the ever-increasing population, food production has to be increased simultaneously. Use of latest techniques in plant genetic engineering could help to increase the production of food crops. A range of different expression platforms have been used for heterologous production of foreign proteins having pharmaceutical, industrial, and agricultural applications (Demain and Vaishnav 2009). Among different production platforms, plants have been used as highly economical and scalable production systems for the expression of recombinant proteins, enzymes, and valuable metabolites. Plastid transformation technology has attracted strong interest for different applications in plant biology owing to several advantages compared to conventional nuclear transformation (Adem et al. 2017). The high ploidy number of chloroplast genome and compartmentalization of proteins allow high levels of foreign protein accumulation. Recombinant proteins are reported to accumulate up to several folds in total leaf soluble protein in plastids (Daniell et al. 2009a, b; Oey et al. 2009; Bock 2015). Plastid genomes are inherited through the maternal parent and thus provide a strong level of biological containment thus avoiding several ethical concerns. Integration of transgene proceeds via homologous recombination and is therefore highly precise and predictable (Verma and Daniell 2007). Another important feature of plastid transformation is that genetic machinery in chloroplast is devoid of gene silencing and other epigenetic mechanisms that interfere with stable transgene expression (Wani et al. 2010; Bock 2014). Recent advancement in plastid transformation is reviewed in (Kumar et al. 2017). In this chapter, we highlight the impact of chloroplast genome engineering on various plant biology applications. Advancement of chloroplast engineering for improving agronomic traits and plastids as platform for production high-value proteins for biofuel production is discussed.

4.2 Recent Developments in Plastid Genome Sequencing

Plant chloroplast genome (plastome) is a short double-stranded circular DNA of ~100–250 kb in size. Plastome of land plants is highly conserved with two ~25 kb inverted repeat (IR) regions that are separated from the rest of the genome into large single copy (LSC) and small single-copy regions (SSC) (Zhang et al. 2018a, b). In general, plastome includes approximately 130 genes encoding transcripts involved in carbon fixation (photosynthesis), transcription, and translation. Although plastid genome is highly conserved in plants, genome size varies across different species (Table 4.1) (Daniell et al. 2016).

Table 4.1 List of important plants with annotated plastome

Sl no	Name of plant	Common name	Genome size (Mb)	References
01	*Amomum compactum*	Cardamom	0.163	Wu et al. (2018)
02	*Ananas comosus*	Pineapple	0.159	Nashima et al. (2015)
03	*Arabidopsis thaliana*	Thale cress	0.154	Sato et al. (1999)
04	*Artemisia annua*	Sweet wormwood	0.150	Shen et al. (2017)
05	*Brassica napus*	Canola	0.152	Hu et al. (2011)
06	*Camellia sinensis*	Tea	0.157	Dong et al. (2018)
07	*Cannabis sativa*	Marijuana/hemp	0.153	Oh et al. (2016) and Vergara et al. (2016)
08	*Capsicum annuum*	Pepper	0.156	Raveendar et al. (2015)
09	*Catharanthus roseus*	Periwinkle	0.154	Ku et al. 2013
10	*Cicer arietinum*	Chickpea	0.125	Jansen et al. (2008)
11	*Citrus sinensis*	Orange	0.160	Bausher et al. (2006)
12	*Cocos nucifera*	Coconut	0.154	Huang et al. (2013)
13	*Coffea arabica*	Coffee	0.155	Samson et al. (2007)
14	*Cucumis sativus*	Wild cucumber	0.155	Liu et al. (2016)
15	*Cuscuta reflexa*	Giant dodder	0.121	Funk et al. (2007)
16	*Cyamopsis tetragonoloba*	Clusterbean	0.152	Kaila et al. (2017)
17	*Daucus carota*	Carrot	0.155	Ruhlman et al. (2006)
18	*Drimys granatenis*	–	0.160	Cai et al. (2006)
19	*Eucalyptus globulus*	Tasmanian bluegum	0.160	Steane (2005)
20	*Glycine max*	Soybean	0.152	Saski et al. (2005)
21	*Hevea brasiliensis*	Rubber	0.161	Tangphatsornruang et al. (2011)
22	*Hordeum vulgare*	Barley	0.135	Middleton et al. (2014)
23	*Jasminum nudiflorum*	Winter jasmine	0.165	Lee et al. (2007)
24	*Lactuca sativa*	Lettuce	0.152	Timme et al. (2007)
25	*Lilium longiflorum*	Tiger lily	0.152	Kim et al. (2017)
26	*Lotus japonicas*	Wild legume	0.150	Kato et al. (2000)
27	*Malus hupehensis*	Wild apple	0.160	Zhang et al. (2018a, b)
28	*Manihot esculenta*	Cassava	0.161	Daniell et al. (2008)
29	*Medicago truncatulata*	Barrel medic	0.124	Gurdon and Maliga (2014)
30	*Musa acuminate*	Banana	0.169	Martin et al. (2013)
31	*Musa balbisiana*	Wild banana	0.169	Niu et al. 2018
32	*Nicotiana tabacum*	Tobacco	0.155	Shinozaki et al. 1986
33	*Nymphaea alba*	Water lily	0.159	Goremykin et al. 2004
34	*Oryza minuta*	Wild rice	0.135	Asaf et al. 2017

(continued)

Table 4.1 (continued)

Sl no	Name of plant	Common name	Genome size (Mb)	References
35	*Oryza sativa*	Rice	0.134	Yu et al. 2017
36	*Panax ginseng*	Ginseng	0.155	Nguyen et al. 2018
37	*Pelargonium* x *hortorum*	Geranium	0.217	Chumley et al. 2006
38	*Piper coenoclatum*	–	0.160	Cai et al. 2006
39	*Rosmarinus officinalis*	Rosemary	0.152	Chen and Hua (2018)
40	*Saccharum* spp.	Sugarcane	0.141	Hoang et al. 2015
41	*Salvia miltiorrhiza*	Red sage	0.151	Qian et al. (2013)
42	*Secale cereale*	Rye	0.135	Middleton et al. (2014)
43	*Solanum lycopersicum*	Tomato	0.155	Wu (2016)
44	*Solanum tuberosum*	Potato	0.155	Chung et al. (2006)
45	*Sorghum bicolor*	Sorghum	0.140	Saski et al. (2007)
46	*Spinacia oleracea*	Spinach	0.150	Schmitz-Linneweber et al. (2001)
47	*Triticum aestivum*	Common wheat	0.135	Middleton et al. (2014)
48	*Vigna radiata*	Mungbean	0.151	Tangphatsornruang et al. (2010)
49	*Vitis vinifera*	Grapes	0.160	Jansen et al. (2006)
50	*Zea mays*	Maize	0.140	Maier et al. (1995)
51	*Zingiber officinale*	Ginger	0.162	Vaughn et al. (2014)

As mentioned above, plastome lacks recombination and exhibits a uniparental (maternal) inheritance. Conventionally, plastome sequencing was based on cloning of plastid DNA fragments to generate DNA libraries followed by long-range PCR. Alternatively a large set of DNA primers were used to amplify and sequence overlapping DNA fragments in plastid genome. With the rapid advancement in next-generation sequencing technology, it is becoming increasingly faster and cheaper to sequence and assemble plastomes. Although different sequencing platforms are available, Illumina and PacBio systems are widely used for sequencing of plastid genome and subsequent assembly (Chen et al. 2015; Lin et al. 2015; Jackman et al. 2016). However, several groups have reported reads generated through PacBio are often low (English et al. 2012). This was later improved by use of latest sequencing chemistry together with a hierarchical genome assembly process algorithm (Daniell et al. 2016).

Since the advancement in sequencing technology, several research groups across the world sequenced plastid genomes of different plants (Table 4.1). Till date, ~650 plant species have their plastid genome sequence publically available in GenBank database. Plastome of two rice varieties (*Oryza sativa* indica and *O. sativa* japonica) were sequenced, and authors reported that divergence of plastome of both varieties occurred approximately 86,000 to 200,000 years ago (Tang et al. 2004). Tong et al. (2016) reported variation in chloroplast genome of rice ecotypes from Asia and Africa. Sequencing of plastid genomes of wild rice *Oryza australiensis* and *Zizania latifolia* gave insights about evolution of rice varieties (Wu and Ge 2016; Zhang et al. 2016). Plastomes of wheat, rye, barley, and other *Triticeae* species were sequenced using Roche/454 technology. Sequence alignments revealed exchange of genetic material by translocation of segment of plastome to the nuclear genome specific to rye/wheat lineage (Middleton et al. 2014). Kim et al. (2014) reported AT-rich plastid genome in orchid *Cypripedium japonicum* and proposed importance of AT residues in effective splicing of the plastid genome. Plastome analysis of banana *Musa acuminate* revealed IR/SSC expansion occurred independently multiple times in monocots during course of evolution (Martin et al. 2013).

In addition to cereals, plastome of vegetables and horticultural crops has been sequenced (reviewed in Rogalski et al. 2015). Comparisons of chloroplast genome structure between tomato (*Solanum lycopersicum*) and a wild diploid potato species (*Solanum bulbocastanum*) revealed that, at gene order, these genomes are identical, and this conservation extends even to more distantly related genera (like tobacco and *Atropa*) (Daniell et al. 2006). Analysis of carrot plastid genome revealed presence of several dispersed direct and inverted repeats scattered throughout coding and noncoding regions (Ruhlman et al. 2006). A comparative plastome analysis of grapes from different geographical locations revealed insights in origin and molecular basis of evolution of *Vitis* (Pipia et al. 2017).

Parasitic plants are interesting host to understand the evolutionary adaptation and ~1% of all known angiosperm species being parasitic plants (Westwood et al. 2010). Plastid genome analysis of several parasitic plants revealed evolutionary reduction in genome size is associated with gene loss (Funk et al. 2007; Gruzdev et al. 2016). It has been proposed that DNA in plastid loci could be horizontally acquired from its host as a result of parasitism (Molina et al. 2014). This is further supported by that finding that plasmodesmatal continuity between partners allows movement of genetic material (Birschwilks et al. 2006; Roney et al. 2007; Talianova and Janousek 2011). Thus comparative plastome analysis revealed information about simplification of plastid gene expression machinery in parasitic plants. The availability of the complete plastid genome sequence for several plant species facilitated generation of novel plastid expression vectors for efficient foreign gene expression in plants through utilization of endogenous flanking sequences and regulatory elements.

4.3 Plastid Bioreactors for Molecular Farming

Advancement in genetic engineering has revolutionized the use of therapeutically and pharmaceutically valuable proteins in a variety of clinical treatments (Grevich and Daniell 2005). In industrial scale, an ideal expression platform should produce safe and biologically active protein (with appropriate post-translational modification) at the lowest cost. Chloroplast engineering offers great promise for both agriculture and pharmaceutical industries, for production of recombinant proteins of interest in plants is emerging as an economical and safe alternative to bacterial, yeast, or animal expression platform (Daniell 2006). Several proteins have been produced in transplastomic plants in the last two decades (Tables 4.2 and 4.3).

Table 4.2 Selected list of vaccine antigens and therapeutic proteins produced in plants by chloroplast genome engineering

Sl no	Name of enzyme	Gene source	Host	References
01	Anti-HIV microbicide griffithsin	*Griffithsia* sp	Tobacco	Hoelscher et al. (2018)
02	Human factor VIII (FVIII)	Human	Tobacco	Kwon et al. (2018)
03	Tetravalent EDIII antigen (EDIII-1-4)	Dengue virus	Lettuce	van Eerde et al. 2018
04	CTB- factor IX	Human	Lettuce	Herzog et al. (2017)
05	CTB-VP1	Poliovirus	Tobacco	Chan et al. (2016)
06	Capsid protein (VP1)	Poliovirus	Lettuce	Daniell et al. (2016)
07	Envelope protein domain III-based antigens	Dengue virus	Tobacco	Gottschamel et al. (2016)
08	gp120 and gp41	HIV	Tobacco	Rubio-Infante et al. (2015)
09	Protective antigen	*Bacillus anthracis*	Tobacco	Gorantala et al. (2014)
10	HPV-16 L1	Human papillomavirus	Tobacco	Hassan et al. (2014)
11	E7 antigen	Human papillomavirus	Tobacco	Morgenfeld et al. (2014)
12	ESAT-6 (6 kDa early secretory antigenic target), Mtb72F (a fusion polyprotein from two TB antigens, Mtb32 and Mtb39), and LipY (a cell wall protein)	*Mycobacterium tuberculosis*	Lettuce/ tobacco	Lakshmi et al. (2013)
13	E protein (junction site of domains I and II)	Dengue virus	Lettuce	Maldaner et al. (2013)
14	Human granulocyte colony-stimulating factor (hG-CSF)	Human	Lettuce	Sharifi Tabar et al. (2013)
15	CTB-proinsulin	Human	Tobacco	Boyhan and Daniell (2011)
16	Premembrane (prM) and truncated envelope (E) protein	Dengue virus	Lettuce	Kanagaraj et al. (2011)
17	Malarial vaccine antigens apical membrane antigen-1 (AMA1) and merozoite surface protein-1 (MSP1) fused with cholera toxin-B subunit	*Plasmodium falciparum*	Lettuce/ tobacco	Davoodi-Semiromi et al. (2010)
18	Insulin-like growth factor	Human	Tobacco	Daniell et al. (2009b)
19	Envelope protein A27L	Vaccinia virus	Tobacco	Rigano et al. (2009)
20	$Pr55^{gag}$	HIV-1	Tobacco	Scotti et al. (2009)

(continued)

Table 4.2 (continued)

Sl no	Name of enzyme	Gene source	Host	References
21	p24	HIV-1	Tobacco	McCabe et al. (2008)
22	p24-Nef	HIV-1	Tobacco/ tomato	Zhou et al. (2008)
23	Viral capsid antigen	Epstein-Barr virus	Tobacco	Lee et al. (2006)
24	SARS-CoV spike protein	Synthetic gene corresponding to homologue protein in SARS virus	Lettuce/ tobacco	Li et al. (2006)
25	E2 protein	Hepatitis E virus	Tobacco	Zhou et al. (2006)

Table 4.3 List of industrially important enzymes/biomaterials produced by plastome engineering

Sl no	Name of enzyme	Gene source	Host	References
01	Polyhydroxybutyrate	*Ralstonia eutropha*	Tomato	Mozes-Koch et al. (2017)
02	β-Glucosidase, xylanase, and endoglucanase	*Trichoderma reesei*	Tobacco	Castiglia et al. (2016)
03	Cellulases and polygalacturonase	*Chaetomium globosum/ Paenibacillus* sp./ *Phanerochaete chrysosporium*	Tobacco	Longoni et al. (2015)
04	Xylanase	*Bacillus* sp.	Tobacco	Pantaleoni et al. (2014)
05	Cellulase	*Thermotoga maritima*	Tobacco	Jung et al. (2013)
06	β-1,4-endoglucanase	*Pyrococcus horikoshii*	Tobacco	Nakahira et al. (2013)
07	Cutinase and swollenin	*Fusarium solani/T. reesei*	Tobacco	Verma et al. (2013)
08	Agglutinin	*Pinellia ternata*	Tobacco	Jin et al. (2012)
09	β-Mannanase	*T. reesei*	Tobacco	Agrawal et al. (2011)
10	Cellulase	*Thermobifida fusca*	Tobacco	Petersen and Bock (2011)
11	Cellulase	*Thermobifida fusca*	Tobacco	Gray et al. (2009)
12	Chitinase	*Brassica juncea*	Tobacco	Guan et al. (2008)
13	Choline monooxygenase	*Beta vulgaris*	Tobacco	Zhang et al. (2008)
14	Cellulase	*Acidothermus cellulolyticus*	Tobacco	Jin et al. (2003)
15	Bioelastic protein-based polymers (PBP)	PBP gene (synthetic)	Tobacco	Guda et al. (2000)

Targeted transformation to plastid genome was thought to be highly challenging in plants as a typical leaf contains more than 2000 copies of the plastid genome with more than 100 chloroplasts (reviewed in Kumar et al. 2017). Nevertheless, the success achieved in green algae *Chlamydomonas* was also achieved in model plant tobacco (Svab et al. 1990; Svab and Maliga 1993). Since then, these two platforms remained the models for plastid transformation for production of several therapeutic proteins and industrially important enzymes. Nevertheless, progress of plastid transformation in agriculturally relevant crops like cereals and monocots is in its infancy (Maliga and Bock 2011; Rigano et al. 2012; Bock 2014). The availability of sequences of plastome, generation of novel expression vectors and development of plastid transformation protocols in crop plants extended use of chloroplast engingeering in economically important species.

Human papillomavirus (HPV) is the causative agent of cervical cancer, and HPV-E7 antigen is one of the major candidates for therapeutic vaccine production. For heterologous production of E7 in tobacco plastids, the expression of E7 as a translational fusion to β-glucuronidase enzyme was attempted. In addition, redirection of E7 into thylakoid lumen was also tried. The use of β-glucuronidase as a fusion protein turned out to be a successful strategy for improving E7 accumulation, and recombinant proteins accumulated ~40 times relative to unfused E7 (Morgenfeld et al. 2014). A high-risk HPV-16 candidate therapeutic vaccine (LALF32–51-E7) was developed by plastid targeting and resulted in 27-fold higher expression compared to cytosolic targeting in *Nicotiana benthamiana*. The authors proposed plastids-based production could be a more affordable therapeutic vaccine for HPV-16 (Yanez et al. 2018). Basic fibroblast growth factor (bFGF) accelerates cell proliferation and differentiation and hence possesses wide clinical applications. A codon-optimized bFGF gene was transformed to tobacco chloroplasts, and recombinant protein accumulation was observed (Wang et al. 2015). Serum antibodies developed in hemophilia B patients against coagulation factor IX (FIX) is highly challenging to eliminate due to nephrotic syndrome after continued infusion. Su et al. (2015) fused FIX with a transmucosal carrier (CTB) in lettuce chloroplast, and recombinant proteins accumulated up to ~1 mg/g. Moreover it was also demonstrated that feeding transgenic lettuce to hemophilia B mice delivered CTB-FIX effectively and not only induced LAP(+) regulatory T-cells but also suppressed IgE formation, anaphylaxis against FIX.

Toxoplasma gondii is an obligate intracellular parasite that causes toxoplasmosis. SAG1 is the main surface antigen in *T. gondii* and proposed as a promising vaccine candidate to produce anti-*T. gondii* vaccine. Transplastomic tobacco expressing SAG1 accumulated ~0.1–0.2 μg protein. Further, transplastomic plants expressing a 90-kDa heat shock protein of *Leishmania infantum* (LiHsp83) fused to SAG1 resulted in antigen accumulation (up to 500-fold). Subsequent oral immunization of fusion protein elicited increase in levels of SAG1-specific antibodies (Albarracín et al. 2015). Similarly, Del et al. (2012) expressed *T. gondii* GRA4 antigen in tobacco chloroplast, and immunization elicited mucosal immune response resulting in production of specific IgA, interferon (IFN-γ), and interleukin (IL-4 and IL-10). Interleukin-2 (IL-2) is a T lymphocyte-derived cytokine. Tobacco expressing human interleukin-2 (targeted to plastids) induced in vitro proliferation of IL-2-dependent murine T lymphocytes (Zhang et al. 2014).

There are approximately 36.9 million people living worldwide with acquired immunodeficiency syndrome (AIDS) in 2017 (UNAIDS). About 5000 new infections are being reported every day. Although human immunodeficiency virus (HIV) causes one of the most deadly infectious diseases, attempts to develop an effective vaccine remain unsuccessful till date. C4V3 is a protein known to induce systemic and mucosal immune responses during HIV infection. Rubio-Infante et al. (2012) expressed a synthetic gene encoding a C4V3 in tobacco plastids. The authors demonstrated that plant-derived C4V3 elicited both systemic and mucosal antibody responses in BALB/c mice. Further, CD4+ T cell proliferation responses were also reported. The authors strongly proposed plant chloroplasts as biofactories to produce HIV candidate vaccines. Human transforming growth factor-β3 (TGFβ3) has high therapeutic value and is used to reduce scarring during wound healing. A synthetic gene with codons optimized for plastid expression resulted in accumulation of the 13-kDa TGFβ3 polypeptide by 75-fold in tobacco (Gisby et al. 2011).

Bacillus anthracis is the causative agent of anthrax, and vaccines are limited as most of them are potentially reactogenic (Gorantala et al. 2011). Gorantala and coworkers expressed domain IV of protective antigen gene [PA(dIV)] from *B. anthracis* in tobacco plastids leading to more than 5% of total soluble protein accumulation. Further, mice challenged with *B. anthracis* and immunized with adjuvanted plant PA(dIV) exhibited 60% and 40% protection upon intraperitoneal and oral immunizations, respectively. Expression of plague F1-V fusion antigen in chloroplasts resulted in ~15% of the total soluble protein. Mice were immunized with F1-V extracts and subsequently exposed to an inhaled aerosolized *Yersinia pestis*. It was interesting to note that 88% of the oral F1-V mice survived aerosolized *Y. pestis*, while all control mice died within 3 days (Arlen et al. 2008). Soria-Guerra et al. (2009) expressed a fusion DPT protein encoding immune-protective epitopes of *Corynebacterium diphtheriae*, *Bordetella pertussis*, and *Clostridium tetani* in tobacco chloroplasts. Transplastomic lines accumulated recombinant proteins, and mice orally immunized with leaf extract accumulated IgG and secretory antibodies specific to DPT toxin in serum and mucosal tissues.

Human thioredoxin 1 (hTrx1) is a stress-responsive protein that functions as an antioxidant during oxidative stress. Lettuce expressing *hTrx1* in chloroplast accumulated upto 1% total soluble protein and recombinant protein protected mouse insulinoma line 6 cells from peroxide-induced damage (Lim et al. 2011). Daniell et al. (2009a, b) reported expression of insulin-like growth factor-1 (IGF-1) in transgenic tobacco chloroplasts, and IGF-1 accumulation reached close to 11 total soluble proteins in transplastomic lines. Transplastomic lettuce and tobacco lines expressing cholera toxin B subunit-human proinsulin (CTB-Pins) fusion protein accumulated up to ~16% and 2.5%, respectively, in tobacco and lettuce. Oral administration of CTB-Pins extract exhibited decreased infiltration of cells characteristic of lymphocytes (insulitis). Further, increased expression of IL-4 and IL-10 was observed in the pancreas of CTB-Pins-treated mice (Ruhlman et al. 2007).

4.4 Plastid as a Biofactory for Industrially Important Enzymes, Metabolites, and Enzyme Cocktails for Biofuel Production

Overproduction of industrially important enzymes through conventional nuclear transformation has met with little success as many reports suggest that recombinant enzymes can negatively affect transgenic plant growth and development. A gene encoding thermostable xylanase enzyme from *Bacillus* sp. when overexpressed in tobacco resulted in accumulation of xylanase up to 6% of the total soluble protein (Leelavathi et al. 2003). Similarly, Kim et al. (2011) demonstrated GH10 xylanase Xyl10B from *Thermotoga maritima* expressed in plastids exhibited high stability and catalytic activities. Further, endoglucanases like endo-β-1,4-xylanase and β-glucosidase expressed in tobacco plastids accumulated more than 75% of total soluble proteins. Subsequent bioconversion experiments confirmed that plastid-derived enzymes were able to hydrolyze industrially pretreated giant reed biomass (Castiglia et al. 2016). *Pyrococcus horikoshii* hyperthermostable archaeal β-1,4-endoglucanase expressed in tobacco plastids produced high levels of active enzymes that were even recovered from dry tissues (Nakahira et al. 2013). Another study reported transplastomic plants expressing four different thermostable cell wall-degrading enzymes from *Thermobifida fusca* accumulated up to 40% of total soluble protein. However, transplastomic lines exhibited pigment-deficient phenotypes (Petersen and Bock 2011). Klinger et al. (2015) proposed that enzymes of prokaryotic origin are efficiently expressed in plants than in a bacterial production platform. Espinoza-Sánchez et al. (2015) expressed pectin lyase and manganese peroxidase in tobacco plastids and recombinant enzymes exhibited improved enzyme activity. In another study, it was demonstrated that thermostable cellulases (*Cel6A* and *Cel6B*) from *Thermobifida fusca* expressed in tobacco chloroplasts were able to hydrolyze crystalline cellulose (Yu et al. 2007).

Plastome has been successfully engineered to produce important biomaterials. Bioelastic protein-based polymers (PBP) have huge industrial applications. These polymers are often used in soft tissue augmentation and regeneration. A synthetic PBP was targeted to both nucleus and plastids. It was confirmed that PBP transcripts accumulated up to 100-fold higher in transplastomic lines compared to nuclear transgenic plants (Guda et al. 2000). Despite the diversion of major metabolic pathway intermediate, metabolic engineering using chloroplast genomes led to production of several bioplastics. Bacterial genes encoding the polyhydroxybutyrate (PHB) pathway encoding enzymes were expressed in tobacco plastome. Transplastomic lines produced ~18.8% dry weight PHB (Bohmert-Tatarev et al. 2011). Another study reported expression of *Ralstonia eutropha* polyhydroxybutyrate operon in plastids and transplastomic lines produced biodegradable PHB (Mozes-Koch et al. 2017). Engineering chloroplast genome resulted in advancement of different biotechnological applications including production of thermostable industrial enzymes, biomaterials, and immunologicals. Nonetheless, how many of these will be approved for commercial level production and will reach market is still a matter of debate.

One of the important applications of plastid transformation is for the expression of enzyme for biofuel production. With the growing interest in biofuels, enzyme cocktails that can digest lignocellulosic biomass into fermentable sugars have gained much attention. A major breakthrough in biofuel production was the report from Verma et al. (2010) about chloroplast-derived enzyme cocktails for the production of fermentable sugars from different sources of lignocellulosic materials. The authors reported that cost of plastid-derived endoglucanase was estimated to be ~3000-fold lower than for the same commercially available recombinant enzymes (Verma et al. 2010). Verma et al. (2013) demonstrated treatment of cotton fiber with plastid produced cutinase exhibited enlarged segments and the intertwined inner fibers due to activity of cutinase. In addition, recombinant protein showed improved esterase and lipase activity in addition to its cutinase activity. Endo- β-mannanase enzyme cocktail mixture derived from chloroplast exhibited 20% more glucose equivalents formation from pinewood compared to cocktail without mannanase (Agrawal et al. 2011). Verma et al. (2010) reported plastid-derived enzyme cocktails yielded more than 3625% glucose from substrates like citrus peel and pine wood than commercially available enzyme cocktails. Transplastomic plants accumulating cell wall-degrading enzymes will be an alternate and cheap renewable source of enzymes for the production of cellulosic ethanol in the near future.

4.5 Updates on Plastid Transformation toward Improving Agricultural Traits

Chloroplast genome engineering has led to stable integration and expression of transgenes to confer valuable agronomic traits (Daniell et al. 2002; Maliga and Block 2011; and Jin and Daniell 2015). Poage et al. (2011) demonstrated expression of mitochondrial superoxide dismutase and *E.coli* glutathione reductase through plastid transformation in tobacco. The transformed lines exhibited increased radical scavenging activity thereby increasing tolerance to heavy metal stress and UV-B radiation. Expression of genes involved in secondary metabolites production has been demonstrated through plastid transformation. Tocochromanol pathway is introduced to plastids as single gene constructs into tobacco and tomato plants which resulted in tenfold higher accumulation of tocochromanol compared to controls (Lu et al. 2013). Plastomic transformation of acetolactate synthase (ALS) gene which catalyzes the first step in branched chain amino acid biosynthesis into the *Arabidopsis thaliana* conferred resistance against herbicides such as pyrimidinylcarboxylate and imidazolinon (Shimizu et al. 2008). Overexpression of cry9Aa2 (insecticidal protein) in chloroplasts of tobacco plants conferred resistance against *Phthorimaea operculella* (potato tuber moth), but this higher expression of protein delayed plant growth and development (Chakrabarti et al. 2006). Similarly, expression of cry1Ac protein through chloroplast transformation in rice resulted in enhanced resistance against common pests of rice skipper and caterpillar (Kim et al. 2009). In another study, transformation of *Pseudomonas pyrrocinia* chloroperoxidase when expressed in tobacco plastids not only resulted in accumulation of recombinant protein but also improved resistance toward *Alternaria Alternata*, *Aspergillus flavus*, *Fusarium* sp., and *Verticillium* sp. (Ruhlman et al. 2014).

Fathi Roudsari et al. (2009) generated improved glyphosate-resistant tobacco plants through biolistic transformation of chloroplasts by introducing a mutated herbicide-tolerant gene coding for EPSP synthase. p-Hydroxybenzoic acid (pHBA) is the major monomer in liquid crystal polymers. *E. coli ubiC* encoding chorismate pyruvate-lyase (CPL) was transformed to tobacco plastome. Total CPL accumulated up to 35% of total soluble protein and was 250 times higher compared to nuclear integration events. Authors reported that CPL integration to chloroplast genome provides a proof of concept of the high-flux potential of shikimate pathway for chorismate biosynthesis and also as a cost-effective route to pHBA production (Viitanen et al. 2004).

Astaxanthin is a rich antioxidant and occurs naturally in certain algae that underlie the red coloration of salmon and other organisms. Expression of different genes including β-carotene ketolase, β-carotene hydroxylase, and isopentenyl diphosphate isomerase from marine bacteria in lettuce plastome led to accumulation of astaxanthin fatty acid esters (Harada et al. 2014).Vitamin E (α, β, γ, and δ-tocopherols) are lipid-soluble antioxidants. Tocopherol biosynthetic machinery in plastids utilized precursors derived from two different metabolic pathways, homogentisic acid, an intermediate of shikimate pathway, and phytyldiphosphate, intermediate from methylerythritol phosphate pathway (Lushchak and Semchuk 2012). Expression of γ-tocopherol methyltransferase (γ-TMT) and tocopherol cyclase (TC) in tobacco and lettuce plastids resulted in improved α-tocopherol levels (Yabuta et al. 2013). Plastome expression of homogentisate phytyltransferase (HPT), TC, and γ-TMT confirmed HPT as the rate-limiting enzymatic step in tocopherol biosynthesis (Lu et al. 2013). Wurbs et al. (2007) demonstrated plastid expression of a bacterial lycopene β-cyclase in tomato-mediated conversion of lycopene to β-carotene and resulted in fourfold enhanced provitamin A content in fruits. Apel and Bock (2009) produced transplastomic tomato by overexpressing β-cyclase that resulted in an increase of up to 50% in provitamin A. Further, transplastomic plants accumulated ~50% increase in total carotenoid, and this could be possible by enhancing the flux through the pathway in chromoplasts. Overall, these findings highlight the potential of chloroplast engineering for production of high-value metabolites in plastid biofactories.

RNA interference (RNAi) technology was used for the first time to engineer the chloroplast genome by Jin et al. (2015). A lepidopteran chitin synthase (Chi), cytochrome P450 monooxygenase (CYP450), and V-ATPase were used as RNAi targets by the above group of researchers. In insects feeding assay in leaves of transplastomic lines, CYP450, Chi, and V-ATPase siRNAs, transcript levels were reduced to undetectable levels in insect midgut. In addition, net weight of the larvae and their growth and pupation were significantly reduced indicating success of technology. Zhang et al. (2015) introduced dsRNA via the chloroplast genome to target the insect β-actin gene and to subsequently elicit resistance against potato beetle. Transgenic potato lines producing dsRNAs targeted against the β-actin gene of the Colorado potato beetle, a deadlier agricultural pest, were successful. Chloroplast expression of long dsRNAs provided protection of potato lines without application of chemical pesticides. A construct encoding ~200-nucleotide duplex-stemmed-hairpin (hp) RNAs, targeting the cholinesterase gene of *Helicoverpa armigera* (cotton bull worm), was targeted to both nuclear and plastid genome of *N. benthamiana*. Undiced, full-length hairpin RNAs (hpRNA) accumulated in *N. benthamiana* and conferred strong protection against *H. armigera* herbivory (Bally et al. 2016). Overall, successful expression of dsRNAs through plastome engineering opens the great possibility of using RNAi approaches to confer desired agronomic traits in plants.

4.6 Concluding Remarks and Future Focus

Engineering of plastid genome has taken a huge leap in plant biology owing to several advantages over nuclear genome engineering. In addition, recent availability of plastid genome sequence of many plants resulted in broadening the host range just from tobacco and lettuce to other crops like tomato, carrot, potato, cotton, soybean, etc. Despite the several advancements made in plastid transformation, there are still several major concerns that need to be addressed. One of the main disadvantages of plastome engineering is related to post-translational modification. Plastids lack the necessary machinery to glycosylate proteins, and absence of glycosylation is one of the major bottlenecks for the proper folding and functioning of glycoproteins like antibodies. Another major concern is stability of recombinant proteins in plastids. Chloroplast stroma contains several proteases that could degrade recombinant proteins produced. The expression of transgenes in non-green plastids is not as efficient as in green plastids (chloroplasts), and challenges related to poor gene expression in non-green plastid are still a major concern. Although the last two decades witnessed expression of a large number of vaccine antigens and therapeutic proteins in plastids (Table 4.2), hardly any chloroplast-made proteins completed clinical trials. This is due to the strict regulatory and IP issues related to the use of plant-made proteins for human consumption. Nevertheless, addressing some of the above concerns could lead to a new paradigm in plastome engineering and could contribute for the development of sustainable production of therapeutic proteins and industrial enzyme in plastids for human consumption and commercial exploitation, respectively.

Acknowledgment The authors thank financial support to Department of Biotechnology, Bharathiar University, under DST- PURSE scheme.

References

Adem M, Beyene D, Feyissa T (2017) Recent achievements obtained by chloroplast transformation. Plant Methods 13:30

Agrawal P, Verma D, Daniell H (2011) Expression of *Trichoderma reesei* β-mannanase in tobacco chloroplasts and its utilization in lignocellulosic woody biomass hydrolysis. PLoS One 6:e29302

Albarracín RM, Becher ML, Farran I, Sander VA, Corigliano MG, Yácono ML, Pariani S, López ES, Veramendi J, Clemente M (2015) The fusion of *Toxoplasma gondii* SAG1 vaccine candidate to *Leishmania infantum* heat shock protein 83-kDa improves expression levels in tobacco chloroplasts. Biotechnol J 10:748–759

Apel W, Bock R (2009) Enhancement of carotenoid biosynthesis in transplastomic tomatoes by induced lycopene-to-provitamin A conversion. Plant Physiol 151:59–66

Arlen PA, Singleton M, Adamovicz JJ, Ding Y, Davoodi-Semiromi A, Daniell H (2008) Effective plague vaccination via oral delivery of plant cells expressing F1-V antigens in chloroplasts. Infect Immun 76:3640–3650

Asaf S, Waqas M, Khan AL, Khan MA, Kang SM, Imran QM, Shahzad R, Bilal S, Yun BW, Lee IJ (2017) The complete chloroplast genome of wild rice (*Oryza minuta*) and its comparison to related species. Front Plant Sci 8:304

Bally J, McIntyre GJ, Doran RL, Lee K, Perez A, Jung H, Naim F, Larrinua IM, Narva KE, Waterhouse PM (2016) *In-plant* protection against *Helicoverpa armigera* by production of long hpRNA in chloroplasts. Front Plant Sci 7:1453

Bausher MG, Singh ND, Lee SB, Jansen RK, Daniell H (2006) The complete chloroplast genome sequence of *Citrus sinensis* (L.) Osbeck var 'Ridge Pineapple': organization and phylogenetic relationships to other angiosperms. BMC Plant Biol 6:21

Birschwilks M, Haupt S, Hofius D, Neumann S (2006) Transfer of phloem-mobile substances from the host plants to the holoparasite *Cuscuta* sp. J Exp Bot 57:911–921

Bock R (2014) Genetic engineering of the chloroplast: novel tools and new applications. Curr Opin Biotechnol 26:7–13

Bock R (2015) Engineering plastid genomes: methods, tools, and applications in basic research and biotechnology. Annu Rev Plant Biol 66:31–33

Bohmert-Tatarev K, McAvoy S, Daughtry S, Peoples OP, Snell KD (2011) High levels of bioplastic are produced in fertile transplastomic tobacco plants engineered with a synthetic operon for the production of polyhydroxybutyrate. Plant Physiol 155:1690–1708

Boyhan D, Daniell H (2011) Low-cost production of proinsulin in tobacco and lettuce chloroplasts for injectable or oral delivery of functional insulin and C-peptide. Plant Biotechnol J 9:585–598

Cai Z, Penaflor C, Kuehl JV, Leebens-Mack J, Carlson JE, de Pamphilis CW, Boore JL, Jansen RK (2006) Complete plastid genome sequences of *Drimys*, *Liriodendron*, and Piper: implications for the phylogenetic relationships of magnoliids. BMC Evolut Biol 6:77

Castiglia D, Sannino L, Marcolongo L, Ionata E, Tamburino R, De Stradis A, Cobucci-Ponzano B, Moracci M, La Cara F, Scotti N (2016) High-level expression of thermostable cellulolytic enzymes in tobacco transplastomic plants and their use in hydrolysis of an industrially pretreated *Arundo donax* L. biomass. Biotechnol Biofuels. 9:154

Chakrabarti SK, Lutz KA, Lertwiriyawong B, Svab Z, Maliga P (2006) Expression of the cry9Aa2 B.t. gene in tobacco chloroplasts confers resistance to potato tuber moth. Transgenic Res 15:481–488

Chan HT, Xiao Y, Weldon WC, Oberste SM, Chumakov K, Daniell H (2016) Cold chain and virus-free chloroplast-made booster vaccine to confer immunity against different poliovirus serotypes. Plant Biotechnol J 14:2190–2200

Chen C, Hua W (2018) The complete chloroplast genome of rosemary (*Rosmarinus officinalis*). Mitochondrial DNA Part B. Resources 4:147–148

Chen X, Li Q, Li Y, Qian J, Han J (2015) Chloroplast genome of *Aconitum barbatum* var. puberulum (Ranunculaceae) derived from CCS reads using the PacBio RS platform. Front Plant Sci 6:42

Chumley TW, Palmer JD, Mower JP, Fourcade HM, Calie PJ, Boore JL, Jansen RK (2006) The complete chloroplast genome sequence of *Pelargonium x hortorum*: organization and evolution of the largest and most highly rearranged chloroplast genome of land plants. Mol Biol Evol 23:2175–2190

Chung H-J, Jung JD, Park H-W, Kim J-H, Cha HW, Min SR, Jeong WJ, Liu JR (2006) The complete chloroplast genome sequences of Solanum tuberosum and comparative analysis with Solanaceae species identified the presence of a 241-bp deletion in cultivated potato chloroplast DNA sequence. Plant Cell Rep 25:1369–1379

Daniell H (2006) Production of biopharmaceuticals and vaccines in plants via the chloroplast genome. Biotechnol J 1:1071–1079

Daniell H, Khan MS, Allison L (2002) Milestones in chloroplast genetic engineering: an environmentally friendly era in biotechnology. Trends Plant Sci 7:84–91

Daniell H, Lee SB, Grevich J, Saski C, Quesada-Vargas T, Guda C, Tomkins J, Jansen RK (2006) Complete chloroplast genome sequences of *Solanum bulbocastanum*, *Solanum lycopersicum* and comparative analyses with other Solanaceae genomes. Theor Appl Genet 112:1503–1518

Daniell H, Wurdack KJ, Kanagaraj A, Lee SB, Saski C, Jansen RK (2008) The complete nucleotide sequence of the cassava (*Manihot esculenta*) chloroplast genome and the evolution of atpF in Malpighiales: RNA editing and multiple losses of a group II intron. Theor Appl Genet 116:723–737

Daniell H, Ruiz G, Denes B, Sandberg L, Langridge W (2009a) Optimization of codon composition and regulatory elements for expression of human insulin like growth factor-1 in transgenic chloroplasts and evaluation of structural identity and function. BMC Biotechnol 9:33

Daniell H, Singh ND, Mason H, Streatfield SJ (2009b) Plant-made vaccine antigens and biopharmaceuticals. Trends Plant Sci 14:669–679

Daniell H, Lin CS, Yu M, Chang WJ (2016) Chloroplast genomes: diversity, evolution, and applications in genetic engineering. Genome Biol 17:134–163

Davoodi-Semiromi A, Schreiber M, Nalapalli S, Verma D, Singh ND, Banks RK, Chakrabarti D, Daniell H (2010) Chloroplast-derived vaccine antigens confer dual immunity against cholera and malaria by oral or injectable delivery. Plant Biotechnol J 8:223–242

Del L, Yácono M, Farran I, Becher ML, Sander V, Sánchez VR, Martín V, Veramendi J, Clemente M (2012) A chloroplast-derived *Toxoplasma gondii* GRA4 antigen used as an oral vaccine protects against toxoplasmosis in mice. Plant Biotechnol J 10:1136–1144

Demain AL, Vaishnav P (2009) Production of recombinant proteins by microbes and higher organisms. Biotechnol Adv 27:297–306

Dong M, Liu S, Xu Z, Hu Z, Ku W, Wu L (2018) The complete chloroplast genome of an economic plant, *Camellia sinensis* cultivar Anhua, China. Mitochond DNA B Resour. https://doi.org/10.1080/23802359.2018.1462124

English AC, Richards S, Han Y, Wang M, Vee V, Qu J, Qin X, Muzny DM, Reid JG, Worley KC, Gibbs RA (2012) Mind the gap: upgrading genomes with Pacific biosciences RS long-read sequencing technology. PLoS One 7:e47768

Espinoza-Sáncheza EA, Álvarez-Hernándeza MH, Torres-Castillob JA, Rascón-Cruzc Q, Gutiérrez-Díeza A, Zavala-Garcíaa F, Sinagawa-García SR (2015) Stable expression and characterization of a fungal pectinase and bacterial peroxidase genes in tobacco chloroplast. Electron J Biotechnol 18:161–168

Fathi Roudsari M, Salmanian AH, Mousavi A, Hashemi SH, Jafari M (2009) Regeneration of glyphosate-tolerant *Nicotiana tabacum* after plastid transformation with a mutated variant of bacterial aroA gene. Iran J Biotechnol 7:247–253

Funk HT, Berg S, Krupinska K, Maier UGand Krause K (2007) Complete DNA sequences of the plastid genomes of two parasitic flowering plant species, *Cuscuta reflexa* and *Cuscuta gronovii*. BMC Plant Biol 7:45

Gisby MF, Mellors P, Madesis P, Ellin M, Laverty H, O'Kane S, Ferguson MW, Day A (2011) A synthetic gene increases TGFβ3 accumulation by 75-fold in tobacco chloroplasts enabling rapid purification and folding into a biologically active molecule. Plant Biotechnol J 9:618–628

Gorantala J, Grover S, Goel D, Rahi A, Jayadev Magani SK, Chandra S, Bhatnagar R (2011) A plant based protective antigen [PA(dIV)] vaccine expressed in chloroplasts demonstrates protective immunity in mice against anthrax. Vaccine 29:4521–4533

Gorantala J, Grover S, Rahi A, Chaudhary P, Rajwanshi R, Sarin NB, Bhatnagar R (2014) Generation of protective immune response against anthrax by oral immunization with protective antigen plant-based vaccine. J Biotechnol 176:1–10

Goremykin VV, Hirsch-Ernst KI, Wölfl S, Hellwig FH (2004) The chloroplast genome of *Nymphaea alba*: whole-genome analyses and the problem of identifying the most basal angiosperm. Mol Biol Evol 21:1445–1454

Gottschamel J, Lössl A, Ruf S, Wang Y, Skaugen M, Bock R, Clarke JL (2016) Production of dengue virus envelope protein domain III-based antigens in tobacco chloroplasts using inducible and constitutive expression systems. Plant Mol Biol 91:497–512

Gray BN, Ahner BA, Hanson MR (2009) High-level bacterial cellulase accumulation in chloroplast-transformed tobacco mediated by downstream box fusions. Biotechnol Bioeng 102:1045–1054

Grevich JJ, Daniell H (2005) Chloroplast genetic engineering: recent advances and future perspectives. Crit Rev Plant Sci 24:83–107

Gruzdev EV, Mardanov AV, Beletsky AV, Kochieva EZ, Ravin NV, Skryabin KG (2016) The complete chloroplast genome of parasitic flowering plant *Monotropa hypopitys*: extensive gene losses and size reduction. Mitochond DNA B Resour. https://doi.org/10.1080/23802359.2016.1155090

Guan Y, Ramalingam S, Nagegowda D, Taylor PW, Chye ML (2008) *Brassica juncea* chitinase BjCHI1 inhibits growth of fungal phytopathogens and agglutinates Gram-negative bacteria. J Exp Bot 59:3475–3484

Guda C, Lee S-B, Daniell H (2000) Stable expression of biodegradable protein-based polymer in tobacco chloroplasts. Plant Cell Rep 19:257–262

Gurdon C, Maliga P (2014) Two distinct plastid genome configurations and unprecedented intraspecies length variation in the accd coding region in *Medicago truncatula*. DNA Res 21:417–427

Harada H, Maoka T, Osawa A, Hattan J, Kanamoto H, Shindo K, Otomatsu T, Misawa N (2014) Construction of transplastomic lettuce (*Lactuca sativa*) dominantly producing astaxanthin fatty acid esters and detailed chemical analysis of generated carotenoids. Transgenic Res 23:303–315

Hassan SW, Waheed MT, Müller M, Clarke JL, Shinwari ZK, Lössl AG (2014) Expression of HPV-16 L1 capsomeres with glutathione-S-transferase as a fusion protein in tobacco plastids: an approach for a capsomere-based HPV vaccine. Hum Vaccin Immunother 10(10):2975–2982

Herzog RW, Nichols TC, Su J, Zhang B, Sherman A, Merricks EP, Raymer R, Perrin GQ, Häger M, Wiinberg B, Daniell H (2017) Oral tolerance induction in hemophilia B dogs fed with transplastomic Lettuce. Mol Ther 25:512–522

Hoang NV, Furtado A, McQualter RB, Henry RJ (2015) Next generation sequencing of total DNA from sugarcane provides no evidence for chloroplast heteroplasmy. New negatives. Plant Sci 1-2:33–45

Hoelscher M, Tiller N, Teh AY, Wu GZ, Ma JK, Bock R (2018) High-level expression of the HIV entry inhibitor griffithsin from the plastid genome and retention of biological activity in dried tobacco leaves. Plant Mol Biol 97:357–370

Hu Z-Y, Hua W, Huang S-M, Wang H-Z (2011) Complete chloroplast genome sequence of rapeseed (*Brassica napus* L.) and its evolutionary implications. Genet Resour Crop Evol 58:875–887

Huang YY, Matzke AJ, Matzke M (2013) Complete sequence and comparative analysis of the chloroplast genome of coconut palm (*Cocos nucifera*). PLoS One 8:e74736

Jackman SD, Warren RL, Gibb EA, Vandervalk BP, Mohamadi H, Chu J, Raymond A, Pleasance S, Coope R, Wildung MR, Ritland CE, Bousquet J, Jones SJ, Bohlmann J, Birol I (2016) Organellar genomes of white spruce (*Picea glauca*): assembly and annotation. Genome Biol Evol 8:29–41

Jansen RK, Kaittanis C, Saski C, Lee SB, Tomkins J, Alverson AJ, Daniell H (2006) Phylogenetic analyses of Vitis (Vitaceae) based on complete chloroplast genome sequences: effects of taxon sampling and phylogenetic methods on resolving relationships among rosids. BMC Evol Biol 6:32

Jansen RK, Wojciechowski MF, Sanniyasi E, Lee SB, Daniell H (2008) Complete plastid genome sequence of the chickpea (*Cicer arietinum*) and the phylogenetic distribution of rps12 and clpP intron losses among legumes (Leguminosae). Mol Phylogenet Evol 48:1204–1217

Jin S, Daniell H (2015) The engineered chloroplast genome just got smarter. Trends Plant Sci 20:622–640

Jin R, Richter S, Zhong R, Lamppa GK (2003) Expression and import of an active cellulase from a thermophilic bacterium into the chloroplast both *in vitro* and *in vivo*. Plant Mol Biol 51:493–507

Jin S, Zhang X, Daniell H (2012) *Pinellia ternata* agglutinin expression in chloroplasts confers broad spectrum resistance against aphid, whitefly, lepidopteran insects, bacterial and viral pathogens. Plant Biotechnol J 10:313–327

Jin S, Singh ND, Li L, Zhang X, Daniell H (2015) Engineered chloroplast dsRNA silences cytochrome p450 monooxygenase, V-ATPase and chitin synthase genes in the insect gut and disrupts *Helicoverpa armigera* larval development and pupation. Plant Biotechnol J 13:435–446

Jung S, Lee DS, Kim YO, Joshi CP, Bae HJ (2013) Improved recombinant cellulase expression in chloroplast of tobacco through promoter engineering and 5′ amplification promoting sequence. Plant Mol Biol 83:317–328

Kaila T, Chaduvla PK, Rawal HC, Saxena S, Tyagi A, Mithra SVA, Solanke AU, Kalia P, Sharma TR, Singh NK, Gaikwad K (2017) Chloroplast genome sequence of clusterbean (*Cyamopsis tetragonoloba* L.): genome structure and comparative analysis. Genes. 8:pii: E212. https://doi.org/10.3390/genes8090212

Kanagaraj AP, Verma D, Daniell H (2011) Expression of Dengue-3 premembrane and envelope polyprotein in lettuce chloroplasts. Plant Mol Biol 76:323–333

Kato T, Kaneko T, Sato S, Nakamura Y, Tabata S (2000) Complete structure of the chloroplast genome of a legume, *Lotus japonicus*. DNA Res 7:323–330

Kim EH, Suh SC, Park BS, Shin KS, Kweon SJ, Han EJ, Park SH, Kim YS, Kim JK (2009) Chloroplast-targeted expression of synthetic cry1Ac in transgenic rice as an alternative strategy for increased pest protection. Planta 230:397–405

Kim JY, Kavas M, Fouad WM, Nong G, Preston JF, Altpeter F (2011) Production of hyperthermostable GH10 xylanase Xyl10B from *Thermotoga maritima* in transplastomic plants enables complete hydrolysis of methylglucuronoxylan to fermentable sugars for biofuel production. Plant Mol Biol 76:357–369

Kim JS, Kim HT, Kim J-H (2014) The largest plastid genome of monocots: a novel genome type containing at residue repeats in the slipper orchid *Cypripedium japonicum*. Plant Mol Biol Report 33:1210. https://doi.org/10.1007/s11105-014-0833-y

Kim JH, Lee SI, Kim BR, Choi IY, Ryser P, Kim NS (2017) Chloroplast genomes of *Lilium lancifolium*, *L. amabile*, *L. callosum*, and *L. philadelphicum*: molecular characterization and their use in phylogenetic analysis in the genus *Lilium* and other allied genera in the order Liliales. PLoS One. 12:e0186788

Klinger J, Fischer R, Commandeur U (2015) Comparison of *Thermobifida fusca* cellulases expressed in *Escherichia coli* and *Nicotiana tabacum* indicates advantages of the plant system for the expression of bacterial cellulases. Front Plant Sci 6:1047

Ku C, Chung WC, Chen LL, Kuo CH (2013) The complete plastid genome sequence of Madagascar periwinkle *Catharanthus roseus* (L) G. Don: plastid genome evolution, molecular marker identification, and phylogenetic implications in asterids. PLoS One 8(6):e68518

Kumar SR, Anunanthini P, Sathishkumar R (2017) Plastome engineering: yesterday, today, and tomorrow. In: Mohan C (ed) Sugarcane biotechnology: challenges and prospects, Springer, pp 139–154

Kwon KC, Sherman A, Chang WJ, Kamesh A, Biswas M, Herzog RW, Daniell H (2018) Expression and assembly of largest foreign protein in chloroplasts: Oral delivery of human FVIII made in lettuce chloroplasts robustly suppresses inhibitor formation in haemophilia a mice. Plant Biotechnol J 16:1148–1160

Lakshmi PS, Verma D, Yang X, Lloyd B, Daniell H (2013) Low cost tuberculosis vaccine antigens in capsules: expression in chloroplasts, bio-encapsulation, stability and functional evaluation *in vitro*. PLoS One 8:e54708

Lee MY, Zhou Y, Lung RW, Chye ML, Yip WK, Zee SY, Lam E (2006) Expression of viral capsid protein antigen against Epstein-Barr virus in plastids of *Nicotiana tabacum* cv. SR1. Biotechnol Bioeng 94:1129–1137

Lee HL, Jansen RK, Chumley TW, Kim KJ (2007) Gene relocations within chloroplast genomes of *Jasminum* and *Menodora* (Oleaceae) are due to multiple, overlapping inversions. Mol Biol Evol 24:1161–1180

Leelavathi S, Gupta N, Maiti S, Ghosh A, Reddy VS (2003) Overproduction of an alkali- and thermo-stable xylanase in tobacco chloroplasts and efficient recovery of the enzyme. Mol Breed 11:59–67

Li HY, Ramalingam S, Chye ML (2006) Accumulation of recombinant SARS-CoV spike protein in plant cytosol and chloroplasts indicate potential for development of plant-derived oral vaccines. Exp Biol Med 231:1346–1352

Lim S, Ashida H, Watanabe R, Inai K (2011) Production of biologically active human thioredoxin 1 protein in lettuce chloroplasts. Plant Mol Biol 76:335–344

Lin CS, Chen JJ, Huang YT, Chan MT, Daniell H, Chang WJ, Hsu CT, Liao DC, Wu FH, Lin SY, Liao CF, Deyholos MK, Wong GK, Albert VA, Chou ML, Chen CY and Shih MC (2015). The location and translocation of ndh genes of chloroplast origin in the Orchidaceae family. Scientific Reports.5:9040

Liu B, Zhang D, Gao LZ (2016) The complete chloroplast genome sequence of *Cucumis sativus* var. Hardwickii, the wild progenitor of cultivated cucumber. Mitochond DNA A DNA Mapp Seq Anal 27:4627–4628

Longoni P, Leelavathi S, Doria E, Reddy VS, Cella R (2015) Production by tobacco transplastomic plants of recombinant fungal and bacterial cell-wall degrading enzymes to be used for cellulosic biomass saccharification. BioMed Research Intl 289759. https://doi.org/10.1155/2015/289759

Lu Y, Rijzaani H, Karcher D, Ruf S, Bock R (2013) Efficient metabolic pathway engineering in transgenic tobacco and tomato plastids with synthetic multigene operons. Proc Nat Acad Sci USA 110:E623–E632

Lushchak V, Semchuk NM (2012) Tocopherol biosynthesis: chemistry, regulation and effects of environmental factors. Acta Physiol Plant 34:1607–1628

Maier RM, Neckermann K, Igloi GL, Kössel H (1995) Complete sequence of the maize chloroplast genome: gene content, hotspots of divergence and fine tuning of genetic information by transcript editing. J Mol Biol 251:614–628

Maldaner FR, Aragão FJ, dos Santos FB, Franco OL, da Rocha Queiroz Lima M, de Oliveira RR, Vasques RM, Nagata T (2013) Dengue virus tetra-epitope peptide expressed in lettuce chloroplasts for potential use in dengue diagnosis. Appl Microbiol Biotechnol 97:5721–5729

Maliga P, Bock R (2011) Plastid biotechnology: food, fuel, and medicine for the 21st century. Plant Physiol 155:1501–1510

Martin G, Baurens FC, Cardi C, Aury JM, D'Hont A (2013) The complete chloroplast genome of banana (*Musa acuminata*, Zingiberales): insight into plastid monocotyledon evolution. PLoS One 8(6):e67350

McCabe MS, Klaas M, Gonzalez-Rabade N, Poage M, Badillo-Corona JA, Zhou F, Karcher D, Bock R, Gray JC, Dix PJ (2008) Plastid transformation of high-biomass tobacco variety Maryland Mammoth for production of human immunodeficiency virus type 1 (HIV-1) p24 antigen. Plant Biotechnol J 6:914–929

Middleton CP, Senerchia N, Stein N, Akhunov ED, Keller B, Wicker T, Kilian B (2014) Sequencing of chloroplast genomes from wheat, barley, rye and their relatives provides a detailed insight into the evolution of the Triticeae tribe. PLoS One 9:e85761

Molina J, Hazzouri KM, Nickrent D, Geisler M, Meyer RS, Pentony MM, Flowers JM, Pelser P, Barcelona J, Inovejas SA, Uy I, Yuan W, Wilkins O, Michel CI, Locklear S, Concepcion GP, Purugganan MD (2014) Possible loss of the chloroplast genome in the parasitic flowering plant *Rafflesia lagascae* (Rafflesiaceae). Mol Biol Evol 31:793–803

Morgenfeld M, Lentz E, Segretin ME, Alfano EF, Bravo-Almonacid F (2014) Translational fusion and redirection to thylakoid lumen as strategies to enhance accumulation of human papillomavirus E7 antigen in tobacco chloroplasts. Mol Biotechnol 56:1021–1031

Mozes-Koch R, Tanne E, Brodezki A, Yehuda R, Gover O, Rabinowitch HD, Sela I (2017) Expression of the entire polyhydroxybutyrate operon of *Ralstonia eutropha* in plants. J Biol Eng 11:44

Nakahira Y, Ishikawa K, Tanaka K, Tozawa Y, Shiina T (2013) Overproduction of hyperthermostable β-1,4-endoglucanase from the archaeon *Pyrococcus horikoshii* by tobacco chloroplast engineering. Biosci Biotechnol Biochem 77:2140–2143

Nashima K, Terakami S, Nishitani C, Kunihisa M, Shoda M, Takeuchi M, Urasaki N, Tarora K, Yamamoto T, Katayama H (2015) Complete chloroplast genome sequence of pineapple (*Ananas comosus*). Tree Genet Genomes 11:60

Nguyen VB, Giang NL, Waminal NE, Park H-S, Kim N-H, Jang W, Lee J, Yang T-J (2018) Comprehensive comparative analysis of chloroplast genomes from seven *Panax* species and development of an authentication system based on species-unique single nucleotide polymorphism markers. J Gingseng Res. https://doi.org/10.1016/j.jgr.2018.06.003

Niu Y-F, Gao C-W, Liu J (2018) The complete chloroplast genome sequence of wild banana, *Musa balbisiana* variety 'Pisang Klutuk Wulung' (Musaceae). DNA Res 3:460–461

Oey M, Lohse M, Kreikemeyer B, Bock R (2009) Exhaustion of the chloroplast protein synthesis capacity by massive expression of a highly stable protein antibiotic. Plant J 57:436–445

Oh H, Seo B, Lee S, Ahn DH, Jo E, Park JK, Min GS (2016) Two complete chloroplast genome sequences of *Cannabis sativa* varieties. Mitochond DNA A DNA Mapp Seq Anal 27:2835–2837

Pantaleoni L, Longoni P, Ferroni L, Baldisserotto C, Leelavathi S, Reddy VS, Pancaldi S, Cella R (2014) Chloroplast molecular farming: efficient production of a thermostable xylanase by *Nicotiana tabacum* plants and long-term conservation of the recombinant enzyme. Protoplasma 251:639–648

Petersen K, Bock R (2011) High-level expression of a suite of thermostable cell wall-degrading enzymes from the chloroplast genome. Plant Mol Biol 76:311–321

Pipia I, Kunelauri N, Gogniashvili M, Kotaria N, Kotorashvili A, Lacombe T, Tabidze V(2017). Complete plastid genomes of South Caucasian, European and Mediterranean Basin wild grapevines (*Vitis vinifera* subsp. *sylvestris*). Acta Horticult 1188. ISHS 2017. https://doi.org/10.17660/ActaHortic.2017.1188.47

Poage M, Le Martret B, Jansen MA, Nugent GD, Dix PJ (2011) Modification of reactive oxygen species scavenging capacity of chloroplasts through plastid transformation. Plant Mol Biol 76:371–384

Qian J, Song J, Gao H, Zhu Y, Xu J, Pang X, Yao H, Sun C, Li X, Li C, Liu J, Xu H, Chen S (2013) The complete chloroplast genome sequence of the medicinal plant *Salvia miltiorrhiza*. PLoS One 8:e57607

Raveendar S, Na YW, Lee JR, Shim D, Ma KH, Lee SY, Chung JW (2015) The complete chloroplast genome of *Capsicum annuum* var. glabriusculum using Illumina sequencing. Molecules 20:13080–13088

Rigano MM, Manna C, Giulini A, Pedrazzini E, Capobianchi M, Castilletti C, Di Caro A, Ippolito G, Beggio P, De Giuli MC, Monti L, Vitale A, Cardi T (2009) Transgenic chloroplasts are efficient sites for high-yield production of the vaccinia virus envelope protein A27L in plant cellsdagger. Plant Biotechnol J 7:577–591

Rigano M, Scotti N, Cardi T (2012) Unsolved problems in plastid transformation. Bioengineered 3:329–333

Rogalski M, do Nascimento Vieira L, Fraga HP, Guerra MP (2015) Plastid genomics in horticultural species: importance and applications for plant population genetics, evolution, and biotechnology. Front Plant Sci 6:586

Roney J, Khatibi P, Westwood J (2007) Cross-species translocation of mRNA from host plants into the parasitic plant dodder. Plant Physiol 143:1037–1043

Rubio-Infante N, Govea-Alonso DO, Alpuche-Solís ÁG, García-Hernández AL, Soria-Guerra RE, Paz-Maldonado LM, Ilhuicatzi-Alvarado D, Varona-Santos JT, Verdín-Terán L, Korban SS, Moreno-Fierros L, Rosales-Mendoza S (2012) A chloroplast-derived C4V3 polypeptide from the human immunodeficiency virus (HIV) is orally immunogenic in mice. Plant Mol Biol 78:337–349

Rubio-Infante N, Govea-Alonso DO, Romero-Maldonado A, García-Hernández AL, Ilhuicatzi-Alvarado D, Salazar-González JA, Korban SS, Rosales-Mendoza S, Moreno-Fierros L (2015) A plant-derived multi-hiv antigen induces broad immune responses in orally immunized mice. Mol Biotechnol 57:662–674

Ruhlman T, Lee SB, Jansen RK, Hostetler JB, Tallon LJ, Town CD, Daniell H (2006) Complete plastid genome sequence of *Daucus carota*: implications for biotechnology and phylogeny of angiosperms. BMC Genomics 7:222

Ruhlman T, Ahangari R, Devine A, Samsam M, Daniell H (2007) Expression of cholera toxin B-proinsulin fusion protein in lettuce and tobacco chloroplasts: oral administration protects against development of insulitis in non-obese diabetic mice. Plant Biotechnol J 5:495–510

Ruhlman TA, Rajasekaran K, Cary JW (2014) Expression of chloroperoxidase from *Pseudomonas pyrrocinia* in tobacco plastids for fungal resistance. Plant Sci 228:98–106

Samson N, Bausher MG, Lee SB, Jansen RK, Daniell H (2007) The complete nucleotide sequence of the coffee (*Coffea arabica* L.) chloroplast genome: organization and implications for biotechnology and phylogenetic relationships amongst angiosperms. Plant Biotechnol J 5 339–353

Saski C, Lee SB, Daniell H, Wood TC, Tomkins J, Kim HG, Jansen RK (2005) Complete chloroplast genome sequence of *Gycine max* and comparative analyses with other legume genomes. Plant Mol Biol 59:309–322

Saski C, Lee SB, Fjellheim S, Guda C, Jansen RK, Luo H, Tomkins J, Rognli OA, Daniell H, Clarke JL (2007) Complete chloroplast genome sequences of *Hordeum vulgare*, *Sorghum bicolor* and *Agrostis stolonifera*, and comparative analyses with other grass genomes. Theor Appl Genet 115(4):571–590

Sato S, Nakamura Y, Kaneko T, Asamizu E, Tabata S (1999) Complete structure of the chloroplast genome of *Arabidopsis thaliana*. DNA Res 6:283–290

Schmitz-Linneweber C, Maier RM, Alcaraz JP, Cottet A, Herrmann RG, Mache R (2001) The plastid chromosome of spinach (Spinacia oleracea): complete nucleotide sequence and gene organization. Plant Mol Biol 45:307–315

Scotti N, Alagna F, Ferraiolo E, Formisano G, Sannino L, Buonaguro L, De Stradis A, Vitale A, Monti L, Grillo S, Buonaguro FM, Cardi T (2009) High-level expression of the HIV-1 Pr55gag polyprotein in transgenic tobacco chloroplasts. Planta 229:1109–1122

Sharifi Tabar M, Habashi AA, Rajabi Memari H (2013) Human granulocyte colony-stimulating factor (hG-CSF) expression in plastids of *Lactuca sativa*. Iran Biomed J 17:158–164

Shen X, Wu M, Liao B, Liu Z, Bai R, Xiao S, Li X, Zhang B, Xu J, Chen S (2017) Complete chloroplast genome sequence and phylogenetic analysis of the medicinal plant *Artemisia annua*. Molecules 22 pii:E1330

Shimizu M, Goto M, Hanai M, Shimizu T, Izawa N, Kanamoto H, Tomizawa K, Yokota A, Kobayashi H (2008) Selectable tolerance to herbicides by mutated acetolactate synthase genes integrated into the chloroplast genome of tobacco. Plant Physiol 14:1976–1983

Shinozaki K, Ohme M, Tanaka M, Wakasugi T, Hayashida N, Matsubayashi T, Zaita N, Chunwongse J, Obokata J, Yamaguchi-Shinozaki K, Ohto C, Torazawa K, Meng BY, Sugita M, Deno H, Kamogashira T, Yamada K, Kusuda J, Takaiwa F, Kato A, Tohdoh N, Shimada H, Sugiura M (1986) The complete nucleotide sequence of the tobacco chloroplast genome: its gene organization and expression. EMBO J 5:2043–2049

Soria-Guerra RE, Alpuche-Solís AG, Rosales-Mendoza S, Moreno-Fierros L, Bendik EM, Martínez-González L, Korban SS (2009) Expression of a multi-epitope DPT fusion protein in transplastomic tobacco plants retains both antigenicity and immunogenicity of all three components of the functional oligomer. Planta 229:1293–1302

Steane DA (2005) Complete nucleotide sequence of the chloroplast genome from the Tasmanian blue gum, *Eucalyptus globulus* (Myrtaceae). DNA Res 12:215–220

Su J, Zhu L, Sherman A, Wang X, Lin S, Kamesh A, Norikane JH, Streatfield SJ, Herzog RW, Daniell H (2015) Low cost industrial production of coagulation factor IX bioencapsulated in lettuce cells for oral tolerance induction in hemophilia B. Biomaterials 70:84–93

Svab Z, Maliga P (1993) High-frequency plastid transformation in tobacco by selection for a chimeric aadA gene. Proc Nat Acad Sci USA 90:913–917

Svab Z, Hajdukiewicz P, Maliga P (1990) Stable transformation of plastids in higher plants. Proc Nat Acad Sci USA 87:8526–8530

Talianova M, Janousek B (2011) What can we learn from tobacco and other Solanaceae about horizontal DNA transfer? Am J Bot 98:1231–1242

Tang J, Xia H, Cao M, Zhang X, Zeng W, Hu S, Tong W, Wang J, Wang J, Yu J, Yang H, Zhu L (2004) A comparison of rice chloroplast genomes. Plant Physiol 135:412–420

Tangphatsornruang S, Sangsrakru D, Chanprasert J, Uthaipaisanwong P, Yoocha T, Jomchai N, Tragoonrung S (2010) The chloroplast genome sequence of mungbean (Vigna radiata) determined by high-throughput pyrosequencing: structural organization and phylogenetic relationships. DNA Res 17:11–22

Tangphatsornruang S, Uthaipaisanwong P, Sangsrakru D, Chanprasert J, Yoocha T, Jomchai N, Tragoonrung S (2011) Characterization of the complete chloroplast genome of *Hevea brasiliensis* reveals genome rearrangement, RNA editing sites and phylogenetic relationships. Gene 475:104–112

Timme RE, Kuehl JV, Boore JL, Jansen RK (2007) A comparative analysis of the *Lactuca* and *Helianthus* (Asteraceae) plastid genomes: identification of divergent regions and categorization of shared repeats. Am J Bot 94:302–312

Tong W, Kim TS, Park YJ (2016) Rice chloroplast genome variation architecture and phylogenetic dissection in diverse *oryza* species assessed by whole-genome resequencing. Rice 9:57

van Eerde A, Gottschamel J, Bock R, Hansen KEA, Munang'andu HM, Daniell H, Liu Clarke J (2018) Production of tetravalent dengue virus envelope protein domain III based antigens in lettuce chloroplasts and immunologic analysis for future oral vaccine development. Plant Biotechnol J 17:1408. https://doi.org/10.1111/pbi.13065

Vaughn JN, Chaluvadi SR, Tushar, Rangan L, Bennetzen JL (2014) Whole plastome sequences from five ginger species facilitate marker development and define limits to barcode methodology. PLoS One 9(10):e108581

Vergara D, White KH, Keepers KG, Kane NC (2016) The complete chloroplast genomes of *Cannabis sativa* and *Humulus lupulus*. Mitochond DNA A DNA Mapp Seq Anal. 27:3793–3794

Verma D, Daniell H (2007) Chloroplast vector systems for biotechnology applications. Plant Physiol 145:1129–1143

Verma D, Kanagaraj A, Jin S, Singh ND, Kolattukudy PE, Daniell H (2010) Chloroplast-derived enzyme cocktails hydrolyse lignocellulosic biomass and release fermentable sugars. Plant Biotechnol J 8:332–350

Verma D, Jin S, Kanagaraj A, Singh ND, Daniel J, Kolattukudy PE, Miller M, Daniell H (2013) Expression of fungal cutinase and swollenin in tobacco chloroplasts reveals novel enzyme functions and/or substrates. PLoS One 8:e57187

Viitanen PV, Devine AL, Khan MS, Deuel DL, Van Dyk DE, Daniell H (2004) Metabolic engineering of the chloroplast genome using the *Echerichia coli ubiC* gene reveals that chorismate is a readily abundant plant precursor for p-hydroxybenzoic acid biosynthesis. Plant Physiol 136:4048–4060

Wang YP, Wei ZY, Zhong XF, Lin CJ, Cai YH, Ma J, Zhang YY, Liu YZ, Xing SC (2015) Stable expression of basic fibroblast growth factor in chloroplasts of tobacco. Intl J Mol Sci 17:pii: E19

Wani SH, Haider N, Kumar H, Singh NB (2010) Plant plastid engineering. Curr Genomics 11:500–512

Westwood JH, Yoder JI, Timko MP, dePamphilis CW (2010) The evolution of parasitism in plants. Trends Plant Sci 15:227–235

Wu Z (2016) The completed eight chloroplast genomes of tomato from *Solanum* genus. Mitochond DNA A DNA Mapp Seq Anal. 27:4155–4157

Wu Z, Ge S (2016) The whole chloroplast genome of wild rice (*Oryza australiensis*). Mitochond DNA A DNA Mapp Seq Anal. 27:1062–1063

Wu ML, Li Q, Xu J, Li XW (2018) Complete chloroplast genome of the medicinal plant *Amomum compactum*: gene organization, comparative analysis, and phylogenetic relationships within Zingiberales. Chin Med 13:10

Wurbs D, Ruf S, Bock R (2007) Contained metabolic engineering in tomatoes by expression of carotenoid biosynthesis genes from the plastid genome. Plant J 49:276–288

Yabuta Y, Tanaka H, Yoshimura S, Suzuki A, Tamoi M, Maruta T, Shigeoka S (2013) Improvement of vitamin E quality and quantity in tobacco and lettuce by chloroplast genetic engineering. Transgenic Res 22:391–402

Yanez RJR, Lamprecht R, Granadillo M, Torrens I, Arcalís E, Stöger E, Rybicki EP, Hitzeroth II (2018) $LALF_{32\text{-}51}$ -E7, a HPV-16 therapeutic vaccine candidate, forms protein body-like structures when expressed in *Nicotiana benthamiana* leaves. Plant Biotechnol J 16:628–637

Yu L-X, Gray BN, Rutzke CJ, Walker LP, Wilson DB, Hanson MR (2007) Expression of thermostable microbial cellulases in the chloroplasts of nicotine-free tobacco. J Biotechnol 131:362–369

Yu Y, Lee H-O, Chin JH, Park YH, Yoo S-C (2017) The complete chloroplast genome sequence of *Oryza sativa* aus-type variety Nagina-22 (Poaceae). Mitochond DNA Part B Resour 2:819–820

Zhang J, Tan W, Yang XH, Zhang HX (2008) Plastid-expressed choline monooxygenase gene improves salt and drought tolerance through accumulation of glycine betaine in tobacco. Plant Cell Rep 27:1113–1124

Zhang XH, Keating P, Wang XW, Huang YH, Martin J, Hartmann JX, Liu A (2014) Production of functional native human interleukin-2 in tobacco chloroplasts. Mol Biotechnol 56:369–376

Zhang J, Khan SA, Hasse C, Ruf S, Heckel DG, Bock R (2015) Pest control. Full crop protection from an insect pest by expression of long double-stranded RNAs in plastids. Science 347:991–994

Zhang D, Li K, Gao J, Liu Y, Gao L-Z (2016) The complete plastid genome sequence of the wild rice *Zizania latifolia* and comparative chloroplast genomics of the rice tribe oryzeae, poaceae. Frontiers in. Plant Sci. https://doi.org/10.3389/fevo.2016.00088

Zhang H, Jin J, Moore MJ, Yi T, Li D (2018a) Plastome characteristics of Cannabaceae. Plant Diversity 40:127–137

Zhang X, Rong C, Qin L, Mo C, Fan L, Yan J, Zhang M (2018b) Complete chloroplast genome sequence of *Malus hupehensis*: genome structure, comparative analysis, and phylogenetic relationships. Molecules 23:pii: E2917

Zhou YX, Lee MY, Ng JM, Chye ML, Yip WK, Zee SY, Lam E (2006) A truncated hepatitis E virus ORF2 protein expressed in tobacco plastids is immunogenic in mice. World J Gastroenterol 12:306–312

Zhou F, Badillo-Corona JA, Karcher D, Gonzalez-Rabade N, Piepenburg K, Borchers AM, Maloney AP, Kavanagh TA, Gray JC, Bock R (2008) High-level expression of human immunodeficiency virus antigens from the tobacco and tomato plastid genomes. Plant Biotechnol J 6:897–913

New-Generation Vectors for Plant Transgenics: Methods and Applications

5

Venkidasamy Baskar, Sree Preethy Kuppuraj, Ramkumar Samynathan, and Ramalingam Sathishkumar

Abstract

Transgenic development is the establishment of novel traits into the plants to enhance its quality. Foreign gene introduction into the nuclear or chloroplast genomes in plants is achieved through the DNA-carrying elements known as plasmids or vectors. Plant genetic engineering will be productive only if we develop small, easy-to-handle, and simple to use *Agrobacterium* binary vectors. Most of the new-generation vectors were derived from conventional vectors, such as pBIN and pCAMBIA series. Conventional vectors are larger in size making it difficult for cloning as it decreases the ability of gene integration. For the functional characterization of genes, it requires comprehensive genetic analysis, which includes overexpression, downregulation (antisense/RNAi), promoter analysis, subcellular localization studies, and gene complementation analysis. These high-throughput functional genetic approaches rely on efficient cloning strategies and new-generation vectors. Ancient cloning procedures based on the restriction and digestion are cumbersome and require large time. Modern cloning approaches were established with the newly arrived next-generation vector systems that will be helpful to reduce the cloning difficulty and to increase cloning efficiency methods. The major purpose of the expression vectors is to achieve high protein expression, which is normally driven by strong promoters. Modern, stable, and transient expression vector systems were established to enhance the expression of the foreign gene and to reduce the complexity of gene construct preparation. This chapter describes the various new-generation vectors and their potential application in the field of plant genetic engineering.

Authors Venkidasamy Baskar and Sree Preethy Kuppuraj are equally contributed to this chapter.

V. Baskar (✉) · S. P. Kuppuraj · R. Samynathan · R. Sathishkumar
Plant Genetic Engineering Laboratory, Department of Biotechnology, Bharathiar University, Coimbatore, Tamil Nadu, India

R. Sathishkumar et al. (eds.), *Advances in Plant Transgenics: Methods and Applications*, https://doi.org/10.1007/978-981-13-9624-3_5

Keywords
Plasmids · Vectors · Cloning · Promoters · Gene expression · Genetic engineering

5.1 Introduction

The introduction of desired character in the plants is the basic principle of genetic engineering. Various methods were employed to transfer the foreign gene into the targeted host plants. Among them, an indirect method using the *Agrobacterium tumefaciens*-mediated genetic transformation is widely used in different varieties of plants including mono- and dicots. The direct plant transformation techniques such as particle gun bombardment, microinjection, and electroporation are also utilized for the gene transfer in plants. To achieve this, the target gene should clone to a suitable vector followed by transfer to plants using direct or indirect plant transformation method. The conventional binary vectors needed for transformation in plants have certain demerits. These include larger size, low copy number, restricted host multiplication, least selectable markers, low promoter availability, and difficult to cloning.

The larger size reduces the cloning efficiency along with the integration ability into the host genome. Furthermore, the use of restriction and digestion-based cloning in the earlier form of binary vectors decreased the cloning ability due to these binary vectors which contain very few restriction enzymes in the multiple cloning sites. The lower-size binary vectors can be capable of used or the transfer of large numbers of DNA into plants. Latest alteration of binary vectors attributes various user-friendly traits such as a wide choice of cloning sites, large *E.coli* copy numbers, enhanced strain compatibility, numerous plant selection markers, and enhanced plant transformation potential. The high copy number vector and ability to replicate in *E.coli* are easy to manipulate for cloning applications. Moreover, the vector DNA size influences the in vitro recombination potential (Wang et al. 2013). Furthermore, numerous vectors are designed for a specific purpose. For example, tissue-specific (spatial), stress/chemical inducible (temporal), ectopic expression, knockdown/–out, transient, and stable expression vectors. Some of the viral-based expression vectors are designed to produce the valuable proteins and enzymes in plant systems rapidly in a transient manner. Gene stacking or gene pyramiding is an essential approach in plant metabolic engineering. Gene stacking is referred to as the overexpression of multiple genes in a metabolic pathway, in order to enhance or modify metabolic flux with the goal of improving/modifying the metabolic outcome (Que et al. 2010). Gene stacking has been achieved in two ways: (1) crossing lines possessing independent transgenes and the identification of plants possessing both genes of interest and (2) transformed again to the plants earlier have the transgene. This could be done by the use of existing vectors with various selection methods (e.g., antibiotics).

Recently, the third option has been adopted in which the transformation is achieved by a single gene construct carrying two or three target genes. The advantage of this system is the integration of two or more foreign genes through a single transformation event. Moreover, the targeted gene modification vectors have arrived recently for modification of the genome in plants. In this chapter, the details of new-generation vectors used for various applications in plant genetic engineering were described in detail.

5.2 GATEWAY-Compatible Binary Vectors

Site-specific recombination permits the bacteriophage to incorporate either within or out of the bacterial chromosome. This mechanism is utilized to develop the GATEWAY-compatible vectors (Katzen 2007). Perhaps, it could overcome the restriction and ligation-based cloning systems. GATEWAY protocols completely rely on BP and LR clonase reactions (Hartley et al. 2000). The BP clonase II enzyme reaction mixture is composed of phage integrase and the integration host factor. This mix transports the gene of interest to the donor vector (pDONR) flanked by two *attB* sites (21–25 bp). The recombination of *attB* and *attP* sites results in the insertion of gene of interest to the backbone of the donor that generates an entry clone (pENTR). In addition, entry clones also produced incorporation of DNA segments in vectors in which multiple cloning sites are flanked by *attL* sites. LR clonase II mix carried out the LR reaction and includes integrase, integration host factor, and the phage excisionase.

The LR clonase mix mediated the transfer of target gene flanked by two *attL* sites (entry clone) into the destination or binary vector (pDEST) consisting two attR sites, which result in the formation of unique expression clone (pEXPR) and again flanked by attB sites. The general mechanism of GATEWAY cloning strategy was shown in Fig. 5.1. In order to perform directional cloning, engineered variants of the original *attB*, *attP*, *attL*, and *attR* sites were developed. Therefore, *attB1* will react specifically with *attP1*, but not with *attP2* and *attP3* (Cheo et al. 2004; Sasaki et al. 2004). It facilitates the ease of cloning without any difficulties, which used to pace in the conventional cloning systems.

GATEWAY-mediated binary vector sets are used for various purposes in plant functional genomics, such as multiple gene transformation, fusion protein expression (Earley et al. 2006), and/or knockout of transgene (Chen et al. 2006).

5.3 Promoter-Reporter (or Native Promoter-Gene Fusion) Constructs

Kurtis and Grossniklaus generated the different sets of plant GATEWAY binary vectors (pMDC series) for various applications, such as promoter-reporter analysis, gene complementation assays, inducible gene expression, subcellular localization studies, and constitutive ectopic overexpression analysis. pMDC107, pMDC110, and pMDC111 contained *GFP6* in all three reading frames which are subjected for promoter-reporter analysis to produce the fused product. Similarly, GUS gene was placed in all frames into the pMDC162, pMDC163, and pMDC164 for promoter-reporter analysis and the fusion of *GUS* gene for subcellular localization analysis as well.

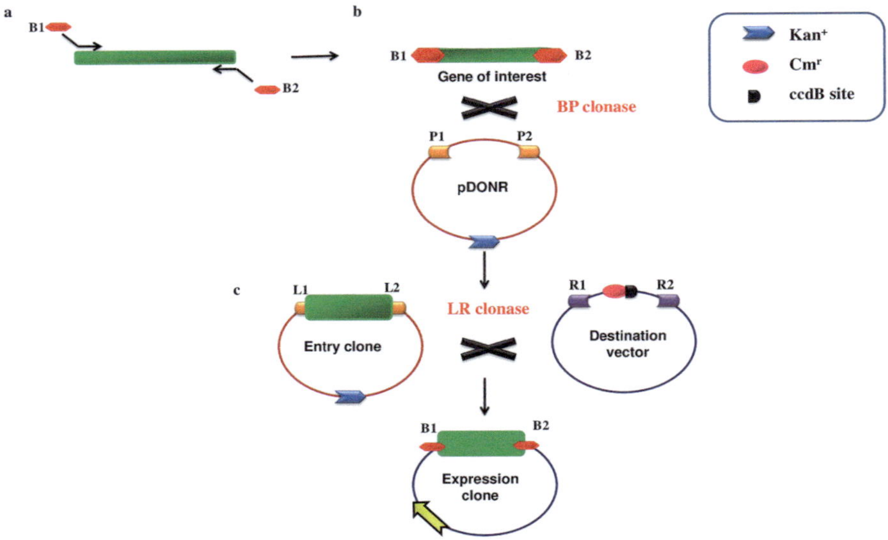

Fig. 5.1 Schematic representation of GATEWAY-based recombination-mediated cloning strategy: (**a**) Generation of PCR product of a gene of interest anchored with the *attB* sites. **(b)** Recombination between the *attB* sites containing gene of interest and *attP* containing donor vector (pDONR) catalyzed by the BP clonase resulting to produce novel entry vector (pENTR) which possesses *attL* sites. (**c**) *attL* sites containing entry vector recombined with the *attR* sites containing destination vector catalyzed by the LR clonase result in the production of expression clone. Blue arrow – kanamycin selection gene; pink circle – *Cm*^{r}, chloramphenicol resistance gene; black arrow – *ccdB*, cytotoxic protein (negative selection). *LB* left border, *RB* right border

5.4 Destination Vectors for Analysis of Subcellular Localization of Proteins

There are several sequence-validated reporter entry clones that are present for LR cloning fluorescence tags [*GFP*, cyan fluorescent protein (*CFP*), yellow fluorescent protein (*YFP*), or red fluorescent protein (*RFP*)], epitope tags [*Myc*, *HA*, or *FLAG*], or enzyme tags [GUS or luciferase (*LUC*)] of ORFs expressing N- or C-terminal fusions (Karimi et al. 2007). Similarly, several destination vectors with C- or N-terminal protein fusions with *GFP* or *GUS* were developed for the investigation of subcellular localization of particular proteins. The destination vectors such as pMDC43, pMDC44, and pMDC45 were generated for GFP C-terminal fusions, while pMDC83, pMDC84, pMDC85, pMDC139, pMDC140, and pMDC141 were developed for N-terminal fusions. The functional activity of these vectors was analyzed in the biolistic bombardment experiments on onion epidermal cells (Varagona et al. 1992). The *att* recombination did not affect the ability of the vectors in the subcellular localization. The diagrammatic representation of GATEWAY binary expression vectors for promoter-reporter and subcellular localization was shown in Fig. 5.2.

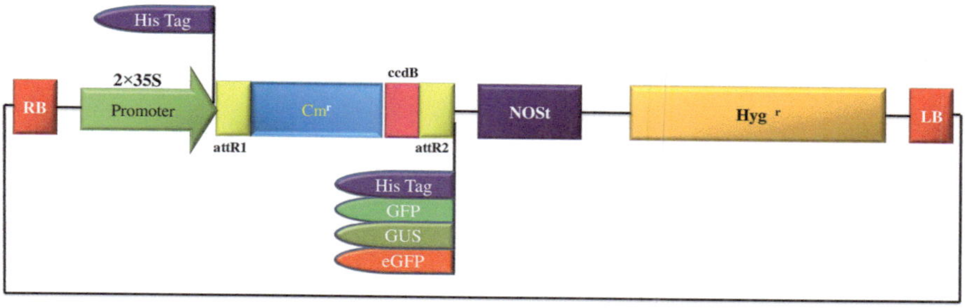

Fig. 5.2 Schematic representation of GATEWAY reporter vectors harboring dual 35S promoter showing the recombination sites (*attR1* and *attR2*), N-terminally and C-terminally tagged sites. N-terminal Histidine (His) tag, C-terminal His-tag, *GFP* Green fluorescent protein, *GUS* β-glucuronidase, *eGFP* enhanced green fluorescent protein. *Cm*^{r} chloramphenicol resistance gene, *ccdB* gene cytotoxic protein, *NOSt* NOS terminator, *Hyg*^{r} Hygromycin resistance gene. *LB* left border, *RB* right border

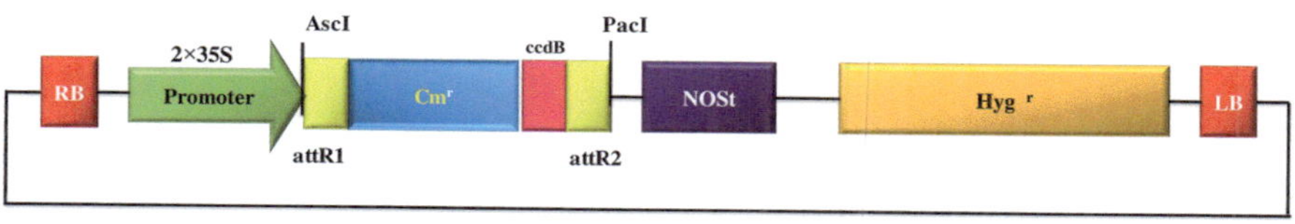

Fig. 5.3 Schematic representation of GATEWAY constitutive overexpression vector harboring a dual 35S promoter showing the recombination sites (*attR1* and *attR2*) flanked by *AscI* and *PacI* enzymes. *Cm*^{r} chloramphenicol resistance gene, *ccdB* gene, cytotoxic protein, *NOSt* NOS terminator, *Hyg*^{r} hygromycin resistance gene. *LB* left border, *RB* right border

5.5 Destination Vectors for Constitutive Ectopic Gene Expression

The overexpression (OE) GATEWAY binary vectors contain the sequences of *R1* (*attR1* recombination attachment site), *Cmr* (chloramphenicol acetyltransferase gene), ccdB (negative selection marker), and *R2* (*attR2* recombination attachment site). The binary vectors pIPKb001 to pIPKb010 are obtained from constitutive *ZmUbi1*, *OsAct1*, and dual *CaMV 35S* promoter and the epidermis-specific *TaGstA1* promoters. The multiple cloning site 1 (MCS1) region is present in the upstream of the GATEWAY destination cassettes of pIPKb001 and pIPKb006 vectors to facilitate cloning desired additional promoters. Similarly, pMDC32 GATEWAY-compatible binary vector bearing the dual *CaMV 35S* promoter was also routinely used for the ectopic constitutive expression analysis in most of the plants. The schematic diagram of constitutive expression binary vector was represented in Fig. 5.3.

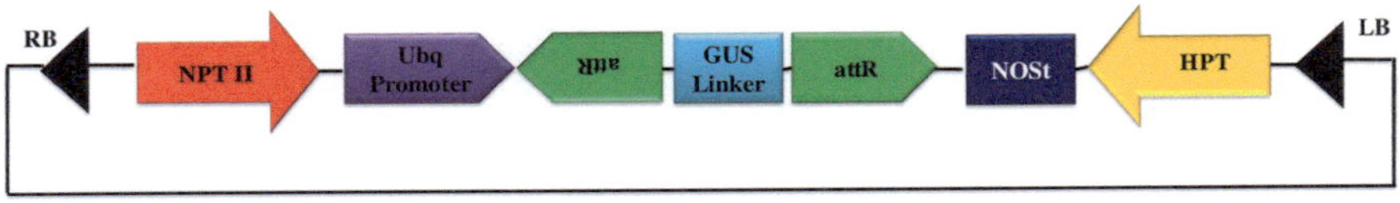

Fig. 5.4 Schematic representation of GATEWAY constitutive downregulation (RNAi) vector – pANDA. *Cm*^{r} chloramphenicol resistance gene, *ccdB* gene cytotoxic protein, *Ubq* maize *ubiquitin 1* promoter, GUS linker, *NOSt* NOS terminator, *HPT*, hygromycin phosphotransferase gene, *NPTII* neomycin phosphotransferase gene. *LB* left border, *RB* right border

5.6 Destination Vectors for Gene Silencing

The hairpin RNA (hpRNA) mediates the production of small interfering RNA that acts as efficient RNA interference (RNAi) inducers in plant (Smith et al. 2000). The sequence selected for silencing, cloned in an entry clone (*attL1*-target-*attL2*), is shifted to a terminal vector with the help of LR clonase reaction that contains two individual GATEWAY cassettes that were separated with an intron sequence (*attR1-ccdB-attR2*-intron-*attR2-ccdB-attR1*). These cloned GATEWAY cassettes are similar, but the recombination sites such as *attR1* and *attR2* are reversed with each other, respectively. Therefore, the sequences of two copies of the gene of interest are placed head-to-head that leads to the formation of hpRNA expression clone. There are several binary destination vectors that are available for gene silencing.

For instance, there are several monocot- and dicot-specific vectors along with the various plant selectable markers, and fluorescent reporters are available (https://gateway.psb.ugent.be/search/index/silencing/any). The diagrammatic representation of simple RNAi vector was shown in Fig. 5.4 (Miki and Shimamoto 2004).

5.7 Constructs for Complementation Analysis of Mutant Plant Lines

The T-DNA binary vectors, namely, pMDC99, pMDC100, and pMDC123, can be utilized for the complementation analysis in mutant backgrounds, and they possess different plant selection markers. It is needed for the fast cloning of larger fragments. For example, the genomic DNA fragment derived from *Arabidopsis* with 12 kb size was cloned between the att recombination sites (Norbert Huck, personal communication) with the help of *E.coli* DH5a. The cloning of larger fragments using the bacterial strains, namely, Stbl2 (Invitrogen, Carlsbad, CA), stabilized the large DNA fragments. The single-step process is familiar among all the vector series, and therefore the gene to be studied can be cloned by utilizing the same approach into numerous vectors that are constructed to facilitate the illumination of gene function.

5.8 Destination Vectors for Inducible Gene Expression

The constitutive ectopic expression of genes attributes a potential method for gene functional studies, while the ubiquitous expression may cause lethality or conceal tissue-specific effects. The advent of inducible vectors can resolve this issue. The heat shock promoter (the Gmhsp17.3B promoter region of SHS3252 plasmid) incorporated in the GATEWAY binary vector pMDC30 can facilitate the induction of transgenes after heat shock.

Similarly, there is a report about chemical-inducible construct called estrogen-inducible ectopic expression vector (PER8) by Zuo et al. (2000). The estrogen-inducible GATEWAY-compatible binary vector pMDC7 was obtained from the PER8 vector. The transgene expression was activated by the transgenic plants treated with the 17-ß-estradiol (Zuo et al. 2002).

5.9 GATEWAY Binary Vectors for Cereals

Crop plant genetic transformation utilized the transgenes for the targeted trait improvement and to know the role of respective genes. The lack of vector systems and the primitive efficiency of transformation in cereals have long been hindered the genetic transformation in cereals. Several previously designed GATEWAY cloning binary vectors are not applicable for monocotyledons, due to restricted role of the promoters used for target gene expression and the plant screening marker (Wesley et al. 2001; Curtis and Grossniklaus 2003; Tzfira et al. 2005), while they are not applicable for monocotyledons, due to restricted role of the promoters used target gene expression the plant screening marker. Some GATEWAY-based binary vectors are also established for the monocotyledonous plants (Miki and Shimamoto 2004).

However, these vectors have certain demerits in the convenient and comprehensive modification with reference to the promoters and the plant screening marker to tailor derivatives for additional specific approaches. A series of GATEWAY-compatible binary vectors were developed by Himmelbach et al. (2007), for transgene overexpression and gene silencing (double-stranded RNA interference) purposes. These vectors also consist of cluster of functionally determined promoters and enable faster integration of foreign gene through GATEWAY-based recombination. Further, these particular vectors provide specialty to desire plant selection markers, cassette orientation, and integration of additional promoters to activate specific gene expression. Furthermore, the transient expression vectors (barley), along with the vectors that can transform both the mono- and dicotyledonous plants, are also developed.

5.10 GATEWAY-Compatible MicroRNA Vectors

Posttranscriptional RNA-mediated silencing of the target RNA in plants is achieved with the help of two different classes of small RNAs, which may include silencing of microRNAs (miRNAs) and transacting small interfering RNAs (tasiRNAs) (Chapman and Carrington 2007; Martínez de Alba et al. 2013). Silencing via amiRNA is more specific and can be induced by the regulation of specific promoters. Small RNA-mediated silencing is the ideal and widely used method for the selective regulation of gene expression. The present methods to create amiRNA or syn-tasiRNA vector constructs required high cost and are difficult to produce in large scale as well as for multiplexing analysis. Carbonell et al. (2014) developed a system to produce rapid, low-cost, and a high-throughput capacity to produce amiRNA and multiplexed syn-tasiRNA gene construct of potential silencing of gene in wide range of plant varieties like *Arabidopsis thaliana* and other plant species. The amiRNA or syn-tasiRNA target genes generated by the two overlapping and partially complementary oligonucleotides are ligated to a zero background *BsaI/ccdB*-based expression vector.

This expression vector for amiRNA or syn-tasiRNA cloning possesses an altered version of *Arabidopsis MIR390a* or *TAS1c* precursors, respectively, in which a portion of the endogenous segment has been replaced by a *ccdB* cassette flanking between two *BsaI* sites. Their results also showed that a single *AtTAS1c*-mediated construct expressing multiple unique syn-tasiRNAs-induced switching of multiple target genes leads to knockdown phenotypes. Similarly, Schwab et al. (2006) reported the amiRNA precursor amplicon generated via the addition of *attB1* and *attB2* recombination sites and cloned into an entry clone (*attL1-amiRNA-attL2*), followed by translocation to the desired vector through standard or MultiSite LR clonase reaction, for expression in diverse tissues, at various developmental times or upon induction.

5.11 Gene Stacking/Multigene Cloning System

The establishment of combination of numerous target genes in the same plant is considered as a difficult process. In order to solve this issue, MultiSite GATEWAY recombination cloning methods are developed for cloning multiple genes. The accessibility of multiple recombination site series were adopted to insert multiple DNA segments concurrently in a single LR clonase reaction which results in the development of an expression clone with two or more contiguous DNA segments in a pre-defined order as well as orientation. This specific MultiSite GATEWAY technique consisting of more than two att series was utilized for the development of plant binary destination vectors (Karimi et al. 2005). Every foreign gene should be introduced into a separate entry clone (like *attL1-gene1-attL2*, *attL4-gene2-attL3*, or *attL6-gene3- attL5*) for multigene cloning and match a distinct GATEWAY destination cassette.

Two or more binary cassettes are all inserted in the T-DNA part of a plant binary vector and are each bound by a different promoter and terminator regulatory sequences (Karimi et al. 2007). Chen et al. (2006) reported the vector system in which the inclusions of multiple transgenes are introduced within the same binary destination vector through sequential stages of LR recombination containing two different entry clones. Exposito-Rodriguez et al. (2017) developed a GATEWAY-compatible vector, namely, pGEMINI, for the introduction of two individual cDNAs at the same time through a one-step LR-clonase reaction. In this vector, back-to-back promoters separate the two inverted recombination with the intention that two individual cDNAs could be cloned in a single reaction from two distinct entry clones. They concluded that this vector was determining the function of a variety of genes in transgenic plants via the transient or stable transformation procedures.

5.12 Tissue-Specific and Stress-Inducible Binary Vectors

Developing transgenic plant with desirable traits is one of the major approaches in genetic engineering. Plant growth and environmental responses influence the expression of various genes. These genes took part in the regulation of developmental events in plants. The cell- and tissue-specific (spatial) gene expression analysis has been used to find out the regulation of many crucial biological processes such as cell-autonomous and non-cell-autonomous regulation of photoperiodic flowering, ovule development, and the tissue-specific response of phytochrome (Groß-Hardt et al. 2002; An et al. 2004; Warnasooriya and Montgomery 2009). In order to study the tissue- or cell type-specific regulation of these genes, the cloning of desired promoters or using two-component systems can be used. In the two-component systems, the application of transcription factor (TF) along with the desired promoter, subjected with the inducers (17-ß-estradiol or ethanol), stimulates the TF to induce the target promoter to permit the temporal-mediated gene induction and spatial regulation provided by the tissue-specific promoters (Deveaux et al. 2003; Maizel and Weigel 2004; Brand et al. 2006; Jia et al. 2007).

Although these systems were influential in understanding the spatial and temporary functions of genes, they have a drawback in higher-order mutants such as unwieldy tissue-specific gene rescue. Depending on the requirement, the gene can be transferred with the particular promoter that can express always or at an appropriate time in a particular tissue. Various types of promoters available from different sources can be utilized for the specific purposes. Based on the activities, there are three types of promoters. The constitutive promoters always control the gene expression in all parts of the plants. Cauliflower mosaic (CaMV) 35S, actin, and ubiquitin are utilized for the strong constitutive transgene expression (Simpson et al. 1986a; Shuai et al. 2002; Laplaze et al. 2007). Some of the promoters, such as maize alcohol 12 dehydrogenase 1 (*Adh1*), sugarcane bacilliform badnavirus (ScBV), maize ubiquitin (*Ubi1*and *Ubi2*), and rice actin (*Act1*), were used for the monocot transformations (Mähönen et al. 2000; Bonke et al. 2003; Thole et al. 2014).

Tissue-specific promoters control the gene expression in various tissues at each developmental stage. In some cases, the constitutive expression (over- or downregulation) leads to plant lethality, which can be solved by the chemical-mediated regulation of gene expression. However, inducible promoters govern the expression of a gene at definite conditions induced by different stimuli, such as chemicals and plant hormones, and different biotic and abiotic stresses promoters (reviewed by Shah et al. 2015). *PR-1a* promoter was activated by the pathogen infection and chemical inducers like benzothiadiazole (Gorlach et al. 1996). Similarly, the herbicide-tolerant benzenesulfonamide safener was shown to induce the maize *In2–2* (inducible gene 2–2) promoter (Veylder et al. 1997). Similar to the tomato wound, inducible hydroxy-3-methylglutaryl CoA reductase 2 (*HMGR2*) promoter, MeGA-PharM (Mechanical Gene Activation Post Harvest Manufacturing), was developed as a commercial inducible promoter system. It permits expression in pharmaceutical proteins encoding transgene as soon as the leaves are harvested and shredded (Ma et al. 2003). The utilization of endogenous inducible promoters possesses few drawbacks such as leaky expression, low-level induction, and the possibility for the stimulation of other endogenous genes by the same inducer, while the other finding utilizing the cloning of desired individual promoters is a longtime process.

In order to reduce the limitations associated with the current approaches, the GATEWAY-based destination vectors was introduced by Michniewicz et al. (2015) for the expression of tissue-specific system to promote the special gene function analysis. These vectors facilitate the faster, uncomplicated construct preparation for the investigation of a tissue-specific expression analysis target gene the higher-order mutant backgrounds. A multiple cloning site from the GATEWAY destination vectors (pEarleyGate100 and pEarleyGate104) was replaced with the cauliflower mosaic virus 35S (*CaMV35S*) promoter to assist the desired promoters. This results in the generation of pMCS:GW and pMCS:YFP-GW vectors. They created many promoter bearing GATEWAY-compatible tissue-specific plant expression vectors. These vectors contain unique promoters that facilitate to investigate the ubiquitous expression and cell type-specific expression such as root and shoots, expression in specific tissues of roots, and generalized expression in root and shoots (Michniewicz et al. 2015). For ubiquitous expression studies, Ubiquitin10 (*UBQ10*) alone or with the YFP tags was cloned into the pMCS:GW and pMCS:YFP-GW to establish pUBQ10:GW and pUBQ10:YFP-GW, respectively.

The upstream regulation site of chlorophyll a/b binding protein 1 (*CAB1*) was cloned into the pMCS:GW and pMCS:YFP-GW vectors for shoot-specific expression, and as a result, it aids in the generation of pCAB1p:GW and pCAB1p:YFP-GW binary vectors. *CAB1* is a light-regulated gene specifically expressed in the tissues of root and stem of peas and tobacco, whereas not in the root tissues (Simpson et al. 1986a, b; Ha and An 1998). The promoter region of *Adh1* was shown to express genes specifically in the root tissues. The upstream/promoter region of ADH1 was cloned into the pMCS:GW and pMCS:YFP-GW vectors that results in the formation of named pADH1:GW and pADH1:YFP-GW vectors. Correspondingly, Scarecrow (SCR) upstream region was cloned into the pMCS:GW and pMCS:YFP-GW vectors to generate pSCR:GW and pSCR:YFP-GW.

SCR revealed tissue-specific expression in the shoot as well as in root tissues (Malamy and Benfey 1997; Di Laurenzio et al. 1996). The Cobra Like1 (*COBL1*) showed expression in various parts of the root tissues and leaf vascular tissues and hydathodes (Brady et al. 2007). The regulatory region (687-bp) was cloned into pMCS:GW and pMCS:YFP-GW resulting in pCOBL1:GW and pCOBL1:YFP-GW. Likewise, the construction of pEXP7:GW and pEXP7:YFP-GW was possible by cloning *EXPANSIN 7* regulatory region with root hair expression to pMCS:GW and pMCS:YFP-GW vectors. These vectors are utilized to study the genes expressed in the root hairs (Cho and Cosgrove 2002). The Lateral Organ Boundaries-Domain 16 (*LBD16*) gene exhibits dominant expression in the root tissues (Shuai et al. 2002). The promoter region covering 1.5 kb was inserted to pMCS:GW and pMCS:YFP-GW to produce pLBD16:GW and pLBD16:YFP-GW. The tissue-specific expression of Wooden Leg/Cytokinin Response1/*Arabidopsis* Histidine Kinase4 (*WOL/CRE1/AHK4*) in root tissues such as vascular cylinder and pericycle was used to clone to pMCS:GW and pMCS:YFP-GW resulting in the generation of pWOL:GW and pWOL:YFP-GW. All these constructs were examined for the predicted tissue-specific expression analysis in *Arabidopsis*. Authors concluded that all the generated GATEWAY-compatible binary constructs exhibit the tissue-specific expression. Earlier studies described various chemical inducible or repressible systems in plants. However, currently, only less number of cloning and application-specific modification of inducible expression systems are available. Schlücking et al. (2013) constructed a series of 57 vectors for stable and transient expression in plants. They included the synthetically optimized XVE expression cassette in all the vectors to permit the ß-estradiol-induced gene expression.

Moreover, the vectors also harbor with the various reporter genes, such as GUS, GFP, mCherry, or with HA and StrepII epitope tags, and possess a multiple cloning sites which facilitate versatile and easy cloning strategies. Besides the vectors with various reporter genes, the vectors are constructed in such a manner that the promoters can be replaced with any tissue-specific promoters for tissue-specific expression to modulate the expression ranges of inducible transgenes. Transient (*Nicotiana benthamiana* leaves) and stable (*Arabidopsis*) transformation were performed to determine the role of designed vectors. Furthermore, the success of gene transformation and expression strictly depends on the selection of a suitable promoter.

5.13 Vectors for Marker-Free Transgenics

The marker gene is used to detect the transgenic plants, whereas their presence after the development of the transgenic plants may lead to unpredicted pleiotropic effects in addition to environmental or biosafety problems. Different methods have been reported to eliminate the SMGs, and most of them were patented and unable to access easily. Recently, Matheka et al. (2013) described the marker-free plants produced using the double right border (DRB) vector generated by conventional cloning methods. The bar gene (gene of interest) was inserted in between the two copies of T-DNA right border sequences to generate DRB vector pMarkfree5.0, and a pMarkerfree5.0 vector was developed from the pMarkfree3.0 by inserting both *gus* and *nptII* genes. Next to the left border sequence, the ß-glucuronidase (gus) and SMG (*nptII* gene) genes were cloned. These vectors have been transformed to tobacco, and 55.6% kanamycin-resistant plants were obtained.

The cotransformation efficiency of both *nptII* and bar transgenes which was distributed among each other using the vector was 66.7%. Subsequently, the *nptII* and bar transgenes were co-expressed and segregated independently, and this was examined by leaf bleach and Basta assays in the transgenic lines. This indicates that the cotransformation using pMarkfree5.0 helps in the separation of transgenes in transgenic plants. They concluded that DRB system was useful to establish SMG-free plants through the separation target gene from SMG. Furthermore, this system is practiced to generate clean plants possessing genes of agronomic trait.

5.14 Plant Gene Targeting Vectors

The traditional methods of plant genetic transformation, such as through *Agrobacterium tumefaciens*-mediated particle bombardment and DNA transfer to protoplast, help in producing transgenic plants in which the transgenes are introduced at random sites in the genome of the plants. On the other hand, latest techniques utilized for the introduction of target gene at the required predetermined locations eliminate many issues coupled with current gene transfer techniques. Many other techniques were evolved for the gene targeting purpose in recent times which include zinc-finger nucleases (ZFNs), transcription activator-like effector nucleases (TALENs), and clustered regularly interspaced short palindrome repeats (CRISPR).

ZFNs are the restriction enzymes containing zinc finger (ZF) domains that help to identify a specific sequence of DNA complexed with the nuclease domain of restriction enzyme *Fokl* (Kim et al. 1996). ZF domains can be engineered to focus on the distinctive sequences of DNA. Especially in the eukaryotic systems, ZFNs were utilized to engineer the endogenous genome loci (Carroll 2008). There are 30 amino acids included in 1 module of DNA binding, and triplet (3) nucleotide was recognized in the integrating module of DNA binding which permits the recognition of 9–18 bp of DNA sequences (Liu et al. 1997). ZFNs were also utilized to establish breaks in the site-specific chromosome, particularly when it is lacking pre-modified sites for the target (Bibikova et al. 2003; Britt and May 2003). The synthetic domain of ZFNs complexed with a cleavage domain of *Fok1* was generally established for the targeted genome modification (TGM) (Urnov et al. 2010; Carroll 2011).

The modifications of the endogenous genes in numerous plant species, such as *Arabidopsis*, soybean, maize, and tobacco, were reported using the ZFNs (reviewed in Sanagala et al. 2017). The inadequate quantity of available target sites added effects on context dependency involving both limited targeting efficiency and specificity, and the effects of frequent off-target produced partly by the nonspecific binding of DNA are the major limitations in the utilization of ZFN (De Francesco 2011). The use of TALENs is an alternative method to ZFNs, specifically for the modification of specific genome, and possesses the enhanced capacity for precise manipulation of the genome (Christian et al. 2010). TALENs also contain an engineered oriented domain of transcription activator-like effectors (TALE) DNA binding and cleavage domain of *FokI*.

The customizable TALE DNA binding domain includes assured but roughly indistinguishable tandem repeated arrays which could target any available sequence on the basis of simple RVD (repeat variable di-residue) code for the recognition of specific nucleotides (Bogdanove et al. 2010; Bogdanove and Voytas 2011). The TALEN-mediated genome modification was accepted in most of the eukaryotes including plants. TALENs are the combination of both cleavage domains of FokI and binding domains of the DNA derived from TALE proteins. The DNA binding domains composed of repeats of 33–35 amino acids, with each domain that identifies a single base pair, and, therefore, at least four types of module in the DNA binding domain are considered necessary to identify each base, such as A, T, G, and C.

In addition, comparative to ZFNs, TALENs are considered as large in recognizing the nucleotides of similar number. The high performance of TALENs was reported for various human cell lines and animal species, whereas the target genome modification mediated by TALENs was elucidated in three different plant species, namely, rice, tobacco, and *Arabidopsis*. Furthermore, several studies utilized TALENs for generating mutations primarily for NHEJ (reviewed by Sanagala et al. 2017). Kusano et al. (2016) developed a simple GATEWAY-mediated vectors for TALEN genes called Emerald–GATEWAY TALEN system for plant genome editing. The entry vectors (pPlat plasmids) were designed to produce the concerned TALEN gene using Platinum Gate TALEN kit, and a destination plasmid (pDual35SGw1301) was also generated to permit on both the DNA strands for the recruitment using GATEWAY technology.

The transformant potato cells were developed by utilizing the TALEN gene constructs, and they observed a site-specific mutation at the desired region of the Granule-bound starch synthase (*GBSS*) gene. Furthermore, they concluded that this system functions effectively when utilized as a suitable method for plant genome editing. ZFNs and TALENs contain artificial DNA-binding proteins fused to *FokI* endonuclease. DNA-binding domains are designed to identify and bind to the desired DNA region, and FokI endonuclease cuts the flanking DNA site. The DNA is cleaved as a double-strand break (DSB), which induced the DNA repairing process that results in a short insertion or deletion via nonhomologous end joining (NHEJ).

NHEJ-mediated introduction of frameshift in the open reading frame results in the disruption of gene of interest that leads to the generation of knockout mutant. Genome engineering methodology with ZFNs and TALENs has been now replaced by other recent approaches utilizing Clustered Regularly Interspaced Short Palindromic Repeats (CRISPR)/CRISPR-associated (Cas) type II prokaryotic system (Sorek et al. 2013). This system makes use of Cas9 nuclease and a designed single guide RNA (sgRNA) which helps in providing the targeted sequence for nucleic acid (reviewed by Sanagala et al. 2017). Several studies are carried out to determine the specificity of CRISPR/Cas system in human cells and in vitro. Most

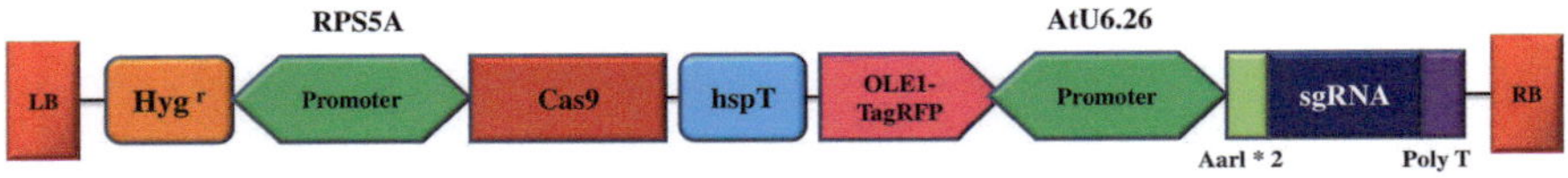

Fig. 5.5 Schematic representation of efficient CRISPR/*Cas9* vector for *Arabidopsis thaliana*, pKAMA-ITACHI Red (pKIR1.1), contains RIBOSOMAL PROTEIN S5A (RPS5A) promoter to express *Cas9* and *AtU6.26* promoter along with red fluorescent protein OLE1-TagRFP. An sgRNA (single guide RNA) with a polyT end and *AarI* insertion at another end. *Hyg*^{r} hygromycin resistance gene, *hspT* heat shock protein terminator. *LB* left border, *RB* right border

of the results specify that 3′-end of the guide segment inside the sgRNA predominantly established the target specificity of CRISPR/Cas system.

For instance, the mismatches between the guide sequence and DNA target of the sgRNA placed inside the last 8–10 bp of 20 bp sequence in target chiefly remove the identification of the target by *Cas9*, while mismatches toward 5′ target end are more resistant (Cong et al. 2013; Nekrasov et al. 2013). The specific alterations of the genomic fractions by removing, adding, and editing the DNA segments are known as genome engineering. The simple, versatile, and high-efficient CRISPR/*Cas9* is a widely used popular technique for genome engineering. The application of CRISPR/*Cas9* in inducing mutations in *Arabidopsis* is lower because stimulation for gene mutation is activated at later developmental stages in cells, and it requires lots of time, efforts, and plant species to acquire the desired plant species with the targeted gene knocked out. A piece of the target RNA, namely, guide RNA (gRNA), and an enzyme called *Cas9* are the key elements of CRISPR/*Cas9*, where *Cas9* functions as a nuclease and breaks DNA strands at particular site in the genome, resulting in the addition or removal of the DNA sequences at that position.

The length of gRNA is around 20 bases predesigned RNA sequence located within the RNA scaffold. The RNA scaffold attached to DNA and the predesigned RNA sequence guide the *Cas9* protein to desired site of the genome, and thus *Cas9* can restrict at the targeted position. Some CRISPR/Cas9-based binary vectors have been reported for plants which possess low efficiency. Tsutsui and Higashiyama (2017) developed a highly efficient CRISPR/*Cas9*-mediated gene knockout vectors called pKAMA-ITACHI. The Ribosomal Protein S5 A (*RPS5A*) promoter present in the pKAMA-ITACHI Red (pKIR) vector helps to drive *Cas9*, and it sustains highly indispensable expression at each stage of development that begins from the egg cell to meristematic cells. Their results explained that pKIR-induced mutations were transmitted to the germ cell line of the T1 generation. pKIR-mediated null phenotypes for few genes such as Phytoene desaturase 3 (*PDS3*), Agamous (*AG*), and Duo pollen 1 (*DUO1*) were observed in the T1 generation. Additionally, in few lines, 100% of the T2 plants possess *adh1* null phenotype which illustrates the heritable mutations done by pKIR effectively. Cas9-free T2 mutant varieties were derived by eliminating fluorescent marker expression in pKIR. Authors suggested that pKIR system could be used as potential genome engineering tool for *Arabidopsis thaliana*. The structural representation of efficient CRISPR/Cas9 binary vector, pKAMA-ITACHI Red (pKIR1.1), for *A. thaliana* was shown in Fig. 5.5.

5.15 Seamless Construct Preparation

The complete regulation of nucleotides flanking the insert in the final construct possesses an advantage in the achievement of plant transformation vectors. The application of type IIS and type IIB restriction endonucleases (REs) reported by Kronbak et al. (2014) for vector linearization, along with sequence and ligase-independent cloning (SLIC), conquers this problem and endorses seamless gene insertion in vectors. They generated two vectors where spacer was placed in between the promoter and terminator. The type IIS and type IIB REs neglect their own recognition segment from the vector, whereas spacer-removal linearization (SRL) does not leave undesired short sequences. Both plant transformation vectors, such as pAUrumII and pAUrumIII, were produced for SRL in combination with SLIC and possess a spacer having the recognition sites for both type IIS and IIB REs, respectively. These vectors contain green fluorescent protein cloned in it. Because of the imperfect linearization, contamination resulted in the transformation by the traces of pAUrumIII, but this problem was eliminated with the type IIS linearized pAUrumII vector. Both the GFP constructs, pAUrumII-GFP and pAUrumIII-GFP, were active when observed under in vitro conditions, on wheat and barley endosperm cells for transient GFP expression. Finally, they proposed that SRL system could be used for a wide range of plant transformation vectors, and it helps in the optimal expression induced by the availability of functional sequences over redundant cloning site remnants.

5.16 Vectors for Virus-Induced Gene Silencing (VIGS)

Virus-induced gene silencing (VIGS) is a quick and potential reverse genetic method for functional genomics in which the phenotypes are excited due to the repression of an endogenous gene present in plants which can be examined in a short period. The plant's gene belongs to plant development, cellular signaling, and regulation of metabolism, and disease resistance can be characterized by the VIGS (Purkayastha and Dasgupta 2009). Furthermore, few plant's two-component gene silencing systems, like the satellite virus-induced silencing system (SVISS), had been utilized (Gossele et al. 2002; Zhou and Huang 2012; Liou et al. 2014). In VIGS-mediated silencing system, viral vectors carrying the target gene segment to develop dsRNA trigger RNA-mediated gene silencing.

Different viruses have been adapted to silence the foreign gene proficiently in a sequence-dependent manner. However, VIGS is a rapid method for the functional genomics and doesn't require a stable transformation. Thus, it is considered as less time-consuming and cost-effective. There are certain limitations in adopting the VIGS; they are viral infection yielding symptoms on its own, and infection is generally confined to particular tissues, but this limitation might be alleviated by the extent of the silencing signals. There are several VIGS vectors which have been utilized for GATEWAY-compatible cloning production. Zhao et al. (2015) reported the generation of tobacco ringspot virus (TRSV)-based vectors for the transient expression of foreign genes for the analysis of endogenous particular role of these genes in plants using virus-induced gene silencing in a broad range of plants.

Moreover, VIGS system may also be helpful for the functional genomic investigation of some fruit trees, such as both peach and cherry that reported hosts for TRSV. The integration of GFP gene between the TRSV movement protein (MP) and coat protein (CP) regions led to increase in-frame expression of the RNA2-encoded viral polyprotein. The introduction of sense and antisense orientations of VIGS targets gene (phytoene desaturase) (*PDS*) into the *SnaBI* site that was produced by mutating the sequence following the CP stop codon. VIGS of PDS results in the photobleaching phenotypes and greatly decreased in PDS mRNA levels in silenced plants.

In recent times, many virus-based vectors are standardized to knock down more than one host plant, namely, Tobacco rattle virus (TRV)-derived viral vectors, which are developed for *Arabidopsis* and *N. benthamiana*. The gene of interest was cloned into the pTRV2 vector. pTRV1 contains TRV1-based cassette (RNA-dependent RNA

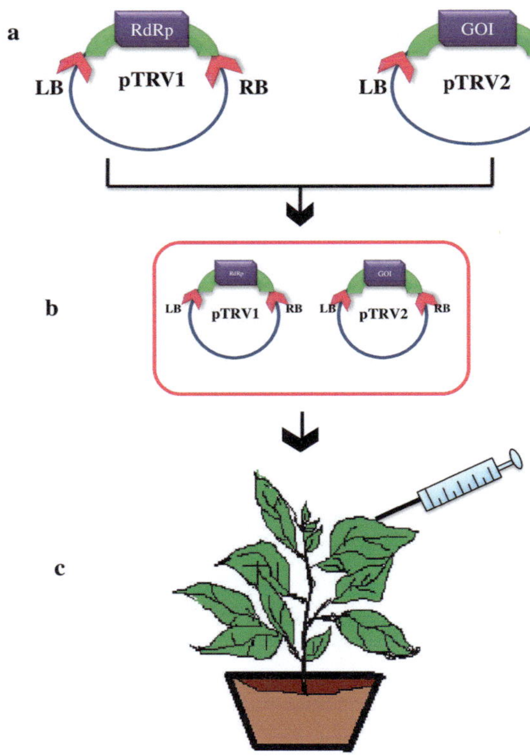

Fig. 5.6 Schematic representation of virus-induced gene silencing system using pTRV (Tobacco rattle virus) vectors. (**a**) pTRV1 encodes recombination genes (RdRp, RNA-dependent RNA polymerase); pTRV2 harboring gene of interest. (**b**) pTRV1 and pTRV2 vectors were mobilized into *Agrobacterium tumefaciens*. (**c**) Agro-inoculated into *Nicotiana benthamiana* plants

polymerase gene, movement protein, etc.), left border (LB) and right border (RB) site for plant transformation. The TRV silencing in plants is usually carried out by *A. tumefaciens*-based inoculation techniques. pTRV1 and pTRV2 are located into LB and RB sites individually. Gene of interest was cloned into the pTRV1, for targeted gene silencing. The plasmids then introduced into *A. tumefaciens*, which are then used for agro-inoculation. Agro-inoculation is inoculated on the plant or seedling via a toothpick and a syringe and through vacuum infiltration. The schematic representation of virus-induced gene silencing system using pTRV (Tobacco rattle virus) vectors was shown in Fig. 5.6. TRV-based gene silencing was utilized to numerous plants from different family members. In addition to dicotyledons, viral silencing vectors are also developed for monocots. Barley stripe mosaic virus (BSMV)-mediated VIGS is one among the viral vectors for gene silencing in barley and wheat.

5.17 Virus-Based Expression Vectors

Virus-based binary expression vectors are recognized as routine tools for protein production and induction of RNA silencing in many herbaceous plants. These vectors are mainly used to transform the existing plants. Moreover, viral vectors help to prevent the plants from the continuously emerging pathogenic and pest's attack with the incorporation of genes in the plant system. Further, viral vectors stand as an alternative tool in promoting transgenic approaches (Dawson and Folimonova 2013). Another main advantage of such vectors is that nothing is permanently added to the environment.

Besides, these vectors are also considered as an emerging tool for the production of many value-added products. For example, GUS and jellyfish GFP marker proteins and several essential pharmaceutical proteins, namely, commercial antibodies, antigens, and vaccines, have been produced in plants by using the plant virus-based expression systems (Gleba et al. 2007; Zhang et al. 2013). Here, it ensures low cost and bulk production and is highly active, as it utilizes the cheap system for the production using these essential viral vectors. Presently, significant efforts were achieved to use plant viral vectors as production platforms for vaccine antigens and antibodies for disease control and analysis (Gleba et al. 2004; Canizares et al. 2005; Lico et al. 2008).

Over and above these strategies in viral vectors, there is another strategy, which has been used over years, and it relies on *Agrobacterium*-mediated delivery of a stripped down vector in infiltrated leaves and has proven to be successful for the bulk production of a number of pharmaceutically essential proteins (Gleba et al. 2005; Gleba et al. 2007; Gleba and Giritch 2011). In a relatively short period of time, these viral replicons produce large number of copies.

5.18 Tobacco Mosaic Virus-Based Transient Vector

Tobacco mosaic virus (TMV) is also called *Tombusvirus* and was the first one to be examined for vector development. TMV is a rod-shaped, single-stranded plus-sense RNA virus. Four proteins are expressed by TMV from three open reading frames (ORFs). Two viral genes responsible for the viral movement protein and the capsid protein are expressed through separate sub-genomic promoters. Modification of TMV has been facilitated to the expression of target gene either by replacing a viral gene (such as the coat protein [CP] gene) with the foreign gene (Scholthof et al. 1996) or by introducing an additional sub-genomic promoter (Dawson et al. 1989; Donson et al. 1991; Pogue et al. 1998) into the viral genome to direct the expression of an inserted foreign gene.

As a result, foreign genes and other DNA fragments are introduced between the plasmids to permit the preferred recombination to take place. Protein expression can also be enhanced up to 25-fold by the co-expression of the RNA silencing suppressor gene of tomato bushy stunt virus known as *P19*. For example, Lindbo (2007) co-inoculated plants with a TMV-based vector and a viral suppressor of RNA silencing; this resulted in the expression of remarkable levels of recombinant proteins (ranging from 600–1200 μg GFP/g of infiltrated tissue) following post-infection of 1 week.

The gene expression can be achieved high by introducing the open reading frame of the target gene closer to the 3′-end of the TMV RNA, as reported by many researchers. An enhanced agroinfection-compatible TMV vector without TMV CP gene coding sequence is referred to as TMV RNA-based overexpression (TRBO) vector. This specialized vector has a number of significant progresses such as (1) higher efficiency of agroinfection, (2) increased expression of recombinant protein, and (3) failure to form virus particles during infection/replication cycle. Compared to *P19*-enhanced agroinfiltration transient expression system, the new expression vector produces 100 times more recombinant protein as described above. Another few examples of TMV-based viral vectors may include pPZP3425, pJLTRBO, pEAQ-HT-GFP, and pBY030-2R (Sainsbury et al. 2009; Huang et al. 2009). The vectors pPZP5025 and pPZPTRBO are primarily constructed depending on the customized high-copy number pPZP family vector pPZP500 (Ali et al. 2012).

The vector pPZP5025 was constructed by insertion of the ß-glucuronidase (GUS) cassette from pPZP3425 as a HindIII fragment into the same site of pPZP500. The vector pPZPTRBO contains the TMV cassette from pJLTRBO and was developed as follows: the cauliflower mosaic virus (CaMV) terminator along with the ribozyme segment from pJLTRBO has incorporated into pPZP500. The TMV cassette having the enhanced 35S promoter, replicase gene, movement protein gene and polylinker was restriction digested from pJLTRBO and ligated to the same site of pPZP500-Ter to produce pPZPTRBO. The GFP coding sequence was inserted into pPZPTRBO and pJLTRBO to make pPZPTRBO-GFP and pJLTRBO-GFP, correspondingly. The vector pBin61-*P19* (Voinnet et al. 2003) was used to express the *P19* RNA silencing suppressor.

5.19 Cowpea Mosaic Virus-Derived Transient Vector

The pEAQ vector series is an efficiently used vector for infiltration studies and the special element Cowpea mosaic virus (CPMV-HT) cassette into which a target gene directly cloned that helps to enhance the expression levels transiently. Transfer of CPMV-HT cassette into pEAQspecialK produces pEAQ-HT. The GFP gene was then inserted within the polylinker of pEAQ-HT to produce untagged GFP 5′ (HisGFP) and 3′ (GFPHis) His-tag fusions from constructs pEAQ-HT-GFP, pEAQ-HT-GFPHis, and pEAQ-HT-HisGFP, respectively, and their level of expression was evaluated by agro-infiltrating constructs into *N. benthamiana* leaves using pEAQspecialK-GFP-HT as a control (Sainsbury and Lomonossoff 2008). Untagged GFP was expressed from pEAQ-HT-GFP to a level (in excess of 1.5 g / kg fresh weight of tissue) even higher than that obtained with pEAQspecialK-GFP-HT as determined by spectrofluorometry. The incorporation of a His-tag reduced the level of GFP detected by spectrofluorometry, although the levels were still in excess of 0.5 g / kg fresh weight of tissue. The decrease was more obvious for N-terminal compared with C-terminal His-tag. pEAQ-HT vector provides an extremely rapid and easy approach for protein expression in plant tissue requiring only 2–3 weeks for cloning and expression. However, traditional restriction enzyme-based cloning is required, which in most cases represents a bottleneck for high-throughput expression platforms.

To alleviate this, a series of three GATEWAY-compatible pEAQ-HT destination vectors were created, which composed of dual-mutated (A115G / U162C) 5' UTR of pEAQ-HT assembled to maximize the expression of an unfused protein (pEAQ-HT-DEST1) or a protein with either an N- (pEAQ-HT-DEST2) or C-terminal (pEAQ-HT-DEST3) His-tag (Sainsbury and Lomonossoff 2008). A GFP entry clone having both start and stop codons was used in GATEWAY recombination (LR) reactions with pEAQ-HT-DEST1 and pEAQ-HT-DEST2 destination vectors, ensuing in producing the gene constructs of both untagged GFP and N-terminally His-tagged GFP (HisGFP) gene expression. pEAQ-HT-DEST3 was used in LR reactions with GFP entry vectors with and without stop codon, thus yielding another untagged construct (GFPstop) and a C-terminally His-tagged construct (GFPHis), respectively. The schematic structural representation of GATEWAY-compatible pEAQ-HT DEST 3 expression vector was shown in Fig. 5.7.

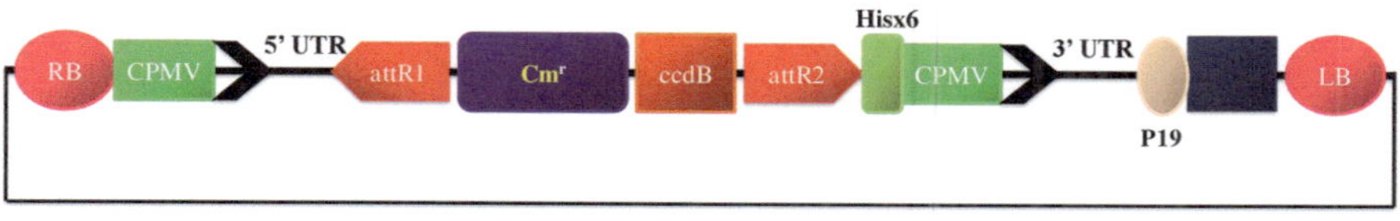

Fig. 5.7 Diagrammatic representation of GATEWAY-compatible pEAQ-HT DEST 3 expression vector. Cowpea mosaic virus (CPMV-HT) allows the rapid production of protein without viral replication; *p19*, suppressor gene; *attR1* and *attR2* GATEWAY recombination sites; *Cm^r*, chloramphenicol resistance gene; *ccdB*, *E.coli* lethal gene; Hisx6, C-terminally His-tagged GFP; 5' UTR and 3' UTR, 5′ and 3′ untranslated regions, black arrows, promoter sequences; navy blue boxes, terminator sequences; LB, left border; RB, right border

5.20 pSIM24 Vector (Mirabilis Mosaic Virus Promoter)

Vector technology has evolved over time, and new generation vectors with enhanced cloning and delivery approaches has been introduced such as pGreen vectors (Hellens et al. 2000). In plant cells, pGD or pSITE vectors are considered as an appropriate system for the stable introduction or transient expression of numerous auto-fluorescent fusion proteins (Goodin et al. 2002; Chakrabarty et al. 2007): the pCLEAN binary vector system (Thole et al. 2007); the pHUGE binary vector system (Untergasser et al. 2012); and binary bacterial artificial chromosome BIBAC vectors (Takken et al. 2004). The TMV RNA-based vector pJLTRBO (Lindbo 2007) and its derivative pPZPTRBO (Shah et al. 2013) were utilized to express recombinant proteins in plants in the absence of RNA silencing inhibitor P19. pEAQ-HT vector is integrated to a *P19* expression cassette which produces comparable expression (Sainsbury et al. 2009). It is reported that a large amount of recombinant protein could be produced by bean yellow dwarf virus single-stranded DNA-based vector, pBY030-2R (Huang et al. 2009). However, the pMAA-Red vector is suitable for unproblematic high expression of target gene in transgenic *Arabidopsis* (Ali et al. 2012). The existing low molecular weight and versatile plant destination expression vectors are still inadequate in plant biotechnology.

Henceforth, pSiM24, a low molecular weight destination vector, which provides wide cloning sites, high copy number, and is active in the transient, affords a reliable transformation if necessary. The T-DNA portion was extracted from pBTdna-rbcT-KanR. The non-T-DNA portion was derived from pBAmpR-ColEI-oriV-trfA. pSi binary vector was constructed by the ligation of two portions composed of T-DNA and non-T-DNA. Mirabilis mosaic virus (M24) complete transcript promoter (Dey and Maiti 1999) was inserted into pSi, and the final product was called as pSiM24. NCBI database is maintaining the complete full-length sequence of pSiM24 (7081 bp) (GenBank accession no. KF032933).

5.21 Concluding Remarks

The success of plant genetic engineering relies on the successful introduction of desirable traits into the plants. Efficient vector systems and easy cloning methods are essential for the successful development of transgenic plants. The larger size and restriction pattern of cloning used in the conventional vectors decreased the cloning efficiency and the integration of the transgene into the host genome. The advancement in vector construction and cloning strategies results in the generation of small size, high copy number, multiple host replication, broad varieties of selection markers, and elimination of restriction-based cloning systems that facilitate the ease of handling and decreased the difficulties in cloning. Several GATEWAY-compatible vectors are designed for various purposes, such as single or multiple gene cloning, over-expression, knockdown, promoter-reporter fusion, gene localization, induced gene expression analysis, etc.

Targeted gene modification using CRISPR/Cas-based vectors is also available for efficient target genome modification in higher plants. Viral-based vectors for transient, stable, and knockdown expression were used for the rapid production of valuable proteins and for functional genomics studies. Most of the vectors are specific to a group of plants that is either monocots or dicots or particular plant families. Therefore, vectors should be developed in such a way that they can be efficiently used for a diverse group of plant families. Although these advanced cloning strategies and vector systems reduced the cloning difficulties and enhanced the functional characteristics, their availability is restricted. Thus, it is essential to eliminate the IPR and technical barriers associated with the next-generation vectors and cloning strategies. Due to these advancements in plant gene expression vectors, now plant production platform has become a potential alternative for the large-scale production of commercially significant proteins.

Acknowledgments This study was supported by a grant (Sanction No. PDF/2016/000750) from the Department of Science and Technology, Science and Engineering Research Board, Government of India, and (Sanction No. No.F.4-2/2006 (BSR)/BL/16-170541) from the D. S. Kothari Postdoctoral Fellowship and was also supported by UGC-SAP, DST-FIST, and DST-PURSE schemes.

References

Ali MA, Shah KH, Bohlmann H (2012) pMAA-Red: a new pPZP-derived vector for fast visual screening of transgenic Arabidopsis plants at the seed stage. BMC Biotech:12

An H, Roussot C, Suárez-López P, Corbesier L, Vincent C, Piñeiro M, Coupland G (2004) CONSTANS acts in the phloem to regulate a systemic signal that induces photoperiodic flowering of Arabidopsis. Development 131(15):3615–3626

Bibikova M, Beumer K, Trautman JK, Carroll D (2003) Enhancing gene targeting with designed zinc finger nucleases. Science 300(5620):764–764

Bogdanove AJ, Voytas D (2011) TAL effectors: customizable proteins for DNA targeting. Science 333:1843–1846

Bogdanove AJ, Schornack S, Lahaye T (2010) TAL effectors: finding plant genes for disease and defense. Curr Opin Plant Biol 1(3):394–401

Bonke M, Thitamadee S, Mähöonen AP, Hauser MT, Helariutta Y (2003) APL regulates vascular tissue identity in Arabidopsis. Nature 426:181–186

Brady SM, Song S, Dhugga KS, Rafalski JA, Benfey PN (2007) Combining expression and comparative evolutionary analysis. COBRA Gene Fam Plant Physiol 143(1):172–187

Brand L, Hörler M, Nüesch E, Vassalli S, Barrell P, Yang W (2006) A versatile and reliable two-component system for tissue-specific gene induction in Arabidopsis. Plant Physiol 141:1194–1204

Britt AB, May GD (2003) Re-engineering plant gene targeting. Trends Plant Sci 8:90–95

Canizares MC, Nicholson L, Lomonossoff GP (2005) Use of viral vectors for vaccine production in plants. Immunol Cell Biol 83:263–270

Carbonell A, Takeda A, Fahlgren N, Johnson SC, Cuperus JT, Carrington JC (2014) New generation of artificial MicroRNA and synthetic trans-acting small interfering RNA vectors for efficient gene silencing in Arabidopsis. Plant Physiol 165(1):15–29

Carroll D (2008) Progress and prospects: zinc-finger nucleases as gene therapy agents. Gene Ther 15:1463–1468

Carroll D (2011) Genome engineering with zinc-finger nucleases. Genetics 188:773–782

Chakrabarty R, Banerjee R, Chung SM, Farman M, Citovsky V (2007) pSITE vectors for stable integration or transient expression of autofluorescent protein fusions in plants: probing Nicotiana benthamiana-virus interactions. Mol Plant-Microbe Interact 20:740–750

Chapman EJ, Carrington JC (2007) Specialization and evolution of endogenous small RNA pathways. Nat Rev Genet 8:884–896

Chen QJ, Zhou HM, Chen J, Wang XC (2006) A Gateway-based platform for multigene plant transformation. Plant Mol Biol 62:927–936

Cheo DL, Titus SA, Byrd DRN, Hartley JL, Temple GF, Brasch MA (2004) Concerted assembly and cloning of multiple DNA segments using in vitro site-specific recombination: functional analysis of multi-segment expression clones. Genome Res 14:2111–2120

Cho HT, Cosgrove DJ (2002) Regulation of root hair initiation and expansin gene expression in Arabidopsis. Plant Cell 14(12):3237–3253

Christian M, Cermak T, Doyle EL, Schmidt C, Zhang F, Hummel A (2010) Targeting DNA double-strand breaks with TAL effector nucleases. Genetics 186:757–761

Cong L, Ran FA, Cox D, Lin S, Barretto R, Habib N (2013) Multiplex genome engineering using CRISPR/Cas systems. Science 339:819–823

Curtis MD, Grossniklaus U (2003) A gateway cloning vector set for high-throughput functional analysis of genes in planta. Plant Physiol. 133(2):462–469

Dawson WO, Folimonova SY (2013) Virus-based transient expression vectors for woody crops: a new frontier for vector design and use. Annu Rev Phytopathol 51:321–337

Dawson WO, Lewandowski DJ, Hilf ME, Bubrick P, Raffo AJ, Shaw JJ, Grantham GL, Desjardins PR (1989) A tobacco mosaic virus-hybrid expresses and loses an added gene. Virology 172:285–292

De Francesco L (2011) Move over ZFNs. Nat Biotechnol 29:681–684

Deveaux Y, Peaucelle A, Roberts GR, Coen E, Simon R (2003) Mizukami Y (2003) the ethanol switch: a tool for tissue-specific gene induction during plant development. Plant J 36:918–930

Dey N, Maiti IB (1999) Structure and promoter/leader deletion analysis of mirabilis mosaic virus (MMV) full-length transcript promoter in transgenic plants. Plant Mol Biol 40:771–782

Di Laurenzio L, Wysocka-Diller J, Malamy JE, Pysh L, Helariutta Y, Freshour G (1996) The SCARECROW gene regulates an asymmetric cell division that is essential for generating the radial organization of the Arabidopsis root. Cell 86:423–433

Donson J, Kearney CM, Hilf ME, Dawson WO (1991) Systemic expression of a bacterial gene by a tobacco mosaic virus-based vector. Proc Natl Acad Sci U S A 88:7204–7208

Earley KW, Haag JR, Pontes O, Opper K, Juehne T, Song K, Pikaard CS (2006) Gateway-compatible vectors for plant functional genomics and proteomics. Plant J 45:616–629

Exposito-Rodriguez M, Laissue PP, López-Calcagno PE, Mullineaux PM, Raines CA, Simkin AJ (2017) Development of pGEMINI, a plant gateway destination vector allowing the simultaneous integration of two cDNA via a single LR-Clonase reaction. Plan Theory 6:55

Gleba YY, Giritch A (2011) Plant viral vectors for protein expression. In: Caranta C, Aranda MA, Tepfer M, Lopez-Moya JJ (eds) Recent advances in plant virology. Caister Academic Press, Norfolk, pp 387, 412 pp–412

Gleba Y, Marillonnet S, Klimyuk V (2004) Engineering viral expression vectors for plants: the "full virus" and the "deconstructed virus" strategies. Curr Opin Plant Biol 7:182–188

Gleba Y, Klimyuk V, Marillonnet S (2005) Magnifection–a new platform for expressing recombinant vaccines in plants. Vaccine 23(17–18):2042–2048

Gleba Y, Klimyuk V, Marillonnet S (2007) Viral vectors for the expression of proteins in plants. Curr Opin Biotechnol 18:134–141

Goodin MM, Dietzgen RG, Schichnes D, Ruzin S, Jackson AO (2002) pGD vectors: versatile tools for the expression of green and red fluorescent protein fusions in agroinfiltrated plant leaves. Plant J 31:375–383

Gorlach J, Volrath S, Beiter GK, Hengym G, Beckhove U (1996) Benzothiadiazole, a novel class of inducers of systemic acquired resistance, activates gene expression and disease resistance in wheat. Plant Cell 8:629–643

Gossele V, Fache I, Meulewaeter F, Cornelissen M, Metzlaff M (2002) SVISS – a novel transient gene silencing system for gene function discovery and validation in tobacco plants. Plant J 32:859–866

Groß-Hardt R, Lenhard M, Laux T (2002) WUSCHEL signaling functions in interregional communication during Arabidopsis ovule development. Genes Dev 16:1129–1138

Ha SB, An G (1998) Identification of upstream regulatory elements involved in the developmental expression of the Arabidopsis thaliana cab1 gene. Proc Natl Acad Sci U S A 85:8017–8021

Hartley JL, Temple GF, Brasch MA (2000) DNA cloning using in vitro site specific recombination. Genome Res 10:1788–1795

Hellens RP, Edwards EA, Leyland NR, Bean S, Mullineaux PM (2000) pGreen: a versatile and flexible binary Ti vector for agrobacterium-mediated plant transformation. Plant Mol Biol 42:819–832

Himmelbach A, Zierold U, Hensel G, Riechen J, Douchkov D, Schweizer P, Kumlehn J (2007) A set of modular binary vectors for transformation of cereals. Plant Physio 145:1192–1200

Huang Z, Chen Q, Hjelm B, Arntzen C, Mason H (2009) A DNA replicon system for rapid high-level production of virus-like particles in plants. Biotechnol Bioeng 103:706–714

Jia H, Van Loock B, Liao M, Verbelen JP, Vissenberg K (2007) Combination of the ALCR/alcA ethanol switch and GAL4/VP16-UAS enhancer trap system enables spatial and temporal control of transgene expression in Arabidopsis. Plant Biotechnol J 5:477–482

Karimi M, De Meyer B, Hilson P (2005) Modular cloning in plant cells. Trends Plant Sci 10:103–105

Karimi M, Depicker A, Hilson P (2007) Recombinational cloning with plant gateway vectors. Plant Physiol 145:1144–1154

Katzen F (2007) Gateway recombinational cloning: a biological operating system. Expert Opin Drug Discov 2:571–589

Kim YG, Cha J, Chandrasegaran S (1996) Construction of a Z-DNA-specific restriction endonuclease. Proc Natl Acad Sci U S A 93:1156–1160

Kronbak R, Ingvardsen CR, Madsen CK, Per Langkjær Gregersen PL (2014) A novel approach to the generation of seamless constructs for plant transformation. Plant Meth 10:10

Kusano H, Onodera H, Kihira M, Aoki H, Matsuzaki H, Hiroaki Shimada H (2016) A simple gateway-assisted construction system of TALEN genes for plant genome editing. Sci Rep 6:30234

Laplaze L, Benkova E, Casimiro I, Maes L, Vanneste S, Swarup R (2007) Cytokinins act directly on lateral root founder cells to inhibit root initiation. Plant Cell 19:3889–3900

Lico C, Chen Q, Sant L (2008) Viral vectors for production of recombinant proteins in plants. J Cell Physiol 216:366–377

Lindbo JA (2007) TRBO: a high-efficiency tobacco mosaic virus RNA-based overexpression vector. Plant Physiol 145(4):1232–1240

Liou MR, Huang YW, Hu CC, Lin NS, Hsu YH (2014) A dual gene-silencing vector system for monocot and dicot plants. Plant Biotechnol J 12:330–343

Liu Q, Segal DJ, Ghiara JB, Barbas CF (1997) Design of polydactyl zinc-finger proteins for unique addressing within complex genomes. Proc Natl Acad Sci U S A 94:5525–5530

Ma JK, Drake PM, Christou P (2003) The production of recombinant pharmaceutical proteins in plants. Nat Rev Gen 4:794–805

Mähönen AP, Bonke M, Kauppinen L, Riikonen M, Benfey PN, Helariutta Y (2000) A novel two-component hybrid molecule regulates vascular morphogenesis of the Arabidopsis root. Genes Dev 14(23):2938–2943

Maizel A, Weigel D (2004) Temporally and spatially controlled induction of gene expression in Arabidopsis thaliana. Plant J 38:164–171

Malamy JE, Benfey PN (1997) Analysis of SCARECROW expression using a rapid system for assessing transgene expression in Arabidopsis roots. Plant J 12:957–963

Martínez de Alba AE, Elvira-Matelot E, Vaucheret H (2013) Gene silencing in plants: a diversity of pathways. Biochim Biophys Acta 1829:1300–1308

Matheka JM, Anami S, Gethi J, Omer RA, Alakonya A, Machuka J, Runo S (2013) A new double right border binary vector for producing marker-free transgenic plants. BMC Res Not 6:448

Michniewicz M, Frick EM, Lucia C, Strader LC (2015) Gateway-compatible tissue-specific vectors for plant transformation. BMC Res Not 8:63

Miki D, Shimamoto K (2004) Simple RNAi vectors for stable and transient suppression of gene function in rice. Plant Cell Physiol 45:490–495

Nekrasov V, Staskawicz B, Weigel D, Jones JD, Kamoun S (2013) Development and validation of a clinical cancer genomic profiling test based on massively parallel DNA sequencing. Nat Biotechnol 31:691–693

Pogue GP, Lindbo JA, Dawson WO, Turpen TH (1998) Tobamovirus transient expression vectors: tools for plant biology and high-level expression of foreign proteins in plants. In: Gelvin SB, Schilperoort RA (eds) Plant molecular biology manual. Springer, Dordrecht

Purkayastha A, Dasgupta I (2009) Virus-induced gene silencing: a versatile tool for discovery of gene functions in plants. Plant Physiol Biochem 47:967–976

Que Q, Chilton MM, de Fontes CM, He C, Nuccio M, Zhu T, Wu Y, Chen JS, Shi L (2010) Trait stacking in transgenic crops. Challenges and opportunities. GM Crops 1:220–229

Sainsbury F, Lomonossoff GP (2008) Extremely high-level and rapid transient protein production in plants without the use of viral replication. Plant Physiol 148(3):1212–1218

Sainsbury F, Thuenemann EC, Lomonossoff GP (2009) pEAQ: versatile expression vectors for easy and quick transient expression of heterologous proteins in plants. Plant Biotechnol J 7:682–693

Sanagala R, Moola AK, Diana RKB (2017) A review on advanced methods in plant gene targeting. J Genetic Eng Biotech 15:317–321

Sasaki Y, Sone T, Yoshida S, Yahata K, Hotta J, Chesnut JD, Honda T, Imamoto F (2004) Evidence for high specificity and efficiency of multiple recombination signals in mixed DNA cloning by the Multisite Gateway system. J Biotechnol 107:233–243

Schlücking K, Edel KH, Köster P, Drerup MM, Eckert C, Steinhorst L, Waadt R, Batistic O, Kudla J (2013) A new β-estradiol-inducible vector set that facilitates easy construction and efficient expression of transgenes reveals CBL3-dependent cytoplasm to tonoplast translocation of CIPK5. Mol Plant 6(6):1814–1829

Scholthof HB, Scholthof KB, Jackson AO (1996) Plant virus gene vectors for transient expression of foreign proteins in plants. Annu Rev Phytopathol 34:299–323

Schwab R, Ossowski S, Riester M, Warthmann N, Weigel D (2006) Highly specific gene silencing by artificial microRNAs in Arabidopsis. Plant Cell 18:1121–1133

Shah KH, Almaghrabi B, Bohlmann H (2013) Comparison of expression vectors for transient expression of recombinant proteins in plants. Plant Mol Biol Reporter 31:1529–1538

Shah SH, Jan SA, Ahmad N, Khan SU (2015) Use of different promoters in transgenic plant development: current challenges and future perspectives. American-Eurasian J Agric Environ Sci 15(4):664–675

Shuai B, Reynaga-Pena CG, Springer PS (2002) The lateral organ boundaries gene defines a novel, plant-specific gene family. Plant Physiol 129:747–761

Simpson J, Schell J, Van Montagu M, Herrera-Estrella L (1986a) Light-inducible and tissue-specific pea lhcp gene expression involves an upstream element combining enhancer- and silencer-like properties. Nature 323:551–554

Simpson J, Vanm M, Herrera Estrella L (1986b) Photosynthesis-associated gene families: differences in response to tissue-specific and environmental factors. Science 233:34–38

Smith NA, Singh SP, Wang MB, Stoutjesdijk PA, Green AG, Waterhouse PM (2000) Total silencing by intron-spliced hairpin RNAs. Nature 407:319–320

Sorek R, Awrence CM, Wiedenheft B (2013) CRISPR-mediated adaptive immune systems in bacteria and archaea. Annu Rev Biochem 82:237–266

Takken FLW, van Wijk R, Michielse CB, Houterman PM, Ram AFJ (2004) A one-step method to convert vectors into binary vectors suited for agrobacterium-mediated transformation. Curr Genet 45:242–248

Thole V, Worland B, Snape JW, Vain P (2007) The pCLEAN dual binary vector system for agrobacterium-mediated plant transformation. Plant Physiol 145:1211–1219

Thole JM, Beisner ER, Liu J, Venkova SV, Strader LC (2014) Abscisic acid regulates root elongation through the activities of auxin and ethylene in *Arabidopsis thaliana*. G3: Genes, Genomes, Genetics 4(7):1259–1274

Tsutsui H, Higashiyama T (2017) pKAMA-ITACHI vectors for highly efficient CRISPR/Cas9-mediated gene knockout in *Arabidopsis thaliana*. Plant Cell Physiol 58(1):46–56

Tzfira T, Tian GW, Lacroix B, Vyas S, Li J, Leitner-Dagan Y, Krichevsky A, Taylor T, Vainstein A, Citovsky V (2005) pSAT vectors: a modular series of plasmids for autofluorescent protein tagging and expression of multiple genes in plants. Plant Mol Biol 57:503–516

Untergasser A, Bijl GJ, Liu W, Bisseling T, Schaart JG (2012) One-step agrobacterium mediated transformation of eight genes essential for rhizobium symbiotic signaling using the novel binary vector system pHUGE. PLoS One 7:e47885

Urnov FD, Rebar EJ, Holmes MC, Zhang HS, Gregory PD (2010) Genome editing with engineered zinc finger nucleases. Nat Rev Genet 11:636–646

Varagona MJ, Schmidt RJ, Raikhel NV (1992) Nuclear localization signal(s) required for nuclear targeting of the maize regulatory protein Opaque-2. Plant Cell 10:1213–1227

Veylder LD, Van MM, Inze D (1997) Herbicide safener inducible gene expression in *Arabidopsis thaliana*. Plant Cell Physiol 38:568–577

Voinnet O, Rivas S, Mestre P, Baulcombe D (2003) Retracted: an enhanced transient expression system in plants based on suppression of gene silencing by the p19 protein of tomato bushy stunt virus. The Plant J 33(5):949–956

Wang CT, Yin XL, Kong XX, Li WS, Ma L (2013) A series of TA-based and zero-background vectors for plant functional genomics. Plos One:8, e59576

Warnasooriya SN, Montgomery BL (2009) Detection of spatial-specific phytochrome responses using targeted expression of biliverdin reductase in Arabidopsis. Plant Physiol 149:424–433

Wesley SV, Helliwell CA, Smith NA, Wang M, Rouse DT, Liu Q, Gooding PS, Singh SP, Abbott D, Stoutjesdijk PA (2001) Construct design for efficient, effective and high-throughput gene silencing in plants. Plant J 27:581–590

Zhang H, Wang L, Hunter D, Voogd C, Joyce N, Davies K (2013) A *Narcissus mosaic* viral vector system for protein expression and flavonoid production. Plant Methods 9:28

Zhao F, Hwang US, Lim S, Yoo RH, Igori D, Lee SH, Lim HS, Moon JS (2015) Complete genome sequence and construction of infectious full-length cDNA clones of tobacco ringspot Nepovirus, a viral pathogen causing bud blight in soybean. Virus Genes 51:163–166

Zhou X, Huang C (2012) Virus-induced gene silencing using begomovirus satellite molecules. Methods Mol Biol 894:57–67

Zuo J, Niu QW, Chau NH (2000) An estrogen receptor-based transactivator XVE mediates highly inducible gene expression in plants. Plant J 24:265–273

Zuo J, Niu QW, Frugis G, Chau NH (2002) The WUSCHEL gene promotes vegetative-to-embryonic transition in Arabidopsis. Plant J 30:349–359

Recent Developments in Generation of Marker-Free Transgenic Plants

6

Rupesh Kumar Singh, Lav Sharma, Nitin Bohra, Sivalingam Anandhan, Eliel Ruiz-May, and Francisco Roberto Quiroz-Figueroa

Abstract

A plant modified through artificial insertion of a foreign DNA into its genome is referred to as "genetically modified plant" or a "transgenic" plant. The selection of the transgenic tissues during the genetic transformation process is based on the constitutively expressed marker gene(s) coding for reporters, such as those conferring resistance against antibiotics and/or herbicides. In this direction, *Agrobacterium*-mediated genetic co-transformation is arguably the most commonly used technique to transfer the gene(s) of interest as well as the marker gene(s). However, the latter is purposeless once a transgenic tissue has been

R. K. Singh (✉)
Centro de Química de Vila Real (CQ-VR), Universidade de Trás-os-Montes e Alto Douro, Vila Real, Portugal
e-mail: rupesh@utad.pt

L. Sharma
CITAB – Centre for the Research and Technology of Agro-Environmental and Biological Sciences, University of Trás-os-Montes and Alto Douro, UTAD, Vila Real, Portugal

N. Bohra
School of Genetics and Biotechnology, University of Trás-os-Montes and Alto Douro, UTAD, Vila Real, Portugal

Department of Biotechnology, National Institute of Technology, Warangal, Telangana, India

S. Anandhan
ICAR- Directorate of Onion and Garlic Research, Rajgurunagar, Pune, Maharashtra, India

E. Ruiz-May
Red de Estudios Moleculares Avanzados, Instituto de Ecología A. C., ClusterBioMimic®, Xalapa, Veracruz, Mexico

F. R. Quiroz-Figueroa
Instituto Politécnico Nacional, Centro Interdisciplinario de Investigación para el Desarrollo Integral Regional Unidad Sinaloa (CIIDIR-IPN Unidad Sinaloa), Laboratorio de Fitomejoramiento Molecular, Guasave, Sinaloa, México

R. Sathishkumar et al. (eds.), *Advances in Plant Transgenics: Methods and Applications*, https://doi.org/10.1007/978-981-13-9624-3_6

selected. Although these marker genes are important for screening purposes, they exhibit safety concerns for the environment as well as among consumers. At times, commercial transgenic plants transfer these gene(s) to the weeds or other organisms, leading to the development of resistance among nontarget plants. Moreover, the escape of such gene could affect the wild relatives or land races via gene flow. Therefore, in order to maintain sustainability, removing the marker gene(s) from a transgenic crop is of utmost importance, prior to its commercialization. Hitherto, several methodologies have been evolved for the development of a marker-free transgenic crop. In the present summary, we discuss the merits and the shortcomings of the *Agrobacterium*-mediated genetic co-transformation. In addition, we review the recent developments among other approaches and their impacts and suggest directions for their maximum utilization in the near future.

Keywords
Marker-free · Transgenic plant · *Agrobacterium*-mediated genetic co-transformation · Gene flow

6.1 Introduction

A "transgenic" plant is a genetically modified plant developed by using various approaches to incorporate a foreign gene(s) into the plant genome. The incorporated gene has a different origin, and it confers its properties to the host plant. This process is called as "genetic engineering" of the plant. *Agrobacterium* is a ubiquitous soil bacterium capable of inserting a DNA fragment into the host plant. *Agrobacterium*-mediated genetic co-transformation is arguably the most popular technique used for genetic engineering among researchers, due to its various benefits, such as (a) the development of disease resistance in susceptible crops, (b) increased tolerance toward various environmental stress conditions, and (c) for the production of important bio-compounds by the incorporation of newer traits in the host crop genome.

The process of genetic modification employs a candidate gene(s) (gene(s) of interest) which improve the plant by imparting a desired property and a marker gene(s) which facilitates the screening of a modified/transformed cell within a selection medium. The marker gene is usually a small DNA fragment coding for the resistance against specific herbicides or antibiotics. The transformed cells, tissues, or regenerated shoots harboring this resistance gene are able to survive in the presence of the corresponding antibiotic or herbicide, whereas the non-transformed cells subsequently bleach out and die. Using a marker gene is instrumental in distinguishing the transformed cells from non-transformed ones, as the recovery of the transformed cells is very less, as in most of the cases, the transformation efficiency is low.

Different marker genes have been used for the genetic transformation in the plant of interest. Neomycin phosphotransferase II (*nptII*) gene has been widely used for the screening in medium supplemented with kanamycin. The antibiotic kanamycin interferes with the protein translation of a normal plant cell leading to its death (Herrera-Estrella et al. 1983; Bevan et al. 1983). However, the transformed cell survives as the *npt* inactivates the antibiotic by its phosphorylation (Goodwin et al. 2004). Another commonly used marker gene is the phosphinothricin acetyltransferase (*pat*) gene isolated from the bacterium *Streptomyces hygroscopicus* (De Block et al. 1987; Miflin and Lea 1977). A marker gene is crucial for the screening of the transformed cells; its use raises concerns pertaining to their integration in a natural ecosystem (Darbani et al. 2007), as listed in the subsequent section.

6.2 Insights into the Urgency for a Marker-Free Genetic Transformation

As hinted above, there are many consumer and environmental concerns regarding the presence of a marker gene in a transformed plant cell.

- *Consumer's concern*

  The incorporated marker gene from the bacterial origin, conferring antibiotic resistance, may affect the intestinal microbiota after the consumption of the transformed crop. The gene can also be shared among other microorganism within the ecosystem, imparting unwanted antibiotic resistance.
- *Environmental concerns.*

  Similarly, the horizontal gene flow of the herbicide resistance gene from the genetically modified plants to the environment can also take place. This may cause serious concerns regarding the crop protection, especially the excessive use of an herbicide

The removal of the marker gene is also important in terms of tackling the regulatory issues on the use of transgenic crops. Furthermore, a stepwise marker removal approach allows the researchers to investigate the role of other genes of interests through retransformation, which is not possible otherwise as the previous incorporated marker hinders the selection of a retransformed plant.

Some reports have suggested that the presence of the markers is maybe not as deleterious as perceived. The transfer of the antibiotic resistance genes from transgenic plants to animals would require a series of steps. For a successful horizontal transfer, the transgene needs itself to be separated from the plant genome and to be inserted into host genome by passing through the gastrointestinal tract of the animal system. Hence, it needs to be protected from the endonucleases, stay stable within the host genome to be functional, and express its protein/product within the host system. This theory still remains unclear as various investigations report some contradictions. One report demonstrated that the mice cells can harbor small amount on plasmid DNA, after being fed on a high quantity of modified product (Schubbert et al. 1998); however, a stable expression of the marker gene is not proved (FAO 2000). Certain limitations have also been reported in transferring the plant DNA into a bacterial/animal system with stable expression in laboratory condition

(Nielsen et al. 1998; Kay et al. 2002). Some other studies have highlighted that the horizontal transfer of the antibiotic resistance gene in a pathogenic bacterium occurs in small proportions, which can be eliminated by the modern plant biotechnology tools (National Research Council 2000; Royal Society 1998). Nonetheless, vertical gene transfer of an herbicide resistant gene is another concern as it may cause serious damage to the natural ecosystem by imparting herbicide resistance among plants (Dale et al. 2002; Jianru et al. 2002).

In a nutshell, the absence of marker gene in transgenic plants is a better option in order to avoid any possible damage or concerns. The modern plant biotechnology methods provide significant advances in this direction. Either the antibiotic/herbicide resistant genes can be eliminated from a modified plant system or a marker, other than those imparting resistance, can be used for the purpose. Some recent approaches to generate marker-free transgenic plants in order to maintain the sustainability of the ecosystem are discussed here.

6.3 Approaches to Develop a Marker-Free Transgenic Plant

Segregation in further generations was one of the first approaches to eliminate marker gene from transgenic plants after the co-transformation (Depicker et al. 1985; McKnight et al. 1987; De Block and Debrouwer 1991). Another system developed was a recombination between defined excision sites in chromosomes called as site-specific recombination and became a very useful tool for marker elimination (Dale and Ow 1991; Gleave et al. 1999). MATVS (multi-auto-transformation vector system) was also a powerful tool for marker-free transgenic development (Ebinuma et al. 1997; Ebinuma and Komamine 2001). Transposon-based methods (Cotsaftis et al. 2002; Goldsbrough et al. 1993) and homologous recombination also emerged in this series (Puchta 2000; Zubko et al. 2000). We summarize some of such methodologies in this chapter.

6.3.1 Two-Vector System

Using a two-vector system, during the genetic manipulation, is one of the most favorable co-transformation methods in the development of the marker-free transgenic plants (Wakita et al. 1998; Shah et al. 2008; Xu et al. 2017). One vector harbors the gene of interest, whereas another vector carries the marker gene. The explants are cocultivated with an *Agrobacterium* carrying both the vectors in separate cultures. Three combinations are possible post transformation using this approach:

(a) One set of transformants harboring only the marker gene.
(b) Another set with only the candidate gene.
(c) A third possibility may also arise where both the candidate gene and the marker gene are present but at different loci inside the host plant genome.

The transformants exhibiting the first and the third possibility are screened in a selective medium. They are further distinguished by a gene-specific PCR to identify the transformants with the third combination. Segregation separates this third combination again into three different categories. The progenies segregate with the clusters of (a) the candidate gene only, (b) the marker gene only, and (c) both the genes. Transgenic lines harboring only the candidate gene are obtained as the final product and are denominated as the marker-free transgenic lines. Due to the wide use of this approach, a detailed stepwise methodology is described below.

Shah et al. (2008) used a two-vector system to co-transform a candidate gene and a marker gene, in the summer squash *Cucurbita pepo* L. cv. Australian green. *Agrobacterium tumefaciens* strain LBA 4404 carrying pCAMBIA 0390 plasmid with cbf1 gene, under the control of stress-inducible promoter rd29A, and pCAMBIA 2200 plasmid, with selectable marker, neomycin phosphotransferase II (*nptII*) gene, under the control of constitutive promoter CaMV 35S, were used in the study. Sterilized shoot tips from in vitro germinated plantlets were used as the explants. A single colony of *Agrobacterium* was inoculated into a 50 ml yeast extract mannitol medium, supplemented with kanamycin (50 mg/l) and rifampicin (25 mg/l) and grown overnight at 28 °C. The bacterial culture was centrifuged and the cells were harvested and re-suspended, until the optical density between 0.15 and 0.18 at 580 nm is achieved. The explants were dipped in the bacterial culture for 20–30 minutes and blot dried. Explants were cocultivated in the dark for 48 hours at 28 °C in a culture medium (MS medium supplemented with 0.05 mg/l BAP). The explants were rinsed properly after cocultivation with sterile water, in order to remove all the bacterial strains, followed by the washing with 250 mg/l cefotaxime, in order to get rid of the bacterial overaccumulation in the later subcultures. MS medium supplemented with 0.05 mg/l BAP and 50 mg/l kanamycin was used to subculture the cocultivated explants until the shoot proliferation. The shoots were cut and transferred in the same medium to achieve an elongation of up to 4–5 cm. Later, the explants were transferred in the MS medium supplemented with 0.5 mg/l IAA for root induction. Well rooted plants were taken out and washed with running tap water and transferred in the potted soil. Plantlets were maintained in the controlled condition for acclimatization for 5 days and were transferred later in a greenhouse. The plants were allowed to grow under normal condition, until the flowering and the fruiting. Subsequently, the seeds were collected, and the next-generation seeds were propagated, followed by one more generation for the segregation analysis. The plants harboring only the candidate gene were selected after F2 generation, and confirmed with gene-specific PCR, and the southern blotting, and denoted as marker- free transgenic plants.

Some recent studies have also adopted a similar strategy. Khidr and Nasr (2018) reported a co-transformation strategy in a cucumber by using gfp (green fluorescent protein) as the candidate gene, inserted within a plasmid, and *nptII* gene as the selectable marker on another plasmid. These two genes were co-transformed in the cotyledonary explants of cucumber. Both the genes were incorporated on the unlinked loci in the host genome and supposed to be segregated in the future generation. Another study, conducted by Rajadurai et al. (2018), was to obtain insect-resistant and marker-free transgenic rice, by using the candidate gene cry2AX1 in one cassette and the selectable marker gene in another. Both the genes were co-integrated and later underwent natural homozygous recombination/segregation in order to produce the marker-free transgenic lines.

6.3.2 Transposon-Mediated Co-Transformation

The transposons are the DNA fragments which are unstable and can change their positions within the genome. They are also referred to as "jumping genes," and this phenomenon was discovered by Barbara McClintock in 1940. Later, it became very useful for the generation of the marker-free transgenic plants. In this method, the marker gene is inserted onto a transposable element, which can be moved to different positions within the host plant genome. This tagged marker is then co-transformed with gene of interest, and later, the separation of marker gene is assisted by segregation. This method is very useful for the elimination of the marker gene, as well as to generate transformants with the candidate gene at different locus, in the host plant. The approach seems promising for the transformation of the recalcitrant plant species, which remains a major challenge for the plant biotechnologists. However, this method has a lot of limitations in its application, as different plants have varying efficiency for the transposable elements. Moreover, the selection of a marker-free transgenic can be quite difficult and time-consuming (Darbani et al. 2007; Goldsbrough et al. 1993; Gorbunova and Levy 1999; Perl et al. 1993).

6.3.3 Site-Specific Recombination

The recombination between a target excision site is called as site-specific recombination method and is generally carried out using the excisionase enzyme. This technique has been very useful in the elimination of a marker gene from the genome of a transgenic plant. Different types of site-specific recombination strategies have been developed to meet various challenges in modern biotechnology. Recently, a multi-auto-transformation vector system (MATVS) was used to develop a marker-free transgenic tobacco plant, where RDR6 (RNA-dependent RNA polymerase 6) was suppressed in posttranscriptional modification event (Mikami et al. 2018). In another report, a Cre-Lox-based system was applied in barley for the auto-removal of a selectable marker gene (Éva et al. 2018). In this system, the Cre enzyme acts on the specific sites under a cold-inducible promoter from wheat (wcs120). The transgenic plants were exposed to a low temperature at different developmental stages, followed by an exposure to a higher temperature, in order to induce recombination and selection of marker-free transgenics in future generations. Another site-specific recombination was reported by employing pMAT21-wasabi defensing multi-auto-transformation vector-based R/RS definition. Isopentenyl transferase (ipt) gene was used for the positive selection, and a MAT cassette was developed with wasabi defensing gene from *Wasabia japonica* (a Japanese horseradish exhibiting antimicrobial proteins). Later, the marker gene was eliminated, and the marker-free transgenic plants were recovered in an eggplant (Darwish et al. 2014). Woo et al. (2015) used a FLP/FRT-based system in rice, where a FLP recombinase target the FRT site, using an oxidative stress-inducible construct, to improve the tocopherol content in seeds and to excise the marker gene.

6.3.4 Negative Selection

Although many selectable marker gene transformation and elimination methodologies are available, an approach guiding the negative selection always exhibits immense potential. In this method, the transformed plants can be selected for the absence of a marker gene. Some negative selection markers have been developed on the basis of substrates or the conversion of the required substrates in a medium (Erikson et al. 2004), although thousands of PCR reactions to screen is a major drawback of this technique.

6.3.5 Unconventional Markers

Marker genes based on the antibiotic or herbicide resistance are conventional and mainly of bacterial origin. This may lead to serious ethical and environmental issues, as discussed before. The genes from a plant origin, which do not pose resistance against antibiotic or herbicides, can be a potential substitute as the marker genes. These markers from the plants are receiving an increased attention from the researchers in the recent times.

A few approaches have been implemented in this direction as well. The amount of mannose was studied for the selection of the transgenic plants on the basis of *Escherichia coli* phosphomannose isomerase (PMI) enzyme as positive selection system in sugar beet (*Beta vulgaris* L.). The non-transformed plants were eliminated by increasing the dose of mannose for up to 10 g/l in the selection medium (Joersbo et al. 1998). In another example, *Agrobacterium*-mediated co-transformation in the immature embryos of maize, with PMI as the selectable marker, recovered 30% transgenic lines (Negrotto et al. 2000). Many investigations have reported the use of PMI gene as an efficient positive selection marker in *Zea mays* (Wright et al. 2001), wheat (Reed et al. 2001), rice (Lucca et al. 2001), pearl millet (Okennedey et al. 2004), and others (Jaiwal et al. 2002; Sonntag et al. 2004; Mentewab and Stewart 2005). Phosphomannose isomerase is nontoxic to plants; however, upon phosphorylation, it changes to mannose-6-phosphate and inhibits the ATP production by hindering the cellular glycolysis (Goldsworthy and Street 1965; Loughman 1966). This, in turn, inversely affects the photosynthetic mechanism of the plant, at the transcriptional level (Jang and Sheen 1997). Hence, the cells grow much slower and result in a very low organogenesis in mannose supplemented medium (Malca et al. 1967; Pego et al. 1999). Only the cells expressing PMI can convert mannose-6-phosphate into fructose-6-phosphate for glycolysis.

Usually the PMIs were reported in the bacteria and the fungi; however, recently, some soybean species are found to carry this gene (Chiang and Kiang 1988; Proudfoot et al. 1994), limiting the use of PMIs as a marker for positive selection in the higher plants. Moreover, some recent reports suggested that the PMIs may be present in some *Arabidopsis* spp. and cabbage (Maruta et al. 2008; Wang et al. 2014a). However, four novel PMIs were recently reported by Hu et al. (2016) from algae and rice, and their corresponding proteins expressions were analyzed. This recent study demonstrated a successful transgenic plants selection under the mannose pressure and verified the use of PMIs as a potential positive selection marker in the higher plants (Hu et al. 2016).

6.4 Recent Contributions Toward Marker-Free Transgenic Plant Development

During the last decades, several other methods have been developed in order to meet the challenges posed by genetic engineering for sustainable crop development. Recent reports in this domain are summarized in Table 6.1.

6.5 Conclusions and Future Perspectives

With an ever-expanding human population, the conventional breeding of crop plants is alone not sufficient to meet the food requirement across the globe. With the advent of biotechnology methods, genetically modified crops are being made which can tolerate adverse conditions, such as the competition for land use, soil erosion, salination, drought, cold, pollution, and changing climate. However, the regulations on the genetically modified crops are strong, and they affect the production of different transgenic crops worldwide. Especially, big markets like European Union have imposed strong rules and limitations on the commercialization and the import of genetically modified foods. Therefore, the development of environmentally safer transgenic crops is the need of the hour.

A major role of plant biotechnology is to improve the crops and, at the same time, minimize the future risk hazards. Several important crops have been transformed with value-added traits to meet various benefits toward an improved quality production. However, the presence of antibiotic or herbicide resistance marker is a major obstacle and needs to be excised from the host plant. The possibilities of the horizontal and the vertical transmission of these genes within the environment cannot be neglected. In this direction, different methods have evolved to generate marker-free transgenic plants, and some of the important approaches were discussed in this chapter. The excision-based marker elimination approaches have been popular; however, the transgenic cells are genetically unstable. The site-specific recombination methods are useful; however, their implementation is time-consuming, and they require an extended skill set. However, a previous work on *Agrobacterium*-mediated genetic co-transformation on summer squash demonstrated an easier and

Table 6.1 Recent advancements in the methodologies for a marker-free genetic transformation of crops

Plant	Method	References
Algae and microalgae	Modular and marker-free chloroplast expression system	Bertalan et al. (2015)
	Precision excision by piggyback transposase	Kasai et al. (2017)
	Dihydropteroate synthase targeted to the chloroplast	Tabatabaei et al. (2018)
	Combined genome editing and marker recycling technology	Verruto et al. (2018)
	Episomal CRISPR system	Poliner et al. (2018)
Arabidopsis	Bsar-R174K, a non-antibiotic selectable marker susceptible to l-lysine and D-alanin	Kuan et al. (2018)
Apple	Chemically inducible R/RS recombinase	Righetti et al. (2014)
Apricot	Chemical-inducible Cre-LoxP system	Petri et al. (2012)
Barley	Cre-lox system-mediated marker gene excision	Éva et al. (2018)
Berry crops	Chimeric genes from its own species as marker genes	Palomo-Ríos et al. (2018)
Eggplant	R/RS site-specific recombination system	Darwish et al. (2014)
Grapevine	Binary vector with a heat-shock-inducible system	Costa et al. (2016)
Maize	Selectable marker independent transformation	Mookkan et al. (2017)
Oil palm	Gfp protein as visual marker	Ghulam and Na'imatulapidah (2018)
Orange	Negative selection marker	De Oliveira et al. (2015)
	Cre/LoxP system	Zou et al. (2013)
	Embryo-specific expression of a visual reporter gene as marker	Dutt et al. (2018)
Pear	Chemically inducible R/RS recombinase	Righetti et al. (2014)
Pigeon pea	Negative selection	Ganguly et al. (2018)
Potato	Cre-lox system through application of the regulatory sequences from cowpea mosaic virus	Kopertekh et al. (2018)
	cis gene stacking	Jo et al. (2014)
Rice	Fluorescence marker gene-mediated approach	He et al. (2018)
	A gene as the negative selection marker for gene targeting	Osakabe et al. (2014)
	Cre-lox-mediated marker elimination and gene reactivation	Terada et al. (2010)
	CRISPR-based targeted nucleotide substitutions	Shimatani et al. (2018)
	Bsar-R174K, a non-antibiotic selectable marker susceptible to l-lysine and D-alanin	Kuan et al. (2018)
	Biolistic co-transformation and segregation approach	Pérez-Bernal et al. (2017)

(continued)

Table 6.1 (continued)

Plant	Method	References
	Agrobacterium-mediated co-transformation system with a twin T-DNA binary vector	Xu et al. (2017)
	Biolistic particle-mediated co-transformation	Feng et al. (2017)
	Targeted base editing in rice and tomato using a CRISPR-Cas9 cytidine deaminase fusion	Shimatani et al. (2017)
	CRISPR/Cas9-mediated precise excision	Srivastava et al. (2017)
	Marker excision system	Zhu et al. (2017)
	RNAi-mediated	Ahmed et al. (2017)
		Wang et al. (2014b)
	Cointegrating vector T-DNA and a binary vector T-DNA in a *Agrobacterium tumefaciens* strain	Sripriya et al. (2008)
	Sequential *Agrobacterium*-mediated co-transformation	Rao et al. (2011)
	Marker excision by piggyBac transposition	Nishizawa-Yokoi et al. (2015)
	Based on strong PNS using diphtheria toxin A-fragment as a negative marker	Shimatani et al. (2015)
	FLP/FRT-mediated spontaneous auto-excision	Woo et al. (2015)
	Single gene targetting	Oliva et al. (2014)
	2 T-DNA binary vector system	Breitler et al. (2004)
Soybean	Conditional negative selectable marker	Shao et al. (2015)
Tobacco	Homoplasmic transplastomic plants crossing with a nuclear-transgenic plants expressing a plastid-targeted Cre recombinase	Hoelscher et al. (2018)
	Cre/lox system	Chakraborti et al. (2008)
	Cre-LoxP system	García-Almodóvar et al. (2014)
Tomato	Targeted base editing in rice and tomato using a CRISPR-Cas9 cytidine deaminase fusion	Shimatani et al. (2017)
Watermelon	Marker-free constructs based on nucleocapsid protein gene	Holkar et al. (2018)
Wheat	Induction of targeted deletions using customized meganuclease	Youssef et al. (2018)
	Zinc finger nuclease-mediated precision genome editing	Ran et al. (2018)
	Gene editing studies	Zhang et al. (2016)
	Agrobacterium-mediated co-transformation strategy	Wang et al. (2017)
	Cre/lox system	Mészáros et al. (2015)
Universal method for all plant types	Recombinant DNA constructs employing site-specific recombination	Gilbertson et al. (2018)

(continued)

Table 6.1 (continued)

Plant	Method	References
Universal method for various plant types	Excision of selectable marker genes	Hare and Chua (2002)
Universal method for various plant types	Marker excision system using an animal-derived piggyBac transposon	Nishizawa-Yokoi et al. (2016)

feasible approach to develop marker-free transgenic crop. Furthermore, homozygous stable lines for candidate genes can be achieved using such approaches. Nonetheless, immaterial to the approach taken, the evaluation of the future generations of the transgenic crops in terms of stability and agronomic parameters is indispensable. Another important aspect of such evaluations is the time frame for observations. The adverse effects of the transgenic crops should be studied in a subchronic manner, i.e., observations under 90 days only. Moreover, long-term effect of transgenic crop products should be investigated for the consumer's safety.

Acknowledgments The authors would like to acknowledge the support from Projeto NORTE-01-0145-FEDER000017- INTERACT/ VitalityWINE, cofinanced by FEDER/Programa NORTE 2020, and Plataforma de inovação da vinha e do vinho-innovine&wine, Norte-01-0145-FEDER000038. Postdoctoral research grant (BPD/UTAD/INNOVINE&WINE/ 424/2016) to RKS is also acknowledged. Financial support (PEst-OE/QUI/UI0616/2014) provided to the Research Unit in Vila Real by Fundaçãopara a Ciência e Tecnologia (FCT), Portugal, and COMPETE is also acknowledged. Assistances from the project UID/AGR/04033/2013 and National Funds by FCT (Portuguese Foundation for Science and Technology) and the European Investment Funds by FEDER/COMPETE/POCI Operacional Competitiveness and Internationalization Programme under the Project POCI-01-0145-FEDER-006958 are also recognized. Chemistry center of Vila Real (CQ-VR) is gratefully acknowledged.

Conflict of Interest The authors declare that there are no conflicts of interest.

References

Ahmed MMS, Bian S, Wang M, Zhao J, Zhang B, Liu Q, Zhang C, Tang S, Gu M, Yu H (2017) RNAi-mediated resistance to rice black-streaked dwarf virus in transgenic rice. Transgenic Res 26:197–207

Bertalan I, Munder MC, Weiß C, Kopf J, Fischer D, Johanningmeier U (2015) A rapid, modular and marker-free chloroplast expression system for the green alga *Chlamydomonas reinhardtii*. J Biotechnol 195:60–66

Bevan MW, Flavell RB, Chilton MD (1983) A chimeric antibiotic resistance gene as a selectable marker for plant cell transformation. Nature 304:184–187

Breitler JC, Meynard D, Van Boxtel J, Royer M, Bonnot F, Cambillau L, Guiderdoni E (2004) A novel 2 T-DNA binary vector allows efficient generation of marker-free transgenic plants in three elite cultivars of rice (*Oryza sativa* L.). Transgenic Res 13:271–287

Chakraborti D, Sarkar A, Mondal HA, Schuermann D, Hohn B, Sarmah BK, Das S (2008) Cre/lox system to develop selectable marker free transgenic tobacco plants conferring resistance against sap sucking homopteran insect. Plant Cell Rep 27:1623–1633

Chiang YC, Kiang YT (1988) Genetic analysis of mannose-6-phosphate isomerase in soybeans. Genome 30:808–811

Costa DL, Piazza S, Campa M, Flachowsky H, Hanke MV, Malnoy M (2016) Efficient heat-shock removal of the selectable marker gene in genetically modified grapevine. Plant Cell Tissue Organ Cult 124:471–481

Cotsaftis O, Sallaud C, Breitler JC, Meynard D, Greco R, Pereira A, Guiderdoni E (2002) Transposon-mediated generation of T-DNA- and marker-free rice plants expressing a Bt endotoxin gene. Mol Breed 10:165–180

Dale EC, Ow DW (1991) Gene transfer with subsequent removal of the selection gene from the host genome. Proc Natl Acad Sci U S A 88:10558–10562

Dale PJ, Clarke B, Fontes MG (2002) Potential for the environmental impact of transgenic crops. Nat Biotechnol 20:567–574

Darbani B, Eimanifar A, Stewart CN, Camargo WN (2007) Methods to produce marker-free transgenic plants. Biotechnol J 2:83–90

Darwish NA, Khan RS, Ntui VO, Nakamura I, Mii M (2014) Generation of selectable marker-free transgenic eggplant resistant to *Alternariasolani* using the R/RS site-specific recombination system. Plant Cell Rep 33:411–421

De Block M, Debrouwer D (1991) Two T-DNA's cotransformed into *Brassica napus* by a double *Agrobacterium tumefaciens* infection are mainly integrated at the same locus. Theor Appl Genet 82:257–263

De Block M, Botterman J, Vandewiele M, Dockx J, Thoen C, Gosselé V, Movva NR, Thompson C, Montagu MV, Leemans J (1987) Engineering of herbicide resistance in plants by expression of a detoxifying enzyme. EMBO J 6:2513–2518

De Oliveira ML, Stover E, Thomson JG (2015) The *codA* gene as a negative selection marker in citrus. Springer Plus 17:p264

Depicker A, Herman L, Jacobs A, Schell J, Montague MV (1985) Frequencies of simultaneous transformation with different T-DNAs and their relevance to the agrobacterium plant cell interaction. Mol Gen Genet 201:477–484

Dutt M, Zambon FT, Erpen L, Soriano L, Grosser J (2018) Embryo-specific expression of a visual reporter gene as a selection system for citrus transformation. PLoS One 13:e0190413

Ebinuma H, Komamine A (2001) Mat (Multi-Auto-Transformation) vector system. The oncogenes of *Agrobacterium* as positive markers for regeneration and selection of marker-free transgenic plants. In Vitro Cell Dev Biol Plant 37:103–113

Ebinuma H, Sugita K, Matsunaga E, Yamakado M (1997) Selection of marker-free transgenic plants using the isopentenyl transferase gene. Proc Natl Acad Sci U S A 94:2117–2121

Erikson O, Hertzberg M, Nasholm T (2004) A conditional marker gene allowing both positive and negative selection in plants. Nat Biotechnol 22:455–458

Éva C, Téglás F, Zelenyánszki H, Tamás C, Juhász A, Mészáros K, Tamás L (2018) Cold inducible promoter driven Cre-lox system proved to be highly efficient for marker gene excision in transgenic barley. J Biotechnol 265:15–24

FAO, Food and Agriculture Organization/World Health Organization (2000) Safety aspects of genetically modified foods of plant origin. Report of a joint FAO/WHO consultation on foods derived from biotechnology. World Health Organization, Geneva

Feng D, Wang Y, Wu J, Lu T, Zhang Z (2017) Development and drought tolerance assay of marker-free transgenic rice with OsAPX2 using biolistic particle-mediated co-transformation. Crop J 5:271–281

Ganguly S, Ghosh G, Purohit A, Sreevathsa R, Chaudhuri RK, Chakraborti D (2018) Effective screening of transgenic pigeonpea in presence of negative selection agents. Proc Natl Acad Sci India Sect B 88:1565

García-Almodóvar RC, Petri C, Padilla IMG, Burgos L (2014) Combination of site-specific recombination and a conditional selective marker gene allows for the production of marker-free tobacco plants. Plant Cell Tissue Organ Cult 116:205–215

Ghulam KAP, Na'imatulapidah AM (2018) Green fluorescent protein as a visual selection marker for oil palm transformation. Ind Crop Prod 115:134–145

Gilbertson LA, Huang S, Malver T (2018) *Recombinant DNA constructs employing site – specific recombination* United States Patent 9856485

Gleave AP, Mitra DS, Mudge SR, Morris BA (1999) Selectable marker-free transgenic plants without sexual crossing: transient expression of Cre recombinase and use of the conditional lethal dominant gene. Plant Mol Biol 40:223–235

Goldsbrough AP, Lastrella CN, Yoder JI (1993) Transposition mediated re-positioning and subsequent elimination of marker genes from transgenic tomato. Nat Biotechnol 11:1286–1292

Goldsworthy A, Street HE (1965) The carbohydrate nutrition of tomato roots: VIII. The mechanism of the inhibition by D-mannose of the respiration of excised roots. Ann Botany-London 29:45–58

Goodwin J, Pastori G, Davey M, Jones H (2004) Transgenic plants: methods and protocols. Method Mol Biol 286:191–202

Gorbunova V, Levy AA (1999) How plants make ends meet: DNA double-strand break repair. Trends Plant Sci 4:263–269

Hare PD, Chua NH (2002) Excision of selectable marker genes from transgenic plants. Nat Biotechnol 20:575–580

He Y, Zhu M, Wang L, Wu J, Wang Q, Wang R, Zhao Y (2018) Programmed self-elimination of the CRISPR/Cas9 construct greatly accelerates the isolation of edited and transgene-free rice plants. Mol Plant 11:1210–1213

Herrera-Estrella L, De Block M, Messens E, Hernalsteens JP, Montagu MV, Schell J (1983) Chimeric genes as dominant selectable markers in plant cells. EMBO J 2:987–995

Hoelscher M, Tiller N, Teh AYH, Wu GZ, Ma JKC, Bock R (2018) High-level expression of the HIV entry inhibitor griffithsin from the plastid genome and retention of biological activity in dried tobacco leaves. Plant Mol Biol 97:357–370

Holkar SK, Mandal B, Jain RK (2018) Development and validation of marker-free constructs based on nucleocapsid protein gene of watermelon bud necrosis orthotospovirus in watermelon. Curr Sci 114:1742–1747

Hu L, Li H, Qin R, Xu R, Li J, Li L, Wei P, Yang J (2016) Plant phosphomannose isomerase as a selectable marker for rice transformation. Sci Rep 6:25921

Jaiwal PK, Sahoo L, Singh ND, Singh RP (2002) Strategies to deal with the concern about marker genes in transgenic plants: some environmentally friendly approaches. Curr Sci 83:128–136

Jang JC, Sheen J (1997) Sugar sensing in higher plants. Trends Plant Sci 2:208–214

Jianru Z, Niu Q, Ikeda Y, Chua N (2002) Marker-free transformation: increasing transformation frequency by the use of regeneration-promoting genes. Curr Opin Biotechnol 13:173–180

Jo KR, Kim CJ, Kim SJ, Kim TY, Bergervoet M, Jongsma M, Visser RGF, Jacobsen E, Vossen JH (2014) Development of late blight resistant potatoes by cisgene stacking. BMC Biotechnol 14:50

Joersbo M, Donaldson I, Kreiberg J, Petersen SG, Brunstedt J, Okkels FT (1998) Analysis of mannose selection used for transformation of sugar beet. Mol Breed 4:111–117

Kasai Y, Matsuzaki K, Ikeda F, Yoshimitsu Y, Harayama S (2017) Precise excision of a selectable marker gene in transgenic *Coccomyxa* strains by the piggyBac transposase. Algal Res 27:152–161

Kay E, Vogel TM, Bertolla F, Nalin R, Simonet P (2002) In situ transfer of antibiotic resistance genes from transgenic (transplastomic) tobacco plants to bacteria. Appl Environ Microbiol 68:3345–3351

Khidr YA, Nasr MI (2018) Generation of transgenic marker-free cucumber plants by co-transformation strategy. Egyptian J Genet Cytol 47:29–43

Kopertekh L, Krebs E, Guzmann F (2018) Improvement of conditional Cre-lox system through application of the regulatory sequences from cowpea mosaic virus. Plant Biotechnol Rep 12:127–137

Kuan YC, Thiruvengadam V, Lin JS, Liu JH, Chen TJ, Wu HM, Wang WC (2018) Broad-specificity amino acid racemase, a novel non-antibiotic selectable marker for transgenic plants. Plant Biotechnol Rep 12:27–38

Loughman BC (1966) The mechanism of absorption and utilization of phosphate by barley plants in relation to subsequent transport to the shoot. New Phytol 65:388–397

Lucca P, Ye X, Potrykus I (2001) Effective selection and regeneration of transgenic rice plants with mannose as selective agent. Mol Breed 7:43–49

Malca I, Endo RM, Long MR (1967) Mechansim of glucose counteraction of inhibition of root elongation by galactose, mannose and glucosamine. Phytopathology 57:272–278

Maruta T, Yonemitsu M, Yabuta Y, Tamoi M, Ishikawa T, Shigeoka S (2008) *Arabidopsis* phosphomannose isomerase 1, but not phosphomannose isomerase 2, is essential for ascorbic acid biosynthesis. J Biol Chem 283:28842–28851

McKnight TD, Lillis MT, Simpson RB (1987) Segregation of genes transferred to one plant cell from two separate *Agrobacterium* strains. Plant Mol Biol 8:439–445

Mentewab A, Stewart CNJ (2005) Overexpression of an *Arabidopsis thaliana* ABC transporter confers kanamycin resistance to transgenic plants. Nat Biotechnol 23:1177–1180

Mészáros K, Éva C, Kiss T, Bányai J, Kiss E, Téglás F, Láng L, Karsai I, Tamás L (2015) Generating marker-free transgenic wheat using minimal gene cassette and cold-inducible Cre/lox system. Plant Mol Biol Report 33:1221–1231

Miflin BJ, Lea PJ (1977) Amino acid metabolism. Annu Rev Plant Physiol 28:299–329

Mikami T, Saeki Y, Hirai S, Shimokawa M, Umeyama Y, Kuroda Y, Kodama H (2018) Transformation efficiency and transgene expression level in marker-free RDR6-knockdown transgenic tobacco plants. Plant Biotechnol Rep 12:389–397

Mookkan M, Nelson-Vasilchik K, Hague J, Zhang ZJ, Kausch AP (2017) Selectable marker independent transformation of recalcitrant maize inbred B73 and sorghum P898012 mediated by morphogenic regulators BABY BOOM and WUSCHEL2. Plant Cell Rep 36:1477–1491

National Research Council (US) Committee on Genetically Modified Pest-Protected Plants (2000) 'Genetically modified pest-protected plants: science and regulation', Washington D.C: National Academy Press, p 1–246

Negrotto D, Jolly M, Beer S, Wenck AR, Hansen G (2000) The use of phosphomannose- isomerase as a selectable marker to recover transgenic maize plants (*Zea mays* L.) via *Agrobacterium* transformation. Plant Cell Rep 19:798–803

Nielsen KM, Bones AM, Smalla K, Elsas JDV (1998) Horizontal gene transfer from transgenic plants to terrestrial bacteria–a rare event? FEMS Microbiol Rev 22:79–103

Nishizawa-Yokoi A, Endo M, Ohtsuki N, Saika H, Toki S (2015) Precision genome editing in plants via gene targeting and piggyBac-mediated marker excision. Plant J 81:160–168

Nishizawa-Yokoi A, Endo M, Osakabe K, Saika H, Toki S (2016) Precise marker excision system using an animal-derived piggyBac transposon in plants. Plant J 77:454–463

Okennedey MM, Burger JT, Botha FC (2004) Pearl millet transformation system using the positive selectable marker gene phosphomannose isomerase. Plant Cell Rep 22:684–690

Oliva N, Chadha-Mohanty P, Poletti S, Abrigo E, Atienza G, Torrizo L, Garcia R, Dueñas CJ, Poncio MA, Balindong J, Manzanilla M, Montecillo F, Zaidem M, Barry G, Hervé P, Shou H, Slamet-Loedin IH (2014) Large-scale production and evaluation of marker-free indica rice IR64 expressing phytoferritin genes. Mol Breed 33:23–37

Osakabe K, Nishizawa-Yokoi A, Ohtsuki N, Osakabe Y, Toki S (2014) A mutated cytosine deaminase gene, codA (D314A), as an efficient negative selection marker for gene targeting in rice. Plant Cell Physiol 55:658–665

Palomo-Ríos E, Quesada MA, Matas AJ, Pliego-Alfaro F, Mercado JA (2018) The history and current status of genetic transformation in berry crops. In: Hytönen T, Graham J, Harrison R (eds) The genomes of rosaceous berries and their wild relatives. Compendium of Plant Genomes. Springer, East Malling, pp 139–160

Pego JV, Weisbeek PJ, Smeekens SCM (1999) Mannose inhibits *Arabidopsis* germination via a hexokinase-mediated step. Plant Physiol 119:1017–1024

Pérez-Bernal M, Delgado M, Cruz A, Abreu D, Valdivia O, Armas R (2017) Marker-free transgenic rice lines with a defensin gene are potentially active against phytopathogenic fungus *Sarocladiumoryzae*. Acta Phytopathologica et Entomologica Hungarica 52:135–144

Perl A, Galili S, Shaul O, Ben-Tzvi I, Galili G (1993) Bacterial dihydrodipicolinate synthase and desensitized aspartate kinase: two novel selectable markers for plant transformation. Nat Biotechnol 11:715–718

Petri C, López-Noguera S, Wang H, García-Almodóvar C, Alburquerque N, Burgos L (2012) A chemical-inducible Cre-LoxP system allows for elimination of selection marker genes in transgenic apricot. Plant Cell Tissue Organ Cult 110:337–346

Poliner E, Takeuchi T, Du ZY, Benning C, Farré EM (2018) Nontransgenic marker-free gene disruption by an episomal CRISPR system in the oleaginous microalga, *Nannochloropsis oceanica*. ACS Synth Biol 7:962–968

Proudfoot AEI, Payton MA, Wells TNC (1994) Purification and characterization of fungal and mammalian phosphomannose isomerases. J Protein Chem 13:619–627

Puchta H (2000) Removing selectable marker genes: taking the shortcut. Trends Plant Sci 5:273–274

Rajadurai G, Kalaivani A, Varanavasiyappan S, Balakrishnan N, Udayasuriyan V, Sudhakar D, Natarajan N (2018) Generation of insect resistant marker-free transgenic rice with a novel cry2AX1 gene. Electron J Plant Breeding 9:723–732

Ran Y, Patron N, Kay P, Wong D, Buchanan M, Cao Y, Sawbridge T, Davies JP, Mason J, Webb SR, Spangenberg G, Ainley WM, Walsh TA, Hayden MJ (2018) Zinc finger nuclease-mediated precision genome editing of an endogenous gene in hexaploid bread wheat (*Triticumaestivum*) using a DNA repair template. Plant Biotechnol J 16:2088–2101

Rao MVR, Parameswari C, Sripriya R, Veluthambi K (2011) Transgene stacking and marker elimination in transgenic rice by sequential *Agrobacterium*-mediated co-transformation with the same selectable marker gene. Plant Cell Rep 30:1241–1252

Reed J, Privalle L, Powell ML, Meghji M, Dawson J, Dunder E, Sutthe J, Wenck A, Launis K, Kramer C, Chang YF, Hansen G, Wright M (2001) Phosphomannose isomerase: an efficient selectable marker for plant transformation. In Vitro Cell Dev Biol Plant 37:127–132

Righetti L, Djennane S, Berthelot P, Cournol R, Wilmot N, Loridon K, Vergne E, Chevreau E (2014) Elimination of the *nptII*marker gene in transgenic apple and pear with a chemically inducible R/Rs recombinase. Plant Cell Tissue Organ Cult 117:335–348

Royal Society (1998) Genetically modified plants for food use. The Royal Society, London

Schubbert R, Hohlweg U, Renz D, Doerfler W (1998) On the fate of orally ingested foreign DNA in mice: chromosomal association and placental transmission to the fetus. Mol Gen Genet 259:569–576

Shah P, Singh NK, Khare N, Rathore M, Anandhan S, Arif M, Singh RK, Das SC, Ahmed Z, Kumar N (2008) *Agrobacterium* mediated genetic transformation of summer squash (*Cucurbita pepo* L. cv. Australian green) with cbf-1using a two vector system. Plant Cell Tissue Organ Cult 95:363–371

Shao M, Michno JM, Hotton SK, Blechl A, Thomson J (2015) A bacterial gene *codA* encoding cytosine deaminase is an effective conditional negative selectable marker in *Glycine max*. Plant Cell Rep 34:1707–1716

Shimatani Z, Yokoi AN, Endo M, Toki S, Terada R (2015) Positive–negative-selection-mediated gene targeting in rice. Front Plant Sci 5:748

Shimatani Z, Kashojiya S, Takayama M, Terada R, Arazoe T, Ishii H, Teramura H, Yamamoto T, Komatsu H, Miura K, Ezura H, Nishida K, Ariizumi T, Kondo A (2017) Targeted base editing in rice and tomato using a CRISPR-Cas9 cytidine deaminase fusion. Nat Biotechnol 35:441–443

Shimatani Z, Fujikura U, Ishii H, Matsui Y, Suzuki M, Ueke Y, Taoka KI, Terada R, Nishida K, Kondo A (2018) Inheritance of co-edited genes by CRISPR-based targeted nucleotide substitutions in rice. Plant Physiol Biochem 131:78–83

Sonntag K, Wang Y, Wallbraun M (2004) A transformation method for obtaining marker-free plants based on phosphomannose isomerase. Acta Universitatis Latviensis Biol 676:223–226

Sripriya R, Raghupathy V, Veluthambi K (2008) Generation of selectable marker-free sheath blight resistant transgenic rice plants by efficient co-transformation of a cointegrate vector T-DNA and a binary vector T-DNA in one *Agrobacterium tumefaciens* strain. Plant Cell Rep 27:1635–1644

Srivastava V, Underwood JL, Zhao S (2017) Dual-targeting by CRISPR/Cas9 for precise excision of transgenes from rice genome. Plant Cell Tissue Org Cult 129:153–160

Tabatabaei I, Bosco CD, Bednarska M, Ruf S, Meurer J, Bock R (2018) A highly efficient sulfadiazine selection system for the generation of transgenic plants and algae. Plant Biotechnol J 7(3):638–649

Terada R, Nagahara M, Furukawa K, Shimamoto M, Yamaguchi K, Iida S (2010) Cre-lox mediated marker elimination and gene reactivation at waxy locus created in the rice genome based on strong positive-negative selection. Plant Biotechnol J 27:29–37

Verruto J, Francis K, Wang Y, Low MC, Greiner J, Tacke S, Kuzminov F, Lambert W, McCarren J, Ajjawi I, Bauman N, Kalb R, Hannum G, Moellering ER (2018) Unrestrained markerless trait stacking in *Nannochloropsis gaditana* through combined genome editing and marker recycling technologies. Proc Natl Acad Sci U S A 115:7015–7022

Wakita Y, Otani M, Iba K, Shimada T (1998) Co-integration, co-expression and co-segregation of an unlinked selectable marker gene and NtFAD3 gene in transgenic rice plants produced by particle bombardment. Genes Genet Syst 73:219–226

Wang XH, Zhang S, Hu D, Zhao X, Li Y, Liu T, Wang J, Hou X, Li Y (2014a) BcPMI2, isolated from non-heading Chinese cabbage encoding phosphomannoseisomerase, improves stress tolerance in transgenic tobacco. Mol Biol Rep 41:2207–2216

Wang Y, Zhang L, Li Y, Liu Y, Han L, Zhu Z, Wang F, Peng Y (2014b) Expression of Cry1Ab protein in a marker-free transgenic Bt rice line and its efficacy in controlling a target pest, *Chilo suppressalis* (Lepidoptera: Crambidae). Environ Entomol 43:528–536

Wang K, Liu H, Du L, Ye X (2017) Generation of marker-free transgenic hexaploid wheat via an *Agrobacterium*-mediated co-transformation strategy in commercial Chinese wheat varieties. Plant Biotechnol J 15:614–623

Woo HJ, Qin Y, Park SY, Park SK, Cho YG, Shin KS, Lim MH, Cho HS (2015) Development of selectable marker-free transgenic rice plants with enhanced seed tocopherol content through FLP/FRT-mediated spontaneous auto-excision. PLoS One 10:e0132667

Wright M, Dawson J, Dunder E, Suttie J, Reed J, Kramer C, Chang Y, Novitzky R, Wang H, Artim-Moore L (2001) Efficient biolistic transformation of maize (*Zea mays* L.) using the phosphomannoseisomerase gene, pmi, as the selectable marker. Plant Cell Rep 20:429–436

Xu M, Zhao S, Zhang Y, Yin H, Peng X, Cheng Z, Yang Z, Zheng J (2017) Production of marker-free transgenic rice (*Oryza sativa* L.) with improved nutritive quality expressing AmA1. Iran J Biotechnol 15:102–110

Youssef D, Nihou A, Partier A, Tassy C, Paul W, Rogowsky PM, Beckert M, Barret P (2018) Induction of targeted deletions in transgenic bread wheat (*Triticumaestivum* L.) using customized meganuclease. Plant Mol Biol Report 36:71–81

Zhang Y, Liang Z, Zong Y, Wang Y, Liu J, Chen K, Qiu JL, Gao C (2016) Efficient and transgene-free genome editing in wheat through transient expression of CRISPR/Cas9 DNA or RNA. Nat Commun 7:12617

Zhu Q, Yu S, Zeng D, Liu H, Wang H, Yang Z, Xie X, Shen R, Tan J, Li H, Zhao X, Zhang Q, Chen Y, Guo J, Chen L, Liu YG (2017) Development of "purple endosperm rice" by engineering anthocyanin biosynthesis in the endosperm with a high-efficiency transgene stacking system. Mol Plant 10:918–929

Zou X, Peng A, Xu L, Liu X, Lei T, Yao L, He Y, Chen S (2013) Efficient auto-excision of a selectable marker gene from transgenic citrus by combining the Cre/loxP system and ipt selection. Plant Cell Rep 32:1601–1613

Zubko E, Scutt C, Meyer P (2000) Intrachromosomal recombination between attP regions as a tool to remove selectable marker genes from tobacco transgenes. Nat Biotechnol 18:442–445

Applications of Genome Engineering/ Editing Tools in Plants

7

Chakravarthi Mohan, Priscila Yumi Tanaka Shibao, and Flavio Henrique Silva

Abstract

The advent of engineered nucleases such as zinc finger nucleases (ZFNs), transcription activator-like effector nucleases (TALENs) and clustered regularly interspaced short palindromic repeats/Cas9 (CRISPR/Cas9) has revolutionized targeted genome editing. CRISPR/Cas9-based editing system has surpassed its predecessors owing to its simplicity, versatility and efficiency. Therefore, it has become the most promising genome-editing tool in recent years which is evident through the increasing number of publications and in several organisms. This technology has profound applications in areas of functional genomics and crop improvement. Recent studies have proved that multiplex genome editing is possible not only in model crops but in major crops too. Unlike transgenic crops which yield random insertions of target genes, genome-editing tools enable targeted gene insertion at a specified locus (knock-in), deletion of desired genes from the genome (knockout) and also genome modification (replacement). In this context, this chapter describes in detail the various applications of genome-editing technologies in crop improvement and highlights how this tool has outwitted transgenic technology in recent times.

Keywords

Base editing · CRISPR/Cas9 · Genome editing · Knock-in · Knockout · TALENs · ZFNs

C. Mohan · P. Y. T. Shibao · F. H. Silva (✉)
Molecular Biology Laboratory, Department of Genetics and Evolution, Federal University of São Carlos, São Carlos, SP, Brazil
e-mail: dfhs@ufscar.br

R. Sathishkumar et al. (eds.), *Advances in Plant Transgenics: Methods and Applications*, https://doi.org/10.1007/978-981-13-9624-3_7

7.1 Introduction

Targeted genome editing relies on the use of site-specific nucleases that create precise modification at specific locations which are subsequently repaired by the cellular repair mechanisms of the host cell. Till date, three major programmable nucleases have been developed for genome editing, namely, zinc finger nucleases (ZFNs; Pabo et al. 2001), transcription activator-like effector nucleases (TALENs; Boch et al. 2009; Moscou and Bogdanove 2009) and RNA-guided nucleases (RGNs) from the clustered regularly interspaced short palindromic repeat (CRISPR) and CRISPR-associated proteins (Cas) system (Cong et al. 2013). Among these, ZFNs and TALENs require two different artificial proteins, comprising a DNA-binding protein domain and the C-terminal FokI nuclease domain for the targeted mutagenesis, and they have been used successfully in plants. However, these methods require exclusive design and are time consuming. On the contrary, CRISPR/Cas9 has emerged as the most efficient tool of choice for genome modification, owing to its versatility, design flexibility and efficiency. In recent times, it has been successfully reported in several plant species. With the advent of biotechnology, an array of transgenics has been developed for several traits. The era of genome editing has further enhanced and fastened the production of next-generation resistant crops which would benefit the increasing human population.

Efficient genome editing using CRISPR/Cas9 system in plants requires effective delivery of Cas9/sgRNA complex into the target cells. Unlike transgenic crops which yield random insertions of target genes, genome-editing tools enable targeted gene insertion at a specified locus (knock-in), deletion of desired genes from the genome (knockout) and also genome modification (replacement). Moreover, genome-edited crops differ from their transgenic counterparts in that they can be screened in the T_1 generation thereby, selecting the mutated crops that possess only the desired mutation and not the vector sequences. Due to this, GE crops do not fall upon the stringent regulations that the transgenic crops go through. In this context, this chapter highlights the various genome-editing systems and discusses the various applications of CRISPR-based genome engineering technique.

7.2 Genome-Editing Systems in a Nutshell

Genome editing, in the shape of transgenic breeding, has surfaced in the 1990s to directly overcome some of the most common plant domestication problems, as biotic and abiotic stresses and low yield of commercial products. However, only in the twenty-first century, the postgenomic era has emerged and revolutionized several fields in general and crop improvement in particular. Besides the availability of several crop genomes, which are more than 120 different species up to date, and transcriptomes, there is an alarming amount of data produced from targeted genome editing. Currently, the use of nucleases has allowed the targeted mutagenesis in several crops. With the aid of nucleases that can cleave specific double-strand DNA regions and the use of the own cell repair mechanism of damaged DNA, genome editing has reached a new level. Herein we will describe the different nucleases and their mechanism of action.

The double-stranded break (DSB) recruits DNA repair factors from the own cell that result in either homologous recombination (HR) repair or nonhomologous end joining (NHEJ), which are endogenous repairing mechanisms. HR uses a similar or homologous DNA as template for accurate repair. In this case, a DNA donor is provided along with the GE cassettes, and the DNA donor must possess homologous regions with appropriate length that can be recognized by the HR machinery. NHEJ does not require a DNA donor. Instead, once the DSB is made, cell machinery is able to resect part of the DNA and ligate the resulting ends, leading to deletion of part of the DNA that was broken (Sonoda et al. 2006). Depending on the aiming of gene edition, one pathway is desirable in detriment of the other. For instance, if the aiming is targeted mutation or gene insertion, HR is the desirable pathway. The main tools for targeted genome editing will be summarized below.

7.3 Zinc Finger Nucleases (ZFNs)

Zinc finger nucleases are synthetic enzymes that consist of a Cys_2-His_2 zinc finger and FokI domain fused together, so the zinc finger (ZF) domain can target the non-specific FokI domain to the desired DNA sequence (Kim et al. 1996). The dimerization of ZFN plays a key role in the process, and each ZFN binds to the DNA in a reverse configuration in a 5- to 6-bp long sequence flanking and allowing the FokI domain to digest the sequence in between, creating a DSB. It is possible to add different ZF domains, resulting in a heterodimer, which will increase site specificity to large genomes, as plants. ZFN constructions with heterodimers should be designed with a linker to enhance specificity. The use of linkers assists to improve the correct binding of ZFN with double-strand DNA, and the linkers' length and composition can be designed with the aid of bioinformatic tools considering the DNA sequence (CG content) and its secondary and tertiary structure (Moore et al. 2001; Cathomen and Keith Joung 2008; Anand et al. 2013). Currently, the ZFN can be either bought or screened from natural libraries.

After the DSB, the cell machinery will be responsible to the DNA repair, which can happen in a homologous or nonhomologous way. If HR is desired, it is necessary that an exogenous DNA sequence, called 'donor DNA', will be the template to the repair machinery, adding the desired sequence to be modified or added to the genome (Porteus and Carroll 2005; Weinthal et al. 2010). It is possible to buy one tailored-made nuclease screened trough traditional phage display techniques. Some research groups have also used the one- and two-hybrid technique and domain shuffling techniques to select the best ZF domain (Greisman and Pabo 1997; Hurt et al. 2003; Mani et al. 2005). Also, there are online tools available to assist the scientist to decide the best ZF sequence for their purpose. However, the main disadvantage from this system is still the high cost and effort to obtain the Zn domains, which led to the development of easier technologies.

7.4 Transcription Activator-Like Effector Nucleases (TALENs)

To overcome low specificity and to facilitate the nuclease screening, the transcription activator-like effector nucleases (TALENs) have arisen. These proteins are naturally found in *Xanthomonas* genus bacteria transcription activator-like effector (TALE) and a nonspecific FokI enzyme domain. TALEs are responsible for its pathogenicity against plants. After entering the cell, TALEs can reach the nucleus and activate gene expression by binding to specific effector DNA sequences that are often responsible for increasing the plant susceptibility to the pathogen. TALEs comprise a central domain of tandem repeats, which are the ones to be engineered for specificity, nuclear localization signals and acidic transcriptional activation domain (Boch et al. 2009). The central region may contain up to 30 repeats with 33–35 amino acid motifs that are largely conservative, with the exception at positions 12 and 13, named the repeat-variable diresidue (RVD), responsible for the TAL specificity. This specificity allows targeted modifications, since different RVDs recognize different DNA base pairs leading to a simple cypher that directs the DNA binding (Moscou and Bogdanove 2009). Each TAL repeat interacts with one base pair of the targeted DNA with different levels of preference, being specific for one base pair, two base pairs or unspecific, which allow the researches to create a full sequence of TALENs to interact in a more or less strong way to the DNA. Besides, TALENs must possess a minimum number of repeats to assure the strength of binding, and such strength must consider the positive and negative effect, which is the sum of attraction strength between the TALENs that interact with the DNA and the ones which does not match (Scholze and Boch 2010).

However, it was only when TALENs connected to the FokI enzyme were dimerized, the first attempt to gene editing was made (Christian et al. 2010). The principle of this gene editing technique combines the TALENs' precise direction, the FokI double-stranded break to the host cell HR or NHEJ machinery to create the desired mutation. It is relatively easy to design the appropriate TALEN once the target is determined, and this can be done with the aid of bioinformatics programs and tools (Cermak et al. 2011).

TALENs, as ZFNs, need chimeric nucleases connected to FokI domain and function as dimers. In addition, both systems need to be experimentally validated, and the DSB created is predicted by the linker length, which may cause off-target cleavage. Thus, both systems cannot be used to create more than one edition at a time (Bortesi and Fischer 2014). Therefore, new technologies have arisen to overcome some of these issues.

7.5 CRISPR/Cas9 System

Clustered regularly interspaced palindromic repeats (CRISPR) associated with the Cas9 nuclease, being called CRISPR/Cas9, is the recent update in genome editing that initiated in 2013. Unlike the previous ZFN and TALENs that use protein to interact with the targeted DNA sequence, CRISPR/Cas9 uses RNA as the guide that can bind to the DNA. CRISPR systems were first observed as adaptive immune system in bacteria and algae against the invasion of exogenous pathogens (Barrangou and Horvath 2012), and their potential to genome editing was readily hypothesized. Indeed, not long after the first observations, it was possible to prospect its use for such purpose. The CRISPR/Cas system is composed by a CRISPR RNA, Cas9 endonuclease protein and two RNA molecules and a trans-activating crRNA (tracrRNA) (Karvelis et al. 2013). CRISPR loci are composed of noncontiguous repeat, varying from 23 to 47 base pairs, separated by sequences called spacers that can vary from 21 to 72 bp (Horvath and Barrangou 2010).

The two RNA molecules act as a guide RNA (gRNA) that dictates the specificity and direction for the endonuclease cleavage (Mali et al. 2013). Furthermore, it is possible to combine more than one RNA molecule to produce a more specific RNA guide (gRNA) to the Cas9 nuclease (Mali et al. 2013). Since the gRNA is a simple molecule, the CRISPR/Cas9 system emerged as the simplest platform to engineer genomes among all gene editing methods (Sander and Joung 2014). Cas9 nuclease is responsible for creating the DSB, and host repair machinery is responsible for finalizing the editing process. NHEJ is the desired repair mechanism for gene silencing, while HR is the most desirable for gene editing as long as a template DNA sequence (donor template) is provided (Sander and Joung 2014). Unlike the previous systems, CRISPR/Cas9 enables multiplex gene editing thus the low-cost, rapidity, the fact it does not need protein engineering, effectiveness, easiness and broad applicability make this system the first choice of research. This system has effortlessly surpassed the other systems. Since this chapter is confined mainly to CRISPR/Cas9 applications, other cas variants are not discussed.

7.6 Delivery Systems

After deciding the best gene editing method and designing the appropriate editing cassette, an appropriate delivery system must be selected for crop transformation. Regarding mammalian cells, there are plenty of delivery systems that can be used: viral (e.g. adeno-associated virus (AAV) and *Lentivirus*), nonviral and physical systems (e.g. microinjection, electroporation and through DNA nano structures) (Liu et al. 2017).

For delivery in plants, two techniques that are widely used are biolistic transformation and *Agrobacterium*-mediated transformation (Ma et al. 2016a, b). Biolistic transformation can be successfully carried out using embryogenic calli as explants (e.g. rice, maize and many plant species). Agroinfiltration is carried out with callus or immature embryos (rice and maize) and floral dip in *Arabidopsis thaliana* (Ma et al. 2016a, b). On the other hand, recent efforts have been made to produce GE plants using viral delivery systems, and geminiviruses (family *Geminiviridae*) and *Tobacco rattle virus* (*Tobravirus*, family *Virgaridae*) are the outstanding choices for such purpose (Zaidi and Mansoor 2017). However, viral delivery system is still very restricted, mainly due to the small size of the viral cargo; for instance, Cas9 coding sequence (longer than 4 kb) is longer than the viral genome of geminivirus (3 kb) (Zaidi and Mansoor 2017). It is also possible to combine two different techniques (e.g. first transfecting protoplasts via agroinfiltration containing a Cas9 expression cassette and then transfecting these lines via biolistic cassettes coding the gRNAs to obtain the desired mutation).

In addition, it is possible to produce GE protoplasts without the introduction of exogenous DNA, only by transfecting preassembled complexes of purified Cas9 protein and gRNAs into different plant protoplasts (Woo et al. 2015).

7.7 Multiplex Genome Engineering

Multiplex genome engineering is defined as simultaneous engineering of several parts of the genome. Thanks to the simplicity of generating different gRNA, CRISPR/Cas9 platform has enabled the possibility of multiplex genome editing. This platform allowed the production of entire gRNA libraries that can be successfully scanned to obtain several mutation loci and also multiplexing only by delivering several gRNA in a single cell and using the cell machinery to produce and process tRNA (Cong et al. 2013; Xie et al. 2015; Thompson et al. 2017).

Multiplex genome editing was simultaneously reported by two different groups in 2013, proving it was feasible to be accomplished in different eukaryotic organisms (Cong et al. 2013; Mali et al. 2013). Gene editing in different loci was successfully obtained using gRNAs for each loci, designed protospacers under only one promoter. Thus, by using the designed gRNA with the aid of bioinformatics tools, HR events were enhanced in detriment of NHEJ, which often leads to unspecific and undesired mutations. It is possible to create different types of constructs aiming multiplex gene editing; one may create polycistronic gene under one single promoter (e.g. Pol III) or to co-inject several gene cassettes containing different single-guide RNAs to obtain multiplex editing (Wang et al. 2013; Xie et al. 2015).

However, due to the necessity of establishing cell lines expressing CRISPR/Cas9 system, plant genome editing relies on co-transfection of plasmid coding for Cas9 system along with that coding for the gRNAs. Co-transfection of plant codon-optimized Cas9 (pcoCas9) and different gRNAs was successfully achieved in *Arabidopsis thaliana*, tobacco and rice protoplasts which prove the feasibility of this kind of technology in plants (Gao et al. 2013; Li et al. 2013). There are also toolkits designed to create different transgenic plants (e.g. maize and *A. thaliana*) and protoplasts with multiplex gene editing (Xing et al. 2014; Ma et al. 2015; Ma and Liu 2016).

Generating gene editing cassettes relies basically in two different strategies: traditional cloning inserts into binary vector, which can be done by sequential rounds of cloning or with the use of multiple cloning enzymes that result in compatible sticky ends, or using the Golden Gate cloning system. In the first approach, it is possible to create up to six different sgRNAs into a vector (Zhang et al. 2015), and it is very time consuming. The latter approach uses the Golden Gate assembly technology (Engler et al. 2008) which relies on the presence of BsaI sites in the inserts and plasmids that are further digested and ligated in one single tube reaction. In Golden Gate assembly, several gRNAs are digested and ligated, generating gene cassettes targeting several loci; then, they are transfected together with another gene cassette coding for Cas9.

7.8 Merits of GE Technology Over the Dwindling Transgenic Approach

Transgenic approach shows high merits, and it is still widely used in the entire world. However, GE technique is rapidly increasing and surely will surpass transgenic approach because of its several advantages, fast results, possibility to edit, delete and insert genes, targeted mutagenesis, studying gene expression, function and regulation and for disease control, which are discussed in the application section. Specifically, CRISPR/Cas9-based GE is very simple, easy to design and versatile (Georges and Ray 2017). Although both genome editing and genetic engineering use plant transformation, they are strikingly different in several ways. Genome engineering/editing enables researchers for precise gene targeting, while genetic engineering always leads to random insertions of transgenes. Using engineered nucleases, it is possible to stack multiple genes in a simpler and efficient manner, whereas it's way more complex using transgenesis.

Unlike genetic engineering, precise base editing/substitution is possible using genome editing. Using ribonucleoproteins, it is possible to obtain DNA-free genome-edited events that are identical to their nontransformed counterparts. However, it is impossible with transgenic technology. Genome editing offers uniform gene expression throughout the plant genome and even in polyploid plants with complex genomes, whereas it is not feasible with genetic engineering. It is possible to obtain genome-edited lines devoid of foreign gene sequences in subsequent generations by segregation, while transgenic approach generates plants containing foreign genes that are stably inherited unless a marker-free approach is utilized. Altogether, genome editing using CRISPR/Cas9 enables rapid, efficient and versatile gene targeting and outwits transgenic approach in both functional genomics and crop improvement research. Moreover, it is assumed that they may not be regulated as GM crops since generally, no foreign gene is present in such genome-edited events unlike genetically modified crops.

7.9 Genome Engineering Applications in Crop Plants

Genome engineering technologies have led to precise manipulation of plant genomes which has invariably resulted in an array of applications that fall under two broad categories – functional genomics and crop improvement. However, recent studies have employed GE in various other applications such as live cell imaging, gene drives and CRISPR interference (CRISPRi), creating mutant libraries, and many are yet to be explored. Although the other editing technologies such as ZFNs and TALENs can also achieve precise targeting, the simplicity, versatility and multiplexing ability of CRISPR/Cas9 or cpf1 systems have enabled their widespread utility in editing plant genomes. In this context, this section of the chapter describes the various accomplishments of genome editing using ZFNs, TALENs and CRISPR/Cas9 systems in crop plants.

7.10 Functional Genomics

7.10.1 Loss of Function/Knockout

Loss of function tests are the simplest and widely carried out to gain understanding of a gene function. Using genome-editing tools, it is now facile to precisely knock out a target gene from a plant genome. With the simplicity and versatility of CRISPR/Cas9 system, gene knockouts have become its primary application in recent years. A simple search in Pubmed database with keywords 'CRISPR plant' yielded a whopping 818 results as of February 27, 2018, reiterating the diverse applications of this technique in plants. CRISPR/Cas9 knockout of the phytoene desaturase (*PDS*) gene that encodes for a key enzyme in carotenoid biosynthesis pathway and whose mutation causes photo bleaching or albino phenotypes has been optimized in a wide variety of plant species including *Arabidopsis thaliana* (Li et al. 2013), tobacco (Li et al. 2013; Nekrasov et al. 2013), rice (Shan et al. 2013), wheat (Upadhyay et al. 2013), tomato (Pan et al. 2016), potato (Wang et al. 2015), soybean (Du et al. 2016), petunia (Zhang et al. 2016a), sweet orange (Jia and Wang 2014), apple (Nishitani et al. 2016), grape (Nakajima et al. 2017), watermelon (Tian et al. 2017), banana (Kaur et al. 2018) and cassava (Odipio et al. 2017). Similarly, a mutant nonfunctional green fluorescent protein (GFP) that contained target site was also used for CRISPR/Cas9 knockout experiments in tobacco and rice (Jiang et al. 2013). Further, several key genes have been knocked out in a variety of plants either to decipher the gene function or for achieving specific trait improvement and have been described in Sect. 6.2. Recently, the function of a long noncoding RNA lncRNA1459 in tomato fruit ripening was elucidated using CRISPR/cas9 system (Li et al. 2018b).

7.10.2 Large Fragment Deletions

Large chromosomal deletions are useful in studies on gene clusters and noncoding RNA. TALENs and CRISPR have been employed to create large size deletions in plants (Christian et al. 2013; Shan et al. 2013). Zhou et al. (2014) produced large chromosomal deletions of 115–245 kb with high frequency using CRISPR/Cas9 in three gene clusters of rice and verified two of them in T_0 plants. A large genomic fragment consisting of three tandemly arranged CBF genes involved in cold acclimation was successfully deleted using CRISPR system (Zhao et al. 2016b). Large deletions have also been reported recently in *Arabidopsis* and tobacco (Ordon et al. 2017).

7.10.3 Gain of Function/Knock-in

Apart from gene knockouts, it is also feasible to insert desired genes into plant genome using genome-editing systems thereby modifying certain phenotypic traits. Shukla et al. (2009) used ZFNs to insert an herbicide-resistant gene in maize and also verified the gene transmission in T_1 generations. In the same year, targeted transgene integration in tobacco using ZFNs was also reported successfully (Cai et al. 2009). Using CRISPR/Cas9, a kanamycin-resistant gene was precisely inserted into the *ADH1* locus of *Arabidopsis* (Schiml et al. 2014), whereas hygromycin resistance gene was inserted in soybean by HR-mediated CRISPR/Cas9 system (Li et al. 2015). In addition, reporter genes have also been inserted via TALENs in tobacco and wheat protoplasts (Zhang et al. 2013; Wang et al. 2014). Recently, CRISPR/Cas9 was used to insert a bialaphos-resistant gene in maize (Svitashev et al. 2015). Although HR-mediated knock-in mechanism is precise and accurate, it needs to be improved further since it has lower frequency rate compared to NHEJ method.

7.10.4 Precise Base Editing

Precise base editing enables the creation or removal of a single-nucleotide variation which determines the genetic diversity within a species and thereby has extensive applications in genetics, therapeutics and agriculture. The base editor technology is a novel approach that utilizes CRISPR/Cas9 and directs cytidine deaminase activity to specific genomic loci thereby introducing precise cytidine to thymidine (C to T) alterations in organisms (Komor et al. 2016). Unlike genome editing, there is no necessity for DSB or donor DNA templates in this method. In addition, it yields reduced superfluous insertion/deletion mutations. Recently, multiple herbicide-resistance point mutations were induced in rice and tobacco by base editing technique (Shimatani et al. 2017). In another study, Zong et al. (2017) reported targeted conversion of cytosine to thymine with frequencies of up to 43.48% in rice, wheat and maize plants. This technique proves to be a promising alternative to HDR-mediated base replacement and would facilitate rapid and precise breeding in plants. Recently, cytosine deaminases with enhanced efficiencies and from several organisms have been reported for precise base editing (Lu and Zhu 2017; Ren et al. 2018).

7.10.5 Change of Function/Gene Replacement

Gene replacement or change of function is possible only when a donor with a similar terminal to that of target gene is available. Gene replacement was successfully demonstrated in tobacco and *Arabidopsis* (Weinthal et al. 2013) where the authors co-delivered a donor molecule with ZFN recognition sites (a promoter less hygromycin B phosphotransferase encoding gene) and an acceptor molecule (GFP), and the results showed that GFP gene sequences were eliminated completely and hygromycin resistance was recovered. Zhao et al. (2016a) utilized CRISPR/Cas9 and obtained 0.8% frequency of gene replacement in T_0 *Arabidopsis*. The replaced eGFP gene was also expressed actively in roots and leaves of T_1 generations. In rice, intron-mediated site-specific replacement of the endogenous 5-enolpyruvylshikima te-3-phosphate synthase (EPSPS) gene was reported with a frequency of 2%, and the replacements were also found to be inherited in the T_1 generation (Li et al. 2016).

7.10.6 Targeting miRNA

MicroRNAs (miRNAs) are short noncoding RNA molecules typically of 22 nucleotides found in several organisms and are involved in the regulation of diverse processes (Bartel 2009). Studies on the role of miRNAs have been increasing due to the advance in sequencing technologies. However, the functions of many miRNAs remain to be elucidated. CRISPR/Cas9 system can be efficiently employed to mutate miRNAs in order to reveal their regulatory networks in plants. Recently, Jacobs et al. (2015) used CRISPR/Cas9 to target two miRNAs – miR151 and miR1509 – and could obtain >95% frequencies confirming the applicability of this approach in studying the roles of miRNAs. Similarly, Zhou et al. (2017) targeted rice miRNAs via CRISPR to study the complex regulation and functions of such microRNAs.

7.10.7 Transcriptional Regulation

Precise regulation of transcription is a powerful tool in cellular engineering. Transcriptional activator-like effectors (TALEs) have been exploited earlier to generate chimeric sequence-specific transcriptional repressors in *Arabidopsis* (Mahfouz et al. 2012). Qi et al. (2013) demonstrated that a catalytically inactive Cas9 protein (dCas9) when combined with guide RNA can create a DNA recognition complex that can repress the transcription of the selected target genes. This simple method can be applied on genome-wide scale to perturb transcription process. Piatek et al. (2015) modified this method to target plant genome transcriptional regulation. They designed transcriptional activators and repressor and successfully proved that targeted transcriptional regulation is possible in tobacco. Using multiple gRNAs, dCas9 can be fused with activators or repressors or other functional domains and can be used effectively in controlling gene expression or methylation. This technique of using CRISPR/dCas9 opens new avenues and will defer answers for several questions pertaining to plant gene regulation.

7.10.8 Verification of DNA Motifs

Recently, CRISPR/Cas9 tool has been employed to verify the DNA motifs of *Arabidopsis thaliana* in their genomic context. Li et al. (2018a, b, c) disrupted CTCTGYTY motifs in DNA sequence-specific H3K27 demethylase gene (REF6) obtained through chromatin immunoprecipitation sequencing (ChIP-seq) and observed the loss of occupancy of the target gene. This strategy can be successfully applied for in vivo verification of DNA motifs identified by ChIP-seq in plants. In addition, the utility of CRISPR/Cas9 in epigenome editing and DNA labelling have also been reported in human cells (Hilton et al. 2015; Ma et al. 2016a). However, these domains are yet to be explored in the case of plants.

7.11 Crop Improvement

Apart from functional genomics, genome-editing tools have also been successfully exploited in order to create crop plants with higher yield, resistant to biotic and abiotic stresses, improved quality and enhanced nutritional value. This section of the chapter reviews the recent studies on the exploitation of genome engineering in crop improvement.

7.11.1 Yield

Quality traits such as yield are regulated by multiple gene loci and thereby difficult to manipulate using conventional breeding methods. CRISPR/Cas9 system has been used to target negative regulators of grain weight leading to enhanced production. Xu and his colleagues (2016) applied CRISPR/Cas9 to mutate three major genes that negatively regulated grain weight in rice and could obtain mutations with higher efficiencies reiterating that multiplex genome editing is a promising technique to understand quantitative traits and in generation and pyramiding of desired alleles. In a similar study, *Hd2*, *Hd4* and *Hd5* genes that negatively control heading date trait in rice were targeted via CRISPR/Cas9 system, and early maturing rice could be obtained (Li et al. 2017). Earlier, Shan et al. (2015) reported the first successful multiplex gene knockout in crop plants wherein they used TALENs to disrupt *OsBADH2*, *OsCKX2* and *OsDEP1* – genes related to rice quality and yield; the desired mutations were acquired in successive generations. Similar work in bread wheat was carried out where *TaGASR7* a gene that negatively regulates grain weight was disrupted (Zhang et al. 2016b). Apart from mutations in coding genes that vary protein expression, *cis*-acting regulatory elements in promoters also affect the gene expression. Rodriguez-Leal et al. (2017) created novel *cis*-regulatory alleles for three genes that regulate fruit size, inflorescence architecture and plant growth habit in tomato by applying CRISPR/Cas9-mediated multiplex genome editing and achieved stabilized promoter alleles in successive generation through segregation thereby confirming the efficiency of CRISPR/Cas9 system in enhancing crop yield.

7.11.2 Biotic Stress

Plants being sessile are constantly exposed to several biotic stresses that lead to severe losses in crop yield. Disease-related genes have been edited in plants via genome-editing technologies. Li et al. (2012) used TALENs to disrupt a specific susceptibility gene *OsSWEET14* in rice in order to confer resistance to bacterial blight disease caused by *Xanthomonas oryzae*. In another study, both TALENs and CRISPR were used to successfully target three homoeoalleles encoding mildew resistance locus (MLO) proteins in the hexaploid bread wheat, and the resultant mutant lines exhibited broad-spectrum resistance to powdery mildew (Wang et al. 2014). These findings have established the possibility of multiplex genome editing in polyploid crops as well. A similar study was reported by Nekrasov et al. (2017) where the authors targeted MLO genes through CRISPR/Cas9 system using double sgRNA approach and generated tomato resistant to powdery mildew in a short span of 10 months. *CsLOB1* was previously revealed to confer susceptibility for citrus canker, a deadly disease in sweet orange caused by *Xanthomonas citri* (Hu et al. 2014). Recently, two independent research groups applied CRISPR technology to knock out *CsLOB1* thereby generating canker-resistant citrus varieties (Jia et al. 2017; Peng et al. 2017). Similarly, CRISPR/Cas9-mediated knockout of *OsERF922* gene gave way to mutant lines displaying enhanced resistance to rice blast disease (Wang et al. 2016).

Apart from bacterial and fungal pathogens, crop yield is also affected by various viral diseases. Mahfouz group delivered sgRNAs specific to tomato yellow leaf curl virus (TYLCV) in tobacco plants. When the plants were challenged with TYLCV, delayed or decreased viral DNA was observed thereby eliminating or attenuating the TYLCV symptoms of infection (Ali et al. 2015). Thus, this report demonstrated the successful in planta viral interference against TYLCV. Similarly, Baltes et al. (2015) could engineer resistant tobacco plants against bean yellow dwarf virus through CRISPR/Cas9 technique. Eukaryotic translation initiation factor *eIF4E* plays a significant role in plant potyvirus interaction (Mazier et al. 2011). CRISPR was used to knock out the *eIF4E* gene in cucumber. T3 progenies exhibited resistance to various viruses such as cucumber vein yellowing virus, zucchini yellow mosaic virus and papaya ring spot mosaic virus (Chandrasekaran et al. 2016). Similarly, Pyott et al. (2016) introduced point mutations in *eIF4E* of Arabidopsis to obtain transgene-free lines resistant to turnip mosaic virus. These results establish the potential of genome-editing tools in developing virus-resistant crops.

7.11.3 Abiotic Stress

In maize, *ARGOS8* is a negative regulator of ethylene response with relatively lower endogenous expression, and its overexpression enhanced grain yield under drought-stressed conditions (Shi et al. 2015). The same research group used CRISPR/Cas9 to generate novel variants of *ARGOS8* with elevated constitutive transcript expression and had higher grain yield in field under drought stress (Shi et al. 2017). The results obtained in this study have successfully proved the efficacy of genome editing to generate novel allelic variations with the aim of developing crops with enhanced abiotic stress tolerance. Osakabe et al. (2016) designed a genome-editing system based on Cas9 driven by a tissue-specific promoter and a tru-RNA complex which can generate mutations without any off-target effects in plant genomes. They verified the system by mutating the *OST2* gene yielding enhanced stomatal response in *Arabidopsis* thereby demonstrating the felicity of genome-editing tools in developing stress-resistant crop plants. In another study, *OsAnn3*, a rice annexin gene which is involved in protecting plants from environmental stresses, was knocked out via CRISPR/cas9 system (Shen et al. 2017). The capacity to withstand cold reduced in the mutant lines revealing the function of *OsAnn3* in cold tolerance. Unlike biotic stress, genome editing for abiotic stress tolerance is still in its infancy, and we hope that sustained research in the near future would give way to more stress-tolerant crop plants.

Genome-editing tools have also been employed in developing herbicide-resistant plants. Earlier, zinc finger nucleases were used to target acetolactate synthase genes (*ALS*) in order to bestow resistance to sulphonylurea and imidazolinone herbicides (Townsend et al. 2009). Recently, Butler et al. (2015) used CRISPR/Cas9 to target *ALS* gene in potato and were able to generate targeted mutations in stable plants. Multiple point mutations were introduced in rice *ALS* gene using CRISPR/cas9-mediated homologous recombination (Sun et al. 2016). Hahn et al. (2017) applied CRISPR/Cas9 to disrupt the alternative visual marker genes *BAR* (bialaphos resistance) and *GL1* (glabrous1) in *Arabidopsis*. The results established the suitability of *GL1* as visual marker, while *BAR* was not suitable since non-chimeric plants could not be obtained.

7.11.4 Augmented Nutritional Value

Improving the nutritional properties of plants has been possible by transgenesis and RNA silencing approaches. With the advent of genome engineering, it is now facile to develop plants with enhanced nutrition. Clasen et al. (2016) knocked out the vacuolar invertase (*VInv*) gene in potato through TALENs, and the resulting mutant lines exhibited reduced levels of reducing sugars which could be beneficial for storage and processing. Sun et al. (2017) used CRISPR/Cas9 to target the starch-branching enzyme genes *SBEI* and *SBEII* in rice wherein the *SBEII* mutants displayed higher amylose content and resistant starch thereby providing improved health benefits. CRISPR/cas9-mediated knockout of fatty acid desaturation 2 (*FAD2*) genes in *Arabidopsis* and *Camelina sativa* increased the oleic acid content of Camelina seeds from 16% to 50% of fatty acid composition with decreased long-chain polyunsaturated fatty acids such as linoleic acid and linolenic acid (Jiang et al. 2017). Hanania et al. (2017) knocked out the β(1,2)- xylosyltranferase (*XylT*) and α(1,3)-fucosyltransferase (*FucT*) genes via CRISPR/Cas9 in tobacco BY2 cell lines. Transformation of the knocked out lines with recombinant DNaseI resulted in production of DNaseI free from xylose and/or ficose residues thereby furnishing a viable biofactory for the production of pharmaceutical products.

Metabolic Engineering.

CRISPR/Cas9 technology has also been utilized to manipulate metabolic pathways in plants. Alagoz et al. (2016) knocked out *4'OMT2* a gene that regulates the biosynthesis of benzylisoquinoline alkaloids (BIAs) in opium poppy which resulted in transgenic plants with reduced BIA biosynthesis. Recently, multiplex CRISPR targeting was used to manipulate the γ-aminobutyric acid (GABA) shunt in tomatoes wherein five genes were targeted yielding enhanced GABA content in the mutant lines (Li et al. 2018c). These findings prove the utility of CRISPR-mediated multiplex editing to manoeuvre complex metabolic pathways in crop plants.

7.11.5 DNA-Free Genome Alteration

As CRISPR/Cas9 system uses vectors to deliver Cas9 and gRNAs into plant cells, there is a possibility of insertion of vector sequences and selectable marker genes in the plant genome. This may further lead to gene disruption, mosaic plants and off targeting. Moreover, it may also lead to reduced targeting efficiency. To overcome these issues, Cas9 ribonucleoproteins (RNPs) are currently being used. These RNPs contain pre-integrated Cas9 and gRNAs and are delivered as RNA molecules into plant cells with equal efficacy to plasmid vector-based expression systems. Cas9 RNPs have been successfully targeted into protoplasts of *Arabidopsis*, tobacco, lettuce and rice with up to 46% frequencies in the regenerated plants (Woo et al. 2015). By this method DNA-free genetically edited crop plants with insertions/deletions that were identical to the naturally occurring genetic variations could be acquired. Svitashev et al. (2016) reported the successful biolistic delivery of RNPs in maize embryos. Another study targeted RNPs into protoplasts of grape and apple to obtain powdery mildew-resistant grapes and fire blight-resistant apples, respectively (Malnoy et al. 2016). Liang et al. (2017) described a rapid protocol of introducing RNPs in wheat through which 4–5 independent mutants could be obtained from a transformation of 100 immature embryos. Recently Kim et al. (2017) delivered recombinant cpf1 proteins guided by a single CRISPR RNA (crRNA) in protoplasts of soybean and tobacco establishing that Cpf1-crRNA complex is also an effective DNA-free genome-editing tool. This method of using RNPs is particularly fascinating since it does not introduce any foreign DNA thereby alleviating regulatory concerns about genetically modified crops.

7.12 Other Applications

Continuous research is being carried out worldwide in order to explore the utility of CRISPR/Cas9 technology in different fields some of which are discussed in this section.

7.12.1 CRISPR Interference and Activation

Recently CRISPR system was repurposed to repress gene transcription (Qi et al. 2013; Larson et al. 2013). This system called as CRISPR interference or CRISPRi uses dCas9 to achieve stable and efficient repression of transcription of any gene. CRISPRi is advantageous over RNAi since it can be used to target multiple genes simultaneously. The dCas9 was fused to SRDX repression domain of the ERF transcription factor and acted as transcriptional repressor (Piatek et al. 2015). However, the method has certain limitations such as availability of less target sites, off-target effects and variation between genes which should be addressed for its widespread exploitation in plants. On contrary, CRISPR activation system or CRISPRa can be used to enhance gene expression up to 1000-fold with a single gRNA (Gilbert et al. 2014). Altogether, developing genome-wide CRISPRi and CRISPRa libraries would certainly be a promising solution to explicate functional genomics studies in plants.

7.12.2 Live Cell Imaging

Genome organization is critical for cellular function and maintenance. Knowledge on how genomes are organized inside the nucleus is indispensable to understand gene regulation during developmental stages of organisms. To this context, live cell imaging is a promising option and has been possible earlier through fluorescent labelled zinc finger proteins. In vivo visualization is possible using fluorescent in situ hybridization (FISH). However, it is unfeasible to image live cells as the method necessitates tissue fixation. Recently, Dreissig et al. (2017) developed a CRISPR/dCas9-based imaging technique to visualize telomere repeats in live tobacco leaf cells. Further, they confirmed the possibility of exploiting CRISPR/dCas9 coupled with fluorescence labelled proteins to study DNA-protein interactions in vivo.

7.12.3 Gene Drives

Recently, Cas9 has been used to create gene drives in order to control the spread of vector-borne diseases. Gantz et al. (2015) used CRISPR/Cas9-mediated HR repair to drive target-specific gene conversion at a high efficiency in transgene heterozygotes of *Anopheles stephensi* in order to control malaria. Esvelt et al. (2014) reviewed the potentiality of RNA-guided gene drives that are capable of spreading genomic alteration through sexually reproducing populations. The possibility of using CRISPR-based gene drives to control pest population in agriculture to enhance crop production has been discussed in recent times (Courtier-Orgogozo et al. 2017; Scott et al. 2017).

7.12.4 Chromatin Remodelling

Chromatin remodelling is the dynamic modification of chromatin architecture in order to allow access of the condensed genomic DNA to the transcriptional regulatory machinery and thereby control gene expression. However, there are no broad methodologies to study or modify the chromatin loops. Recently, Morgan et al. (2017) applied CRISPR/Cas9 system (CLOuD9) to modify the chromatin loop structure and thereby alter gene expression. By using this technique, it is possible to elucidate the function of chromatin organization in transcriptional control. Liu et al. (2018) used CRISPR activation to target and remodel endogenous *Oct4* and *Sox2* to generate induced pluripotent stem cells.

7.12.5 Creation of Mutant Libraries

Generation of large scale mutant libraries is inevitable to study both functional genomics and genetic improvement of plants. Creation of mutant libraries through existing methods such as T-DNA insertions and TILLING (targeting-induced local lesions in genomes) is time consuming and labour intensive. CRISPR/Cas9-based mutant libraries have been developed in cultured eukaryotic cells (Shalem et al. 2015). Recently, Meng et al. (2017) applied CRISPR to create a high-quality genome-wide mutant library in rice demonstrating the robustness of this system. Jacobs et al. (2017) transformed pooled CRISPR T-DNAs into tomato and generated mutant libraries rapidly and with minimal transformation efforts.

7.13 Regulating Crops with Edited Genomes

Now that genome editing poses various advantages over transgenic crops, how these crops will be regulated worldwide is not clear yet. In general, two approaches of regulation exist – process-based regulation (e.g. European Union, Brazil) that considers the techniques used to develop new crops and product-based regulation (e.g. the United States, Canada) that focuses mainly on the final product and the risks it poses. Since the genetic changes introduced by GE tools are precise, researchers advocate a product-based regulation for the genome-edited crops. Huang et al. (2016) have recommended five-step preliminary principles for the regulation of genome-edited crops which would benefit mankind if adopted. Stringent regulation will in turn affect the development costs and delay the commercialization of GE crops. Moreover, public acceptance and consumption of genome engineered crop plants also play a critical role. It is thus necessary for the scientific community to convey the benefits and significance of genome-editing technologies to the general public in a convincing manner to make them embrace this novel advancement for crop improvement. Owing to the potentials of this approach, considerable progress is being carried out by regulatory authorities throughout the world in order to devise a sensible and pragmatic regulation for the genome-edited crops. Earlier this year, the United States Department of Agriculture (USDA) stated that it will not impose further regulations on certain categories of crops developed through genome-editing techniques (USDA 2018). It is important to note that the USDA had already approved the non-browning mushroom, oil-rich Camelina and a corn variety (Wx1 gene knockout) which were developed through CRISPR-Cas9-based genome editing.

7.14 Conclusion

The advent of CRISPR/Cas9-based genome-editing technology has undoubtedly opened several opportunities in both functional genomics and crop improvement research. CRISPR-mediated knock-in and knockout approaches have significant potentials in comprehending gene functions. Advances in next-generation sequencing have facilitated sequencing of a large amount of plant species. Such genome information can be particularly useful for genome-editing applications. Nevertheless, the rapid advances and innovations in genome editing will facilitate development of more stress-tolerant crops in the near future. Currently, several problems such as increasing population, limited cultivable lands, climate change and global food crisis pose a serious threat to mankind, and crop genome editing will certainly be one of the promising solutions in the future. Although GE poses certain limitations, stringent efforts are being taken up by researchers to resolve them, and we hope they would be overcome in the near future.

Acknowledgements CM and PYTS gratefully acknowledge the São Paulo Research Foundation (FAPESP) for the postdoctoral research grant (Proc. 2015/10855-9) and doctoral grant (Proc. No. 2017/16118-1), respectively. FHS is a recipient of a Research Productivity Scholarship from the National Council for Research and Development (CNPq #311745/2013-0).

References

Alagoz Y, Gurkok T, Zhang B, Unver T (2016) Manipulating the biosynthesis of bioactive compound alkaloids for next-generation metabolic engineering in opium poppy using CRISPR-Cas 9 genome editing technology. Sci Rep 6:309–310

Ali Z, Abulfaraj A, Idris A, Ali S, Tashkandi M, Mahfouz MM (2015) CRISPR/Cas9-mediated viral interference in plants. Genome Biol 16:238–249

Anand P, Schug A, Wenzel W (2013) Structure based design of protein linkers for zinc finger nuclease. FEBS Lett 587(19):3231–3235

Baltes NJ, Hummel AW, Konecna E, Cegan R, Bruns AN, Bisaro DM, Voytas DF (2015) Conferring resistance to geminiviruses with the CRISPR–Cas prokaryotic immune system. Nat Plants 1:15145

Barrangou R, Horvath P (2012) CRISPR: new horizons in phage resistance and strain identification. Annu Rev Food Sci Technol 3:143–162

Bartel DP (2009) MicroRNAs: Target recognition and regulatory functions. Cell 136(2):215–233

Boch J, Scholze H, Schornack S, Landgraf A, Hahn S, Kay S, Lahaye T, Nickstadt A, Bonas U (2009) Breaking the code of DNA binding specificity of TAL-type III effectors. Science 326(5959):1509–1152

Bortesi L, Fischer R (2014) The CRISPR/Cas9 system for plant genome editing and beyond. Biotechnol Adv 33(1):41–52

Butler NM, Atkins PA, Voytas DF, Douches DS (2015) Generation and inheritance of targeted mutations in potato (*Solanum tuberosum* L.) using the CRISPR/Cas System. PLoS One 10:e0144591

Cai C, Doyon Y, Ainley W, Miller J, DeKelver R et al (2009) Targeted transgene integration in plant cells using designed zinc finger nucleases. Plant Mol Biol 69:699–709

Cathomen T, Keith Joung J (2008) Zinc-finger Nucleases: The Next Generation Emerges. Mol Ther 16(7):1200–1207

Cermak T, Doyle EC, Christian M, Wang L, Zhang Y, Schmidt C, Baller JA, Somia NV, Bogdanove AJ, Voytas DF (2011) Efficient design and assembly of custom TALEN and other TAL effector-based constructs for DNA targeting. Nucleic Acids Res 39(12):e82

Chandrasekaran J, Brumin M, Wolf D, Leibman D, Klap C, Pearlsman M et al (2016) Development of broad virus resistance in non-transgenic cucumber using CRISPR/Cas9 technology. Mol Plant Pathol 17:1140–1153

Christian M, Cermak T, Doyle EL, Schmidt C, Zhang F, Hummel A, Bogdanove AJ, Voytas DF (2010) Targeting DNA Double-Strand Breaks with TAL Effector Nucleases. Genetics 186(2):757–761

Christian M, Qi Y, Zhang Y, Voytas DF (2013) Targeted mutagenesis of Arabidopsis thaliana using engineered TAL Effector Nucleases (TALENs). G3 (Bethesda) 3:1697–1705

Clasen BM, Stoddard TJ, Luo S, Demorest ZL, Li J, Cedrone F, Tibebu R, Davison S, Ray EE, Daulhac A, Coffman A, Yabandith A, Retterath A, Haun W, Baltes NJ, Mathis L, Voytas DF, Zhang F (2016) Improving cold storage and processing traits in potato through targeted gene knockout. Plant Biotechnol J 14:169–176

Cong L, Ran FA, Cox D, Lin S, Barretto R, Habib N, Hsu PD, Wu X, Jiang W, Marraffini LA, Zhang F (2013) Multiplex genome engineering using CRISPR/Cas systems. Science 339(6121):819–823

Courtier-Orgogoz V, Morizot B, Boete C (2017) Agricultural pest control with CRISPR based gene drive: time for public debate: Should we use gene drive for pest control? EMBO Rep 18:878–880

Dreissig S, Schiml S, Schindele P, Weiss O, Rutten T, Schubert V, Gladilin E, Mette MF, Puchta H, Houben A (2017) Live-cell CRISPR imaging in plants reveals dynamic telomere movements. Plant J 91(4):565–573

Du H, Zeng X, Zhao M, Cui X, Wang Q, Yang H, Cheng H, Yu D (2016) Efficient targeted mutagenesis in soybean by TALENs and CRISPR/Cas9. J Biotechnol 217:90–97

Engler C, Kandzia R, Marillonnet S (2008) A one pot, one step, precision cloning method with high throughput capability. PLoS One 3(11):e3647

Esvelt KM, Smidler AL, Catteruccia F, Church GM (2014) Concerning RNA-guided gene drives for the alteration of wild populations. elife 3:e03401

Gantz VM, Jasinskiene N, Tatarenkova O, Fazekas A, Macias VM, Bier E, James AA (2015) Highly efficient Cas9-mediated gene drive for population modification of the malaria vector mosquito Anopheles stephensi. Proc Natl Acad Sci U S A 112:6736–6743

Georges F, Ray H (2017) Genome editing of crops: a renewed opportunity for food security. In: (Ed.). GM Crops Food 8(1):1–12

Gilbert LA, Horlbeck MA, Adamson B (2014) Genome-scale CRISPR-mediated control of gene repression and activation. Cell 159(3):647–661

Greisman HA, Pabo CO (1997) A general strategy for selecting high-affinity zinc finger proteins for diverse DNA target sites. Science 275(5300):657–661

Hahn F, Mantegazza O, Greiner A, Hegemann P, Eisenhut M, Weber APM (2017) An efficient visual screen for CRISPR/Cas9 activity in Arabidopsis thaliana. Front Plant Sci 8:39

Hanania U, Ariel T, Tekoah Y, Fux L, Sheva M, Gubbay Y, Weiss M, Oz D, Azulay Y, Turbovski A, Forster Y, Shaaltiel Y (2017) Establishment of a tobacco BY2 cell line devoid of plant-specific xylose and fucose as a platform for the production of biotherapeutic proteins. Plant Biotechnol J 15:1120–1129

Hilton IB, D'Ippolito AM, Vockley CM, Thakore PI, Crawford GE, Reddy TE et al (2015) Epigenome editing by a CRISPR-Cas9-based acetyltransferase activates genes from promoters and enhancers. Nat Biotechnol 33:510–517

Horvath P, Barrangou R (2010) CRISPR/Cas, the immune system of bacteria and archaea. Science 327(5962):167–170

Hu Y, Zhang J, Jia H, Sosso D, Li T, Frommer WB, Yang B, White FF, Wang N, Jones JB (2014) Lateral organ boundaries 1 is a disease susceptibility gene for citrus bacterial canker disease. Proc Natl Acad Sci U S A 111:521–529

Huang S, Weigel D, Beachy RN, Li J (2016) A proposed regulatory framework for genome-edited crops. Nat Genet 48(2):109–111

Hurt JA, Thibodeau SA, Hirsh AS, Pabo CO, Joung JK (2003) Highly specific zinc finger proteins obtained by directed domain shuffling and cell-based selection. Proc Natl Acad Sci U S A 14;100(21):12271–12276

Jacobs TB, LaFayette PR, Schmitz RJ, Parrott WA (2015) Targeted genome modifications in soybean with CRISPR/Cas9. BMC Biotechnol 15:16

Jacobs TB, Zhang N, Patel D, Martin GB (2017) Generation of a collection of mutant tomato lines using pooled CRISPR libraries. Plant Physiol 174:2033–2037

Jia H, Wang N (2014) Targeted genome editing of sweet orange using Cas9/sgRNA. PLoS One 9(4):e93806

Jia H, Zhang Y, Orbovic V, Xu J, White F, Jones J, Wang N (2017) Genome editing of the disease susceptibility gene CsLOB1 in citrus confers resistance to citrus canker. Plant Biotechnol J 15:817–823

Jiang W, Zhou H, Bi H, Fromm M, Yang B, Weeks DP (2013) Demonstration of CRISPR/Cas9/sgRNA-mediated targeted gene modification in Arabidopsis, tobacco, sorghum and rice. Nucl Acids Res 41:e188

Jiang WZ, Henry IM, Lynagh PG, Comai L, Cahoon EB, Weeks DP (2017) Significant enhancement of fatty acid composition in seeds of the allohexaploid, *Camelina sativa*, using CRISPR/Cas9 gene editing. Plant Biotechnol J 15:648–657

Karvelis T, Gasiunas G, Siksnys V (2013) Programmable DNA cleavage *in vitro* by Cas9. Biochem Soc Trans 41(6):1401–1406

Kaur N, Alok A, Shivani KN, Pandey P, Awasthi P, Tiwari S (2018) CRISPR/Cas9-mediated efficient editing in phytoene desaturase (PDS) demonstrates precise manipulation in banana cv. Rasthali genome. Func Int Genomics 18(1):89–99

Kim YG, Cha J, Chandrasegaran S (1996) Hybrid restriction enzymes: zinc finger fusions to Fok I cleavage domain. Proc Natl Acad Sci U S A 93(3):1156–1160

Kim H, Kim S-T, Ryu J, Kang B-C, Kim J-S, Kim S-G (2017) CRISPR/Cpf1-mediated DNA-free plant genome editing. Nat Commun 8:14406

Komor AC, Kim YB, Packer MS, Zuris JA, Liu DR (2016) Programmable editing of a target base in genomic DNA without double-stranded DNA cleavage. Nature 33:420–424

Larson MH, Gilbert LA, Wang X, Lim WA, Weis JS (2013) CRISPR interference (CRISPRi) for sequence-specific control of gene expression. Nat Protoc 8:2180–2196

Li T, Liu B, Spalding MH, Weeks DP, Yang B (2012) High-efficiency TALEN-based gene editing produces disease-resistant rice. Nat Biotechnol 30:390–392

Li JF, Norville JE, Aach J, McCormack M, Zhang D, Bush J, Church GM, Sheen J (2013) Multiplex and homologous recombination-mediated genome editing in Arabidopsis and *Nicotiana benthamiana* using guide RNA and Cas9. Nat Biotechnol 31:688–691

Li Z, Liu ZB, Xing A, Moon BP, Koellhoffer JP, Huang L et al (2015) Cas9-guide RNA directed genome editing in soybean. Plant Physiol 169(2):960–970

Li J, Meng X, Zong Y, Chen K, Zhang H, Liu J, Li J, Gao C (2016) Gene replacements and insertions in rice by intron targeting using CRISPR-Cas9. Nat Plants 2:16139

Li X, Zhou W, Ren Y, Tian X, Lv T, Wang Z, Fang J, Chu C, Yang J, Bu Q (2017) High-efficiency breeding of early-maturing rice cultivars via CRISPR/Cas9-mediated genome editing. J Genet Genomics 44(3):175–178

Li C, Chen C, Chen H, Wang S, Chen X, Cui Y (2018a) Verification of DNA motifs in Arabidopsis using CRISPR/Cas9-mediated mutagenesis. Plant Biotechnol J 8:1446–1451

Li R, Fu D, Zhu B, Luo Y, Zhu H (2018b) CRISPR/Cas9-mediated mutagenesis of lncRNA1459 alters tomato fruit ripening. Plant J. (in press)

Li R, Li R, Li X, Fu D, Zhu B, Tian H, Luo Y, Zhu H (2018c) Multiplexed CRISPR/Cas9-mediated metabolic engineering of γ-aminobutyric acid levels in Solanum lycopersicum. Plant Biotechnol J 16:415–427

Liang Z, Chen K, Li T, Yi Zhang Y, Wang Y, Zhao Q, Liu J, Zhang H, Liu C, Ran Y, Gao C (2017) Efficient DNA-free genome editing of bread wheat using CRISPR/Cas9 ribonucleoprotein complexes. Nat Commun 8:14261

Liu C, Zhang L, Liu H, Cheng K (2017) Delivery strategies of the CRISPR-Cas9 gene-editing system for therapeutic applications. J Control Release 266:17–26

Liu P, Chen M, Liu Y, Qi LS, Ding S (2018) CRISPR-based chromatin remodeling of the endogenous Oct4 or Sox2 locus enables reprogramming to pluripotency cell. Stem Cell 22(2):252–261

Lu Y, Zhu JK (2017) Precise editing of a target base in the rice genome using a modified CRISPR/Cas9 System. Mol Plant 10:523–525

Ma X, Liu YG (2016) CRISPR/Cas9-based multiplex genome editing in monocot and dicot plants. Curr Protoc Mol Biol 115:31.6.1–31.6.21

Ma X, Zhang Q, Zhu Q, Liu W, Chen Y, Qiu R, Wang B, Yang Z, Li H, Lin Y, Xie Y, Shen R, Chen S, Wang Z, Chen Y, Guo J, Chen L, Zhao X, Dong Z, Liu YG (2015) A Robust CRISPR/Cas9 system for convenient, high-efficiency multiplex genome editing in monocot and dicot plants. Mol Plant 8(8):1274–1284

Ma H, Tu LC, Naseri A, Huisman M, Zhang S, Grunwald D, Pederson T (2016a) Multiplexed labeling of genomic loci with dCas9 and engineered sgRNAs using CRISP Rainbow. Nat Biotechnol 34(5):528–531

Ma X, Zhu Q, Chen Y, Liu YG (2016b) CRISPR/Cas9 platforms for genome editing in plants: developments and applications. Mol Plant 9(7):961–974

Mahfouz MM, Li L, Piatek M, Fang X, Mansour H, Bangarusamy DK, Zhu J-K (2012) Targeted transcriptional repression using a chimeric TALE-SRDX repressor protein. Plant Mol Biol 78(3):311–321

Mali P, Yang L, Esvelt KM, Aach J, Guell M, DiCarlo JE, Norville JE, Church GM (2013) RNA-guided human genome engineering via Cas9. Science 339(6121):823–826

Malnoy M, Viola R, Jung MH, Koo OJ, Kim S, Kim JS et al (2016) DNA-free genetically edited grapevine and apple protoplast using CRISPR/Cas9 ribonucleoproteins. Front Plant Sci 7:1904

Mani M, Kandavelou K, Dy FJ, Durai S, Chandrasegaran S (2005) Design, engineering, and characterization of zinc finger nucleases. Biochem Biophys Res Commun 335(2):447–457

Mazier M, Flamain F, Nicolaï M, Sarnette V, Caranta C (2011) Knock-down of both eIF4E1 and eIF4E2 genes confers broad-spectrum resistance against Potyviruses in Tomato. PLoS One 6(12):e29595

Meng X, Yu H, Zhang Y, Zhuang F, Song X, Gao S, Gao C, Li J (2017) Construction of a genome-wide mutant library in rice using CRISPR/Cas9. Mol Plant 10(9):1238–1241

Moore M, Klug A, Choo Y (2001) Improved DNA binding specificity from polyzinc finger peptides by using strings of two-finger units. Proc Natl Acad Sci U S A 98(4):1437–1441

Morgan SL, Mariano NC, Bermudez A, Arruda NL, Wu F, Luo Y, Shankar G, Jia L, Chen H, Hu JF, Hoffman AR, Huang CC, Pitteri SJ, Wang KC (2017) Manipulation of nuclear architecture through CRISPR-mediated chromosomal looping. Nat Commun 8:15993

Moscou MJ, Bogdanove AJ (2009) A simple cipher governs DNA recognition by TAL effectors. Science 326:1501

Nakajima I, Ban Y, Azuma A, Onoue N, Moriguchi T, Yamamoto T et al (2017) CRISPR/Cas9-mediated targeted mutagenesis in grape. PLoS One 12(5):e0177966

Nekrasov V, Staskawicz B, Weigel D, Jones JDG, Kamoun S (2013) Targeted mutagenesis in the model plant *Nicotiana benthamiana* using Cas9 RNA-guided endonuclease. Nat Biotechnol 31:691–693

Nekrasov V, Wang C, Win J, Lanz C, Weigel D, Kamoun S (2017) Rapid generation of a transgene-free powdery mildew resistant tomato by genome deletion. Sci Rep 7:482

Nishitani C, Hirai N, Komori S, Wada M, Okada K, Osakabe K, Yamamoto T, Osakabe Y (2016) Efficient genome editing in apple using a CRISPR/Cas9 system. Sci Rep 6:31481

Odipio J, Alicai T, Ingelbrecht I, Nusinow DA, Bart R, Taylor NJ (2017) Efficient CRISPR/Cas9 genome editing of phytoene desaturase in cassava. Front Plant Sci 8:1780

Ordon J, Gantner J, Kemna J, Schwalgun L, Reschke M, Streubel J, Boch J, Stuttmann J (2017) Generation of chromosomal deletions in dicotyledonous plants employing a user-friendly genome editing toolkit. Plant J 89(1):155–168

Osakabe Y, Watanabe T, Sugano SS, Ueta R, Ishihara R, Shinozaki K, Osakabe K (2016) Optimization of CRISPR/Cas9 genome editing to modify abiotic stress responses in plants. Sci Rep 6:26685

Pabo C, Peisach E, Grant R (2001) Design and selection of novel Cys2 His2 zinc finger proteins. Annu Rev Biochem 70:313–340

Pan C, Ye L, Qin L, Liu X, He Y, Wang J, Chen L et al (2016) CRISPR/Cas9-mediated efficient and heritable targeted mutagenesis in tomato plants in the first and later generations. Sci Rep 6:24765

Peng A, Chen S, Lei T, Xu L, He Y, Wu L et al (2017) Engineering canker resistant plants through CRISPR/Cas9-targeted editing of the susceptibility gene CsLOB1 promoter in citrus. Plant Biotechnol J 15(12):1509–1519

Piatek A, Ali Z, Baazim H, Li L, Abulfaraj A, Al-Shareef S (2015) RNA-guided transcriptional regulation in planta via synthetic dCas9-based transcription factors. Plant Biotechnol J 13:578–589

Porteus MH, Carroll D (2005) Gene targeting using zinc finger nucleases. Nat Biotechnol 23(8):967–973

Pyott DE, Sheehan E, Molnar A (2016) Engineering of CRISPR/ Cas9- mediated potyvirus resistance in transgene-free Arabidopsis plants. Mol Plant Path 17:1276–1288

Qi LS, Larson MH, Gilbert LA, Doudna JA, Weissman JS, Arkin AP et al (2013) Repurposing CRISPR as an RNA-guided platform for sequence specific control of gene expression. Cell 152:1173–1183

Ren B, Yan F, Kuang Y, Li N, Zhang D, Zhou X, Lin H, Zhou H (2018) Improved base editor for efficiently inducing genetic variations in rice with CRISPR/Cas9-guided hyperactive hAID mutant. Mol Plant doi 11:623. https://doi.org/10.1016/j.molp.2018.01.005

Rodríguez-Leal D, Lemmon ZH, Man J, Bartlett ME, Lippman ZB (2017) Engineering quantitative trait variation for crop improvement by genome editing. Cell 171(2):470–480

Sander JD, Joung JK (2014) CRISPR-Cas systems for editing, regulating and targeting genomes. Nat Biotechnol 32(4):347–355

Schiml S, Fauser F, Puchta H (2014) The CRISPR/Cas system can be used as nuclease for in planta gene targeting and as paired nickases for directed mutagenesis in Arabidopsis resulting in heritable progeny. Plant J 80:1139–1150

Scholze H, Boch J (2010) TAL effector-DNA specificity. Virulence 1(5):428–432

Scott MJ, Gould F, Lorenzen MD, Grubbs N, Edwards OR, O'Brochta DA (2017) Agricultural production: assessment of the potential use of Cas9-mediated gene drive systems for agricultural pest control. J Respon Innov 5:98–120

Shalem O, Sanjana NE, Zhang F (2015) High-throughput functional genomics using CRISPR-Cas9. Nat Rev Genet 16:299–311

Shan Q, Wang Y, Li J, Zhang Y, Chen K, Liang Z, Zhang K, Liu J, Xi JJ, Qiu JL, Gao C (2013) Targeted genome modification of crop plants using a CRISPR-Cas system. Nat Biotechnol 31(8):686–688

Shan Q, Zhang Y, Chen K, Zhang K, Gao C (2015) Creation of fragrant rice by targeted knockout of the OsBADH2 gene using TALEN technology. Plant Biotechnol J 13:791–800

Shen C, Que Z, Xia Y, Tang N, Li D, He R, Cao M (2017) Knock out of the annexin gene OsAnn3 via CRISPR/Cas9-mediated genome editing decreased cold tolerance in rice. J Plant Biol 60(6):539–547

Shi J, Habben JE, Archibald RL, Drummond BJ, Chamberlin MA, Williams RW, Lafitte HR et al (2015) Overexpression of ARGOS genes modifies plant sensitivity to ethylene, leading to improved drought tolerance in both Arabidopsis and maize. Plant Physiol 169:266–282

Shi J, Gao H, Wang H, Lafitte HR, Archibald RL, Yang M, Hakimi SM, Mo H, Habben JE (2017) ARGOS8 variants generated by CRISPR-Cas9 improve maize grain yield under field drought stress conditions. Plant Biotechnol J 15:207–216

Shimatani Z, Kashojiya S, Takayama M, Terada R, Arazoe T, Ishii H, Teramura H, Yamamoto T, Komatsu H, Miura K (2017) Targeted base editing in rice and tomato using a CRISPR-Cas9 cytidine deaminase fusion. Nat Biotechnol 35:441

Shukla VK, Doyon Y, Miller JC, DeKelver RC, Moehle EA, Worden SE, Mitchell JC et al (2009) Precise genome modification in the crop *Zea mays* using zinc finger nucleases. Nature 459:437–441

Sonoda E, Hochegger H, Saberi A, Taniguchi Y, Takeda S (2006) Differential usage of non-homologous end-joining and homologous recombination in double strand break repair. DNA Repair (Amst) 5(9–10):1021–1029

Sun Y, Zhang X, Wu C, He Y, Ma Y, Hou H, Guo X et al (2016) Engineering herbicide-resistant rice plants through CRISPR/Cas9-mediated homologous recombination of acetolactate synthase. Mol Plant 9:628–631

Sun Y, Jiao G, Liu Z, Zhang X, Li J, Guo X, Du W, Du J, Francis F, Zhao Y et al (2017) Generation of high-amylose rice through CRISPR/Cas9-mediated targeted mutagenesis of starch branching enzymes. Front Plant Sci 8:298

Svitashev S, Young JK, Schwartz C, Gao H, Falco SC, Cigan AM (2015) Targeted mutagenesis, precise gene editing and site-specific gene insertion in maize using Cas9 and guide RNA. Plant Physiol 169:931–945

Svitashev S, Schwartz C, Lenderts B, Young JK, Cigan AM (2016) Genome editing in maize directed by CRISPR-Cas9 ribonucleoprotein complexes. Nat Commun 7:13274

Thompson DB, Aboulhouda S, Hysolli E, Smith CJ, Wang S, Castanon O, Church GM (2017) The future of multiplexed eukaryotic genome engineering. ACS Chem Biol 13(2):313–325

Tian S, Jiang L, Gao Q, Zhang J, Zong M, Zhang H, Ren Y, Guo S, Gong G, Liu F, Xu Y (2017) Efficient CRISPR/Cas9-based gene knockout in watermelon. Plant Cell Rep 36(3):399–406

Townsend JA, Wright DA, Winfrey RJ, Fu F, Maeder ML, Joung JK, Voytas DF (2009) High-frequency modification of plant genes using engineered zinc-finger nucleases. Nature 459:442–445

Upadhyay SK, Kumar J, Alok A, Tuli R (2013) RNA-Guided Genome Editing for Target Gene Mutations in Wheat. G3: Genes, Genomes. Genetics 3(12):2233–2238

USDA (2018). Available at https://www.usda.gov/media/press-releases/2018/03/28/secretary-perdue-issues-usda-statement-plant-breeding-innovation

Wang H, Yang H, Shivalila CS, Dawlaty MM, Cheng AW, Zhang F, Jaenisch R (2013) One-step generation of mice carrying mutations in multiple genes by CRISPR/Cas-mediated genome engineering. Cell 153(4):910–918

Wang Y, Cheng X, Shan Q, Zhang Y, Liu J, Gao C, Qiu JL (2014) Simultaneous editing of three homoeoalleles in hexaploid bread wheat confers heritable resistance to powdery mildew. Nat Biotechnol 32:947–951

Wang S, Zhang S, Wang W, Xiong X, Meng F, Cui X (2015) Efficient targeted mutagenesis in potato by the CRISPR/Cas9 system. Plant Cell Rep 34(9):1473–1476

Wang F, Wang C, Liu P, Lei C, Hao W, Gao Y, Liu YG, Zhao K (2016) Enhanced rice blast resistance by CRISPR/Cas9-targeted mutagenesis of the ERF transcription factor gene OsERF922. PLoS One 11:e0154027

Weinthal D, Tovkach A, Zeevi V, Tzfira T (2010) Genome editing in plant cells by zinc finger nucleases. Trends Plant Sci 15(6):308–321

Weinthal DM, Taylor RA, Tzfira T (2013) Non homologous end joining-mediated gene replacement in plant cells. Plant Physiol 162:390–400

Woo JW, Kim J, Kwon SI, Corvalán C, Cho SW, Kim H, Kim SG, Kim ST, Choe S, Kim JS (2015) DNA-free genome editing in plants with preassembled CRISPR-Cas9 ribonucleoproteins. Nat Biotechnol 33(11):1162–1164

Xie K, Minkenberg B, Yang Y (2015) Boosting CRISPR/Cas9 multiplex editing capability with the endogenous tRNA-processing system. Proc Natl Acad Sci U S A 112(11):3570–3575

Xing HL, Dong L, Wang ZP, Zhang HY, Han CY, Liu B, Wang XC, Chen QJ (2014) A CRISPR/Cas9 toolkit for multiplex genome editing in plants. BMC Plant Biol 14:327

Xu R, Yang Y, Qin R, Hao L, Qiu C, Li L, Wei P, Yang J (2016) Rapid improvement of grain weight via highly efficient CRISPR/Cas9-mediated multiplex genome editing in rice. J Genet Genomics 43:529–532

Zaidi SS, Mansoor S (2017) Viral vectors for plant genome engineering. Front Plant Sci 11(8):539

Zhang Y, Zhang F, Li X, Baller JA, Qi Y, Starker CG, Bogdanove AJ, Voytas DF (2013) TALENs enable efficient plant genome engineering. Plant Physiol 161(1):20–27

Zhang Z, Mao Y, Ha S, Liu W, Botella JR, Zhu JK (2015) A multiplex CRISPR/Cas9 platform for fast and efficient editing of multiple genes in Arabidopsis. Plant Cell Rep 35(7):1519–1533

Zhang Y, Liang Z, Zong Y, Wang Y, Liu J, Chen K et al (2016a) Efficient and transgene-free genome editing in wheat through transient expression of CRISPR/Cas9 DNA or RNA. Nat Commun 7:12617

Zhang B, Yang X, Yang C, Li M, Guo Y (2016b) Exploiting the CRISPR/Cas9 System for targeted genome mutagenesis in Petunia. Sci Rep 6:20315

Zhao Y, Zhang C, Liu W, Gao W, Liu C, Song G, Li WX et al (2016a) An alternative strategy for targeted gene replacement in plants using a dual-sgRNA/Cas9 design. Sci Rep 6:23890

Zhao C, Zhang Z, Xie S, Si T, Li Y, Zhu JK (2016b) Mutational evidence for the critical role of CBF genes in cold acclimation in Arabidopsis. Plant Physiol 171:2744–2759

Zhou H, Liu B, Weeks DP, Spalding MH, Yang B (2014) Large chromosomal deletions heritable small genetic changes induced by CRISPR/Cas9 in rice. Nucleic Acids Res 42:10903–10914

Zhou J, Deng K, Cheng Y, Zhong Z, Tian L, Tang X, Tang A, Zheng X, Zhang T, Qi Y, Zhang Y (2017) CRISPR-Cas9 based genome editing reveals new insights into MicroRNA function and regulation in rice. Front Plant Sci 8:1598

Zong Y, Wang Y, Li C, Zhang R, Chen K, Ran Y, Qiu JL, Wang D, Gao C (2017) Precise base editing in rice, wheat and maize with a Cas9- cytidine deaminase fusion. Nat Biotechnol 35:438–440

High-Throughput Analytical Techniques to Screen Plant Transgenics

8

Furkan Ahmad and Pragadheesh VS

Abstract

Analytical chemistry plays a vital role in the screening of marker compounds in transgenic plants. A transgenic plant, due to the introduction of a new trait, possesses a different chemical composition, from the naturally occurring population, which needs a systematic evaluation to quantify one or more marker compounds. Advancement in the analytical technique aids the high-throughput screening of plant metabolites in large number with less time. Sample preparation is an important task prior to the analysis, which focuses the enrichment of marker compound in the extract having a complex mixture of plant metabolites. Selection of appropriate extraction method based on the target compound and analysis is essential for the biological or chemical assays. Hyphenation of liquid chromatography (LC) and gas chromatography (GC) with mass spectrometry (MS) employs the separation and identification of transgenic plant metabolites in real time. High-resolution mass spectrometry imaging, nuclear magnetic resonance (NMR), and Fourier transform infrared (FT-IR) spectroscopy are the other spectroscopic techniques which are widely used for the screening of new plant metabolites. The chapter "High-throughput analytical techniques to screen plant transgenics" highlights the recent developments on the screening of transgenic plants including the sample preparation, analysis, interpretation, and elucidation of the structure of plant metabolites with future prospective.

F. Ahmad
Department of Natural Products, National Institute of Pharmaceutical Education and Research, SAS Nagar, Punjab, India

P. VS (✉)
Organic and Medicinal Chemistry Division, CSIR-Indian Institute of Chemical Biology, Kolkata, India
e-mail: vspragadheesh@gmail.com

R. Sathishkumar et al. (eds.), *Advances in Plant Transgenics: Methods and Applications*, https://doi.org/10.1007/978-981-13-9624-3_8

Keywords
Analytical chemistry · Transgenic plants · Plant metabolites · Liquid chromatography · Gas chromatography · Nuclear magnetic resonance · Fourier transform infrared spectroscopy

8.1 Introduction

Economically valued plant products often vary in their cost due to various reasons like annual production, climatic factors, and seasonal changes and also due to the low abundance of the potential compounds (Daviet and Schalk 2010). Several methods including organic synthesis of plant products are in progress, still the complex plant products are hard to synthesize through chemical reactions (Kraft and Denizot 2013). Thus metabolic engineering of plant biosynthetic pathway grows considerably with the help of molecular biology tools. Introduction of a foreign (trans) gene from other species or organism or manipulation of the gene in the same organism by biotechnological means makes a new transgenic plant with required traits. Analytical chemistry is imperative in the field of biotechnology as the result of every experiment needs to be confirmed finally by an analytical technique. Transgenic plants are analyzed to assess the change in the metabolites due to the introduction of a new gene in the wild type. Analytical instruments play a major role in the development and screening of plant transgenics by analyzing monoclonal antibody, metabolites, etc. The inclusion of a trans gene makes the plants to generate new proteins that give pest resistance and disease tolerance and also to biosynthesize economically valued products in plants (Anklam et al. 2002).

Among the several transformation methods for the development of transgenics, the simple method like floral dip method produces numerous seeds (Clough and Bent 1998). Production of thousands of seeds as reported in Clough and Bent (1998) makes the screening of thousand plants grown from the respective seeds an intricate task. Thus high-throughput analytical techniques which can screen the metabolites in the plantlets grown from the seeds will be a precious tool in the transgenic plant development.

Analytical techniques are also needed to analyze the transgenic plants to assess the compliance of the transgenic plants with respect to the various regulations. Regulatory authorities are concerned about the safety of genetically modified organism; therefore, well-established analytical techniques are required which also have the ability to detect any unintended effect without alteration of any metabolites in genetically modified plants (Piccioni et al. 2009).

Every metabolomic analysis comprises of three steps: the first one is the sample preparation; instrumental analysis, and data acquisition is the second step, and the third step is data processing and interpretation of results (Kim and Verpoorte 2010). This chapter discusses mainly the analytical techniques used in transgenic plants and throws some light on the sample preparation for the metabolites.

8.2 High-Throughput Chromatography and Mass Spectrometry for the Screening of Transgenic Plants

8.2.1 Sample Preparation for Plant Metabolite Analysis in Chromatography

Sample preparation is the major task in the analysis of any plant metabolites, which separates the analyzable metabolites from other matrices of the plants. Sample preparation methods can also be developed to separate or enhance one group of metabolites from others. The result of any analytical technique can be authentic, only if the extracted sample (in a large sample set) is representative and homogenous. Due to the similarity of the metabolites in plants, the preparation of the good quality sample is a challenging and critical task. Every extraction and analyzing method and even with the sophisticated techniques employed for the detection of the metabolites are not well applied to all the metabolites, and it shows some degree of inclination toward the extraction of some metabolite to another group (Hall 2006).

Extraction of non-polar to mid-polar compounds such as monoterpenoids, sesquiterpenoids, diterpenoids, phenylpropanoids, fatty acid derivatives, hydrocarbons, etc. can be extracted using pentane, hexane, or dichloromethane and analyzed in gas chromatography. Mid-polar to polar compounds such as flavonoids, sterols, some alkaloids, etc. can be extracted using ethyl acetate or chloroform, and polar compounds such as glycosides can be extracted using methanol and/or water. A lot of intermediate compounds based on the functional groups present in the metabolite will get extracted in the solvents of similar polarity. Hence, there is no general rule for the extraction of all plant secondary metabolites. Further, every plant will have a different profile of metabolites, and a method developed for a plant cannot be applied to another plant. Interestingly, the same species of plants collected from the different geographical regions may also have different metabolites and extraction method that need to be optimized separately (Figueiredo et al. 2008; Verma et al. 2013). For the volatile metabolites, hydro- or steam distillation, extraction from the headspace of the sample using solid-phase microextraction (SPME) or solid phase extraction, or direct headspace sampling can be used to avoid the nonvolatile matrix in the analysis. There is no universal solvent which can dissolve all metabolites of the plant to extract completely, and the solubility of every component in the solvent also plays a major role in the extraction. For instance, a low solubility compound requires a lot of solvents to extract it completely, and additionally due to the high dissolution of a polar or non-polar compound, the solvents polarity get altered and further extraction may also get modulated. Due to the un-employability of a general method to different plants, automation cannot be possible or lead to erroneous sample preparation; thus most of the sample preparation is manual and high labor-oriented process. Yet, utmost care needs to be taken in the sample preparation process, and the method should be simple and fast to avoid the degradation of samples during the process of handling large samples manually (Kim and Verpoorte 2010).

8.2.2 Gas Chromatography and Mass Spectrometry

8.2.2.1 Gas Chromatography in Transgenic Plant Analysis

Gas chromatography (GC) is one of the widely used analytical techniques, for the study of volatile metabolites and in the development of plant transgenics. GC is mainly used for the analysis of mono-, sesqui-, and diterpenoid compounds, fatty acid and its derivatives, and other low-boiling compounds of the plants. Simple instrumentation, easy to operate, various methodologies to identify the compounds, effective coupling with a mass spectrometer (MS), availability of several sample introduction techniques (Bonn 2008) made GC and GC-MS a very important instrument in the plant secondary metabolite analysis (Jonsson et al. 2005). Several identification methodologies, viz., mass spectral data search/match, relative retention index, and co-injection of standards (Adams 1995; Toth and Praszna 1998; Zellner et al. 2008), are available to interpret the identity of the metabolites. Further, gas chromatography supports various below-mentioned sample introduction techniques for the volatile metabolites:

1. Thermal desorption (TD) – In TD, the sample is introduced into a thermal desorption tube at room temperature or lower temperature, and the temperature of thermal desorption tube is increased to extract the volatile metabolites from the nonvolatile matrix. The extracted volatile metabolites are trapped in an ultra-low temperature cool injection system (up to −100 °C) till the temperature of the thermal desorption reaches the maximum and ensures the extraction of all volatile metabolites from the sample. The cool injection system is now heated ballistically to a higher temperature to pass all the metabolites into the column for the separation. This is the best method to study the volatiles which may degrade at a higher temperature of the conventional GC inlet systems (Aharoni et al. 2003; Seethapathy et al. 2008; Kallenbach et al. 2014; Nordström et al. 2017).
2. Solid-phase microextraction (SPME) – In SPME, an aliquot of the volatile sample is collected on a thin layer of the solid phase extraction material and desorbed into the GC inlet by heating. The solid phase material comprises of polydimethylsiloxane, divinylbenzene, carboxen, polyacrylate, polyethylene glycol, etc. This method is very widely used in the transgenic plant analysis of terpenoids and other volatile metabolites (Meija et al. 2002; Aharoni et al. 2003).
3. Gas chromatography also compliments other sample introduction techniques such as purge and trap, headspace sampling, direct inlet, etc. (Aharoni et al. 2003).

Numerous researches have been carried by co-expressing the terpene synthase enzymes with isoprenoid pathway carbon flux controlling enzymes to increase the amount of accumulation of terpenoids in transgenic plants (Wu et al. 2006; Daviet and Schalk 2010). Several transgenic plants were screened using GC and GC-MS for the qualitative and quantitative analysis of the metabolites intent to modulate in the transgenic plants. Gas chromatography alone or in combination with mass spectrometry is also used for the analysis of volatile secondary metabolites in hairy root

Table. 8.1 Examples of transgenic plants analyzed for terpenoids using gas chromatography and gas chromatography-mass spectrometry

S. no	Transgenic crops	Metabolites studied	Technique	Column	References
1.	*Nicotiana tabacum*	Terpenoids	GC-MS	Rtx-5	Wu et al. (2006)
2.	*Nicotiana tabacum*	Amorpha-4,11-diene	GC-MS	HP-5MS HP-INNOwax	Wallaart et al. (2001)
3.	*Arabidopsis thaliana*	Linalool and Nerolidol Germacrene A	GC-MS	HP-5MS DB-Wax	Aharoni et al. (2003)
4.	*Mentha piperita*	Menthofuran	GC-MS	AT-1000(PEG)	Mahmoud and Croteau (2001)
5.	*Petunia hybrid*	(*S*)-linalool synthase	GC-MS	HP-5MS	Lucker et al. (2001)
6.	*Eucalyptus grandis x Eucalyptus urophylla*	Terpenoids	GC-MS	HP-5MS	Lucas et al. (2017)
7.	Soybean	Campesterol, stigmasterol, and β-sitosterol	GC	PTE5	Mounts et al. (2009)
8	*Mentha arvensis*	Total monoterpenes	GC	Supelcowax	Diemer et al. (2001)
9.	*Nicotiana tabacum*	Limonene	GC-MS	HP-5MS	Ohara et al. (2003)
10.	*Dianthus caryophyllus*	Linalool and linalool oxide	GC-MS	HP-5MS	Lavy et al. (2002)
11.	*Lycopersicon esculentum*	Linalool and 8-hydroxy linalool	GC	HP-5	Lewinsohn et al. (2001)

cultures (Szarka et al. 2007; Abdelkader and Lockwood 2016). Most of the GC and GC-MS analysis of transgenic plants are to quantify the monoterpenoids and sesquiterpenoid enhancement in the crop varieties (Table 8.1).

8.2.2.2 Applications of Gas Chromatography-Mass Spectrometry in Transgenic Plant

Nicotiana tobaccum **(Tobacco)**

GC-MS was used for the quantification of patchoulol and amorpha-4,11-diene in tobacco plants and found that more than 1000-fold increase in these sesquiterpenoid compounds when the pathway is engineered in tobacco plants (Wu et al. 2006). An amorpha-4,11-diene synthase gene was transformed into tobacco plants to elicit the biosynthesis of amorpha-4,11-diene, a precursor of artemisinin, very effective antimalarial drug (Wallaart et al. 2001). A GC-MS analysis was performed for identification, and the amount of amorpha-4,11-diene was quantified as 0.2–1.7 ng per g leaf tissue in the transgenic plant.

Mentha piperita **(Peppermint)**
When a 1-deoxy-D-xylulose-5-phosphate reductoisomerase cDNA and menthofuran synthase cDNA were transformed to a peppermint plant, two transgenic plant groups were regenerated, and one group exhibited a 50% increase in the yield of essential oil. Further, to analyze the composition of the terpenoids in essential oil, a GC-MS analysis was performed and found no alteration in the monoterpene composition of the essential oil of the transgenic plants in comparison with the wild type (Mahmoud and Croteau 2001).

Arabidopsis thaliana
The heterologous expression of terpenoids in *A. thaliana* was demonstrated by transforming strawberry *FaNES1* gene. Results show that flowering plants of *Arabidopsis* produced linalool and nerolidol, whereas these compounds were not emitted from leaves. SPME was used to sample the volatiles from the vegetative parts of the plant and analyzed in GC-MS, and the results show that plants produce varying levels of linalool and low level of nerolidol. Headspace volatiles produced by the transgenic plants was sampled for a period of 24 hours by solid phase extraction method using Tenax trapping and eluted with pentane/diethyl ether (4:1). In the same study, an enantioselective multidimensional GC-MS was employed to characterize the enantiomer of linalool using a 2,3-di-*O*-ethyl-6-*O*-tert-butyldimethylsilyl-β-cyclodextrin-based chiral column (Aharoni et al. 2003).

Petunia hybrida
When *Petunia hybrida* transgenic plant was generated by transferring (*S*)-linalool synthase gene of *Clarkia breweri*, it was analyzed using a GC by sampling the volatiles by SPME. Linalool was not obtained in the headspace sampling of vegetative tissues, whereas it was detected when the transgenic plant tissues are digested with $CaCl_2$ solution and analyzed in GC-MS (Lucker et al. 2001).

8.2.2.3 High-Throughput GC and GC-MS Screening of Plant Transgenics

Combination of sample preparation method, fast temperature programming, and quick analysis method makes the high-throughput analysis possible for gas chromatography (Gras et al. 2017). Several hyphenation techniques like headspace, thermal desorption, SPME, etc. will also speed up the sample preparation process. For a high-throughput analysis, fast cool down rate and quick equilibration are essential. Modern advanced hardware technologies in chromatography help in the faster cooling rate of 300 °C to 40 °C in 3 minutes. Aharoni et al. (2003) employed SPME for a semiquantitative headspace measurement using an autosampler for exposure of the SPME to the headspace of the sample and GC-MS inlet. This method made possible the analysis of limonene and β-myrcene and screening of around 40 different lines of transgenic plants in overnight. GC/GC-MS technique is not only limited to volatiles; other metabolites such as carbohydrates, steroids, organic acids, phenols, amides, sulfides, and amino acids can also be analyzed in gas chromatography with appropriate chemical derivatization. Several derivatization reagents are commercially available for the silylation, acylation, alkylation, and esterification (Hall 2006). Derivatization not only enhances the volatilization of the metabolites, but it also provides better peak shape and detector response in GC analysis.

8.2.3 Liquid Chromatography and Mass Spectrometry

8.2.3.1 Liquid Chromatography in Transgenic Plant Analysis

Development of transgenic in food crops shows higher importance in comparison with the cash crops as the world's population is increasing at a steady rate, and it is expected to become double by 2050. Food security will be the most important social concern as the same area of cultivable land has to cater the need of expected population (Herrera-estrella 2000). The majority of the transgenic crops developed are to enhance the final products (mainly primary metabolites) of the food crops. High-performance liquid chromatography (HPLC) is widely used to analyze all types of metabolites in plants and has an extensive application in transgenic plant development and screening.

8.2.3.2 Applications of LC and LC-MS in Transgenic Plant Screening

HPLC with ultraviolet (UV) diode array detector (DAD) was used for the qualitative and quantitative measurement of flavonoids in transgenic sugarcane and compared with cultivated sugarcane (Colombo et al. 2006). Thus, HPLC-UV/DAD can be used for the HPLC analysis of flavonoids and its glycosides in transgenic plants. The technique comprises of capillary electrophoresis (CE) along with the time of flight mass spectrometer (TOF MS) which can be successfully employed in high-throughput analysis to tentatively identify metabolites in plants with the support of isotopic pattern and electrophoretic mobility of the compounds (Garcia-Villalba et al. 2008). Normal phase and reverse phase HPLC was used to quantify the tocopherols and phospholipids in genetically modified soybeans and found that more than 25–50 times increase in tocopherols and a meager change in the phospholipid content (Mounts et al. 2009). Another study used a CE-TOF MS in unifying with Fourier transform-ion cyclotron resonance-mass spectrometry (FT-ICR-MS), and various metabolites were identified and the same can be potentially used as a biomarker to identify the transgenic plants from their parental lines (non-transgenic) (Leon et al. 2009).

8.2.3.3 High Throughput Analysis of HPLC, LC-MS and MS Techniques in Plant Transgenics

Liquid chromatography in combination with mass spectrometry (LC-MS) is a versatile technique which can be used to analyze a large group of compounds varying in polarity, size, etc. and has no limitation. For the high-throughput analysis of plant transgenics, liquid chromatography can be used in combination with high sensitive and high-resolution detector like ultraviolet detection or mass detector. Due to the mass accuracy and high-resolution of FT-ICR-MS, a large number of metabolites can be screened using this technique in high-throughput analysis (Leon et al. 2009). Similarly, advancement in liquid chromatography, viz., ultra-performance liquid chromatography (UPLC), operates with smaller columns which in turn reduce the time of analysis and are very effective in high-throughput screening of metabolites in transgenic plants (Boba et al. 2011; Mierziak et al. 2014; Yuan et al. 2015). Development in column chemistry aids the separation of similar compounds with a greater resolution which will further reduce the time of analysis (Hall 2006; Foubert et al. 2010).

8.3 High-Throughput Nuclear Magnetic Resonance (NMR) Spectroscopy for the Screening of Transgenic Plants

8.3.1 Introduction

Nuclear magnetic resonance (NMR) spectroscopy is one of the most popular analytical techniques which is used for the structural determination and quantification of chemical and biological compounds up to micro-molar concentration. NMR spectroscopy gives unbiased results for metabolites.

Recently, NMR spectroscopy as a high-throughput analytical technique has been widely used for several applications such as biosynthetic pathway studies using ^{13}C, metabolite profiling, metabolite dynamics, and structure elucidation of novel compounds. NMR spectroscopy is simple, automated, and non-destructive to the sample which can be used for the further experiments. NMR spectroscopy has a major advantage over several analytical techniques of metabolite analysis due to its incomparable reproducibility and robustness. Keun et al. (2002) have performed a comparison in cross-laboratory with different strengths of NMR magnetic field which showed a coefficient of variation of $<$ 2%. From several years, NMR spectroscopy has been used for qualitative and quantitative assessment of metabolites and requires minimal sample preparation, and the samples do not need any derivatization. The main disadvantage of the NMR technique is relatively low sensitivity toward low-abundance metabolites present in the sample (Zhang et al. 2012; Holmes et al. 2006). High-field strengths of frequency up to 900 MHz, a recent advancement in NMR spectrometers, are available. The more commonly available NMR spectrometers which are ranging from 300 to 600 MHz are mostly used for metabolic investigation. With increasing field strength, NMR gives better resolution and minimum overlapping signals in the spectrum (Krishnan et al. 2005).

^1H NMR spectra of crude extracts are overcrowded with overlapping signals of metabolites, so it is very difficult to assign correct signals, and in most of the cases, it is not feasible. Overlapping of the ^1H NMR signals can be resolved by 2D-NMR experiments. The 2D-NMR techniques can be applied for identification of metabolites which include correlation methods such as ^{1}H-^{1}H correlations like correlation spectroscopy (COSY), total correlation spectroscopy (TOCSY), and ^{13}C-^{1}H correlations like heteronuclear multiple bond correlation (HMBC) and heteronuclear single quantum coherence spectroscopy (HSQC). J-resolved NMR technique is the most popular among 2D-NMR spectroscopy. It is used for peak assignment of metabolites due to its simplicity and short acquisition time. J-resolved NMR gives two different informations, a graph of the chemical shifts against the coupling constants (*J* value). A simplified projection of the ^{1}H spectrum in which each multiplet signal appears as singlet and can be extracted in a reduction in the complexity of the spectrum solves most of the signal overlapping problems. This projected two-dimensional NMR can provide further information for multivariate analysis. The COSY spectrum shows a ^{1}H-^{1}H correlation with mutual spin-spin couplings, whereas TOCSY illustrates the spinning of coupled protons within the compound. It is a very useful analytical technique especially for identification of carbohydrate or amino acid. The HMBC is heteronuclear spectroscopy which shows correlations between carbon and protons at long range. HSQC, heteronuclear spectroscopy, has been found useful particularly in the metabolomics field due to its merit in the stability of ^{13}C chemical shift for the change in pH and concentration.

As already mentioned, the disadvantage of NMR spectroscopy for metabolomics studies are less sensitivity to detect low-abundant metabolites and the complexity of the spectra produced by overlapping signals (which makes the assignment of signals challenging). To overcome these problems, NMR is coupled with HPLC for a better separation of the metabolites. Further development has been implemented in which LC-NMR coupled with MS for simultaneous mass determination.

8.3.2 Sample Preparation for the NMR Analysis

Sample perpetration is a very crucial step for any analytical study. Establishment of a simple extraction method for polar and non-polar metabolites will aid in evading the interfering matrix during analytical measurement. Direct extraction of metabolites from the sample with deuterated water may steer clear of any artifacts formed during the extraction by H_2O and lyophilization process. Most of the water-soluble metabolites like organic acids, amino acids, phenols, amines, etc. show acid-base properties. Existence of protonated and deprotonated metabolites plays an important role for recording NMR spectrum of the sample. A small change in the pH of even 0.1 unit leads to the chemical shift of 0.02 ppm or more in the upfield or downfield in a 600 MHz NMR. It is difficult to compare the water extracts of NMR signals of pH dependence with different pHs. Such type of problems can be minimized by using buffered solutions. The phosphate buffer solution is recommended for NMR spectroscopy due to the capability of exchangeable protons without any additional signals in the ^{1}H NMR spectrum. Metabolites like citric acids and malic acids (organic acids) have the ability to chelate with metals owing to the paramagnetic cations, namely, Fe^{2+} and Mn^{2+}, existing in the plant extracts which results in the broadening of NMR signals. This type of peak broadening can be avoided by the addition of EDTA in the neutral solution which chelates the metal ions and leads to sharp signals in ^1H NMR of citric and malic acids. Precaution should be taken during the extraction and analysis of samples in NMR for the stability of metabolites (Sobolev et al. 2010b; Colquhoun 2007).

8.3.3 Statistical Analysis

NMR experiments in combination with multivariate statistical methods, such as principal component analysis (PCA) and partial least squares-discriminant analysis (PLS-DA), provide the variation in the metabolite levels and the metabolomics network. By employing NMR spectroscopy with chemometrics and multivariate statistics, even a slight variation in the metabolite level can be successfully traced among transgenic plants. Thus NMR spectroscopy can be an authoritative technique to identify metabolites in genetically modified crops (Krishnan et al. 2005; Mattoo et al. 2006).

8.3.4 Application of NMR Spectroscopy for Transgenic Plants

Several kinds of literature are available for metabolomic studies of transgenic plants by NMR spectroscopic techniques. Few transgenic plants which have been evaluated using NMR spectroscopy are discussed here.

8.3.4.1 *Lycopersicon esculentum* (Tomato)

^1H NMR spectroscopy was used to investigate the amount of organic acids, carbohydrates, amino acids, and other compounds in the genetically modified tomatoes and compared with the corresponding control lines by coupling with univariate and multivariate methods. It has been observed that metabolites like glutamine, aspargine, citric acid, phenylalanine, sucrose, and trigonelline were increased in the two- to three fold and flavonoid glycosides of more than ten fold in the transgenic tomato as compared to control tomatoes. It has also been found that amino acids, organic acids, and nuleotides/sides are higher in control tomatoes (La Gall et al. 2003a). Further study was conducted by overexpressing the transcription factors such as C1 and LC in the tomato fruits. From this experiment, the significant amount of flavonoids and flavonoid glycosides was increased in the transgenic ripe tomatoes and identified by LC/NMR (La Gall et al. 2003b). NMR analysis was used for metabolite comparison of isogenic non-GM and GM lines at various locations and seasons. The results indicated that almost 95% of the metabolites have no variation in their concentration (Noteborn et al. 2000).

8.3.4.2 *Solanum tuberosum* (Potato)

The four Desire'e GM plants with altered polyamine metabolism and respective control plants were analyzed by NMR in combination with statistical analysis like PCA and ANOVA. It has been found that proline, trigonelline, and other phenolics were significantly affected, and GM lines show the abnormal phenotype. Few lines of other groups showed many metabolites with slight variation in concentration as compared to the control (Defernez et al. 2004).

8.3.4.3 *Zea mays* (Maize)

The transgenic maize variety 33P67 and non-transgenic seed were investigated employing one-dimensional and two-dimensional NMR spectroscopy. About 40 polar metabolites of various classes have been identified in the maize extracts. Most of the metabolites including ethanol, organic acids, amino acids, sugars, and adenine have been reported for the first time using the ^{1}H NMR spectrum. Both transgenic and non-transgenic seed maize extracts showed the same signals in the ^{1}H spectra and same metabolites (Piccioni et al. 2009).

From GM maize plants, maize seeds containing the Cry1Ab transgene have been identified and categorized by the metabolomic fingerprinting approach. This study explains the differentiation of maize seeds from genetically modified and their control lines, with the help of NMR and multivariate statistical data analysis (Manetti et al. 2004; Manetti et al. 2006). Another study was performed in a similar way by introducing the gene which downregulates the antisense-mediated gene, and the Rpd3 gene was overexpressed in the genome of inbred maize line. ^{1}H NMR spectra and multivariate statistical PLS analysis showed that major differences were found on initial development. This study suggested the role of Rpd3 gene in cell cycle control for the accumulation of the ZmRpd3 transcripts and proteins (Castro and Manetti 2007). An experiment was conducted to compare the metabolite change of two transgenic plants in three different growing seasons. NMR spectra showed significant differences among the samples of 3 years, and further 36 metabolites were identified. It has been found that 13.8- and 6.9-fold increment in the glucose and fructose, respectively, in the Bt plant against the non-GM and the RR plants (Barros et al. 2010).

8.3.4.4 *Oryza sativa* (Rice)

When the metabolic profile of wild-type bred parent line and genetically modified rice with cry1Ab gene from *Bacillus thuringiensis* was studied using NMR spectroscopy, in combination with multivariate analysis utilized for the differentiation of sample, the results show the advantage of supervised over unsupervised statistical analysis for the categorization function (Keymanesh et al. 2009).

8.3.4.5 *Pisum sativum* (Pea)

When six independent lines of a GM pea was transferred with a Ds transposable element and five transgenes, it has been reported that the transgene was lost in two control lines, the null segregant control and the non-transformed pea plant control. Multivariate analysis like linear discriminant analysis (LDA) and PCA did not give any significant difference. This study suggested that the considerable difference was observed in the null segregant group and wild type (Charlton et al. 2004).

8.3.4.6 Wheat

NMR spectra showed a greater influence of metabolites in the polar soluble compounds. A significant variation was observed in transgenic lines (B73–6-1) and parental lines (L88–6). Variations were observed amid the parental and transgenic lines in which high concentration of carbohydrates (maltose and sucrose) and amino acids was found in transgenic line (Baker et al. 2006).

8.3.4.7 *Vitis vinifera* (Grapes)

When metabolic profiling of Silcora and Thompson varieties of grapes was studied after the transfer of one/three copies of the DefH9-iaaM construct, statistically significant differences were observed for metabolites by ^{1}H NMR spectra in the transgenic and corresponding control lines. Major variation was observed in the signals of aromatic protons of tryptophan and indole derivatives. Similarly, significant variation was also found in the organic acid regions. These results suggested that the introduction of large copies of genes leads to a higher result and also alters the entire pattern of targeted metabolites (Picone et al. 2011).

8.3.4.8 *Citrus sinensis* (Sweet Orange)

A study has been reported for investigation of the metabolite modification in transgenic and parental *C. sinensis* which is related to the infection of citrus canker employing ^{1}H NMR with multivariate and univariate statistics. According to this study, few metabolites were found different and occurred at the different stages of infection with Xac. The NT and STX genotypes of citrus leaves showed a significant difference in metabolites (Apparecido et al. 2017).

8.3.4.9 *Lactuca sativa* (Lettuce)

In the study of metabolism of transgenic lettuce leaves using NMR profiles of water-soluble metabolites, three lines in which *E. coli* asparagine synthetase A gene was overexpressed in the wild type, NMR data combining with statistical analyses indicated that the metabolite change was related to pMAC:asnA transgene at a different developmental period of leaves (Sobolev et al. 2007, 2010a). In another study when the Arabidopsis KNAT1 gene was overexpressed and transgenic lettuce was generated, there is no significant difference that was observed in metabolite profile of transgenic lettuce and the wild type by NMR spectra. Significant statistical variation was found in the metabolites at different developmental stages of leaf in the transgenic plant and found that the major role of the transgene is the modification of sugar metabolism (Sobolev et al. 2010b).

8.3.4.10 *Nicotiana tabacum* (Tobacco)

When transgenic and wild-type tobacco was studied, a small variation was monitored in the ^{1}H NMR spectra of the $CHCl_3$ extracts of the different samples. The ^{1}H NMR spectra of the signals of phenylpropanoid were comparatively larger in the control leaves in corresponding to CSA-line leaves. Chlorogenic acids, sugars, and malic acid were the key metabolites responsible for the variation. The principal component analysis showed that no major variation was found in the $CHCl_3$ extracts of transgenic and wild-type tobacco plants. It has also been found that the metabolite level was not equal in the leaves and veins of the wild-type plants and significantly more than the CSA plants (Choi et al. 2004). When tobacco plants are engineered by diverting the carbon flux to plastids from the cytosol, sesquiterpenoids such as patchoulol and amorpha-4,11-diene have increased, and the ^{13}C-labeled carbon in patchoulol was confirmed by NMR spectroscopy (Wu et al. 2006).

8.3.5 Summary

NMR spectroscopy has been widely used for metabolomic studies of the transgenic plant. NMR techniques give unbiased results due to the merit of the technique of having very less chance of metabolite degradation. When NMR data combine with multivariate statistical techniques, it can differentiate the variation even at very low level. In the recent advancement, NMR is available with hyphenated techniques like LC-NMR, LC-NMR-MS, etc. NMR profiling of plant metabolites is highly robust and quantitatively reliable. Thus NMR technique will be of greatest applicability in the future in high-throughput screening of transgenic plants.

8.4 High-Throughput Fourier Transform Infrared (FT-IR) Spectroscopy for the Screening of Transgenic Plants

8.4.1 Introduction

Fourier transform infrared (FT-IR) spectroscopy becomes one of the important analytical techniques to determine the quantitative and qualitative information for a wide range of metabolites in a non-destructive manner. The principle of the infrared spectroscopy is that the sample is irradiated with IR radiation, and the change in the molecular properties of the compound/metabolite leads to the spectral change in the reflected and transmitted radiation. Employing the spectral variation, the molecular properties of the compound such as functional groups, structure, etc. can be determined for the metabolites. IR spectroscopy can also be used for the detection of unstable substances such as non-isolatable intermediates of the chemical and biological processes. Advancement in the instrumentation, application of Fourier transform function, and user-friendly computer software aids the operation and data interpretation from the complex IR data to a simple spectrum and provides the spectral information for specific functionalities in a metabolite. Further, based on the spectral characteristics, IR spectroscopy is divided into near-infrared (NIR) covers between 13,400 and 4000 cm^{-1}, mid-infrared (MIR) covers in the range of 4000 to 400 cm^{-1}, and far-infrared (FIR) up to about 10 cm^{-1} (Cozzolino 2011). All three IR spectroscopies have different applications in the metabolite analysis.

IR spectroscopy can be used for solid, liquid, and gaseous samples. The sample preparation for the IR spectroscopy is simple where solid and liquid samples can be made as a pellet or spread between the plates, respectively, using IR transparent salts such as NaCl, KBr, and CaF_2, whereas gaseous samples are filled in the sample cell with longer pathlength to indemnify the dilution of the sample.

8.4.2 Applications in Transgenic Plant Analysis

Due to its simplicity, NIR spectroscopy was extensively used in the transgenic plant analysis to determine different types of metabolites in transgenic and parental lines.

8.4.2.1 *Lycopersicon esculentum* (Tomato)

NIR spectroscopy in combination with visible spectrum as VIS-NIR was employed to screen antisense LeETR2 transgenic tomatoes and compared with their parental lines and non-transgenic tomatoes. PCA, DA, and PLS-DA were used to categorize tomato lines into two clusters. VIS-NIR technique along with the statistical analysis was successful in the classification of transgenic and non-transgenic lines (Xie et al. 2007).

8.4.2.2 Barley

NIR spectroscopy was used to screen the high-lysine mutants in barley. The NIR technique in combination with multivariate statistical analysis was used to identify the genetic variation in the mutants of barley (Munck et al. 2004).

8.4.2.3 Trees

The robust mini-scale instrumentation of NIR spectroscopy affords the efficient monitoring of the properties of the transgenic trees. The sample preparation was also simple as the wood meal was made as pellet and measured directly in the transmittance NIR spectroscopy. The data from the NIR spectroscopy is in good correlation with the conventional techniques for the measurement of lignin, cellulose, and xylose content (Yamada et al. 2006).

8.5 Conclusion and Future Perspectives

Currently, there is no single analytical technique which gives sufficient information about the structure and functional group of metabolites; thus a comprehensive approach is needed using different analytical techniques to interpret and elucidate the structure of a compound (metabolite). Selection of appropriate extraction method and separation technique is crucial in the plant metabolite analysis. A simple graphical representation of the extraction solvents for various metabolites and corresponding analytical techniques is given in Fig. 8.1. While developing a high-throughput analytical technique for transgenic plant analysis, one must be acquainted with the chemical and physical properties of the metabolite and the limitations of the analytical technique to select the right technique for the transgenic plant metabolites.

In the past years, substantial advancement has occurred in the high-throughput analytical techniques for plant metabolite analysis. In the future, the following

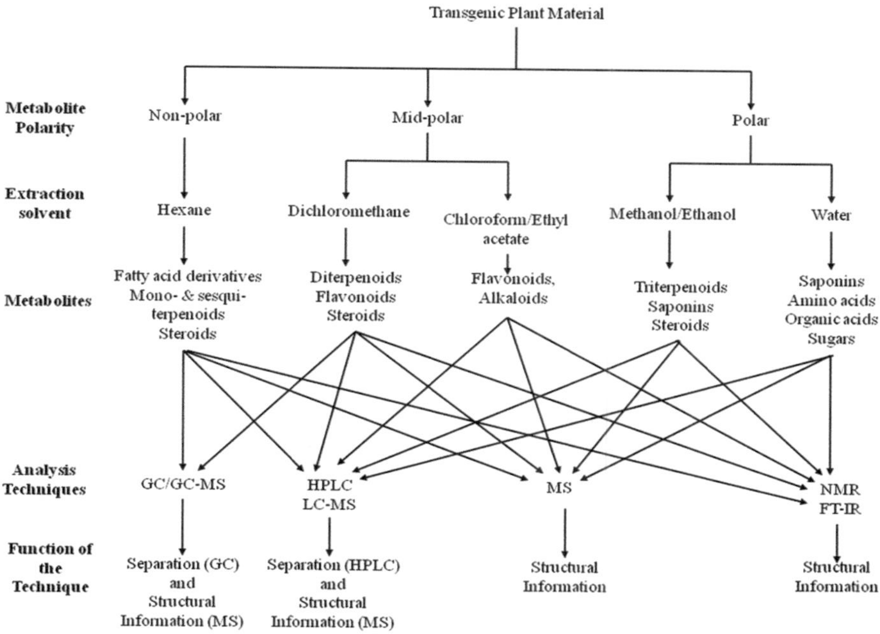

Fig. 8.1 Graphical representation for the extraction solvent and analysis techniques for the metabolites of transgenic plants. **Note:** *GC* gas chromatography, *HPLC* high-performance liquid chromatography, *MS* mass spectrometry, *NMR* nuclear magnetic resonance spectroscopy

advancements are required for the fast, sensitive, and accurate analysis of transgenic plant metabolites:

1. Advanced stationary phases in chromatography which can be applicable to a wide range of metabolites of different polarity and functionality.
2. Universal detection techniques for a different class of metabolites.
3. Development in the hyphenation of sample introduction, separation, and different spectroscopic techniques in tandem mode.
4. Sophisticated analysis software with artificial intelligence to interpret the structure of the compounds by combining two or three independent parameters and make a meaningful judgment for the structural elucidation of the metabolites.
5. Robustness and miniaturization in the instrument for easy portability with self-diagnosing and troubleshooting instrument control software.

References

Abdelkader MSA, Lockwood GB (2016) Essential oils from the plant, hairy root cultures and shoot cultures of Egyptian Anethum graveolens (dill). J Essent Oil Res 28:104–112. https://doi.org/10.1080/10412905.2015.1091790

Adams RP (1995) Identification of essential oil components by gas chromatography/ mass spectroscopy. Allured Publishing, Carol Stream

Aharoni A, Giri AP, Deuerlein S, Griepink F, Kogel W De, Verstappen FWA, Verhoeven HA, Jongsma MA, Schwab W, Bouwmeester HJ (2003) Terpenoid metabolism in wild-type and transgenic arabidopsis Plant Cell 15:2866–2884 https://doi.org/10.1105/tpc.016253.The

Anklam E, Gadani F, Heinze P, Pijnenburg H, Den VEG (2002) Analytical methods for detection and determination of genetically modified organisms in agricultural crops and plant-derived food products. Eur Food Res Technol 214:3–26. https://doi.org/10.1007/s002170100415

Apparecido R, do P, Carlos EF, Lião LM, Vieira LGE, Alcantara GB (2017) NMR-based metabolomics of transgenic and non-transgenic sweet orange reveals different responses in primary metabolism during citrus canker development. Metabolomics 13:20

Baker JM, Hawkins ND, Ward JL, Lovegrove A, Napier JA, Shewry PR, Beale MH (2006) Ametabolomic study of substantial equivalence offield-grown genetically modified wheat. Plant Biotechnol J 4:381–392

Barros E, Lezar S, Anttonen MJ, van Dijk JP, Röhlig RM, Kok EJ, Engel KH (2010) Comparison of two GM maize varieties with a nearisogenic non-GM variety using transcriptomics, proteomics and metabolomics. Plant Biotechnol J 8:436–451

Boba A, Kulma A, Kostyn K, Starzycki M, Starzycka E, Szopa J (2011) The influence of carotenoid biosynthesis modification on the Fusarium culmorum and Fusarium oxysporum resistance in flax. Physiol Mol Plant Pathol 76:39–47. https://doi.org/10.1016/j.pmpp.2011.06.002

Bonn J (2008) Improved techniques for sampling and sample introduction in gas chromatography. Royal Institute of Technology, Stockholm. https://www.diva-portal.org/smash/get/diva2:13747/FULLTEXT01.pdf) dated 24/07/2019

Castro C, Manetti C (2007) A multiway approach to analyze metabonomic data: A study of maize seeds development. Anal Biochem 371:194–200

Charlton A, Allnutt T, Holmes S, Chisholm J, Bean S, Ellis N, Mullineaux P, Oehlschlager S (2004) NMR profiling of transgenic peas. Plant Biotechnol J 2:27–35

Choi HK, Choi YH, Verbernea M, Lefeberc AWM, Erkelensc C, Verpoorte R (2004) Metabolic fingerprinting of wild type and transgenic tobacco plants by 1H NMR and multivariate analysis technique. Phytochemistry 65:857–864

Clough SJ, Bent AF (1998) Floral dip: A simplified method for agrobacterium-mediated transformation of Arabidopsis thaliana. Plant J 16:735–743. https://doi.org/10.1046/j.1365-313X.1998.00343.x

Colombo R, Lanças FM, Yariwake JH (2006) Determination of flavonoids in cultivated sugarcane leaves, bagasse, juice and in transgenic sugarcane by liquid chromatography-UV detection. J Chromatogr A 1103:118–124. https://doi.org/10.1016/j.chroma.2005.11.007

Colquhoun IJ (2007) Use of NMR for metabolic profiling in plant systems. J Pestic Sci 32:200–212

Cozzolino D (2011) Infrared methods for high throughput screening of metabolites: food and medical applications. Comb Chem High Throughput Screen 14:125–131

Daviet L, Schalk M (2010) Biotechnology in plant essential oil production: progress and perspective in metabolic engineering of the terpene pathway. Flavour Fragr J 25:123–127. https://doi.org/10.1002/ffj

Defernez M, Gunning YM, Parr AJ, Shepherd LVT, Davies HV, Colquhoun IJ (2004) NMR and HPLC–UV profiling of potatoes with genetic modifications to metabolic pathways. J Agric Food Chem 52:6075–6085

Diemer F, Caissard J-C, Moja S, Chalchat J-C, Jullien F (2001) Altered monoterpene composition in transgenic mint following the introduction of 4S-limonene synthase. Plant Physiol Biochem 39:603–614. https://doi.org/10.1016/S0981-9428(01)01273-6

Figueiredo AC, Barroso JG, Pedro LG, Scheffer JJC (2008) Factors affecting secondary metabolite production in plants: volatile components and essential oils. Flavour Fragr J 23:213–226. https://doi.org/10.1002/ffj

Foubert K, Cuyckens F, Vleeschouwer K, Theunis M, Vlietinck A, Pieters L, Apers S (2010) Rapid quantification of 14 saponins of Maesa lanceolata by UPLC – MS/MS. Talanta 81:1258–1263. https://doi.org/10.1016/j.talanta.2010.02.018

Garcia-Villalba R, Leon C, Dinelli G, Segura-Carretero A, Fernandez-Gutierrez A, Garcia-Canas V, Cifuentes A (2008) Comparative metabolomic study of transgenic versus conventional soybean using capillary electrophoresis – time-of-flight mass spectrometry. J Chromatogr A 1195:164–173. https://doi.org/10.1016/j.chroma.2008.05.018

Gras R, Hua Y, Luong J (2017) High-throughput gas chromatography for volatile compounds analysis by fast temperature programming and adsorption chromatography. J Sep Sci 40:1979–1984. https://doi.org/10.1002/jssc.201700016

Hall RD (2006) Plant metabolomics: from holistic hope, to hype, to hot topic. New Phytol 169:453–468

Herrera-estrella LR (2000) Genetically modified crops and developing countries. Plant Physiol 124:923–925

Holmes E, Tang H, Wang Y, Seger C (2006) The assessment of plant metabolite profile by NMR-based methodologies. Planta Med 72:771–785

Jonsson P, Johansson AI, Gullberg J, Trygg J, Grung B, Marklund S, Sjöström M, Antti H, Moritz T (2005) High-throughput data analysis for detecting and identifying differences between samples in GC/MS-based metabolomic analyses. Anal Chem 77:5635–5642. https://doi.org/10.1021/ac050601e

Kallenbach M, Oh Y, Eilers EJ, Veit D, Baldwin IT, Schuman MC (2014) A robust, simple, high-throughput technique for time- resolved plant volatile analysis in field experiments. Plant J 78:1060–1072. https://doi.org/10.1111/tpj.12523

Keun HC, Ebbels TM, Antti H, Bollard ME, Beckonert O, Schlotterbeck G, Senn H, Niederhauser U, Holmes E, Lindon JC, Nicholson JK (2002) Analytical reproducibility in 1H NMR-based metabonomic urinalysis. Chem Res Toxicol 15:1380–1386

Keymanesh K, Darvishi MH, Sardari S (2009) Metabolome comparison of transgenic and non-transgenic rice by statistical analysis of FTIR and NMR spectra. Rice Sci 16:119–123

Kim HK, Verpoorte R (2010) Sample preparation for plant metabolomics. Phytochem Anal 21:4–13. https://doi.org/10.1002/pca.1188

Kraft P, Denizot N (2013) Synthesis of a spirocyclic seco structure of the principal vetiver odorant Khusimone. Eur J Org Chem 2013:49–58. https://doi.org/10.1002/ejoc.201201318

Krishnan P, Kruger NJ, Ratcliffe RG (2005) Metabolite fingerprinting and profiling in plants using NMR. J Exp Bot 56:255–265

Lavy M, Zuker A, Lewinsohn E, Larkov O, Ravid U, Vainstein A, Weiss D (2002) Linalool and linalool oxide production in transgenic carnation flowers expressing the Clarkia breweri linalool synthase gene. Mol Breed 9:103–111. https://doi.org/10.1023/A:1026755414773

Le Gall G, Colquhoun IJ, Davis AL, Collins GJ, Verhoeyen ME (2003a) Metabolite profiling of tomato (Lycopersicon esculentum) using 1H NMR spectroscopy as a tool to detect potential unintended effects following a genetic modification. J Agric Food Chem 51:2447–2456

Le Gall G, DuPont MS, Mellon FA, Davis AL, Collins GJ, Verhoeyen ME, Colquhoun IJ (2003b) Characterization and content of flavonoid glycosides in genetically modified tomato (Lycopersicon esculentum) fruits. J Agric Food Chem 51:2438–2446

Leon C, Rodriguez-meizoso I, Lucio M, Garcia-Canas V, Ibanez E, Schmitt-Kopplin P, Cifuentes A (2009) Metabolomics of transgenic maize combining Fourier transform-ion cyclotron resonance-mass spectrometry, capillary electrophoresis-mass spectrometry and pressurized liquid extraction. J Chromatogr A 1216:7314–7323. https://doi.org/10.1016/j.chroma.2009.04.092

Lewinsohn E, Schalechet F, Wilkinson J, Matsui K, Tadmor Y, Nam K, Amar O, Lastochkin E, Larkov O, Ravid U, Hiatt W, Gepstein S, Pichersky E (2001) Enhanced levels of the aroma and flavor compound S-linalool by metabolic engineering of the Terpenoid pathway in tomato fruits. Plant Physiol 127:1256–1265. https://doi.org/10.1104/pp.010293.1256

Lucas AM, Pasquali G, Astarita LV, Cassel E (2017) Comparison of genetically engineered (GE) and non-GE Eucalyptus trees using secondary metabolites obtained by steam distillation. J Essent Oil Res 29:22–31. https://doi.org/10.1080/10412905.2016.1187674
Lucker J, Bouwmeester HJ, Schwab W, Blaas J, Plas LHW Van Der, Verhoeven HA (2001) Expression of Clarkia S -linalool synthase in transgenic petunia plants results in the accumulation of S -linalyl- b - D - glucopyranoside. 27:315–324
Mahmoud SS, Croteau RB (2001) Metabolic engineering of essential oil yield and composition in mint by altering expression of deoxyxylulose phosphate reductoisomerase and menthofuran synthase. Proc Natl Acad Sci 98:8915–8920. https://doi.org/10.1073/pnas.141237298
Manetti C, Bianchetti C, Bizzarri M, Casciani L, Castro C, D'Ascenzo G, Delfini M, di Cocco ME, Lagana A, Miccheli A, Motto M, Conti F (2004) NMR-based metabonomic study of transgenic maize. Phytochemistry 65:3187–3198
Manetti C, Bianchetti C, Casciani L, Castro C, di Cocco ME, Miccheli A, Motto M, Conti F (2006) A metabonomic study of transgenic maize (Zea mays) seeds revealed variations in osmolytes and branched amino acids. J Exp Bot 57:2613–2625
Mattoo AK, Sobolev AP, Neelam A, Goyal RK, Handa AK, Segre AL (2006) Nuclear magnetic resonance spectroscopy-based metabolite profiling of transgenic tomato fruit engineered to accumulate spermidine and spermine reveals enhanced anabolic and nitrogen-carbon interactions. Plant Physiol 142:1759–1770
Meija J, Montes-bayo M, Duc DL Le, Terry N, Caruso JA (2002) Simultaneous Monitoring of volatile selenium and sulfur species from Se accumulating plants (wild type and genetically modified) by GC / MS and GC / ICPMS using solid-phase microextraction for sample introduction. 74:5837–5844
Mierziak J, Wojtasik W, Kostyn K, Tadeusz C, Szopa J, Kulma A (2014) Crossbreeding of transgenic flax plants overproducing flavonoids and glucosyltransferase results in progeny with improved antifungal and antioxidative properties. Mol Breed 34:1917–1932. https://doi.org/10.1007/s11032-014-0149-5
Mounts TL, Abidi SL, Rennick KA (2009) Effect of genetic modification on the content and composition of bioactive constituents in soybean oil. J Am Oil Chem Soc 73:581–586. https://doi.org/10.1007/BF02518111
Munck L, Moller B, Jacobsen S, Sondergaard I (2004) Near infrared spectra indicate specific mutant endosperm genes and reveal a new mechanism for substituting starch with (1→3,1→4)-b-glucan in barley. J Cereal Sci 40:213–222
Nordström K, Dahlbom J, Pragadheesh VS, Ghosh S, Olsson A, Dyakova O, Krishna Suresh S, Olsson S (2017) In situ modeling of multimodal floral cues attracting wild pollinators across environments. Proc Natl Acad Sci 114:13218–13223. https://doi.org/10.1073/pnas.1714414114
Noteborn HP, Lommen A, van der Jagt RC, Weseman JM (2000) Chemical fingerprinting for the evaluation of unintended secondary metabolic changes in transgenic food crops. J Biotechnol 77:103–114
Ohara K, Ujihara T, Endo T, Sato F, Yazaki K (2003) Limonene production in tobacco with Perilla limonene synthase cDNA. J Exp Bot 54:2635–2642. https://doi.org/10.1093/jxb/erg300
Piccioni F, Capitani D, Zolla L, Mannina L (2009) NMR metabolic profiling of transgenic maize with the Cry1A(b) gene. J Agric Food Chem 57:6041–6049
Picone G, Mezzetti B, Babini E, Capocasa F, Placucci G, Capozzi F (2011) Unsupervised principal component analysis of NMR metabolic profiles for the assessment of substantial equivalence of transgenic grapes (Vitis vinifera). J Agric Food Chem 59:9271–9279
Seethapathy S, Tadeusz G, Li X (2008) Passive sampling in environmental analysis. J Chromatogr A 1184:234–253. https://doi.org/10.1016/j.chroma.2007.07.070
Sobolev AP, Segre AL, Giannino D, Mariotti D, Nicolodi C, Brosio E, Amato ME (2007) Strong increase of foliar inulin occurs in transgenic lettuce plants (Lactuca sativa L.) overexpressing the asparagine Synthetase A gene from Escherichia coli. J Agric Food Chem 55:10827–10831
Sobolev AP, Testone G, Santoro F, Nicolodi C, Iannelli MA, Amato ME, Ianniello A, Brosio E, Giannino D, Mannina L (2010a) Quality traits of conventional and transgenic lettuce (Lactuca sativa L.) at harvesting by NMR metabolic profiling. J Agric Food Chem 58:6928–6936

Sobolev AP, Capitani D, Giannino D, Nicolodi C, Testone G, Santoro F, Frugis G, Iannelli MA, Mattoo AK, Brosio E, Gianferri R, D'Amico I, Mannina L (2010b) NMR-metabolic methodology in the study of GM foods. Nutrients 2:1–15

Szarka S, Héthelyi ÉB, Lemberkovics É, Bálványos I, Szőke É, Farkas E, Kuzovkina IN (2007) Essential oil constituents of intact plants and in vitro cultures of tagetes patula L. J Essent Oil Res 19:85–88. https://doi.org/10.1080/10412905.2007.9699235

Toth A, Praszna L (1998) Improvement of the reproducibility of retention indices in temperature-programmed gas - liquid chromatography. Part I. DB-1 column. Flavour Fragr J 13:196–202

Verma RS, Chauhan A, Padalia RC, Jat SK, Thul S, Sundaresan V (2013) Phytochemical diversity of Murraya koenigii (L.) Spreng. From Western Himalaya. Chem Biodivers 10:628–641

Wallaart TE, Bouwmeester HJ, Hille J, Poppinga L, Maijers NCA (2001) Amorpha-4,11-diene synthase: cloning and functional expression of a key enzyme in the biosynthetic pathway of the novel antimalarial drug artemisinin. Planta 212:460–465. https://doi.org/10.1007/s004250000428

Wu S, Schalk M, Clark A, Miles RB, Coates R, Chappell J (2006) Redirection of cytosolic or plastidic isoprenoid precursors elevates terpene production in plants. Nat Biotechnol 24:1441–1447. https://doi.org/10.1038/nbt1251

Xie L, Ying Y, Ying T, Yu H, Fu X (2007) Discrimination of transgenic tomatoes based on visible/near-infrared spectra. Anal Chim Acta 584:379–384

Yamada T, Yeh T-F, Chang H-M, Li L, Kadla JF, Chiang VL (2006) Rapid analysis of transgenic trees using transmittance near-infrared spectroscopy (NIR). Holzforschung 60:24–28

Yuan Y, Qi L, Yang J, Wu C, Liu Y, Huang L (2015) A Scutellaria baicalensis R2R3-MYB gene, SbMYB8, regulates flavonoid biosynthesis and improves drought stress tolerance in transgenic tobacco. Plant Cell Tissue Organ Cult 120:961–972. https://doi.org/10.1007/s11240-014-0650-x

Zellner B d'A, Bicchi C, Dugo P, Rubiolo P, Dugo G, Mondello L (2008) Linear retention indices in gas chromatographic analysis: a review. Flavour Fragr J 29:297–314. https://doi.org/10.1002/ffj

Zhang A, Sun H, Wang P, Han Y, Wang X (2012) Modern analytical techniques in metabolomics analysis. Analyst 137:293–300

Part II

Applications in Genetic Improvement of Plants

Transgenic Technologies and Their Potential Applications in Horticultural Crop Improvement

9

Varsha Tomar, Shashank Sagar Saini, Kriti Juneja, Pawan Kumar Agrawal, and Debabrata Sircar

Abstract

The increasing world population is facing a threatening health crisis in the form of malnutrition or hidden hunger, and simultaneously global demand for food and nutrition is growing day by day. Horticultural crops are widely grown and used globally and play an important role in human dietary nutrition in terms of providing vitamins, edible fibers, essential minerals, and health protective bioactive plant metabolites. Consumption of horticultural crops, mainly vegetables and fruits, is also linked with enhancement of intestinal health and digestion, better body immunity, and reduced risk of several life-threatening diseases, such as diabetes, cardiovascular disease, and many forms of cancer. The growing world population needs engineered horticultural crops to enhance yield, nutrient value, and tolerance toward climate change. As the scientific community debates the advantages and risks of transgenic plants, the proportion of the transgenic field crops has significantly increased worldwide, since last few years, and the public acceptance of transgenic crops continues to demonstrate. Nevertheless, despite the encouraging success story of commercialization of transgenic field crops, the production and marketing of transgenic horticultural crops, such as vegetables, fruits, and ornamentals, are very poor. Transgenic vegetable crops could provide an economically viable solution toward sustainable production of horticultural crops in the coming years to meet the growing demands. However, countries differ in their policies and regulations for acceptance of transgenic horticultural crops. Production of transgenic horticultural crops will be more mean-

V. Tomar · S. S. Saini · K. Juneja · D. Sircar (✉)
Department of Biotechnology, Indian Institute of Technology Roorkee, Roorkee, Uttarakhand, India
e-mail: debsrfbt@iitr.ac.in

P. K. Agrawal (✉)
Krishi Anusandhan Bhawan-I, ICAR, New Delhi, India
e-mail: pawankagrawal@hotmail.com

R. Sathishkumar et al. (eds.), *Advances in Plant Transgenics: Methods and Applications*, https://doi.org/10.1007/978-981-13-9624-3_9

ingful and publically acceptable if their advantages in terms of nutrient and yield and their safety issues are demonstrated clearly. In this chapter, authors presented the latest updates available on plant transgenic technologies and their possible applications in the improvement of horticultural crops, particularly on vegetables and fruits. The future challenges, strategies, and opportunities of transgenic technologies in the improvement of horticultural crops are discussed.

Keywords
Transgenic · Technologies · Horticulture · Safety issues · Commercialization · Bioactive · Globally · Malnutrition · Dietary

9.1 Introduction

Horticultural crops (vegetables, fruits, and ornamentals) constitute an important group among agricultural crops, both in terms of intake and nutrition value. Among horticultural crops, vegetables and fruits are the most important in dietary point of view and are grown throughout the world in more than 200 countries. Vegetables and fruits are the major source of vitamins, minerals, food fibers, and health protective metabolites, especially as source of polyphenols (Dias and Ryder 2011). Consumption of vegetables and fruits is linked with improved gastrointestinal health and reduced risk of several life-threatening diseases such as cardiovascular diseases, diabetes, and several forms of cancer (Keatinge et al. 2010; Dias and Ortiz 2013).

As evident from a survey conducted throughout the world approximately 400 vegetable crops representing approximately 230 genera are cultivated globally (Kays 2011; Dias and Ortiz 2014). In 2017, the total global production of horticulture crops was approximately 7 billion tons which is even more than the production of staple crops. Presently, China is the biggest producer of horticultural crops, followed by India and Brazil. Horticulture contributes about 35% of GDP in agriculture, using only 17% land area. Horticulture production has increased nearly 70% within the last 10 years. The production rate of horticultural crops among top five leading world producers is given in Table 9.1. Vegetables and fruits are mostly marketed fresh; only a few are sold frozen, either whole part or pieces. Consumption of fresh vegetables and fruits just after harvesting is associated with gaining optimum nutrition value.

Table 9.1 Production of horticultural crops among top five leading world producers

Fruits		Vegetables	
Top five producers	Production (Tons)	Top five producers	Production (Tons)
China	137,066,750	China	573,935,000
India	81,285,334	India	262,186,567
Brazil	38,368,678	United States of America	35,947,720
United States of America	26,548,859	Turkey	27,818,918
Indonesia	17,744,411	Iran	23,485,675

Source: FAO (2013)

Recent trend of continuous increase in the world population is alarming for the limited availability of horticultural products in the coming years. By the year 2050, the world population is expected to reach up to 9 billion (Tanuja and Kumar 2017). Due to continuous production, horticultural crops are vulnerable to many abiotic and biotic stresses. The biggest challenge comes from biotic stresses triggered mainly by pathogens and pests. According to an earlier estimate, annual global pre-harvest loss of horticultural crops is around 40%, including 15%, 13%, and 12% damage in production caused by insects, microbial pathogens, and weeds, respectively (Pimentel 1997). In order to protect from pests and pathogen, pesticides have been used since long time; however, pesticide residues are extremely harmful for the environment and human health. Due to increase in awareness, consumers are very concerned about nutritional quality, toxicity, and safety of the food they are consuming, along with the surrounding environment under which those crops are produced. Most consumers prefer pesticide-free organically grown fruits and vegetables. Therefore, the prices for major vegetables and fruits are increasing to maintain high crop quality and safety.

To cope up with rapidly increasing population and their high demand for getting low-cost fresh and nutritious food, there is an urgent need for enhanced production of horticultural crops with high nutrition values and better productivity with improved pest and disease resistance. Conventional plant breeding to enhance the quality of horticultural crops is still very limited mostly because of genetic restrictions. The most serious drawback of conventional plant breeding of fruit and vegetables is requirement of long breeding time to perform several backcrosses for the introgression of desired trait. In several cases, the desirable traits are present in wild cultivars, so introgression of such resistant trait by conventional breeding involves multiple backcrosses, which ultimately reduce the fruit/vegetable quality.

Transgenic technology offers a precise tool for introducing desirable traits in target plants with high accuracy. Traits are controlled by genes, which are the basis of any plant breeding. Earlier during conventional breeding practices, breeders use to select novel phenotypes without knowing much about the plant genotype. Due to tremendous advancement in the area of plant molecular biology, now scientists understand how most of the desirable or undesirable traits are inherited and their underlying gene functions. At present, scientists can precisely manipulate the gene (addition/deletion/modification) encoding a particular trait to develop novel phenotypes using recombinant DNA and genome editing technologies, together known as transgenic breeding technology. Plants developed using these technologies are known as transgenic plants or genetically modified (GM) plants or biotech plants. The key process of transgenic breeding technology is the incorporation of desired foreign genes into the target host plant genome. The biggest advantage of this technology is that its precision, less time requirements, and the improvement of a trait without changing the other genetic constitution of an elite plant genotype to maintain the superiority of the elite germplasm. This is mainly useful for many horticultural plants because they are vastly heterozygous. Our present social trends indicate that consumers are looking for healthy vegetables and fruits with more nutrition, better freshness and taste, and more number of varieties in their daily eating meals. Consumers are also seeking greater ease in availability and lower buyer cost of those fresh horticultural products. Transgenic horticultural crops bring solutions to all these issues by bringing favorable genes, often previously inaccessible, into existing elite cultivars. Despite release of few transgenic horticultural crops in the market, large-scale commercial success of this technology is of great concern due to associated possible risk factors.

This chapter summarizes the latest updates available on plant transgenic technologies and their possible applications in the improvement of horticultural crops with special emphasize on some commercially successful horticultural crops. At the end, future challenges, strategies, and opportunities associated with implication of transgenic technologies in the improvement of horticultural crops are also discussed.

9.2 Transgenic, Cisgenic, and RNA Interference Technology

The basis of transgenic or cisgenic technology relies on the fact that a foreign donor gene encoding desirable trait and regulated by specific genetic regulatory sequences, such as promoter, enhancer, terminator, etc., is transferred to a target host plant (recipient plant) using biotechnology tools. Based on the source from which the donor gene has been taken, the technology is divided into transgenic and cisgenic (Fig. 9.1). If the donor gene and all of its regulatory sequences to be inserted into a target host crop belong to different crops or non-cross breedable species or organisms (such as bacterial or fungal gene), the resulting crop is called "transgenic," and the corresponding donor gene is known as transgene (Sticklen 2015). In contrary, if the foreign gene and all of its regulatory sequences belong to the same crop species or its cross breedable species, then the resulting crop is known as "cisgenic," and the

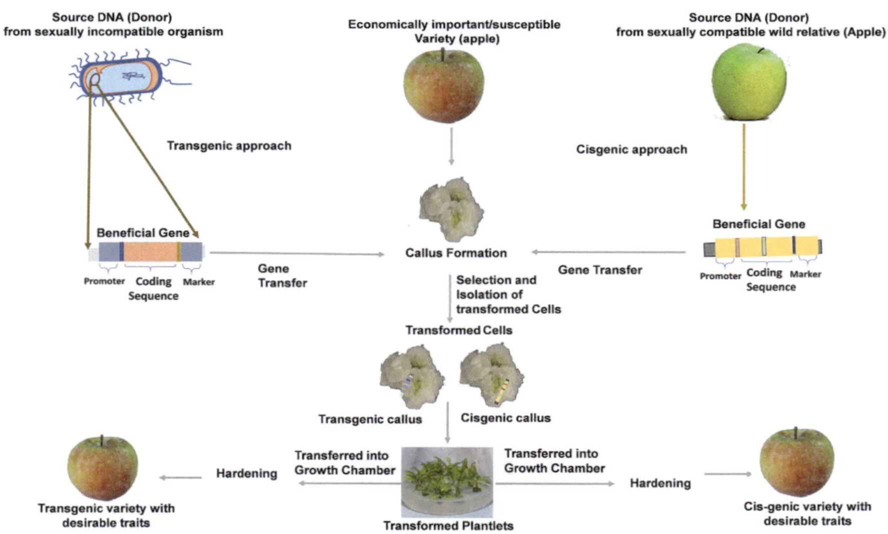

Fig. 9.1 Schematic diagram showing difference between cisgenic and transgenic plants

corresponding gene is called cisgene (Schouten et al. 2006). In developing cisgenic plants, it is important that cisgene and its regulatory sequences must integrate in the recipient host plant in the original-sense orientation.

There is another technology available to develop a separate type of genetically modified crop, known as "intragenic" plants. In this technology the scientist recombines different genes (or part of DNAs), all of which came from the same crop or its sexually compatible cross breedable relatives into a single construct to be introduced inside a host plant. In this case, mostly the promoter comes from one gene and the coding region from another gene (Holme et al. 2013).

RNA interference (RNAi) is another emerging technology to develop genetically modified plants where target gene in the host plant is downregulated by means of inserting the same gene from the same host plant but in antisense orientation. Nevertheless, regulatory sequences in those plants come from the same host plant or from any cross breedable species or from any other organism. Consequently, the RNAi-modified plant could fall under the cisgenic, transgenic, or interagenic groups. Moreover, in case of transgenic, cisgenic, or intragenic technology, there is a requirement of an in vitro selectable marker to select the positive gene transfer. Accordingly, the origin of the selectable marker gene and their regulatory sequences will also fall under transgenic, cisgenic, or intragenic groups.

9.3 Traits Commonly Targeted for Developing Genetically Modified Horticultural Crops

A number of traits have been approved by different regulatory bodies and widely adopted by the companies to focus on those traits for genetic modification. In general, traits affecting production of the crop are preferred over the traits involved in modification of the final product. One of the prime foci is the selection of traits for enhanced resistance against insects and other pathogens. Many vegetable and fruit crops have been genetically modified to include the following traits (Ram and Dasgupta 2005):

1. To enhance resistance toward insect, pests, and diseases.
2. To enhance tolerance toward herbicides and insecticides.
3. To enhance tolerance toward abiotic stresses such as high temperature, cold, salt, light, etc.
4. To enhance nitrogen fixation.
5. To modify end products (qualitative or quantitative).
6. To enhance shelf-life of fruits and vegetables.

The list of approved traits is steadily increasing, and approval differs from country to country based on regulatory guidelines of different countries. Traits approved by the regulatory authority of the United States are listed in Table 9.2 (Clark et al. 2004; Xiong et al. 2015).

Table 9.2 Approved traits for raising transgenic plants in the USA

Approved Traits	Target modification
Insect resistance	*Bt* kurstaki
Herbicide tolerance	Glyphosate, sulfonyl urea, bromoxynil, glufosinate
Virus resistance	Potato leaf roll virus, cucumber mosaic virus, papaya ringspot virus, zucchini yellow mosaic virus, watermelon mosaic virus
Modification of fruit ripening for enhanced shelf-life	ACC synthase, polygalacturonase
Modified oils content	Myristic, oleic acids
Modified product quality	Apples, squash, sweet pepper

9.4 Different Approaches for Developing Transgenic Horticultural Crops

Development of transgenic horticultural crop requires transfer of foreign gene from donor plant to the recipient host plant, the process popularly known as transformation. Currently, there are two main methods available for performing stable transformation of the horticultural crops:

(i) *Agrobacterium*-mediated transformation.
(ii) Particle-bombardment-mediated transformation.

Apart from these two methods, the following two other methods are also available, which are not strictly categorized under transgenic but routinely used to change native genetic backbone of the plants.

(iii) RNA interference.
(iv) Genome editing.

(i) ***Agrobacterium*-Mediated Transformation**: The bacterium *Agrobacterium tumefaciens* naturally infects the wound of a plant to develop the crown gall disease. This is possible because *Agrobacterium* carries the tumor-inducing (Ti) plasmid, which has a virulence (*vir*) region and a T-DNA (transfer DNA), which actually transferred from the bacterium to plant. During the process of transformation, multiple components of the Ti plasmid work together for the effective transfer of the gene of interest into the plant cells. The following components of the Ti plasmid play crucial roles in transferring foreign DNA inside plant cell:

- The left and right border sequences of T-DNA, which distinguish the DNA segment (T-DNA) to be transferred inside the plant genome.
- *vir* genes (virulence genes), whose products are required for transferring the T-DNA from *Agrobacterium* to plant, but *vir* genes themselves are not transferred.

- T-DNA region between the borders, which harbors genes that are incorporated into the nuclear genome of the host plant cell.

Further, the wounded plant cells produce specialized signaling molecule like acetosyringone, which activate the *vir* region of the Ti plasmid. After integration into the plant genome, the T-DNA encodes enzymes that synthesize plant hormones, which in turn destabilize the delicate balance of auxins and cytokinins in the host plant. This results in an uncontrolled division of the infected plant cells that ultimately form a tumor. Scientists have exploited this unique combination of the *vir* region and the T-DNA in Ti plasmid of *Agrobacterium* to transfer desirable traits into horticulture crops. For the transfer of the T-DNA, only specific sequences at its left and right borders are required, while the harmful tumor-inducing genes can be completely discarded. Thus, the harmful gene sequence can be replaced by a gene of interest within the T-DNA, which is transferred into the nuclear genome of host plant cell. This makes the Ti plasmid in *Agrobacterium* a very useful tool for the transformation of horticulture crops. The process of *Agrobacterium*-mediated transformation involves several steps:

(a) Cloning and isolation of the gene of interest from the source organism.
(b) Creating a functional transgenic construct bearing the promoter, gene of interest, terminator, and selectable marker gene.
(c) Insertion of the transgenic construct into the Ti plasmid.
(d) Introduction of the T-DNA-containing-plasmid into host *Agrobacterium* to form transformed *Agrobacterium*.
(e) Co-culture of the transformed *Agrobacterium* with the plant cells to facilitate the transfer of T-DNA into plant chromosome.
(f) Selection of the positively transformed plants using the property of selectable marker.
(g) Regeneration of the transformed cells into genetically modified plants.
(h) Finally testing the functionality of transgene at laboratory and field level.

Figure 9.2 illustrates the *Agrobacterium*-mediated plant transformation process.

Agrobacterium-mediated transformation is a simple and inexpensive biological technique, which can be applied *in planta* as well. The transformation results in either single or low copy number of T-DNA insertions, which prevent homology-dependent gene silencing or rearrangement of inserted genes by recombination. This is advantageous over methods that insert target sequences in multiple copies. However, it can be applied successfully more toward dicot plants than that of monocots. Monocots are generally hard to transform by this method (Katsube-Tanaka and Utsumi 2000); however, several monocots (barley, wheat, rice, and maize) have also been reported to be successfully transformed by this method (Hensel et al. 2009). A list of horticultural crops successfully transformed by this method is given in Table 9.3.

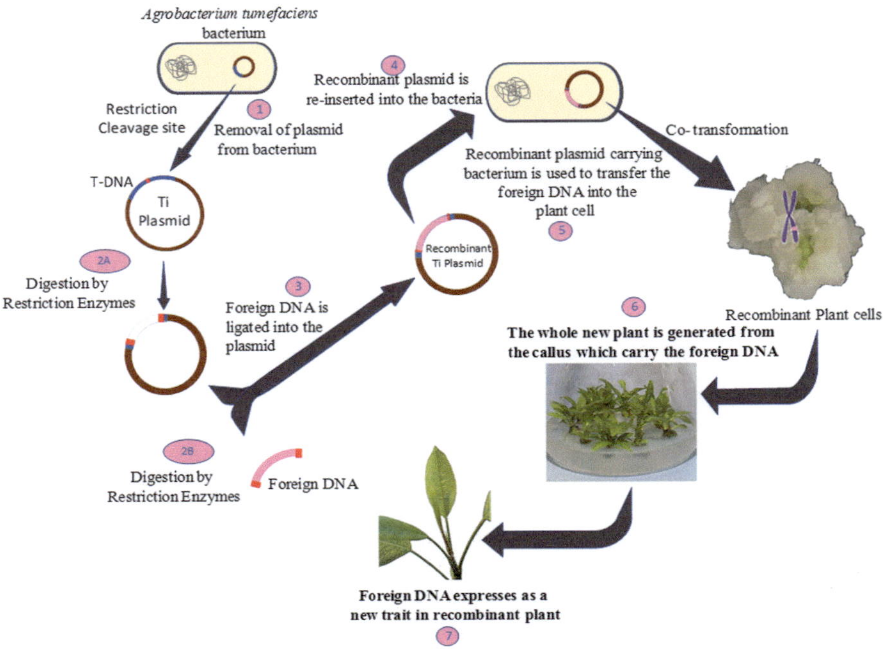

Fig. 9.2 Schematic illustration of the *Agrobacterium*-mediated plant transformation system

Table 9.3 Selected horticultural crops successfully transformed by *Agrobacterium*-mediated transformation

Horticulture crop	Characteristic introduced	Reference
Banana	Disease resistance	Gómez-Lim and Litz (2004)
Citrus	Transformation efficiency test	Mendes et al. (2002)
Grapes	Seedless fruit	Dutt et al. (2007)
Guava	Transformation efficiency test	Mishra et al. (2014)
Lily	Purple flowers	Mori et al. (2004)
Mango	Disease resistance	Gómez-Lim and Litz (2004)
Melon	Resistance to Fusarium wilt	Ntui et al. (2010)
Orange	Virus resistance	Fagoaga et al. (2006)
Papaya	PRSV resistance	Chen et al. (2001)
Papaya	Disease resistance	Azad et al. (2013)
Pineapple	Disease resistance	Gómez-Lim and Litz (2004)
Plum	Virus resistance	Scorza et al. (1994)
Potato	Disease resistance	Khan et al. (2006)
Spinach	Resistant to *Fusarium oxysporum*	Chin et al. (2009)
Strawberry	Disease resistance	Sugaya et al. (2008)
Sunflower	Transformation efficiency test	Rao (1999)

(ii) **Particle-Bombardment-Mediated Transformation**: This is a physicochemical method that bombards micro-carriers containing gene of interest at high speed on plant cell walls using the so called gene gun. This gene gun technique was developed by John Stanford of Cornell University in 1984. It is also called the particle gun method. In this process, micro-carriers made up of gold or tungsten particles are first coated with DNA that contains the gene construct and then propelled forward by helium or CO_2 gases, or an electric discharge, with compressed helium gas being the most widely used accelerant (Katsube-Tanaka and Utsumi 2000). This particle-bombardment method also follows the same procedure as *Agrobacterium*-mediated method till the formation of transgenic construct bearing the promoter, gene of interest, terminator, and selectable marker gene and finally the insertion of the transgenic construct into the Ti plasmid (construct plasmid). Once the construct plasmid is ready, plasmid DNAs are coated on to tungsten or gold particles (micro-projectiles). The coated particles are accelerated with gas pressure and shot using a gene gun into plant tissue kept on a petri plate. The gene construct can be circular or linear plasmid or a linear expression cassette. Factors that affect successful transformation include size and density of micro-carriers, velocity of micro-carriers at the point of impact, nature of plant tissue to be transformed, and suitable pre-culture or pre-treatment of the target plant explants (Anderson et al. 2000). When the micro-projectiles enter inside the plant cell, the transgenes are known to be released from the particle surface and may incorporate into the plant chromosomal DNA. Selectable markers are used to spot the cells that are positively transformed by the transgene. The transformed plant cells are then regenerated into whole plants using tissue culture techniques. Using this method, one can also go for the transformation of organelles like chloroplasts, which enables scientists to develop transplastomic plants. In contrast to plastome, plant mitochondrial genomes have not been transformed so far. Plastome transformation has many advantages, for example, transgenes are normally targeted to a particular locus in the plastome; thereby there are little or no chances of "position effects" that influence gene expression. Moreover, it was reported that in many cases, a very high level of accumulation of recombinant protein was observed when plastomes were transformed instead of nuclear genome; for example, 46% enhanced accumulation of *Bt* Cry2A protein was observed in tobacco when gene was transformed into plastome (De Cosa et al. 2001).

As a whole, this particle-bombardment-mediated transformation method is useful for the transformation of many horticultural crops (Table 9.4). However, this technique requires expensive equipment. In addition, the transformation with the particle gun often gives rise to chimeric plants that consist of both transformed and non-transformed cells. If the reproductive cell line of such chimeras does not contain the target gene, then the next generation of such a plant would not be transformed. Transformation by this method can result in multiple copies of target DNA being inserted randomly anywhere in the plant genome.

Table 9.4 Selected horticultural crops transformed by particle-bombardment method to generate transgenic plants

Horticulture crop	Characteristic introduced	Reference
Orchid	Stable transgenic expression	Kuehnle and Sugii (1992)
Rose	Stable transgenic expression	Marchant (1998)
Chrysanthemum	Stable transgenic expression	Yepes et al. (1995)
Tomato	Resistant to fruit borer	Chaithra et al. (2016)
Banana	Resistance to virus	Ismail et al. (2011)
Cabbage	Plastid transformation	Menq-Jiau Tseng et al. (2014)
Papaya	Virus resistance	Tennant (1994)

Table 9.5 List of selected transgenic horticultural crops developed by RNAi technology

Horticulture crop	Characteristic introduced	Reference
Potato	High amylose starch	Du et al. (2012)
Apple	Reduced oxidative browning	Cao et al. (2004)
Cucumber	Viral resistance	Rodríguez-hernández et al. (2012)
Soybean	High oleic acid, low linoleic acid contents	Zhang et al. (2014)

(iii) **RNA Interference**: RNA interference, or RNAi, is a method of creating transgenic plants by introducing double-stranded RNA (dsRNA) into a plant in order to cause targeted mRNA degradation, which leads to gene silencing. A dsRNA is introduced as a transgene into the target cell. This dsRNA is completely homologous to the target gene. Dicer, a ribonuclease III enzyme, cleaves the dsRNA into double-stranded siRNAs 21–25 nucleotides in length. These siRNAs are integrated into the RNA-induced silencing complex (RISC). Next, a helicase unwinds the RNA duplex in the RISC, thus activating the silencing complex. The activated RISC recognizes and cleaves mRNA that is complementary to the siRNA incorporated into the RISC. This results in mRNA fragments of approximately 22 nucleotides. The siRNA is recycled for the next mRNA degradation. RNA interference, thus, silences genes at the posttranscriptional level. This method has been successfully adopted for raising many transgenic horticultural plants (Table 9.5).

(iv) **Genome Editing**: Conventional breeding techniques for incorporation of desired trait into a horticultural crop lack precision, accuracy, and predictability. Genome editing tools use rationally designed biomolecules (DNA, RNA, or protein) to modify the genetic makeup of crop plants at precise locations with the desired accuracy. Genome editing relies on the process of DNA repair using engineered endonucleases (EENs) which can cleave DNA at a specific sequence because of the presence of a sequence-specific DNA-binding domain or RNA sequence. First, EENs are used to mutate DNA with an addition, deletion, or modification at the desired locus. After DNA cleavage, these nucleases leave a single-stranded break (called a nick) or a double-stranded break. These breaks can be repaired using DNA repair mechanisms, including the non-homologous end joining (NHEJ) process, which is native to the cell and may result in a mutation at the edited site, or the homology directed repair (HDR), in which the cellular machinery incorporates a donor DNA molecule that was provided at the time of DNA editing, leading to gene modification at the target sites. Some of the most popular and majorly used sequence-specific nucleases are zinc finger nucleases (ZFNs), transcription activator-like effector nucleases (TALENs), and the clustered regularly interspaced short palindromic repeats (CRISPR)/Cas9 nuclease system.

(a) **Zinc Finger Nucleases**: Zinc finger domains are DNA-binding motifs in proteins that naturally function as transcription factors. A zinc finger nuclease (ZFN) is a hybrid created by fusing the DNA-cutting nuclease domain of the FokI protein to a zinc finger domain that has been manipulated to bind to a specific DNA sequences. To cut DNA for genome editing, a pair of ZFNs have to work in tandem at the target locus. ZFNs relay on the NHEJ as well as HDR processes of DNA repair to induce mutations.
(b) **Transcription Activator-Like Effector Nucleases**: Like ZFNs, TALENs are also hybrids between DNA-binding and DNA-cutting units. Transcription activator-like effectors (TALEs) are naturally secreted by plant pathogens and bind to promoters of plant genes to suppress its transcription. TALEs can be engineered to specifically bind to any desired site in the target plant genome. When such customized TALEs are hybridized with the FokI nuclease domain, the complex is referred to as TALENs. Like ZFNs, TALENs work in pairs to edit the target genome.
(c) **CRISPR/Cas9**: Clustered regularly interspaced short palindromic repeats (CRISPR)/Cas9 is a bacterial system of innate defense against predators and uses RNA-guided nucleases to specifically cleave foreign DNA sequences. This process has been adapted as a tool for genome editing. The system is comprised of the CRISPR RNA (crRNA) that contains a region complementary to the protospacer adjacent motif (PAM) immediately following the targeted DNA sequence, transactivating crRNA (tracrRNA), and the Cas9 nuclease. In the bacterial system, the foreign DNA is incorporated into the CRISPR cluster. This foreign DNA is used to produce crRNA (~40 nt long). Specificity is conferred by a stretch sequence that is 12 nt upstream of PAM and matches between the RNA and target DNA. The crRNA hybridizes with the tracrRNA to form a guide RNA (gRNA), which binds to and activates the Cas9 nuclease. The gRNA guides the Cas9 to the target DNA site, where it induces double-stranded breaks. The CRISPR/Cas9 system relies on both NHEJ and HDR to ligate the breaks in the DNA.

This system is quite efficient owing to the ability of one Cas9 nuclease to target and alter more than one site by using multiple gRNA. This approach is used when one gRNA is insufficient for disrupting the target gene or even when altering more than one gene simultaneously. This technique has become very popular because of its simplicity, ease of design, specificity of gRNA, ability to target multiple genes,

Table 9.6 List of selected horticultural plants edited by genome editing techniques

Horticulture Crop	Characteristic introduced	Genome editing tool	Reference
Soybean	High oleic acid content	TALENs	Haun et al. (2014)
Potato	Minimized reducing sugars	TALENs	Clasen et al. (2016)
Tomato	High anthocyanin content	TALENs	Čermák et al. (2015)
Potato	High amylopectin content	CRISPR/Cas9	Andersson et al. (2017)
Tomato	Powdery mildew resistance	CRISPR/Cas9	Nekrasov et al. (2017)
Grapefruit	Citrus canker resistance	CRISPR/Cas9	Jia et al. (2017)
Orange	Citrus canker resistance	CRISPR/Cas9	Peng et al. (2017)
Cucumber	Virus resistance	CRISPR/Cas9	Chandrasekaran et al. (2016)
Apple	Stable transgenic expression	ZFNs	Peer et al. (2015)
Fig	Stable transgenic expression	ZFNs	Peer et al. (2015)
Cabbage	Dwarf phenotype	CRISPR/Cas9	Lawrenson et al. (2015)
Carrot	Low anthocyanin biosynthesis	CRISPR/Cas9	Klimek-Chodacka et al. (2018)
Strawberry	Faster growth	CRISPR/Cas9	Zhou et al. (2018)
Watermelon	Albino phenotype	CRISPR/Cas9	Tian et al. (2017)

and low cost compared to ZFNs and TALENs. However, in some cases, the CRISPR/Cas method may also introduce unwanted off-target mutations (Pattanayak et al. 2013). Genome editing has been performed to a number of horticultural plants as listed in the Table 9.6.

9.5 Production of Genetically Modified Horticultural Crops: A Global Scenario

Currently, transgenic technology is used to solve certain specific problems associated with horticultural crops such as enhancement of pest and disease resistance, enhanced drought and salt tolerance, higher tolerance toward metal toxicity, improvement of nutritional value, betterment of herbicide tolerance, etc. The first commercialized transgenic horticultural crop approved was the Flavr Savr tomato, which was approved for commercialization by the US Government on 1994 (Bruening and Lyons 2000). Flavr Savr tomato had modified cell wall genes that made its ripening delayed resulting in better shelf-life and taste. After initial success of Flavr Savr tomato, transgenic papaya with enhanced virus resistance was another big success story among transgenic horticultural crops, which was approved by the USA in 1996 (Azad et al. 2014). At present, a reasonable number of transgenic horticultural crops are already commercialized in different countries, and many more are in the pipeline of final commercialization stage. A list of major horticultural crops raised using transgenic technology is presented in Table 9.7.

Table 9.7 List of transgenic horticultural plants commercialized at various countries

Horticultural crop	Trait(s) modified	Trade name	Manufacturer	Approval country	Commercialization year
Papaya (*Carica papaya*)	Disease resistance	55–1 (Rainbow, SunUp)	Cornell University and University of Hawaii	Canada, USA, Japan	1996
Melon (*Cucumis melo*)	Quality of product	Melon A (NA); Melon B (NA)	Agritope Inc. (USA)	USA	1999
Plum (*Prunus domestica*)	Disease resistance	C-5 (NA)	United States Department of Agriculture - Agricultural Research Service	USA	2007
Potato (*Solanum tuberosum L.*)	Insect resistance	1210 amk (Lugovskoi plus)	Centre Bioengineering, Russian Academy of Sciences	Russia	2007
Potato (*Solanum tuberosum L.*)	Quality of product	AM04–1020 (Starch Potato); G11 (Innate® G Potato)	J.R. Simplot Co.	USA	2014
Potato (*Solanum tuberosum L.*)	Insect and disease resistance	HLMT15–15, –3, –46 (Hi-Lite NewLeaf® Y potato)	Monsanto Company	USA	1998
		RBMT15–101 (NewLeaf® Y Russet Burbank potato)	Monsanto Company	Canada, USA, Australia, Japan, Philippines, Mexico, New Zealand, South Korea	1997, 1998
	Herbicide tolerance and disease resistance	RBMT22–082, –186, –238, –262		Canada, USA, Australia, Japan, Philippines, Mexico, New Zealand,	1997
Eggplant (*Solanum melongena*)	Disease resistance	*Bt* Brinjal Event EE1 (BARI *Bt* Begun-1, –2, –3, –4)	Maharashtra Hybrid Seed Company	Bangladesh	2013

(continued)

Table 9.7 (continued)

Horticultural crop	Trait(s) modified	Trade name	Manufacturer	Approval country	Commercialization year
Chicory (*Cichorium intybus*)	Herbicide tolerance	RM3–3, RM3–4, RM3–5	Bejo Zaden BV (the Netherlands)	USA	1997
Apple (*Malus domestica* cv. Golden Delicious)	Quality of product (non-browning)	GD734	Okanagan Specialty Fruits Incorporated	USA, Canada	2015
Petunia (*Petunia hybrida*)	Quality of product	Petunia-CHS (NA)	Beijing University	China	1998
Squash (*Cucurbita pepo*)	Disease resistance	CZW3 (NA)	Seminis Vegetable Seeds (Canada) and Monsanto Company (Asgrow)	Canada, USA	1994
Sweet pepper (*Capsicum annuum*)	Disease resistance	PK-SP01 (NA)	Beijing University	China	1998
Tomato (*Lycopersicon esculentum*)	Quality of product	1345–4 (NA)	DNA Plant Technology Corporation (USA)	Canada, USA	1995
		B (NA); Da (NA)	Zeneca Plant Science and Peto seed Company	USA, Mexico	1994
		Flavr Savr®	Monsanto Company	Mexico, Canada	1994
	Insect resistance	5345 (NA)	Monsanto Company	Canada, USA	1998

Table 9.8 Increase in global area producing genetically modified crops in last 10 years (since 2006 to 2016)

Year	Area (million hectares)
2006	102.0
2007	143.3
2008	125.0
2009	134.0
2010	148.0
2011	160.0
2012	170.3
2013	175.2
2014	181.5
2015	179.7
2016	185.1

Source: ISAAA (2016).

Flavr Savr tomato was commercialized in 1994 as the first genetically modified crop; thereafter in the last 20 years (1996–2015), a significant number of genetically modified horticultural crops have arrived into the market, and those crops have delivered substantial economic, environmental, health, social, and agronomic benefits to the farmers. The acceptance of genetically modified horticultural crops is steadily increasing among the consumers across the globe. In the last 20 years, approximately 2 billion hectares of land was used for growing genetically modified crops (ISAAA 2016) which comprised of approximately 1.0 billion hectares of genetically modified soybean, 0.6 billion hectares of maize, 0.3 billion hectares of cotton, and 0.1 billion hectares of canola. In the last 10 years, there is significant increase in the total global area (increase in the global area producing genetically modified crop is presented in Table 9.8). In the last few years, genetically modified crops have significantly contributed toward providing food to the current 7.4 billion population. The world population is expected to reach 9.9 billion in 2050 which would require an estimated 50%–70% increase in food production with reduced availability of resources. So together with conventional breeding technology, transgenic crops also provide excellent promises to feed the immensely increasing world population.

At present genetically modified crops are grown in a total of 26 countries, including 19 developing and 7 developed countries. Among the top 10 countries producing genetically modified crops, the USA stands first with production of approximately 72.9 million hectares of genetically modified crops which accounts for approximately 39% of the total global production of transgenic crop. Brazil stands second, followed by Argentina, Canada, India, Paraguay, Pakistan, China, South Africa, and Uruguay (Table 9.9). Transgenic soybean is the main horticultural crop grown

Table 9.9 Area-wise worldwide production of genetically modified crops in the year 2016 by top 10 countries

Country	Production area (million hectares)	Percentage (%) of world production
USA	72.9	39
Brazil	49.1	27
Argentina	23.8	13
Canada	11.1	6
India	10.8	6
Paraguay	3.6	2
Pakistan	2.9	2
China	2.8	2
South Africa	2.7	1
Uruguay	1.3	1

Source: ISAAA (2016).

Table 9.10 Type of genetically modified horticultural crops grown in top four world produces in the year 2016

Country	Type of genetically modified horticultural crop grown (2016)	Total area (million hectares)
USA	Soybean	33.8
	Cotton	3.9
	Canola	0.6
	Sugar beet	0.47
	Alfalfa	8.4
	Papaya	<0.01
	Squash	<0.01
	Potato	<0.01
Brazil	Soybean	33.8
	Cotton	1
Argentina	Soybean	21.1
	Cotton	0.5
Canada	Soybean	2
	Sugar beet	<0.01
	Canola	7.5

Source: ISAAA (2016).

worldwide. The USA is the largest producer of genetically modified soybean, followed by Brazil, Argentina, and Canada. In India, the major genetically modified crop is transgenic mustard, developed by the Delhi University. Currently, this mustard is under environmental release and waiting to be marketed soon (ISAAA 2016). A list of area-wise production of major genetically modified horticultural crops by the top four countries is given in Table 9.10.

9.6 Application of Transgenic Technologies for The Improvement of Horticultural Crops: Commercial Success Story

Genetic modification has been introduced into a number of horticultural crops for the improvement of their traits. We will discuss on some of the most striking commercially successful case studies related to horticultural crops.

9.6.1 Tomato with Enhanced Shelf Life

Flavr Savr™ tomato was the first commercially grown horticultural plant which was developed by Calgene in 1994.This genetically modified tomato bears a *polygalacturonase* (*PG*) gene in antisense orientation, to reduce the ripening of tomatoes to increase their shelf-life. However, Flavr Savr™ tomatoes were not appreciated by the consumers because of inferior quality and therefore rapidly withdrawn from the market (Dias and Ortiz 2014). At present, genes of ethylene biosynthetic pathway, such as 1-aminocyclopropane-1-carboxylate (ACC) synthase and S-adenosyl methionine (SAM) synthase, have been targeted for inducing delayed fruit ripening.

9.6.2 Potato with Enhanced Pest and Disease Resistance

Bt potatoes were developed by putting *Bacillus thuringiensis cry 3A* gene. *Bt* potatoes showed excellent resistance against Colorado potato beetle (*Leptinotarsa decemlineata*), *Potato virus Y*, and *Potato leafroll virus*. These *Bt* potatoes were approved for marketing in the USA in 1995. They are sold in the market in the trade name of NewLeaf® and NewLeaf® Y by the Monsanto Company. This potato was very popular in the USA from 1995 to 1999. Farmers benefited from higher yields and less requirements for insecticides. However, later on, due to lack of proper segregation technology to discriminate between natural and transgenic potato, gradually, people showed negative response toward buying such genetically modified potatoes. Therefore, even after initial success, Monsanto decided to close these genetically potatoes in 2001.

9.6.3 Brinjal (Eggplant) with Enhanced Disease Resistance

Eggplant is vulnerable to attack by many insects, but the most threatening pest is fruit and shoot borer (*Leucinodes orbonalis*), which causes significant losses (up to 70%) (Choudhary and Gaur 2009). Earlier the only control measure for the fruit and shoot borer was insecticide spray. To combat with this problem, Mahyco Company in collaboration with Monsanto developed the first *Bt*-eggplant in the year 2000, which showed excellent resistance against fruit and shoot borer. *Bt*-eggplant was transformed with the *cry1Ac* gene. In the year 2006 and 2009, an expert committee was constituted by the Indian government to look into the biosafety issues and data regarding *Bt*-eggplant. The Genetic Engineering Appraisal Committee (GEAC) of India approved *Bt*-eggplant for commercialization in October 2009. However, in February 2010, the Government of India announced that more time is required before its release to farmers (Jayaraman 2010). However, to reduce pesticide use and enhance yield, Bangladesh recently approved commercialization of four cultivars of *Bt*-eggplant in October 2013 (Shelton et al. 2018).

9.6.4 Papaya with Enhanced Disease Resistance

Papaya production is threatened by a viral disease caused by *Papaya ringspot virus* (PRSV). PRSV infection substantially reduces both fruit yield and quality. To combat with PRSV infection, transgenic papaya was developed by introducing a viral coat protein (CP) gene which provides effective resistance against PRSV strains (Lius et al. 1997). Seeds of such transgenic cultivars "SunUp" and "Rainbow" were available to the farmers in Hawaii since 1998. Cultivation of "Rainbow" cultivar led to a significant increase in papaya production (Lius et al. 1997). Furthermore, farmers are using PRSV-resistant transgenic papaya as a trap crop by growing it on the border around the non-transgenic papaya crop and allowing it to eradicate aphids of PRSV. As a result, the papaya producers can now produce both transgenic and non-transgenic papaya in the same field. The availability of the draft genome of the "SunUp" cultivar makes this papaya cultivar the best-characterized transgenic crop, which would help to overcome regulatory barriers in other countries.

9.6.5 Summer Squash with Multiple Virus Resistance

Summer squash production is badly affected by three viruses, namely, *Watermelon mosaic virus* (WMV), *Cucumber mosaic virus* (CMV), and *Zucchini yellow mosaic virus* (ZYMV). These viruses collectively cause around 60% loss in summer squash production in the USA. In the year 1996, two transgenic summer squash lines were developed by expressing viral coat protein (CP) gene of ZYMV, WMV, and CMV. Such genetically modified summer squash with enhanced resistance against ZYMV, WMV, and CMV was commercialized in the USA in 1996. In subsequent years, several squash cultivars have been developed breeding with the two initially produced transgenic lines. This material is highly resistant to infection by one, two, or all three of the targets (Fuchs et al. 1998). Currently, US farmers are benefitted by growing these virus-resistant summer squash cultivars.

9.6.6 Plum with Resistance Toward Virus Infection

Plum pox virus (PPV) is the causative agent of sharka, one of the devastating diseases of plum and other related stone fruits. The PPV coat protein (CP) gene was transgenically introduced within plum (Ravelonandro et al. 1992) to develop a highly PPV-resistant cultivar "Honey Sweet." To test the PPV resistance, this cultivar was field tested in Europe and tested under greenhouse conditions in the USA (Polak et al. 2008; Dias and Ortiz 2014). Later on transgenic resistance of "Honey Sweet" was transferred to other cultivars through crossbreeding to develop additional PPV-resistant cultivars (Scorza et al. 1998). "Honey Sweet" plum showed excellent fruit quality and high productivity. In the year 2011, "Honey Sweet" got regulatory approval for commercialization in the USA. "Honey Sweet" plum offers an interesting source of virus-resistant germplasm for PPV control worldwide. At present other countries are also in a process to adopt "Honey Sweet" cultivar.

9.6.7 Banana with Resistance Against Xanthomonas Infection

Banana is one of the very important fruit crops grown in the tropical countries with an approximate global production of 130 million tons (FAO 2013). Africa is one of the largest producers of banana. Banana fruit production is heavily affected by wild disease caused by the bacterium *Xanthomonas campestris.* This disease is more prevalent in the African countries. In order to protect bananas from this disease, the International Institute of Tropical Agriculture (IITA) and National Agricultural Research Organization of Uganda developed transgenic bananas expressing the hypersensitivity response-assisting protein (*Hrap*) and the plant ferredoxin-like protein (*Pflp*) genes of *Capsicum annuum.* These genetically modified bananas exhibited enhanced resistance toward *X. campestris.* Currently, those transgenic bananas are under environment testing in Uganda before final commercialization (Dias and Ortiz 2014).

9.7 Nutritional Health Benefits of Transgenic Horticultural Crops

Transgenic horticultural crops are well known for providing enhanced disease resistance, better tolerance to biotic and abiotic stresses, and higher shelf-life. However, transgenic crops can also contribute toward more nutritious and healthy food. We will discuss some of the promising use of transgenic fruits and vegetables in the context of healthy and nutritious food. Tomatoes are a very rich source of carotenoids such as lycopene and β-carotene, which are important health-promoting metabolites. Scientists were trying to enhance lycopene and β-carotene contents of tomatoes using transgenic technology. Earlier Romer et al. (2000) have developed a transgenic tomato by overexpressing bacterial phytoene desaturase gene, which resulted in approximately 45% enhancement in the β-carotene level. In order to combat folate deficiency, transgenic tomatoes were developed with higher levels of folate (Díaz et al. 2004). Lu et al. (2006) developed transgenic cauliflower with enhanced accumulation of β-carotene. Similarly, transgenic *Brassica rapa* was developed by Wahlroos et al. (2004) to produce oilseed with higher proportion of histidine. Similar to folate, calcium (Ca) is also an essential nutrient. Transgenic carrots were developed to increase calcium level, with higher bioavailability (Park et al. 2004). A transgenic lettuce cultivar "Salinas 88" was developed producing very high content of zinc (Zuo 2002). Polyphenols are important nutritional components present mostly in fruits and vegetables, whose dietary intake is directly linked to reduced risk of getting chronic diseases. A transgenic tomato was developed by heterologous expression of flavonoid pathway genes for enhanced production of flavonoids (Schijlen et al. 2006). However, all these examples of transgenic crops given above are under laboratory conditions. Extensive trials and biosafety assessments are required before they actually come into the market.

9.8 Future Challenges, Strategies, and Opportunities of Transgenic Technologies in the Improvement of Horticultural Crop

Although genetically modified horticultural crops have proven to be a powerful tool for management of pest and diseases and several biotic and abiotic stresses with reduced use of pesticides/insecticides and environmental benefits, still their public acceptability is poor, and many countries are still engaged in discussions about the safety issues and potential negative impacts of these crops on the environment and human health. Considering the panic about potential harmful effects of those transgenic plants, different countries have implemented very stringent regulatory systems for the adaptation of these transgenic crops (Ortiz and Smale 2007). People also claim that growing transgenic crops might benefit big seed companies while throbbing small farmers because of the huge monetary investments required for growing these crops successfully. These unwanted concerns are still present despite the highly successful adoption of several transgenic crops like soybeans, canola, cotton, maize, and papaya in many countries of the world. Another reason for poor public acceptance of transgenic crops is that several activists opposed in accepting genetically modified crops. In some cases, these activists have been successful in developing negative attitude among the consumer and government toward adaptation of genetically modified crops. The acceptance for genetically modified crop is very poor in Europe because European governments have inclined to follow the expressed desires by the activists and a majority of consumers opposed to transgenic crops. In contrary, the USA and Canada strongly support farmers and agri-companies to bring genetically modified crops in the market after strict biosafety assessments.

In the last 18 years, there is a steady rise in the global area for the production of transgenic crops. The so-called potential health risks of consuming transgenic crops have been subject to extensive research, but till date as such no dramatic evidence has been found indicating potential health risk from the transgenic crops. Still, many people continue to argue that there are long-term health risks of using transgenic crops, but the kind of risk is yet to be specified. In the year 2009, the Federal

Office of Consumer Protection and Food Safety of Germany and partners published the report "Biological and Ecological Evaluation Towards Long-Term Effects" (popularly known as BEETLE report) which clearly stated that so far, there are no adverse effects on human health reported due to consumption of transgenic crops (Dias and Ortiz 2014; FOCPFS 2009). This report concludes that there are negligible chances of getting adverse health effects in the long run by consuming transgenic crops.

9.9 Summary

Although there are many challenges associated with the use of genetically modified horticultural crops that need to be resolved, we are optimistic those hurdles will be resolved with emerging technologies. The most important point that should be taken into consideration is the public acceptance of new horticultural crops developed using transgenic technology. This is very important to establish clear policies for the generation and use of these genetically modified crops. Ultimately, transgenic technology, in combination with other conventional breeding technologies, will result in more resistant, nutritious, high-yield, tasteful, and healthy vegetables, fruits, and flowers which will fulfill the greater food demand and make our lives healthier. Transgenic crops will succeed in the market if only clear advantages and safety issues are properly demonstrated to both consumers and farmers.

References

Anderson RB, Boynton JE, Dawson J (2000) Sub-micron gold particles are superior to larger particles for efficient biolistic ® transformation of organelles and some cell types. Mol Biol

Andersson M, Turesson H, Nicolia A, Fält AS, Samuelsson M, Hofvander P (2017) Efficient targeted multiallelic mutagenesis in tetraploid potato (*Solanum tuberosum*) by transient CRISPR-Cas9 expression in protoplasts. Plant Cell Rep 36:117–128

Azad M, Rabbani MG, Amin L, Sidik NM (2013) Development of transgenic papaya through agrobacterium -mediated transformation. Int J Genomics:1–5

Azad MAK, Amin L, Sidik NM (2014) Gene technology for papaya ringspot virus disease management. Sci World J 2014:1–11

Bruening G, Lyons JM (2000) The case of the FLAVR SAVR tomato. Calif Agric 54:6–7

Cao YH, Zhang Z, Yao QH, Peng RH, Xiong AS, Li X (2004) Suppression of apple polyphenol oxidase by double-stranded RNA (RNAi). Shi Yan Sheng Wu Xue Bao 37:487–493

Čermák T, Baltes NJ, Čegan R, Zhang Y, Voytas DF (2015) High-frequency, precise modification of the tomato genome. Genome Biol 16:232

Chaithra N, Gowda PHR, Guleria N (2016) Transformation of tomato with Cry2ax1 by biolistic gun method for fruit borer resistance. Int J Agric Environ Biotechnol 8:795

Chandrasekaran J, Brumin M, Wolf D, Leibman D, Klap C, Pearlsman M, Sherman A, Arazi T, Gal-On A (2016) Development of broad virus resistance in non-transgenic cucumber using CRISPR/Cas9 technology. Mol Plant Pathol 17:1140–1153

Chen GC, Ye M, Huang JC, Yu M, Li BJ (2001) Cloning of the papaya ringspot virus (PRSV) replicase gene and generation of PRSV-resistant papayas through the introduction of the PRSV replicase gene. Plant Cell Rep 20:272–277

Chin DP, Bao JH, Mii M (2009) Transgenic spinach plants produced by Agrobacterium-mediated method based on the low temperature-dependent high plant regeneration ability of leaf explants. Plant Biotechnol 26:243–248

Choudhary B, Gaur K (2009) The development and regulation of Bt brinjal in India

Clark D, Klee H, Dandekar A (2004) Despite benefits, commercialization of transgenic horticultural crops lags. Calif Agric 58:89–98

Clasen BM, Stoddard TJ, Luo S, Demorest ZL, Li J, Cedrone F, Tibebu R, Davison S, Ray EE, Daulhac A et al (2016) Improving cold storage and processing traits in potato through targeted gene knockout. Plant Biotechnol J 14:169–176

De Cosa B, Moar W, Lee S-B, Miller M, Daniell H (2001) Overexpression of the Bt cry2Aa2 operon in chloroplasts leads to formation of insecticidal crystals. Nat Biotechnol 19:71–74

Díaz R, Quinlivan EP, Klaus SMJ, Basset GJC, Gregory JF, Hanson AD (2004) Folate bio fortification in tomatoes by engineering the pteridine branch of folate synthesis. Proc Natl Acad Sci 101(38):13720–13725. https://doi.org/10.1073/pnas.0404208101

Dias JS, Ortiz R (2013) Transgenic vegetables for 21st century horticulture. Acta Hortic:15–30

Dias JS, Ortiz R (2014) Advances in transgenic vegetable and fruit breeding. Agric Sci:1448–1467

Dias JS, Ryder E (2011) World vegetable industry: production, breeding, trends. Hortic Rev 38:299–356

Du H, Yang T, Ma C, Feng D, Zhang N, Si H, Wang D (2012) Effects of RNAi silencing of SSIII gene on phosphorus content and characteristics of starch in potato tubers. J Integr Agric 11:1985–1992

Dutt M, Li ZT, Dhekney SA, Gray DJ (2007) Transgenic plants from shoot apical meristems of *Vitis vinifera* L. "Thompson Seedless" via Agrobacterium-mediated transformation. Plant Cell Rep 26:2101–2110

Fagoaga C, Rodrigo I, Conejero V, Hinarejos C, Tuset JJ, Arnau J, Pina JA, Navarro L, Peña L (2006) Increased tolerance to *Phytophthora citrophthora* in transgenic orange plants constitutively expressing a tomato pathogenesis related protein PR-5. Mol Breed 7:175–185

Federal Office of Consumer Protection and Food Safety (Germany) and Partners FOCPFS (2009) Long-Term Effects of Genetically Modified (GM) Crops on health and the environment (Including Biodiversity): prioritization of potential. Risks and delimitation of uncertainties. Federal office of consumer protection of food safety, Berlin

Food and agriculture organization of the United Nations FAO Stat 2013

Fuchs M, Klas FE, McFerson JR, Gonsalves D (1998) Transgenic melon and squash expressing coat protein genes of aphid-borne viruses do not assist the spread of an aphid non- transmissible strain of cucumber Mosaic Virus in the Field. Transgenic Res 7:449–462

Gómez-Lim MA, Litz RE (2004) Genetic transformation of perennial tropical fruits. In Vitro Cell Dev Biol – Plant 40:442

Haun W, Coffman A, Clasen BM, Demorest ZL, Lowy A, Ray E, Retterath A, Stoddard T, Juillerat A, Cedrone F et al (2014) Improved soybean oil quality by targeted mutagenesis of the fatty acid desaturase 2 gene family. Plant Biotechnol J 12:934–940

Hensel G, Kastner C, Oleszczuk S, Riechen J, Kumlehn J (2009) Agrobacterium -mediated gene transfer to cereal crop plants: current protocols for barley, wheat, triticale and maize. Int J Plant Genomics:1–9

Holme IB, Wendt T, Holm PB (2013) Intragenesis and cisgenesis as alternatives to transgenic crop development. Plant Biotechnol J 11:395–407

ISAAA (2016) Increase in global area producing genetically modified crops in last ten years since 2006 to 2016

Ismail RM, El-Domyati FM, Wagih EE, Sadik AS, Abdelsalam AZE (2011) Construction of banana bunchy top nanovirus-DNA-3 encoding the coat protein gene and its introducing into banana plants cv. Williams. J Genet Eng Biotechnol 9:35–41

Jayaraman K (2010) Bt brinjal splits Indian cabinet. Nat Biotechnol 28:296

Jia H, Zhang Y, Orbović V, Xu J, White FF, Jones JB, Wang N (2017) Genome editing of the disease susceptibility gene CsLOB1 in citrus confers resistance to citrus canker. Plant Biotechnol J 15:817–823

Kays SJ (2011) Cultivated vegetables of the world: a multilingual onomastic on, Wageningen Academic Publishers, The Netherlands. https://doi.org/10.3920/978-90-8686-720-2

Keatinge JDH, Waliyar F, Jammadass RH, Moustafa A, Andrade M, Drechsel P, Hughes J'A, Kardivel P, Luther K (2010) Re-learning old lessons for the future of food: by bread alone no longer—Diversifying diets with fruit and vegetables. Crop Sci 50:51–62

Khan RS, Nishihara M, Yamamura S, Nakamura I, Mii M (2006) Transgenic potatoes expressing wasabi defensin peptide confer partial resistance to gray mold (*Botrytis cinerea*). Plant Biotechnol 23:179–183

Klimek-Chodacka M, Oleszkiewicz T, Lowder LG, Qi Y, Baranski R (2018) Efficient CRISPR/Cas9-based genome editing in carrot cells. Plant Cell Rep 37:575–586

Kuehnle A, Sugii N (1992) Transformation of dendrobium orchid using particle bombardment of protocorms. Plant Cell Rep 11

Lawrenson T, Shorinola O, Stacey N, Li C, Østergaard L, Patron N, Uauy C, Harwood W (2015) Induction of targeted, heritable mutations in barley and *Brassica oleracea* using RNA-guided Cas9 nuclease. Genome Biol 16:258

Lius S, Manshardt RM, Fitch MMM, Slightom JL, Sanford JC, Gonsalves D (1997) Pathogen-derived resistance provides papaya with effective protection against papaya ringspot virus. Mol Breed 3:161–168

Lu S, Van Eck J, Zhou X, Lopez AB, O'Halloran DM, Cosman KM, Conlin BJ, Paolillo DJ, Garvin DF, Vrebalov J et al (2006) The cauliflower Or gene encodes a DnaJ cysteine-rich domain-containing protein that mediates high levels of -carotene accumulation. Plant Cell Online 18:3594–3605

Marchant R (1998) Biolistic Transformation of Rose (Rosa hybridaL.). Ann Bot 81:109–114

Mendes BMJ, Boscariol RL, Mourão Filho F de AA, Almeida WAB de (2002) Agrobacterium-mediated genetic transformation of 'Hamlin' sweet orange. Pesq Agrop Brasileira 37:955–961

Mishra M, Jalil SU, Sharma N, Hudedamani U (2014) An Agrobacterium mediated transformation system of guava (*Psidium guajava* L.) with endochitinase gene. Crop Breed Appl Biotechnol 14:232–237

Mori S, Kobayashi H, Hoshi Y, Kondo M, Nakano M (2004) Heterologous expression of the flavonoid 3,5-hydroxylase gene of Vinca major alters flower color in transgenic Petunia hybrida. Plant Cell Rep 22:415–421

Nekrasov V, Wang C, Win J, Lanz C, Weigel D, Kamoun S (2017) Rapid generation of a transgene-free powdery mildew resistant tomato by genome deletion. Sci Rep 7:482

Ntui V, Otang V, Sher Khan R, Bullet D, Chin P, Bullet I, Bullet N, Mii M (2010) An efficient Agrobacterium tumefaciens-mediated genetic transformation of '"Egusi"' melon (Colocynthis citrullus L.)

Ortiz R, Melinda S (2007) Transgenic technology: pro-poor or pro-rich. Chronica Hortic 47:9–12

Park S, Kim C-K, Pike LM, Smith RH, Hirschi KD (2004) Increased calcium in carrots by expression of an Arabidopsis H +/Ca 2+ transporter. Mol Breed 14:275–282

Pattanayak V, Lin S, Guilinger JP, Ma E, Doudna JA, Liu DR (2013) High-throughput profiling of off-target DNA cleavage reveals RNA-programmed Cas9 nuclease specificity. Nat Biotechnol 31:839

Peer R, Rivlin G, Golobovitch S, Lapidot M, Gal-On A, Vainstein A, Tzfira T, Flaishman MA (2015) Targeted mutagenesis using zinc-finger nucleases in perennial fruit trees. Planta 241:941–951

Peng A, Chen S, Lei T, Xu L, He Y, Wu L, Yao L, Zou X (2017) Engineering canker-resistant plants through CRISPR/Cas9-targeted editing of the susceptibility gene CsLOB1 promoter in citrus. Plant Biotechnol J 15:1509–1519

Pimentel D (1997) Techniques for reducing pesticide use. Economic and environmental benefits. Wiley, New York

Polák J, Pívalová J, Kundu JK, Jokes M (2008) Behaviour of transgenic plum pox virus -resistant *Prunus domestica* L. Clone C5 grown in the open field under a high and permanent infection pressure of the Ppv-Rec strain. J Plant Pathol 90:2003–2006

Ram RB, Dasgupta SK (2005) Transgenic vegetable crops. J New Seeds 6:409–428

Rao K (1999) Agrobacterium-mediated transformation of Sunflower (*Helianthus annuus L.*): A Simple Protocol. Ann Bot 83:347–354

Ravelonandro M, Monsion M, Teycheney PY, Delbos R, Dunez J (1992) Construction of a chimeric viral gene expressing plum pox virus coat protein. Gene 120:167–173

Rodríguez-hernández AM, Gosalvez B, Sempere RN, Burgos L, Aranda MA, Truniger V (2012) Melon RNA interference (RNAi) lines silenced for Cm-eIF4E show broad virus resistance. Mol Plant Pathol 13:755–763

Römer S, Fraser PD, Kiano JW, Shipton CA, Misawa N, Schuch W, Bramley PM (2000) Elevation of the provitamin A content of transgenic tomato plants. Nat Biotechnol 18:666–669

Schijlen E, Ric de Vos CH, Jonker H, van den Broeck H, Molthoff J, van Tunen A, Martens S, Bovy A (2006) Pathway engineering for healthy phytochemicals leading to the production of novel flavonoids in tomato fruit. Plant Biotechnol J 4:433–444

Schouten HJ, Krens FA, Jacobsen E (2006) Cisgenic plants are similar to traditionally bred plants: International regulations for genetically modified organisms should be altered to exempt cisgenesis. EMBO Rep 7:750–753

Scorza R, Ravelonandro M, Callahan A, Cordts J, Fuchs M, Dunez J, Gonsalves D (1994) Transgenic plums (Prunus domestica L.) express the plum pox virus coat protein gene. Plant Cell Rep 14

Scorza R, Callahan AM, Levy L, Damsteegt V, Ravelonandro M (1998) Transferring pot virus coat protein genes through hybridization of transgenic plants to produce plum pox virus resistant plums (*Prunus domestica L.*). Acta Hort (472):421–427

Shelton AM, Hossain MJ, Paranjape V, Azad AK, Rahman ML, Khan ASMMR, Prodhan MZH, Rashid MA, Majumder R, Hossain MA et al (2018) Bt Eggplant Project in Bangladesh: History, Present Status, and Future Direction. Front Bioeng Biotechnol 6

Sticklen M (2015) Transgenic, cisgenic, intragenic and subgenic Crops. Adv Crop Sci Technol 03

Sugaya T, Yano M, Sun H-J, Hirai T, Ezura H (2008) Transgenic strawberry expressing the taste-modifying protein miraculin. Plant Biotechnol 25:329–333

Tanaka KT, Utsumi S (2000) Plant Transformation and Transgenic Crops. Food Sci Technol Res 6:241–247

Tanuja P, Kumar AL (2017) Transgenic fruit crops – A review. Int J Curr Microbiol App Sci 6:2030–2037

Tennant PF (1994) Differential protection against papaya ringspot virus isolates in coat protein gene transgenic papaya and classically cross-protected Papaya. Phytopathology 84:1359

Tian S, Jiang L, Gao Q, Zhang J, Zong M, Zhang H, Ren Y, Guo S, Gong G, Liu F et al (2017) Efficient CRISPR/Cas9-based gene knockout in watermelon. Plant Cell Rep 36:399–406

Tseng MJ, Yang MT, Chu WR, Liu CW (2014) Plastid Transformation in Cabbage (*Brassica oleracea* L. var. capitata L.) by the Biolistic Process. Chloroplast Biotechnol:355–366

Wahlroos T, Susi P, Solovyev A, Dorokhov Y, Morozov S, Atabekov J, Korpela T (2004) Increase of histidine content in *Brassica rapa* subsp. oleifera by over-expression of histidine-rich fusion proteins. Mol Breed 14:455–462

Xiong J-S, Ding J, Li Y (2015) Genome-editing technologies and their potential application in horticultural crop breeding. Horticulture Res 2:15019

Yepes L, Mittak V, Pang S-Z, Gonsalves C, Slightom J, Gonsalves D (1995) Biolistic transformation of chrysanthemum with the nucleocapsid gene of tomato spotted wilt virus. Plant Cell Rep 14

Zhang L, Yang X, Zhang Y, Yang J, Qi G, Guo D, Xing G, Yao Y, Xu W, Li H et al (2014) Changes in Oleic Acid Content of Transgenic Soybeans by Antisense RNA Mediated Post transcriptional Gene Silencing. Int J Genomics:1–8

Zhou J, Wang G, Liu Z (2018) Efficient genome editing of wild strawberry genes, vector development and validation. Plant Biotechnol J 16:1868–1877

Zuo X (2002) Expression of the mouse metallothionein mutant -cDNA in the lettuces (*Lactuca sativa L.*). Chin Sci Bull 47:558

Commercial Applications of Transgenic Crops in Virus Management

10

Ashirbad Guria and Gopal Pandi

Abstract

The economically important crops such as cotton, potato, tomato, etc. are infected by various plant viruses and pose major limitations for the production. Plant viral diseases are difficult to manage; it is strongly advised to use insecticides to eradicate the viral pathogens. Alternatively, implementation of virus-free source materials or selection of disease-resistant plants has been recommended. However, above approaches often result in failure, and scientist has proposed the concept of pathogen-derived resistance (PDR) which accelerated the research to acquire virus resistance plant through genetic engineering. Genetic engineering approaches are highly beneficial to confer and control the viral diseases in a wide range of crops. Compared to traditional breeding methods, genetic engineering offers precise and rapid technology to attain the virus-resistant plants. The mechanism behind the PDR can be a protein- or nucleic acid-mediated pathway. Based on available results, it has been observed that protein-mediated resistance confers reasonable protection against wide variety of related viruses while nucleic acid-mediated resistance provides an optimum level of protection which is only against specific virus or closely related strains of a virus. Coat protein-mediated protection has been reported for *Papaya ringspot virus*, *Tobacco mosaic virus*, *Cucumber mosaic virus*, *Alfalfa mosaic virus*, *Potato virus X*, *Potato virus Y* and *Potato leaf roll virus*. In addition, sequences from the viral replicase gene, defective virus movement protein genes and virus antisense RNA have been engineered into plants to obtain virus resistance. This chapter is devoted to the comprehensive description of the various strategies used to produce virus-resistant plants.

A. Guria · G. Pandi (✉)
Department of Plant Biotechnology, School of Biotechnology, Madurai Kamaraj University, Madurai, India
e-mail: pgopal.biotech@mkuniversity.org

R. Sathishkumar et al. (eds.), *Advances in Plant Transgenics: Methods and Applications*, https://doi.org/10.1007/978-981-13-9624-3_10

Keywords
Crops · Plant viral diseases · Pathogen-derived resistance · *Papaya ringspot virus* · *Tobacco mosaic virus* · Virus resistance

10.1 Introduction

With the increasing burden of ever-growing population, considerable efforts have been taken to augment the cultivable area to meet up the world's food requirement. However, such attempt is often hampered by the equally needed industrial growth again for the benefit of human for which land is the first and foremost requirement. Further, based on the population projections by the United Nations and income estimation from the Food and Agricultural Policy Research Institute (FAPRI), global food demand is anticipated to rise persistently for the sufficient supply of the demand by overcoming pressure on the farm lands from urbanisation, climate change and competition from high-value agriculture cereals. As area under cultivation has increased over the years, crops infected with the emerging pest and diseases have also shot up. Farmers have incurred huge investments in excessive usage of pesticides, fungicides, antiviral chemicals, etc. over the years in combating plant diseases. This has though proved them with immediate success but causes long-term post-effect with respect to depletion of natural flora and fauna leading to poorer soil quality, unfit for growing of other crops if proper care is not taken like integrated pest management (IPM) and other cultivation practices. Moreover, pests frequently acquire resistance to particular dosage, convert into super bugs, become persistent in soil and cause sustainable losses to crops and farmer's livelihood. However, recent research on biotechnology and genetic engineering (GE) on plants has given impetus to overcome the issues related with the sustained production. Indeed, recent technological advancement such as genetic modification (GM) of the plants has helped us to increase the food production by minimising the loss due to biotic and abiotic stress.

GM crops have taken a foothold more than three decades back. Although it is not widely accepted across the globe, it does have many success stories in both industrial and developing nations. Presently, a large number of countries across continents are negotiating to bring laws pertinent to the introduction and growth of GM crops. As a result, more number of transgenic plant events is getting approved day by day for agriculturally important crops, trees, fruits and even ornamentals and is expected to cover to a wider range of plants in the future. Area under GM crops has increased significantly to around 190 million ha in 2017 as compared to 1.7 million ha in 1996 by 17 million farmers from 24 countries worldwide (http://www.isaaa.org/resources/infographics/default.asp). In addition to it, 43 countries have adopted the GM crops by importing them. The top 5 GM crop-growing countries have reached close to its adoption saturation, in decreasing order of acreage under cultivation, topped by the USA (~40% of global area under GM crops) followed by Brazil (~27%), Argentina (12.4%), Canada (~7%) and India (6%). Spain, which grows ~95% of biotech crops of Europe, leads among the EU countries, successfully cultivating GM maize (http://www.isaaa.org/resources/infographics/default.asp). Sudan (presently South and North Sudan) and South Africa are the two major African countries actively growing GM plants like cotton, maize and soybean. ~50% of the world's biotech crop area is under the cultivation of GM soybean followed by GM maize (holds ~31% of GM area), GM cotton (12.8%) and GM canola (5.4%) possessing mostly herbicide resistance (HR), insecticide resistance (IR) or combination of both. As of 2013, 57% of the transgenic crop area is cultivated for HR GM crops followed by HR+IR crops (27%) and IR crops (16%). Generally, plant resistance against various phytopathogens including viruses is derived by enhancing the RNA silencing or N or R gene-related pathway (Cillo and Palukaitis 2014). Alternatively, pathogen-derived resistance (PDR) can be employed by expressing the pathogen protein to confer resistance against various species of viruses or producing immunity against particular pathogen by using the nucleic acid-mediated resistance. Utilisation of variety of sequences and subsequent resistance mechanism has been comprehensively described by many researchers across the globe (Cillo and Palukaitis 2014; Kavanagh and Spillane 1995). However, in this chapter, we exclusively focus on the management of the virus resistance via genetic modification (GM) of the plants with more emphasis on the model and the non-model plants.

10.2 Essentiality of Virus Resistant (VR) Transgenic Plant Using GE Approach

Infection by virus may not always bring visible symptoms to the plants. Often, plants show asymptomatic when the viral titre is under the threshold level (Ghosal and Sanfacon 2015). Interaction between virus and resistant plant produced local lesions (Carr et al. 2010; Padmanabhan and Dinesh-Kumar 2014), whereas a susceptible plant tends to produce systemic infection (Ghosal and Sanfacon 2015). In general, there will be reduction in yield and average life-span of plants on post-virus infection. Symptoms are mostly seen on the leaves, while virus also shows striking symptoms on stem, fruits and roots (http://www.agriinfo.in/default.aspx?page=topic&superid=5&topicid=1855). Some common symptoms which one can observed during virus infection are mosaic, yellowing, mottling, shortening of plant height, chlorosis and necrosis of leaves and other developmental abnormalities (Roossinck 2010). Blackening (http://www.agriinfo.in/default.aspx?page=topic&superid=5&topicid=1855) and blistering of leaves by *Potato virus Y* (PVY) along with *Potato virus X* (PVX) (Senanayake and Mandal 2014), rolling and leathery leaves with yellow or reddish coloration, phloem and tuber necrosis by *Potato leaf roll virus* (PLRV) (http://vegetablemdonline.ppath.cornell.edu/NewsArticles/Potato_Virus.htm) are few other markedly known visible symptoms of virus infestation, bronzing by *Potato virus S* (PVS) and *Tomato spotted wilt virus* (TSWV) (http://www.agri-info.in/default.aspx?page=topic&superid=5&topicid=1855) are few other markedly known visible symptoms of virus infestation. Intriguingly, the virus infection also affects and produces visible symptoms on the fruit as in the case of apple fruit pitting by apple groove virus. Stem pitting by *Citrus tristeza virus* (CTV) on kagzi lime, swelling of stem by *Cocoa swollen shoot virus* (CSSV), vein banding by *Beet curly top virus* (BCTV), tumour formation by *Clover wound tumour virus*, oily streaks on petioles and upper part of trunk, ringspots and shoestring appearance on leaves by *Papaya ringspot virus* (PRSV) on papaya (https://www.apsnet.org/edcenter/intropp/lessons/viruses/pages/papayaringspotvirus.aspx), leaf enation by *Tobacco leaf curl virus* (TLCV), *Cotton leaf curl virus* (CLCuV), *Okra enation leaf curl virus* (OELCuV), etc. are some of the other striking symptoms that are visible after virus has established a compatible interaction with its host. On the basis of symptoms, a significant yield reduction across the agricultural hotspots has been noted (Table 10.1).

In general, there is no specific viricide or antiviral chemicals that can kill specific virus without affecting the host. In combination with poor socio-economic condition, farmers find uprooting and burning of virus-infected plants being the only feasible option and hence practiced in many parts of the world even today which causes environmental pollution and loss of habitat. In another way, farmers use insecticide in an attempt to reduce the virus-transmitted insect vectors. But this strategy causes effect on host and nearby-grown non-host plants and degrades soil quality and environment which eventually results in loss and debt with farmers. Despite losses, intense exploration on the infected plant field leads to identification of resistant plant for against the particular strain of virus or relative or diverse virus families. This phenomenon is called as cross-protection which is shown by McKinney against *Tobacco mosaic virus* (TMV) (McKinney 1929). Although mechanism of cross-protection is not understood (Bergua et al. 2016; Folimonova 2013), scientists have developed many agriculturally important crop plants for cross-protection against different viruses, the prominent example being the protection of citrus plant from the CTV infection. Simultaneously, breeders are on the continuous look out for dominant virus-resistant genes which are generally difficult to find and mostly time-consuming. In order to prevent the yield losses by various viruses in a short time, a hypothesis is proposed to confer resistance against viruses which is known as pathogen-derived resistance (PDR) (Hamilton 1980; Sanford and Johnston 1985). By following the footstep of the PDR concept, Robert Beachy's group produced a transgenic tobacco plant by expressing the coat protein of TMV which showed good resistance upon the same virus infection by delaying the infection (Abel et al. 1986). Consequently, various sequences of viruses have been utilised to provide immunity against different kinds of viruses with the promising outcome. These sequences are either proteins which are involved in structural, replication association, movement function or other proteins and variety of RNAs (non-coding RNAs, hairpin RNAs, inverted repeat RNAs, double-stranded RNAs, antisense RNAs, etc.). Alternatively, non-viral sequences (N or R genes, RNases, antiviral inhibitor proteins, plantibodies, etc.), response elicitors and various host-derived genes can be used to enhance the virus resistance. With the rapid progress seen in genome editing technology, researchers may soon come out with VR GM plants developed using CRISPR/Cas9 approach. Benefits of the genetically engineered commercially approved crop are found to be multifold and detailed later in the chapter.

Table 10.1 Estimate of loss by virus attack on field crop

Virus	Crop	Yield loss	Reference(s)
TSWV	Tomato	90%	Van Zijl et al. (1986)
TSWV	Lettuce	90%	Cho et al. (1987)
Barley yellow dwarf virus (BYDV)	Barley	75–80%	Andrews and Paliwal (1986) and Risius et al. (1991a)
BYDV	Wheat	60%	Risius et al. (1991b)
BYDV	Oats	60%	Haber (1990)
Rice yellow mottle virus (RYMV)	Rice	82%	Taylor et al. (1990)
PRSV	Bottle gourd	100%	Singh (1989)
GFLV	Grapevine	70%	Walter (1988)
Cowpea mosaic virus (CPMV)	Cowpea	65%	Thouvenel (1988)
Bean common mosaic virus (BCMV)	Bean	65%	Vishwa and Gutha (1990)
Rice tungro bacilliform virus (RTBV) + *Rice tungro spherical virus* (RTSV)	Rice	85%	Hasannuddin and Hibino (1989)
Maize streak virus (MSV) + MSpV	Maize	100%	Roca de Doyle and Autrey (1992)
Rice gall dwarf virus (RGDV)	Rice	150 thousand tonnes	Oerke et al. (1994)
Soybean mosaic virus (SMV)	Soybean	132 thousand tonnes	Oerke et al. (1994)
Maize dwarf mosaic virus (MDMV)	Maize	2.5 million tonnes	Oerke et al. (1994)
BYDV	Barley	350 thousand tonnes	Oerke et al. (1994) and Pennazio et al. (1996)
Bean golden mosaic virus (BGMV)	Bean	90–280 thousand tonnes/year	Aragao et al. (2013)
Cucumber mosaic virus (CMV)	Peanut	30%	Breuil et al. (2012)
CMV	Capsicum	28–40%	Myti et al. (2014) and Rahman et al. (2016)
CMV	Tomato	15–25%	Montasser et al. (1991), Sikhora et al. (1998) and Jalender et al. (2015)
Watermelon mosaic virus (WMV)	Melon, squash, watermelon	78%, 40%, 35%	Ayazpour and Vahidian (2016)
PLRV	Potato	20 million tonnes/year	Abbas et al. (2016)
PVY + PLRV	Potato	60–75%	Gupta et al. (1985) and Rahman and Akanda (2008)
PVY	Potato	95%	Hossain and Ali 1992 and Rahman and Akanda (2008)
PLRV	Potato	80–90%	Rovere et al. (2001)

Virus-resistant (VR) transgenic crops are grown in less than 1% of global GM crop area and reduced from 40% when VR GM crops were first introduced for cultivation in 1996, either as a single trait or stacked along with IR or/and HR trait(s). It is also grown to confer resistance against a single virus or multiple viruses. Although it holds a small proportion in terms of acreage under cultivation, but yield loss can escalate up to 96% as in *Bhendi yellow vein mosaic virus* (BYVMV)-infected field as reported previously (Pun and Doraiswamy 1999). Several virus-causing diseases proved endemic wiping out entire field, thereby, dwindling national loss tremendously, the classical example being the havoc caused by PRSV in Hawaii which has reduced papaya production by more than 40% within 5 years (Slater et al. 2013). Estimation of loss by virus alone is difficult to obtain as field crop is under multiple abiotic and biotic stresses. However, there are few old reports (Table 10.1) to show the extent of damage caused by virus to stress on the need of having more VR GM crops for cultivation. Until now, there are seven agricultural crops which harbour VR against different viruses and are approved for commercialisation. However, due to varied regulatory terms prevailing in different countries further protested by non-governmental pleas, VR GM crops could not show its benefits to the masses.

10.3 Methods to Develop VR GM Crops

Plant breeders over the years have tirelessly crossbred different varieties with the hope of getting superior quality in terms of yield, quality, resistance against pest and diseases, etc. However, conventional breeding is a time-consuming process which takes at least 12–15 years to get a hybrid ready for commercial cultivation. With the development of several plant genome sequencing initiatives and other advancement in the genetic engineering (cloning, DNA double helix, etc.), both forward genetics and reverse genetics are at interplay. Several methods have since then tried with varied results; some of which are detailed below, in developing virus-resistant genetically modified VR GM plants.

10.4 Coat Protein-Mediated Resistance (CPMR)

Numerous approaches are followed in developing a GM plant against the invading virus to different levels of tolerance or resistance. The most frequently adopted technique utilises the expression of coat protein (CP) gene of a virus cloned in a plant to derive resistance against either the same or related virus called as CPMR. It was first tested on transgenic *N. tabacum* with complete resistance against TMV (Abel et al. 1986) which was inherited across many generations. Interestingly, it is also observed that CP gene of a virus provides protection to related or diverse viruses owing to amino acid sequence homology between them (Kavanagh and Spillane 1995; Stark and Beachy 1989; Nejidat and Beachy 1990). The idea marked the development of virus-free GM plants which could be used for mass cultivation in an attempt to prevent the already existing epidemic caused by viruses. Multiple field trials were conducted to check for the similar agronomic traits present in GM crops as those already present

Table 10.2 Lists of transgenic plants to confer CPMR

CP of virus	Method of introduction	Transgenic plant developed (event)	Resistance against virus	Reference(s)	Approved for commercialisation
TMV	*Agrobacterium* mediated	*N. tabacum*	TMV	Abel et al. (1986)	No
Tomato yellow leaf curl virus (TYLCV)	*Agrobacterium* mediated	Tomato	TYLCV	Kunik et al. (1994)	No
Alfalfa mosaic virus (AlMV)	*Agrobacterium* mediated	*N. tabacum*	AlMV	Loesch-Fries et al. (1987)	No
PRSV	Microparticle bombardment	Papaya (55-1)	PRSV	Gonsalves et al. (2004)	Yes
PRSV	*Agrobacterium* mediated	Papaya (Huanong No. 1)	PRSV	Guo et al. (2009)	Yes
Plum pox virus (PPV)	*Agrobacterium* mediated	*N. benthamiana*, Plum (C-5)	PPV	Ravelonandro et al. (2000)	No, yes
CMV	*Agrobacterium* mediated	Sweet pepper	CMV	Zhu et al. (1996) and Chen et al. (2003)	Yes
CMV	*Agrobacterium* mediated	Tomato	CMV	Yang et al. (1995) and Chen et al. (2003)	Yes
CMV + *Zucchini yellow mosaic virus* (ZYMV) + WMV2	*Agrobacterium* mediated	Squash (CZW3)	CMV + ZYMV + WMV2	Fuchs et al. (1998)	Yes
ZYMV + WMV2	*Agrobacterium* mediated	Squash (ZW20)	ZYMV + WMV2	Fuchs et al. (1998)	Yes

in non-GM crops with high success (Potato-Jongedijk et al. 1992; Tomato-Sanders et al. 1992; Cucumber-Gonsalves et al. 1992; Kavanagh and Spillane 1995, etc.). Table 10.2 shows the inexhaustible list of transgenic plant developed with CP gene of different viruses, some of which are commercialised.

10.5 Replicase Protein-Mediated Resistance (RPMR)

Replicase-mediated protection mechanism is the second most frequently adopted technique (Rovere et al. 2001; Varma et al. 2002). In case of *Geminivirus* and *Luteovirus* families, CP-mediated protection fails to confer adequate level of resistance. In such case, replicase gene construct can be used for obtaining resistant plants which offer protection either protein-mediated or by RNA-mediated, both of which are mentioned separately. Modified or mutant *Rep* REP provides stronger and specific resistance against virus as compared to unmodified full length protein (Rovere et al. 2001). However, there are few exceptions like the high resistance from PVX by the transgenic plant encoding full-length PVX Rep geneprotein (Donson et al. 1993; Braun and Hemenway 1992). Table 10.3 shows some of the examples of transgenic plants developed by RPMR.

10.6 Movement Protein-Mediated Resistance (MPMR)

Scientists have also explored the possible mechanism of using movement protein (MP) gene as a tool to develop VR GM crops to protect against viruses. It is observed that functional movement protein (MP) does not confer any protection against the virus in transgenic-derived plant (Ziegler-Graff et al. 1991). Indeed, geminiviral MP enhances the susceptibility of virus infection in transgenic plants (Hussain et al. 2005; Gopal et al. 2007; Babu et al. 2018). However, a mutant version of MP was shown to resist viruses from infection in GM plant (Lapidot et al. 1993). Resistance is believed to be obtained because of the competition between the dysfunctional MP (from GM plant) and the functional MP from the challenged virus for plasmodesmatal binding sites (Cooper et al. 1995). This method is not explored as much as CPMR or RPMR in spite of the fact that many MPMR-derived plants are actually broad-spectrum resistant offering protection against a wide range of virus families (Baulcombe 1996). None of the MPMR-derived plants are approved for field trial or commercialisation, yet few promising examples are worth noting. Table 10.4 lists down some of the examples of MPMR.

10.7 Nucleic Acid-Mediated Resistance (NAMR)

Apart from protein-mediated resistance, research was also carried out to test the level of resistance from viral RNA transcripts in transgenic plants. In nucleic acid-mediated resistance (NAMR), any portion of DNA or RNA, either normal or mutated, is either cloned in a vector to derive a transgenic plant or added mechanically in combination with the parent virus with the sole motive of activating defence mechanism against virus after infection. There are different types of NAMR like resistance induction by satellite RNA, defective interfering RNA (DI-RNA), antisense technology, dsDNA formation by hairpin construct, viral N gene, etc., some of which are listed in Table 10.5, where it shows appreciable

Table 10.3 Lists of transgenic plants showing RPMR

Virus used	Part of virus	Method of introduction	Transgenic plant developed (event name)	Protection against virus	Approved for commercialisation	References
TMV	Rep gene plus extra sequence	*Agrobacterium* mediated	*N. tabacum*	TMV, *Tomato mosaic virus* (ToMV), *Tobacco mild green mosaic virus* (TMGMV), *Green tomato atypical mosaic virus* (GTAMV), *Ribgrass mosaic virus* (RMV), *Tomato mosaic virus-fruit necrosis strain* (ToMV-FN)	No	Donson et al. (1993)
PLRV[a]	Rep gene	*Agrobacterium* mediated	Potato (RBTM21-350, RBTM21-129, RBTM22-82)	PLRV	Yes	Lawson et al. (2001)
Pea early browning virus (PEBV)	54KDa of rep protein	*Agrobacterium* mediated	*N. benthamiana*	PEBV, *Broad bean yellow band virus* (BBYBV), British PEBV isolate (PGRO R)	No	Farlane and Davies (1992)
TMV	54 KDa of rep protein	Electroporation of protoplasts	*N. tabacum*	TMV	No	Carr et al. (1992)
PLRV	Mutant ORF1	*Agrobacterium* mediated, agroinoculation of leaf disc	Potato	PLRV	No	Sadowy et al. (2001)
PLRV	Mutant ORF2	*Agrobacterium* mediated, agroinoculation of leaf disc	Potato	PLRV	No	Sadowy et al. (2001)

[a]Stacked along with Cry3A to confer resistance against Colorado potato beetle

Table 10.4 MPMR-derived plants showing resistance against virus

Virus used	MP of virus	Method of gene introduction	Transgenic plant developed	Protection against virus	References
TMV	Defective mutant of MP gene	Inoculation	*N. tabacum*	*Tobacco rattle virus* (TRV), *Tobacco ringspot virus* (TRSV), AlMV, *Peanut chlorotic streak virus* (PClSV), CMV	Cooper et al. (1995)
White clover mosaic virus (WClMV)	Mutant triple gene block (TGB)	Mechanical inoculations	*N. clevelandii*, white clover, subterranean clover, pea, cowpea	WClMV	Beck et al. (1991)
WClMV	Mutant TGB of WClMV-O	*Agrobacterium* mediated	*N. benthamiana*	WClMV-O,-M,-J, PVX, *Narcissus mosaic virus* (NMV), PVS	Beck et al. (1994)
PLRV	Mutant ORF4	*Agrobacterium* mediated	Potato	PLRV, PVX, PVY	Tacke et al. (1996)

result. In many of the cases, modified or unmodified constructs of viral replicase gene are taken for introduction into plants.

DI-RNA is actually a deletion mutant of infectious viral genome found in both plants and animals although it is rarely found in plants in nature (Kollar et al. 1993). However, it can be constructed artificially as already done with *Brome mosaic virus* (Marsh et al. 1991). They are known to be of parasitic nature as they compete for the parent virus RNA-dependent RNA polymerase (RdRp)/replicase for its own multiplication which ameliorates high viral titre in the host (Pathak and Nagy 2009). Less viral accumulation has led to symptom attenuation and increased resistance against infection from plant viruses. The only exception is the DI-RNA associated with *Turnip crinkle virus* where symptoms are aggravated further (Li et al. 1989). The classical advantage of symptom suppression has lured the researchers to develop DI-RNA-derived transgenic plant to promote resistance against plant viruses upon infection. The concept is also further explored to produce recombinant proteins in plants and can act as surrogate templates to study viral replication (Patak and Nagy 2009).

Satellite RNA (satRNA) behaves similarly like DI-RNA as they depend on helper virus for its replication. In the contrary to DI-RNA, satRNAs share little or no sequence similarity with parent virus. They are generally <1500 nt and categorised into four types: single-stranded (ss) RNA satellite virus, ss satellite DNA, double-stranded (ds) satellite RNA and ss satellite RNA (Hu et al. 2009; Roossinck et al.

Table 10.5 Transgenic plant developed to confer nucleic acid mediated resistance

Virus	Sequence used	Method of introduction	Transgenic plant developed	Protection from virus	Approved for commercialisation	Reference(s)
BGMV	Sense and antisense RNA of Rep gene	Particle bombardment	Bean	BGMV	Yes	Aragao et al. (2013)
TSWV	Nucleoprotein (N) gene	*Agrobacterium* mediated	Tobacco	TSWV	No	De Haan et al. (1992)
Tomato golden mosaic virus (TGMV)	Antisense DNA of AL1	*Agrobacterium* mediated	*N. tabacum*	TGMV	No	Day et al. (1991)
TMV	Two synthetic non-overlapping antisense DNA of 5′-UTR	*Agrobacterium* mediated	*N. tabacum*	TMV	No	Nelson et al. (1993)
Turnip yellow mosaic virus (TYMV)	Two genes including 3′-UTR	*Agrobacterium* mediated	Rapeseed	TYMV	No	Zaccomer et al. (1993)
Cymbidium ringspot virus (CyRSV)	Defective interfering RNA	*Agrobacterium* mediated	*N. benthamiana*	CyRSV	No	Kollar et al. (1993)
CMV	Satellite RNA	*Agrobacterium* mediated	*N. tabacum*	CMV, *Tomato aspermy virus* (TAV)	No	Harrison et al. (1987)
Tobacco ringspot virus (TobRSV)	Satellite RNA	*Agrobacterium* mediated	Tobacco	TobRSV	No	Gerlach et al. (1987) and kavanagh and spillane (1995)
PLRV	Three different constructs of ORF2b	*Agrobacterium* mediated	Potato	PLRV	No, but field trials conducted	Rovere et al. (2001)

(continued)

Table 10.5 (continued)

Virus	Sequence used	Method of introduction	Transgenic plant developed	Protection from virus	Approved for commercialisation	Reference(s)
African cassava mosaic virus (ACMV)	DNA B	*Agrobacterium* mediated	*N. benthamiana*	ACMV	No	Stanley et al. (1990)
PLRV	Inverted repeat of CP gene	Agroinoculation	Potato	PLRV	No	Orbegozo et al. (2016)
BYDV	Hairpin construct of polymerase	Agroinoculation	Barley	BYDV	No	Wang et al. (2000)
Sorghum mosaic virus (SrMV)	Hairpin construct of CP gene	Agroinoculation	Sugarcane	SrMV	No	Guo et al. (2015)
TLCuNDV	Antisense RNA of AV2	*Agrobacterium* mediated	Tobacco	TLCuNDV	No	Mubin et al. (2007)

1992). The first satRNA was found to be associated with *Tobacco ringspot virus* (ToRSV) in 1969 (Schneider 1969; Roossinck et al. 1992). They are similar to DI-RNA in action by providing symptom attenuation and increased resistance against plant virus infection and hence can also be called as parasite towards virus. Thus, satRNA acts as a biocontrol agent by providing resistance to transgenic-derived host plants from natural or artificial viral infection (Hu et al. 2009). However, symptoms like chlorosis and necrosis do exacerbate upon satRNA associated with CMV infection on host plants like tomato and pepper (Roossinck et al. 1992). In another case, satRNA associated with *Groundnut rosette virus* (GRV) is essential for encapsidation of the parent virus which will ultimately increase the infection process (Robinson et al. 1999).

Antisense technology explores the concept of utilising complementary sequence of a portion of viral gene for generation of a transgenic plant. Infection by the same or related virus results in viral replication inhibition that ultimately leads to suppression of viral titre and augmented resistance to plants (Aragao et al. 2013; Bejarano and Lichtenstein 1992). The strategy is widely explored against both DNA and RNA plant viruses (Bejarano and Lichtenstein 1992). RNA silencing can also be obtained by production of dsRNA by introducing a hairpin construct in transgenic plant. Generation of siRNA will lead to breakage of viral RNA by RNAi upon viral infection (Mitter and Dietzgen 2012). Transgenic tobacco plant with viral N gene is shown to abort viral replication and transcription upon infection by TSWV by altering the mode of viral polypeptide (De Haan et al. 1992).

10.8 Artificial MicroRNA (amiR)-Mediated Resistance

amiR is the third-generation small RNA-mediated resistance technology yet to be explored to its full potential (Khalid et al. 2017). amiR is considered to be of significant interest to medical and agricultural community in silencing viral genes (Vu et al. 2013) by degradation (Zhang et al. 2011). In amiR technology, natural mature miRNA sequence is replaced by a sequence-specific RNA which will be complementary to viral target(s). Upon introduction into transgenic-derived plants, amiR gene transcribes and processes into mature miRNA using the cellular miRNA biogenesis machinery. Thus, this newly derived GM plant can protect itself from the challenged virus upon infection (Khalid et al. 2017). Generally, 21 nucleotide (nt) short sequences are used in amiR-derived resistance mechanism which can reduce off-target host genes as often seen in long hairpin-mediated silencing (Wagaba et al. 2016). Although the mechanism is still at its infancy, Table 10.6 shows some of the recent GM plants developed using amiR technology.

Table 10.6 List showing the transgenic plant developed to confer resistance of virus by amiR strategy

Virus	Small RNA used	Method of introduction	Transgenic plant developed	Protection from virus	Approved for commercialisation	References
CMV	Artificial miRNA targeting 2b (miR2b)	*Agrobacterium* mediated	Tobacco	CMV	No	Qu et al. (2007)
TYMV	amiR-P69	Agroinfiltration	*A.thaliana*	TYMV	No	Niu et al. 2006
Turnip mosaic virus (TuMV)	amiR-HC-Pro	Agroinfiltration	*A. thaliana*	TuMV	No	Niu et al. (2006)
TYMV + TuMV	Pre-amiR-P69/ miR-HC-Pro	Agroinfiltration	*A.thaliana*	TYMV + TuMV	No	Niu et al. (2006)
PVY	605 bp from 3′ of CP	*Agrobacterium* mediated	Potato	PVY	No	Missiou et al. (2004)
Cymbidium mosaic virus (CymMV)	amiR-CymMV-ORSV targeting viral RdRp	*Agrobacterium* mediated leaf disc	*N. benthamiana*	CymMV	No	Petchthai et al. (2018)
Cassava brown streak virus (CBSV)	amiR-CBSV targeting P1, CP, NIb	Agroinfiltration	*N. benthamiana*	CBSV, *Ugandan cassava brown streak virus* (UCBSV)	No	Wagaba et al. (2016)
UCBSV	amiR-UCBSV targeting P1, CP, Nib	Agroinfiltration	*N. benthamiana*	UCBSV, CBSV	No	Wagaba et al. (2016)
Grapevine virus A (GVA)	amiR-1 targeting GVA ORF1	Agroinfiltration	*N. benthamiana*	GVA	No	Roumi et al. (2012)
GVA	amiR-5 targeting GVA ORF5	Agroinfiltration	*N. benthamiana*	GVA	No	Roumi et al. (2012)

CMV	amiR-2a/2b	*Agrobacterium* mediated	Tomato	CMV	No	Zhang et al. (2011)
CMV	amiR-3′UTR	*Agrobacterium* mediated	Tomato	CMV	No	Zhang et al. (2011)
Tomato leaf curl New Delhi virus (ToLCNDV)	amiR-AV1-3 targeting middle region of AV1	Agroinfection	Tomato	ToLCNDV	No	Vu et al. (2013)
ToLCNDV	amiR-AV1-1 targeting common region of AV1 and AV2	Agroinfection	Tomato	ToLCNDV	No	Vu et al. (2013)

10.9 Plant Protection by Expressing Other Proteins

Several other non-viral sequences which are derived from plants or other organisms have been tested by many researchers across the earth and observed plant protection against various types of viruses. Utilising dsRNAse from yeast (Pac1) or *E.coli* (RNase III), several laboratories confirmed the partial protection against CMV, PVY, ToMV, *Potato spindle tuber viroid* (PSTVd), TSWV, *Chrysanthemum stunt viroid* (CSVd), PVX, TMV and *Tobacco etch virus* (TEV) in transgenic plants (Cillo and Palukaitis 2014; Watanabe et al. 1995; Sano et al. 1997; Ogawa et al. 2005; Colalongo et al. 2005; Langenberg et al. 1997). Similarly, many kinds of plant-derived antiviral proteins such as pokeweed antiviral proteins (PAPs) have been expressed transgenically, observed broad-spectrum resistance to various viruses like PVX, PVY and CMV (Lodge et al. 1993). By drawing cues from PAP results, subsequently, inhibitor of virus replication (IVR) has been identified from TMV-infected tobacco plants and showed resistance to many viruses by ectopic expression (Gera and Loebenstein 1983; Akad et al. 2005). Dominant plant R genes can also be introgressed to provide protection by hypersensitive response (HR) in transgenic plant (Cillo and Palukaitis 2014). NB-LRR is the most abundant class of R gene that aids resistance against TMV in *N. glutinosa*, tobacco, and against TMV and ToMV in tomato by HR mechanism (Whitham et al. 1996). Similarly, Rx gene, which is similar to R gene apart from structural exception, provides extreme resistance against avirulent PVX in transgenic *N. benthamiana* (Bendahmane et al. 1999). CRISPR/Cas9 is the current hot cake in genome-editing process which has found its application towards geminivirus-infected plants. Single-guide RNA (sgRNA) of target 20 nt prepared from CP or intergenic region (IR) of monopartite *Cotton leaf curl kokhran virus* (CLuKoV) and bipartite TYLCV is used to derive transgenic plant. CRISPR/Cas9-derived *N. benthamiana* plant having coding CP can obstruct the CLuKoV, MeMV and TYLCV viral interference, but it produces different viral variants by non-homologous end joining (NHEJ) mechanism. However, there is no indel/virus variant formation upon CLuKoV and one strain of TYLCV infection in non-coding IR-derived transgenic plant showing a possible degradation of viral particles which further makes the plant more immune (Ali et al. 2016). In another case, a non-viral protein, agrobacterium VirE2-derived transgenic *N. benthamiana*, efficiently reduces symptom expression and viral titre upon *Sri Lankan cassava mosaic virus* (SLCMV) and *Mungbean yellow mosaic virus* (MYMV) infection and opens scope for protection by other geminiviruses or ssDNA viruses (Resmi et al. 2015).

10.10 Implications of Commercially Cultivated VR GM Food Crops

10.10.1 Virus-Resistant Squash

Squash stocked with the CP of ZYMV, WMV, ZYMV, WMV and CMV are found to be resistant against these viruses and commercially approved for cultivation in the USA in 1994 and 1996, respectively. It shows up to 50-fold increases in yield than the susceptible lines (Fuchs et al. 1998). It covers up to 12% of total US squash production in 2006 and continues to grow even today. Another study estimates the benefit of growers at 24 million USD in 2007 (Johnson et al. 2007; Klas et al. 2011).

10.10.2 Virus-Resistant Potato

PVY and PLRV are the two major viruses affecting potato cultivation worldwide. PLRV replicase gene harbouring transgenic line and PVY-resistant potato line were started growing commercially in 1999 after initial screening and multiple field trials (Kaniewski and Thomas 2004). A report stated a saving of ~150 USD/acre for farmers in Columbia and Idaho after growing PLRV-resistant potato. Moreover, pesticide application could be reduced to 1.8 million pounds/year upon continuous growing of GM potato (Kaniewski and Thomas 2004) apart from providing environmental benefits and being popular with customers. Unfortunately, VR potato was not grown after 2001 because of anti-GM protests although there was no evidence of VR GM potato to be having side effects than non-transgenic cultivars.

10.10.3 Virus-Resistant Papaya

PRSV is a bottleneck in papaya cultivation worldwide. PRSV CP was inserted in transgenic papaya lines, and after several years of testing, it was deregulated in the USA in 1997. Field trial reports of resistant lines quoted a yield of 125,000 pounds/acre/year as compared to 5,000 pounds/acre/year with susceptible lines (Gonsalves et al. 2004). Commercially grown VR papaya started to be available in market from 1998 onwards (USDA-APHIS 2016). Seventy-five percent of papaya grown in Hawaii is VR until 2009 which increases the papaya export to 14 million USD. The export market increases further to 23 million USD in 2012 (VIB 2014) which clearly shows the consumer acceptance towards VR papaya. Increased cultivation of transgenic papaya in Hawaii had further led to the creation of a buffer zone around it against virus load where non-transgenic crop was successfully cultivated without PRSV infection (Fitch 2016). Later, PRSV-CP-resistant papaya was deregulated for cultivation in 2006 in China. VIB2014 reported an estimated 3,500 ha under VR papaya in 2007 which spiked up further to 6,275 ha (~60% of papaya area) in 2012. Efforts are on to create locally available transgenic papaya lines resistant against different potyvirus and its strains to make it available in different countries (Bau et al. 2003; Kung et al. 2009).

10.10.4 Virus-Resistant Plum

PPV, which causes Sharka disease in plum, is a serious threat to stone fruit industry worldwide, particularly in Europe and North America. Around 100 million trees were infected in Europe (Garcia et al. 2014), out of which an estimated 2.3 million trees were destroyed in Spain between 1989 and 2006 costing more than 63 million Euros (Cambra et al. 2006). Similarly, 264,000 trees in Canada and 190,000 trees in Pennsylvania were destroyed as part of the disease eradication programme. Cost of PPV inoculum removal is around 40 million USD in Pennsylvania alone over a 10-year period (Cambra et al. 2006; Scorza et al. 2013). Transgenic VR plum was successfully tested in lab, greenhouse and at multiple field locations for more than 16 years without any PPV infection (Scorza et al. 2013), the result of which is the deregulation of Honeysweet VR plum by the US government. But, because of fear of consumer acceptance after probable protests from activists, it is not yet commercially available anywhere in the world (Lindbo and Falk 2017). Hence, removal of PPV inoculum from plum remains a costly affair spending millions per year on uprooting thousands of trees which could have been avoided by growing VR GM plum.

Brazil and China are the other two countries apart from the USA which have deregulated and cultivated different VR food crops. Transgenic tomato and pepper inserted with TMV-CP were deregulated in China in 1998. However, the extent of cultivation is not mentioned (James 2009; National Academy of Science and Medicine 2016). Likewise, BGMV-resistant bean was deregulated in Brazil in 2011 (Tollefson 2011) although it is not yet commercially grown as of 2017. A report estimated a production loss of up to 280,000 tonnes of bean infected by BGMV alone in Brazil (Faria et al. 2016). Cultivation of VR GM bean could reduce this loss significantly if approved by the government in the future.

10.11 Insights into the Progression of GM Plants

One of the primary concerns to bring a GM food crop to commercial cultivation is the protest from anti-GMO campaigns. In several cases, finding a gene resistant to particular trait is of utmost concern to develop a GM crop. Once the problem is solved by inserting a suitable trait to make a plant transgenic with several rounds of field trials conducted across locations of different climatic conditions, simultaneous protests start magnifying discussing the ill-effects of launching GM crop into market. Even if a GM crop gets approved for commercial planting, prolonged and continuous efforts by anti-GM organisations compel the government to ban the newly established GM crop from further rounds of cultivation. As a result, benefits reaped are short lived and could not be reached to its beneficiaries. On the other hand, more than 800 million people sleep with an empty stomach every day worldwide. 98% of these people belong to developing nations; 75% of them are thriving in rural Asia and African communities. Seventy percent of the food-deprived people belong to farming communities, either having marginal lands or working as agricultural labourers (Roberts 2018). Anti-GMO campaigns were initially started in Europe and were successful. Few other NGOs from the USA also protested by citing the probable damages that GM food crop might cause to human, although there were no scientific evidences till date. Because of continuous street protests, demonstrations, media coverage, etc., common people, with little or no knowledge about GM, get scared which ultimately convinced the African, Asian and Latin American governments to bring laws against the cultivation of GM food crop. It should be pointed out that population is not a constraint in developed countries. Farmers in rich countries account for a mere percentage of total population who have large acreage holdings with advanced machinery for cultivation as opposed to more than 50% of population in developing countries who solely depend on farming-related activities yet being poor at the end of the day. There is no need to opt for GM crop in rich countries as production is enough to sustain everyone. Activists quoting as 'playing with God's gift if adopting GM' ultimately play with religious sentiments of poorer sections and made them scapegoat in these 20 plus years of long battle against GM adoption.

Before GM technology, breeders use to crossbred different species or genera to obtain superior quality traits without knowing the gene being transferred. Further, researchers use radiation technology to bring genetic mutation with the ultimate aim of obtaining new plant varieties. In the 1980s, scientists discovered a natural way of gene transfer by *Agrobacterium* into plants. Human insulin is also being prepared from bacteria and yeast using GM techniques. Celery contains psoralens which cause dermatitis and skin cancer. But it is still consumed as human body can detoxify it. So, growth of celery is continuing without any protest. Of all the above cases, activists did not question for testing the food for probable health-deteriorating contaminants. Therefore, it is unfair to stop adopting only the transgenic crop even though all the major scientific organisations of the world like The Royal Society in London, the British Medical Association, the French Academy of Science and Medicine, the Union of German Academies of Sciences and Humanities, etc. vouch for it (Paarlberg 2014). It is also unfair to believe completely on anti-GM activists who are mostly from nonscientific background and corporate motivated. It is clearly understandable that once a GM crop comes into market, use of pesticides, insecticides, fertilisers and other agricultural inputs will go down which will be a big jolt to the multinational companies. To counteract this, we need proper dissemination of knowledge among the deluded masses to be given by the scientific personnel to make the commoners satisfied with what they are eating. For example, GM papaya is being grown now for over 20 years without a single side effect to its consumers across the world which should be taken as a live evidence for everyone to believe in GM food crops. Golden rice (Ahmad 2018; Roberts 2018) and GM eggplant (Paarlberg 2014) are successfully commercialised in Bangladesh. GMO sweet corn is approved in the Philippines; Indonesia approved GM sugarcane in 2014. GM summer squash and sweet corn are also grown in the USA (Paarlberg 2014). We need people like Patrick Moore and Mark Lynas, who once used to campaign for anti-GM, who revert their stand and understand the need of quick acceptance of

GM food crop in market (Paarlberg 2014). It is trustworthy to believe Nobel Laureate Richard J Roberts and 128 other Nobel laureates who have voted for the adoption of GM food crops by requesting the anti-GM organisations to review their stand in order to deplete hunger and malnutrition-derived diseases from the face of the earth (Roberts 2018). Finally, it is up to governments across the world that should come together in encouraging cultivation of GM food crops.

References

Abbas A, Arif M, Ali M (2016) A review paper on potato leaf roll virus (PLRV) of potato in Pakistan. Asian J Agric Biol 4(3):77–86

Abel PP, Nelson RS, De B, Hoffmann N, Rogers SG, Fraley RT, Beachy RN (1986) Delay of disease development in transgenic plants that express the tobacco mosaic virus coat protein gene. Science 232(4751):738–743

Ahmad R (2018) Wait nearly over for Golden Rice release in Bangladesh. Beta carotene-rich grain key to fight vitamin-A deficiency. United News of Bangladesh

Akad A, Teverovsky E, Gidoni D, Elad Y, Kirschner B, Rav-David D, Czosnek H, Loebenstein G (2005) Resistance to *Tobacco mosaic virus* and *Botrytis cinerea* in tobacco transformed with the complementary DNA encoding an inhibitor of viral replication-like protein. Ann Appl Biol 147:89–100

Ali Z, Ali S, Tashkandi M, Zaidi SS, Mahfouz MM (2016) CRISPR/Cas9-mediated immunity to geminiviruses: differential interference and evasion. Nat Sci Rep 6:26912

Andrews CJ, Paliwal YC (1986) Effects of barley yellow dwarf virus infection and low temperature floading on cold stress tolerance of winter cereals. Can J Plant Pathol 8:311–316

APS. Papaya ringspot virus. https://www.apsnet.org/edcenter/intropp/lessons/viruses/pages/papayaringspotvirus.aspx. Accessed 30 Nov 2018

Aragão FJL, Nogueira EOPL, Tinoco MLP, Faria JC (2013) Molecular characterization of the first commercial transgenic common bean immune to the Bean golden mosaic virus. J Biotechnol 166:42–50

Ayazpour K, Vahidian M (2016) Study of watermelon mosaic virus in cucurbit fields of Jahrom area, Iran. Vegetos 29:4

Babu KSD, Guria A, Karanthamalai J, Srikakulum N, Kumari K, Sharma P, Chandran SA, Barnabas D, Tennyson J, Pandi G (2018) DNA methylation suppression by *Bhendi yellow vein mosaic virus*. Epigenomes 2(2):7

Bau HJ, Cheng YH, Yu TA, Yang JS, Yeh SD (2003) Broad-spectrum resistance to different geographic strains of papaya ringspot virus in coat protein gene transgenic papaya. Phytopathology 93:112–120

Baulcombe DC (1996) Mechanisms of pathogen-derived resistance to viruses in transgenic plants. Plant Cell 8:1833–1844

Beck DL, Guilford PJ, Voot DM, Andersen MT, Forster RLS (1991) Triple gene block proteins of white clover mosaic potex-virus are required for transport. Virology 183:695–702

Beck DL, Van Dolleweerd CJ, Lough TJ, Balmori E, Voot DM, Andersen MT, OBrien IEW, Forster RLS (1994) Disruption of virus movement confers broad spectrum resistance against systemic infection by plant viruses with a triple gene block. Proc Natl Acad Sci U S A 91:10310–10314

Bejarano ER, Lichtenstein CP (1992) Prospects for engineering virus resistance in plants with antisense RNA. Trends Biotechnol 10:383–387

Bendahmane A, Kanyuka K, Baulcombe DC (1999) The Rx gene from potato controls separate virus resistance and cell death responses. Plant Cell 11:781–791

Bergua M, Phelan DM, Bak A, Bloom DC, Folimonova SY (2016) Simultaneous visualization of two Citrus tristeza virus genotypes provides new insights into the structure of multi-component virus populations in a host. Virology 491:10–19

Braun CJ, Hemenway CL (1992) Expressions of amino-terminal portions or full-length viral replicase genes in transgenic plants confers resistance to potato virus X infection. Plant Cell 4:735–744

Cambra M, Capote N, Myrta A, Llacer G (2006) *Plum pox virus* and the estimated costs associated with sharka disease. EPPO Bull 36:202–204

Carr JP, Marsh LE, Lomonossoff GP, Sekiya ME, Zaitlin M (1992) Resistance to tobacco mosaic virus induced by the 54-KDa gene sequence requires expression of the 54-KDa protein. Mol Plant-Microbe Interact 5(5):397–404

Carr JP, Lewsey MG, Palukaitis P (2010) Signaling in induced resistance. Adv Virus Res 76:57–121

Chen ZL, Gu H, Li Y, Su Y, Wu P, Jiang Z, Ming X, Tian J, Pan N, Qu LJ (2003) Safety assessment for genetically modified sweet pepper and tomato. Toxicology 188(2–3):297–307

Cho JJ, Mitchell WC, Mau RFL (1987) Epidemiology of tomato spotted wilt virus disease on crispheed lettuce in Hawaii. Plant Dis 71:505–508

Cillo F, Palukaitis P (2014) Transgenic resistance. In: Loebenstein G, Katis N (eds) Advances in virus research, vol 90. Academic, Burlington, pp 35–146

Colalongo C, Contaldo N, Faccioli G, Pagan P (2005) The use of pac1 gene from *Schizosaccaromyces pombe* to protect potato from potato virus Y (PVY) and potato virus X (PVX) infections. Phytopathol Mediterr 44:226–231

Cooper B, Lapidot M, Heick JA, Dodds JA, Beachy RN (1995) A defective movement protein of TMV in transgenic plants confers resistance to multiple viruses whereas the functional analog increases susceptibility. Virology 206:307–313

Day AG, Bejarano ER, Buck KW, Burrell M, Lichtenstein CR (1991) Expression of an antisense viral gene in transgenic tobacco confers resistance to the DNA virus tomato golden mosaic virus. Proc Natl Acad Sci U S A 88:6721–6725

De Breuil S, Giolitti FJ, Bejerman N, Lenardon SL (2012) Effects of cucumber mosaic virus on the yield and yield components of peanut. J Plant Pathol 94(3):669–673

De Doyle MMR, Autrey LJC (1992) Assessment of yield losses as a result of co-infection by maize streak virus and maize stripe virus in Mauritius. Ann Appl Biol 120:443–450

De Faria JC, Aragao FJL, Souza TLPO, Quintela ED, Kitajima EW, Ribeiro SDG (2016) Golden mosaic of common beans in Brazil: management with a transgenic approach. APS*net* Feature, American Phytopathological Society

De Haan P, Gielen JJL, Prins M, Wijkamp IG, Van Schepen A, Peters D, Van Grinsven MQJM, Goldbach R (1992) Characterization of RNA-mediated resistance to tomato spotted wilt virus in transgenic tobacco plants. Biotechnology 10:1133–1137

Donson J, Kearney CM, Turpen TH, Khan IA, Kurath G, Turpen AM, Jones GE, Dawson WO, Lewandowski DJ (1993) Broad resistance to tobamoviruses is mediated by a modified tobacco mosaic virus replicase transgene. Mol Plant-Microbiol Interact 6(5):635–642

Farlane SAM, Davies JW (1992) Plants transformed with a region of the 201-kilodalton replicase gene from pea early browning virus RNA1 are resistant to virus infection. Proc Natl Acad Sci 89(13):5829–5833

Fitch MMM (2016) Update on gene transfer biotechnology of papaya. Acta Hortic 1111:7–18

Folimonova SY (2013) Developing an understanding of cross-protection by Citrus tristeza virus. Front Microbiol 4:76

Fuchs M, Tricoli DM, Carney KJ, Schlesser M, McFerson JR, Gonsalves D (1998) Comparative virus resistance and fruit yield of transgenic squash with single and multiple coat protein genes. Plant Dis 82:1350–1356

Garcia JA, Glasa M, Cambra M, Candresse T (2014) *Plum pox virus* and sharka: a model potyvirus and a major disease. Mol Plant Pathol 15:226–241

Gera A, Loebenstein G (1983) Further studies of an inhibitor of virus replication from tobacco mosaic virus-infected protoplasts of a local lesion responding tobacco cultivar. Phytopathology 73:111–115

Gerlach WL, Llewellyn D, Haseloff J (1987) Construction of a disease resistance gene using the satellite RNA of tobacco ringspot virus. Nature 328:802–806

Ghoshal B, Sanfaçon H (2015) Symptom recovery in virus-infected plants: revisiting the role of RNA silencing mechanisms. Virology 479–480:167–179

Gonsalves D, Chee P, Provvidenti R, Seem R, Slightom JL (1992) Comparison of coat protein-mediated and genetically derived resistance in cucumbers to infection by cucumber mosaic virus under field conditions with natural challenge inoculations by vectors. Biotechnology 10:1562–1570

Gonsalves D, Gonsalves C, Ferreira S, Pitz K, Fitch M, Manshardt R, Slightom J (2004) Transgenic virus resistant papaya: from hope to reality for controlling *Papaya ringspot virus* in Hawaii. APS*net* Feature, American Phytopathological Society

Gopal P, Kumar PP, Sinilal B, Jose J, Yadunandama AK, Usha R (2007) Differential roles of C4 and βC1 in mediating suppression of post-transcriptional gene silencing: evidence for transactivation by the C2 of *Bhendi yellow vein mosaic virus*, a monopartite begomovirus. Virus Res 123:9–18

Guo J, Yang L, Liu X, Zhang H, Qian B, Zhang D (2009) Applicability of the chymopapain gene used as endogenous reference gene for transgenic Huanong No. 1 papaya detection. J Agric Food Chem 57:6502–6509

Guo J, Gao S, Lin Q, Wang H, Que Y, Xu L (2015) Transgenic sugarcane resistant to sorghum mosaic virus based on coat protein gene silencing by RNA interference. Biomed Res Int 2015:1–9. https://doi.org/10.1155/2015/861907

Gupta BM, Sing BP, Varma HN, Shrivastava KM (1985) Perspective, in plant virology. Printer House India 1:1–132

Haber S (1990) Situation review of barley yellow dwarf in Canada. In: Burnett PA (ed) World perspectives on barley yellow dwarf. CIMMYT, Mexico D.F, pp 7–10

Hamilton RI (1980) Defenses triggered by previous invaders: viruses. In: Horsfall JG, Cowling EB (eds) Plant disease: an advance treatise, vol 5. Academic, New York, pp 279–303

Harrison BD, Mayo MA, Baulcombe DC (1987) Virus resistance in plants that express cucumber mosaic virus satellite RNA. Nature 328:799–802

Hasanuddin A, Hibino H (1989) Grain yield reduction, growth retardation and virus concentration in rice plants infected with tungro-associate viruses. Trop Agric Res Ser 22:56–73

Hossain M, Ali MS (1992) Effect of potato virus Y severities on virus concentration, dilution end point and potato yield. Bangladesh J Plant Pathol 8:27–29

Hu C-C, Hsu Y-H, Lin N-S (2009) Satellite RNAs and satellite viruses of plants. Viruses 1:1325–1350

Hussain M, Mansoor S, Iram S, Fatima AN, Zafar Y (2005) The nuclear shuttle protein of *Tomato leaf curl New Delhi virus* is a pathogenicity determinant. J Virol 79(7):4434–4439

ISAAA http://www.isaaa.org/resources/infographics/default.asp. Accessed 30 Nov 2018

Jalender P, Anitha K, Prasanthi Y, Bhat BN (2015) Effect of cucumber mosaic virus (CMV) on yield and yield attributes of tomato. Res J Agric Sci 6(2):443–446

James C (2009) Global status of commercialized biotech. GM crops 2008. ISAAA, Ithaca

Johnson SR, Strom S, Grillo K (2007) Quantification of the impacts on US agriculture of biotechnology-derived crops planted in 2006. National Center for Food and Agriculture Policy. http://www.ncfap.org

Jongedijk E, de Schutter AJM, Stolte T, Van den Elzen PJM, Comelissen BJC (1992) Increased resistance to potato virus X and preservation of cultivar properties in transgenic potato under field conditions. Biotechnology 10:422–429

Kaniewski WK, Thomas PE (2004) The potato story. AgBioforum 7:41–46

Kavanagh TA, Spillane C (1995) Strategies for engineering virus resistance in transgenic plants. Euphytica 85:149–158

Khalid A, Zhang Q, Yasir M, Li F (2017) Small RNA based genetic engineering for plant viral resistance: application in crop protection. Front Microbiol 8:43

Klas FE, Fuchs M, Gonsalves D (2011) Fruit yield of virus-resistant transgenic summer squash in simulated commercial plantings under conditions of high disease pressure. J Hortic For 3(2):46–52

Kollar A, Dalmay T, Burgyan J (1993) Defective interfering RNA-mediated resistance against cymbidium ringspot tombusvirus in transgenic plants. Virology 193:313–318

Kung YJ, Bau HJ, Wu YL, Huang CH, Chen TM, Yeh SD (2009) Generation of transgenic papaya with double resistance to papaya ringspot virus and papaya leaf-distortion mosaic virus. Phytopathology 99(11):1312–1320

Kunik T, Salomon R, Zamir D, Navot N, Zeidan M, Michelson I, Gafni Y, Czosnek H (1994) Transgenic tomato plants expressing the *Tomato yellow leaf curl virus* capsid protein are resistant to the virus. Biotechnology 12:500–506

Langenberg WG, Zhang L, Court DL, Guinchedi L, Mitra A (1997) Transgenic tobacco plants expressing the bacterial *rnc* gene resist virus infection. Mol Breed 3:391–399

Lapidot M, Gafny R, Ding B, Wolf S, Lucas WJ, Beachy RN (1993) A dysfunctional movement protein of tobacco mosaic virus that partially modifies the plasmodesmata and limits virus spread in transgenic plants. Plant J 4:959–970

Lawson EC, Weiss JD, Thomas PE, Kaniewski WK (2001) New leaf plus russet Burbank potatoes: replicase-mediated resistance to *Potato leaf roll virus*. Mol Breed 7(1):1–12

Li XH, Heaton LA, Morris TJ, Simon AE (1989) *Turnip crinkle virus* defective interfering RNAs intensify viral symptoms and are generated de novo. Proc Natl Acad Sci U S A 86:9173–9177

Lindbo JA, Falk BW (2017) The impact of coat protein-mediated virus resistance in applied plant pathology and basic research. Phytopathology 107:624–634

Lodge JK, Kaniewski WJ, Tumer NE (1993) Broad-spectrum virus resistance in transgenic plants expressing pokeweed antiviral protein. Proc Natl Acad Sci U S A 90:7089–7093

Loesch-Fries LS, Merlo D, Zinnen T, Burhop L, Hill K, Krahn K, Jarvis N, Nelson S, Halk E (1987) Expression of *Alfalfa mosaic virus* RNA 4 in transgenic plants confers virus resistance. EMBO J 6:1845–1851

Marsh LE, Pogue GP, Szybiak U, Connell JP, Hall TC (1991) Non-replicating deletion mutants of *Brome mosaic virus* RNA-2 interfere with viral replication. J Gen Virol 72:2367–2374

McKinney HH (1929) Mosaic diseases in the Canary Islands, West Africa and Gibraltar. J Agric Res 39:557–578

Missiou A, Kalantidis K, Boutla A, Tzortzakaki S, Tabler M, Tsagris M (2004) Generation of transgenic potato plants highly resistant to *Potato virus Y* (PVY) through RNA silencing. Mol Breed 14(2):185–197

Mitter N, Dietzgen RG (2012) Use of hairpin RNA constructs for engineering plant virus resistance. In: Watson J, Wang MB (eds) Antiviral resistance in plants. Methods in molecular biology (Methods and protocols), vol 894. Humana Press, Totowa

Montasser MS, Tousignant ME, Kaper JM (1991) Satellite-mediated protection of tomato against cucumber mosaic virus: I. Greenhouse experiments and simulated epidemic conditions in the field. Plant Dis 75:86–92

Mubin M, Mansoor S, Hussain M, Zafar Y (2007) Silencing of the AV2 gene by antisense RNA protects transgenic plants against a bipartite begomovirus. Virol J 4:10

My Agriculture Information Bank. http://www.agriinfo.in/default.aspx?page=topic&superid=5&topicid=1855. Accessed 30 Nov 2018

Myti S, Shabbir AK, Akhter S, Uddin A, Kamruzzaman M, Faruq MO, Biswas GC (2014) Identification of the most prevalent and spatially disperse virus on chilli at Northern and Eastern part of Bangladesh. Int J Biosci 5:40–49

National Academies of Sciences and Medicine (2016) Genetically engineered crops: experiences and prospects. The National Academies Press, Washington, DC

Nejidat A, Beachy RN (1990) Transgenic tobacco plants expressing a tobacco virus coat protein gene are resistant to some tobamoviruses. Mol Plant-Microbiol Interact 3:247–251

Nelson RS, Roth DA, Johnson JD (1993) Tobacco mosaic virus infection of transgenic *Nicotiana tabacum* plants is inhibited by antisense constructs directed a the 5′ region of viral-RNA. Gene 127:227–232

Niu Q-W, Lin S-S, Reyes JL, Chen K-C, Wu H-W, Yeh S-D, Chua N-H (2006) Expression of artificial microRNAs in transgenic *Arabidopsis thaliana* confers virus resistance. Nat Biotechnol 24:1420–1428

Oerke EC, Dehne HW, Schömbeck F, Weber A (1994) Crop production and crop protection. Estimated losses in major food and cash crops. Elsevier, Amsterdam

Ogawa T, Toguri T, Kudoh H, Okamura M, Momma T, Yoshioka M, Kato K, Hagiwara Y, Sano T (2005) Double-stranded RNA-specific ribonuclease confers tolerance against Chrysanthemum stunt viroid and tomato spotted wilt virus in transgenic chrysanthemum plants. Breed Sci 55:49–55

Orbegozo J, Solorzano D, Cuellar WJ, Bartolini I, Roman ML, Ghislain M, Kreuze J (2016) Marker-free PLRV resistant potato mediated by Cre-loxP excision and RNAi. Transgenic Res 25(6):813–828

Paarlberg R (2014) A dubious success: the NGO campaign against GMOs. GM Crops Food 5(3):223–228

Padmanabhan MS, Dinesh-Kumar SP (2014) The conformational and subcellular compartmental dance of plant NLRs during viral recognition and defense signaling. Curr Opin Microbiol 20:55–61

Pathak KB, Nagy PD (2009) Defective interfering RNAs: foes of viruses and friends of virologists. Viruses 1:895–919

Pennazio S, Roggero P, Conti M (1996) Yield losses in virus-infected crops. Arch Phytopathol Plant Protect 30(4):283–296

Petchthai U, Yee CSL, Wong S-M (2018) Resistance to CymMV and ORSV in artificial microRNA transgenic *Nicotiana benthamiana* plants. Nat Sci Rep 8:9958

Pun KB, Doraiswamy S (1999) Effect of age of okra plants on susceptibility to *Okra yellow vein mosaic virus*. Indian J Virol 15:57–58

Qu J, Ye J, Fang R (2007) Artificial microRNA-mediated virus resistance in plants. J Virol 81:6690–6699

Rahman MS, Akanda AM (2008) Infection pattern of PVY and PLRV in potato tubers obtained from potato plants infected with both the viruses. Agriculturists 6(1&2):118–123

Rahman MS, Akanda AM, Mian IH, Bhuiyan MKA, Hossain MM (2016) New sources of resistance to cucumber mosaic virus in *Capsicum annuum*. J Crop Sci Biotechnol 19(3):249–258

Ravelonandro M, Scorza R, Callahan A, Levy L, Jacquet C, Monsion M, Damsteegt V (2000) The use of transgenic fruit trees as a resistance strategy for virus epidemics: the plum pox (sharka) model. Virus Res 71:63–69

Resmi TR, Hohn T, Hohn B, Veluthambi K (2015) The *Agrobacterium tumefaciens* Ti plasmid virulence gene virE2 reduces *Sri Lankan cassava mosaic virus* infection in transgenic *Nicotiana benthamiana* plants. Virology 7:2641–2653

Risius ML, Gildow FE, Hoffman T (1991a) Barley yellow dwarf virus effects on five winter barley cultivar. In: Barnett PA (ed) Barley yellow dwarf newsletter, vol 4. CIMMYT, Mexico D.F, pp 43–44

Risius ML, Gildow FE, Hoffman T (1991b) Responses of five winter wheat cultivars infected with BYDV. In: Barnett PA (ed) Barley yellow dwarf newsletter, vol 4. CIMMYT, Mexico D.F, p 48

Roberts RJ (2018) The nobel laureates' campaign supporting GMOs. J Innov Knowl 3:61–65

Robinson DJ, Ryabov EV, Raj SK, Roberts IM, Taliansky ME (1999) Satellite RNA is essential for encapsidation of groundnut rosette umbravirus RNA by groundnut rosette assistor luteovirus coat protein. Virology 254:105–114

Roossinck MJ (2010) Lifestyles of plant viruses. Philos Trans R Soc B Biol Sci 365(1548):1899–1905

Roossinck MJ, Sleat D, Palukaitis P (1992) Satellite RNAs of plant viruses: structures and biological effects. Microbiol Rev 56(2):265–279

Roumi V, Afsharifar A, Saldarelli P, Niazi A, Martelli GP, Izadpanah K (2012) Transient expression of artificial microRNAs confers resistance to *Grapevine virus A* in *Nicotiana benthamiana*. J Plant Pathol 94(3):643–649

Rovere CV, Asurmendi S, Hopp HE (2001) Transgenic resistance in potato plants expressing potato leaf roll virus (PLRV) replicase gene sequences is RNA-mediated and suggests the involvement of post-transcriptional gene silencing. Arch Virol 146:1337–1353

Sadowy E, Juszczuk M, David C, Gronenborn B, Hulanicka MD (2001) Mutational analysis of the proteinase function of *Potato leafroll virus*. J Gen Virol 82:1517–1527

Sanders PR, Sammons B, Kaniewski W, Haley L, Layton J, Lavallee BJ, Delannay X, Turner NE (1992) Field resistance of transgenic tomatoes expressing the tobacco mosaic virus or tomato mosaic virus coat protein genes. Phytopathology 82:683–690

Sanford JC, Johnston S (1985) The concept of parasite-derived resistance-deriving resistance genes from the parasite's own genome. J Theor Biol 113(2):395–405

Sano T, Nagayama A, Ogawa T, Ishida I, Okada Y (1997) Transgenic potato expressing a double-stranded RNA-specific ribonuclease is resistant to potato spindle tuber viroid. Nat Biotechnol 15:1290–1294

Schneider IR (1969) Satellite-like particle of tobacco ringspot virus that resembles tobacco ringspot virus. Science 166:1627–1629

Scorza R, Callahan A, Dardick C, Ravelonandro M, Polak J, Malinowski T, Zagrai I, Cambra M, Kamenova I (2013) Genetic engineering of *Plum pox virus* resistance: 'HoneySweet' plum—From concept to product. Plant Cell Tissue Organ Cult 115:1–12

Senanayake DMJB, Mandal B (2014) Expression of symptoms, viral coat protein and silencing suppressor gene during mixed infection of a N–Wi strain of potato virus Y and an asymptomatic strain of potato virus X. Virus Dis 25(3):314–321

Sikora EJ, Gudauskas RT, Murphy JF, Porch DW, Andrianifahanana M, Zehnder GW, Bauske EM, Kemble JM, Lester DF (1998) A multivirus epidemic of tomatoes in Alabama. Plant Dis 82:117–120

Singh SJ (1989) Effect of *Watermelon mosaic virus* 1 on growth and yield of bottlegourd (Lagenaria siceraria). Indian J Virol 5:73–78

Slater A, Scott NW, Fowler MR (2013) Reducing the effects of viral disease. Plant biotechnology; the genetic manipulation of plants, 2nd edn. Oxford University Press, Oxford

Stanley J, Frischmuth T, Ellwood S (1990) Defective viral DNA ameliorates symptoms of geminivirus infection in transgenic plants. Proc Natl Acad Sci U S A 87:6291–6295

Stark DM, Beachy RN (1989) Protection against potyvirus infection in transgenic plants: evidence for broad spectrum resistance. Biotechnology 7:1257–1262

Tacke E, Salamini F, Rohde W (1996) Proceedings of the EAPR virology section meeting, pp 15–22

Taylor DR, Fofie AS, Suma M (1990) Natural infection of rice yellow mottle virus disease on rice in Sierra Leone. Int Rice Res Newsl 15:19

Thouvenel JC (1988) A serious disease caused by cowpea mottle virus in the Ivory Coast. Plant Dis 72:363

Tollefson J (2011) Brazil cooks up transgenic bean. Nature 478:168

USDA-APHIS (2016) Petitions for determination of nonregulated status. USDA-APHIS petitions for determination of nonregulated status. https://www.aphis.usda.gov/aphis/ourfocus/biotechnology/permits-notifications-petitions/petitions/petition-status

Van Zijl JJB, Bosch SE, Coetzee CPJ (1986) Breeding tomatoes for processing in South Africa. Acta Hortic 194:69–75

Varma A, Jain RK, Bhat AI (2002) Virus resistant transgenic plants for environmentally safe management of viral diseases. Indian J Biotechnol 1:73–86

Vegetable MD Online, Cornell University. http://vegetablemdonline.ppath.cornell.edu/NewsArticles/Potato_Virus.htm. Accessed 30th Nov 2018

VIB (2014) Virus resistant papaya in Hawaii: the local papaya industry's life raft. In: Fact series: virus resistant papaya in Hawaii. Ghent, VIB

Vishwa D, Gutha SN (1990) Effect of bean common mosaic virus on yield and yield attributes in french bean. Indian J Pulse Res 3:89–91

Vu TV, Choudhury NR, Mukherjee SK (2013) Transgenic tomato plants expressing artificial microRNAs for silencing the pre-coat and coat proteins of a begomovirus, *Tomato leaf curl New Delhi virus*, show tolerance to virus infection. Virus Res 172:35–45

Wagaba H, Patil BL, Mukasab S, Alicai T, Fauquet CM, Taylor NJ (2016) Artificial microRNA-derived resistance to cassava brown streak disease. J Virol Methods 231:38–43

Walter B (1988) Quelques exemples de la réaction physioloque de la vigne en présence de virus. Bull OIV 687–688:383–389

Wang M-B, Abbott DC, Waterhouse PM (2000) A single copy of a virus-derived transgene encoding hairpin RNA gives immunity to barley yellow dwarf virus. Mol Plant Pathol 1(6):347–356
Watanabe Y, Ogawa T, Takahashi H, Ishida I, Takeuchi Y, Yamamoto M, Okada Y (1995) Resistance against multiple viruses in plants mediated by a double stranded-RNA specific ribonuclease. FEBS Lett 372:165–168
Whitham S, McCormick S, Baker B (1996) The N gene of tobacco confers resistance to tobacco mosaic virus in transgenic tomato. Proc Natl Acad Sci U S A 93:8776–8781
Yang RC, Xu HL, Yu WG, Lu CG, Long MS, Liu CQ, Pan NS, Chen ZL (1995) Transgenic tomato expressing CMV-CP gene and its resistance to CMV. Acta Agric Jiangsu 11:42–46
Zaccomer B, Cellier F, Boyer J-C, Haenni A-L, Tepfer M (1993) Transgenic plants that express genes including the 3′ untranslated region of the turnip yellow mosaic virus (TYMV) genome are partially protected against TYMV infection. Gene 136:87–94
Zhang X, Li H, Zhang J, Zhang C, Gong P, Ziaf K, Xiao F, Ye Z (2011) Expression of artificial microRNAs in tomato confers efficient and stable virus resistance in a cell-autonomous manner. Transgenic Res 20(3):569–581
Zhu YX, Ou-Yang WJ, Zhang YF, Chen ZL (1996) Transgenic sweet pepper plants from agrobacterium mediated transformation. Plant Cell Rep 16:71–75
Ziegler-Graff V, Guilford PJ, Bauloombe DC (1991) Tobacco rattle virus RNA-1 29K gene product potentiates viral movement and also affects symptom induction in tobacco. Virology 182:145–155

A Review on Reed Bed System as a Potential Decentralized Wastewater Treatment Practice

11

Soumya Chatterjee, Anindita Mitra, Santosh K. Gupta, and Dharmendra K. Gupta

Abstract

To develop sustainable societies, wastewater treatment and management are two most significant challenges the world is facing today. A large amount of potable water is being wasted regularly from various sources, including household, in the form of wastewater. Reed bed systems (RBSs) are cost-effective engineered constructed wetlands (CWs) which can be used as an eco-friendly technology for treatment of different wastewaters. Several factors are crucial for effective performance of this sustainable technology, which include design parameters of RBSs, media materials, hydraulic regime, inhabiting microorganisms, biofilms and functionalized plants. The biotic community (especially, microbial assemblages) dwelling RBSs can vary considerably in different environment and wastewater compositions. Thus, RBSs can be an excellent option for decentralized wastewater treatment with water reuse potentiality in the space constrained areas. This review presents recent developments of RBSs and its potential for using as decentralized cost-effective wastewater treatment processes.

Keywords

Constructed wetlands · Cold · Nitrogen · Phosphorus · Metagenomics

S. Chatterjee · S. K. Gupta
Defence Research Laboratory, DRDO, Tezpur, Assam, India

A. Mitra
Bankura Christian College, Bankura, West Bengal, India

D. K. Gupta (✉)
Ministry of Environment Forest and Climate Change (MoEFCC), Indira Paryavaran Bhavan, New Delhi, India

R. Sathishkumar et al. (eds.), *Advances in Plant Transgenics: Methods and Applications*, https://doi.org/10.1007/978-981-13-9624-3_11

11.1 Introduction

The United Nations water and sustainable development programmes state that 'water is a finite and irreplaceable resource that is fundamental to human well-being'. It is only renewable if well managed (http://www.un.org/waterforlifedecade/). Development and implementation of reliable, environment-friendly water renewable technology is an important challenge the world is facing today. Due to rapid urbanization and varied anthropogenic activities, production of wastewater has been increased manyfold in recent days, which created a problem of wide contamination and health hazards. It is a worldwide challenge to clean up the wastewater before releasing into the environment. As the nature and quality of the wastewater may vary considerably, appropriate and affordable wastewater treatment systems are the need of the hour. Conventional wastewater treatment facilities (WTPs) are having limited application for wide range of users due to high establishment, operational and maintenance cost and manpower intensiveness. In an alternative solution, natural treatment processes are significantly gaining popularity throughout the world. In recent past, constructed wetlands (CWs) have been widely considered as a valid technology for different wastewater treatments (Fig. 11.1). CWs are ecological engineered systems with shallow depth of water, formed like channels or ponds, having specific macrophytes and capable of treating wide range of wastewaters (Sánchez 2017). Kathe Seidel (1960s) and Reinhold Kickuth (1970s) of Germany were pioneers in initiating research on these CWs (Kadlec and Wallace 2009). Reed bed systems (RBSs) are specific type of CWs that create further decentralized, non-mechanized, environment-friendly gravel-filled treatment beds planted with reeds,

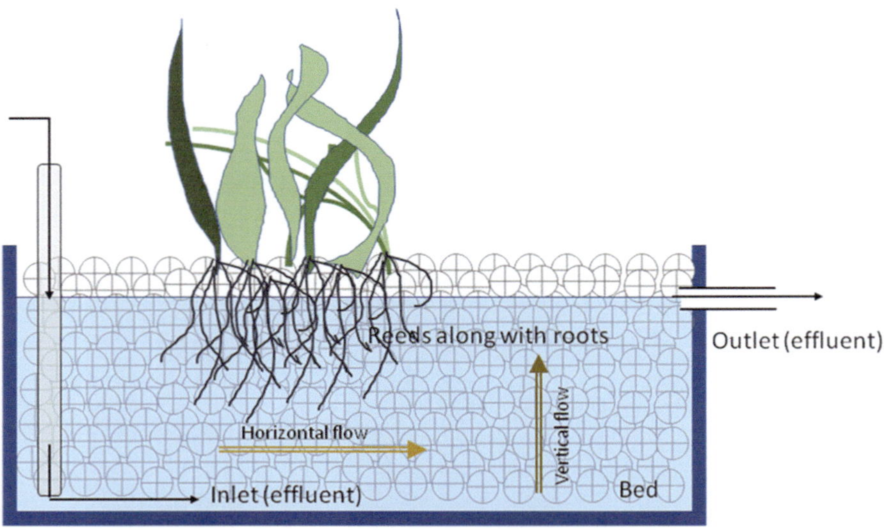

Fig. 11.1 Diagram representing a typical reed bed where water may flow horizontally or vertically through the designed bed and reed plants play a major role in decontaminating water

a solution to maintain intensive mechanized treatment systems. Wastewater streams through the gravels and roots of reed plants and is cleansed by joint activities of microbial components and other edaphic factors.

RBSs were primarily used for domestic wastewater treatments. Because of its ease in construction and maintenance at user level, it is a popular technology and widely used in several places all over the world (Fig. 11.1). Further, RBS can be considered as a technology having least impacts on the community's culture and lifestyles, developing a stable waste management practice (Wong et al. 2009). The structure primarily consists of impermeable basement layer, having gravels and sands as natural filtering media and reed plants (Masciandaro et al. 2015). The systems are natural, environment-friendly, sustainable and having reduced carbon footprint (http://www.reedbedtechnology.com; Rizzo et al. 2018). Basically, RBSs treat wastewater through amalgamation of physical, chemical and biological processes, interacting between the sludge, reeds, microorganisms, gravel and environment (Tiwari et al. 2017; Carvalho et al. 2017). Processes include sedimentation, volatilization, plant uptake, sorption, photodegradation, accretion transpiration flux, microbial biodegradation and transformations, etc. (Kadlec and Wallace 2009; Larsen 2017; Sánchez 2017). The design of the bed area is important for hydraulic conductivity (flow of water) and oxygen saturation, which are critical for bed activity and effluent water treatments (Chen et al. 2016).

Treating different kinds of wastewaters and getting benefits through naturally removing pathogenic microbial load and removal of parasitic helminth eggs by applying the systems are being reported from different corners of the world (Kadlec and Knight 1996; Stott et al. 1999; Karpiscak et al. 2001; Jackson and Jackson 2008; Drewe et al. 2009). Due to evapotranspiration activities of plants, dewatering of sludge is an important functioning of reed bed (Kinsley et al. 2017). Besides the potential of decontaminating a range of pollutants, RBSs play very important role in carbon, nitrogen and sulphur cycling and metal removal (Faulwetter et al. 2009; Larsen 2017). RBSs are also capable of treating pesticides (Vymazal and Brezinova 2015; Lv et al. 2017), herbicides (Elsayed et al. 2014), pharmaceutical residues (Li et al. 2016), etc. Variants of CWs are mostly according to the hydraulic flow and vegetation types. The most common path of hydraulic flow in RBSs can either be vertical or horizontal flow, where water flows either vertically (both down and up flow) or horizontally (surface or subsurface flow). The Netherlands was the first country reported to use RBS as constructed wetlands in the early 1960s (Brix and Schierup 1989). Various countries, like Germany, the USA, Australia, the UK, Thailand, Sweden, China and India, are adopting this technology and stressing on these natural wastewater treatment systems based on aquatic plants and microorganism (Kurniadie 2011; Vymazal 2013; Sánchez 2017). Besides wastewater treatment and pathogen removal, RBSs can also be used to generate electricity by combining it with emerging technologies like microbial fuel cells (MFCs) through bioelectrochemical mechanisms (Arends et al. 2014; Sánchez 2017). In this review, we try to highlight the recent developments related to RBSs functioning, important components, mechanisms of abatement of contaminants, related microbial processes and the scope of the technology for implementation in different climates and conditions.

11.2 Components of Reed Bed

Reed beds are having a variety of different components which act in combination to decontaminate wastewaters through physico-chemical and biological processes.

11.2.1 Media

Bed media strongly affect the efficiency of RBSs. Gravel and sand are the most common materials through which hydraulic regime are created. It also helps the reed growths which are functionally attached through their roots penetrated into the in-between spaces of the media. Therefore, filtration of pollutants is subjected to the quality of the media components, like porosity (interstitial pore spaces), particle size, texture (grades of irregularity) and permeability (Taleno 2012). Kadlec (2009) reported that in natural wetlands, bed components, like soil type and stratification and humic/organic substances, help in ammonia removal, while P is removed by adsorption and ion exchange reactions in soil rich in divalent or trivalent ions (Brix et al. 2001). However, due to limited presence of ion cations, phosphorus removal is generally very limited in RBSs, which can either be augmented by supplementing alternative natural media components like limestone, apatite, dolomite, zeolite (e.g. clinoptilolite-rich tuff), bauxite, laterite or shale or with materials like coal fly ash, furnace slags, burnt oil shale, oyster shell, etc. (Vohla et al. 2011; Valipour and Ahn 2016).

11.2.2 Vegetation

Herbaceous plants, like reeds, rushes or grasses that can sustain higher level of water logging, play a vital role in any CWs for treating wastewaters. The roots, stems and leaves of these marshy plants provide perfect sites for microbial attachments and filtration while liberating oxygen (Valipour et al. 2014a; Valipour and Ahn 2016). Similarly, reed beds can be a designated site for bioremediation (including phytoremediation) where plant-specific species act on amelioration. Selections of plants are crucial, as high growth and high biomass producing, native, ecologically adaptive plants are preferred. Because of their immense activities, although plants are crucial biotic component, however, their role for uptake of contaminants depends on the wastewater qualities and interacting time (Valipour and Ahn 2016).

In CWs submerged, free-floating, floating-leaved and emergent aquatic macrophytes are present. However, in RBSs, emergent aquatic plants having vegetative parts above the water surface but rooted in shallow water of the bed are preferred (http://www.fao.org/). These reed plants take up N and P compounds and other components through the root zone epidermis and vascular bundles and transported to the aerial parts of the plant (Valipour et al. 2014a, b). Several interrelated factors affect the efficiency of plant for uptake of the components, which include the reed bed configuration, amount of wastewater loading range, contaminant concentrations, environmental conditions, hydraulic retention time, contact time, etc. Interestingly, vegetative evaporation-transpiration mechanism of reeds plays a significant role in contaminant adsorption, absorption, uptake, retention and volatilization, which is directly linked to the capability index of water purification of the plants (Valipour et al. 2015; Valipour and Ahn 2016). Age of plants also plays a crucial role in these decontamination processes. Due to more suberization and lignification processes in ageing roots, impurity absorption and oxygen release are less in older plants than younger plants, leading to altered RBSs functioning (Heers 2006; Valipour and Ahn 2016).

11.2.3 Microorganisms/Biofilms

RBS harbours a range of microbial population, including bacteria, algae and fungi, which act individually or form a biofilm attached to bed medium and other objects. Diversified highly active metabolic processes are evident in the beds, where rhizospheres of the plants are mostly accompanied by gram-negative bacteria which help plant in nutrition, growth and environmental stress management generated from the components present in wastewater (Valipour et al. 2014b). Different fungi, including arbuscular mycorrhizal fungi, also play an important role in the sequestration processes (Martínez-Viveros et al. 2010; Gontia-Mishra et al. 2014; Valipour and Ahn 2016).

The efficiency of removal of organic load of RBSs is mostly carried out by obligate aerobic or facultative anaerobic bacteria. Mander et al. (2003) reported that approximately less than 5% P and 10% N removal (which is almost equal to the level of plants' removal efficiency in reeds) takes place by bacteria from soluble phase. Although nutrient removal efficiency may be reduced once the bacterial growth is reaching steady state, however, in natural conditions, various environmental factors including oxygen availability, organic carbon, wastewater loading, etc. affect ammonification and denitrification processes (Valipour and Ahn 2016). However, due to volatile/atmospheric components, N removal in RBSs is more effective than P (Ciria et al. 2005).

11.2.4 Oxygen Availability

Oxygen content in the bed depends upon its structure, media components, porosity, seasonal weather and reeds. Although it differs amongst species, however, it is estimated that aquatic macrophytes transfer 30–40% of atmospheric oxygen to the rhizosphere through its aerenchyma (internal gas space) (Li et al. 2011a). However, this gas exchange and transport is regulated by several factors, like photosynthesis, thermoosmosis, diffusion, pressure gradient-induced convective flow and humidity (Wiessner et al. 2006). Physically, due to the presence of plants (along with plant's leaves and aerial parts), several factors determine some amount of wind flow in RBS, which include Venturi effects, wind stagnation pressures, pressure oscillations (due to turbulence in wind flow) and heat tube effects (condensation and evaporation at reverse ends) (Allen 1997; Valipour and Ahn 2016). In different climatic conditions, especially in cold areas, these wind flow factors characterize the operation of RBSs.

11.3 Types of RBSs

RBSs are a post-treatment system (typically for wastewater after anaerobic treatments). As per the water flow regime, RBSs can either be surface (SF) or subsurface (SSF) flow. Submerged, free-floating, or emergent macrophytes can be present in SF, which has horizontal flow regime. Although easy to construct and can ameliorate wastewater having higher suspended solids, SF requires more space and creates problems of odour and insect breeding (as free water surface is present). On the other hand, SSF can either be horizontal (HSSF) or vertical (VSSF) type, where emergent reeds are used. A thick bacterial biofilm catering microbial activities is supported by the roots submerged within the water and bed medium (Valipour et al. 2015). In HSSF, wastewater flows horizontally through a media-packed 50–100 cm active zone and 5–15 cm below the surface of the media (Tee et al. 2012), while VSSF may either be upflow (UF-VSSF) or downflow (DF-VSSF) having active zone ranging from a minimum of 50 to 100 cm. DF-VSSF is a frequently used system for wastewater treatment as it provides better oxygen transfer, which is good for improved treatment (Zhao et al. 2011). SSF matrix provides better surface area for biofilm growth, and subsequently faster treatment with less space is required. Due to subsurface water accumulation, which gives a thermal insulation, SSF types are good for cold climates. However, SSF system is vulnerable to chocking and clogging, which is relying on porosity (interstitial pore spaces) and particle size of the media. However, hybrid RBSs (coupling parallel HSSF and VSSF, with DF-VSSF and UF-VSSF variants) are more effective for accomplishing higher treatment efficacy, better BOD, suspended solids (SS) and nitrogen removal and less clogging problems (Ávila et al. 2014; Jamshidi et al. 2014).

11.4 Performance Studies

As mentioned, a range of environmental conditions, vegetation types, microbial (especially, bacterial) population, oxygen enrichment potentials and design cum operational plans are responsible for RBSs performance (Valipour and Ahn 2016). Most of the studies with RBS deal primarily with organic load and pollutant removal. Reeds like *Phragmites* sp. are common because of their capacity to improve water quality through nutrient recycling, elaborate root systems with better microbial biofilms, higher growth rate and biomass production, better oxygen transference, etc. (Valipour and Ahn 2016). However, for treating specific wastewater through RBS, hydraulic retention time (HRT) which is related to bed size, space and capital investments is an important criterion, as short HRT may lead to truncated treatment and long HRT may clog the system with higher economic burdens (Valipour and Ahn 2016).

11.5 Removal of Organic Substances

RBSs are good in removal of various organic matters. In HSSF *Phragmites* sp. bed, minimum single day retention is required to treat anaerobically pretreated effluent, which is entirely nutrient rich and comprising of higher pathogen concentration. Valipour et al. (2015) demonstrated that the *Phragmites* sp. planted bio-rack system requires shorter HRT and can act under higher loading rates. The elaborate root system of *Phragmites* sp. augments aerobic microbial growth that contributes higher removal efficiencies of both organic and inorganic substances (Valipour et al. 2015). However, clogging may affect in cold areas where freezing temperature may block the pores and void spaces reducing diffusion of atmospheric air and, therefore, treatment efficiency (Austin and Nivala 2009). It has been reported that an average of 73–93% organic pollutant removal can be achieved in two-stage hybrid systems (Q = 0.04 m3 day^{-1}, HLR = 0.07–0.07 m3/m2 day^{-1}) with varying the influent COD/N ratios (2.5, 5, and 10) (Zhao et al. 2011).

11.5.1 Removal of Nitrogen

Nitrogen removal efficiency varies in RBSs according to the input concentrations of nitrogen and organic components, transfer of oxygen within the system, microbial population concerning nitrogen assimilation, HRT and adsorption factors of the bed media. On an average, 20–99% removal efficiency has been reported for ammonium nitrogen (NH_4-N). However, higher BOD_5 (≥300-mgL^{-1}; representing high concentrations of organic matters) can be a limiting factor for nitrogen, as heterotrophic bacteria present in the bed quickly utilize available oxygen for metabolism of organic carbon, resulting in bed clogging due to sudden excessive growth of biofilms (Ciria et al. 2005; Valipour and Ahn 2016). But addition of carbon source and additional aeration may help RBS in a better way to remove nitrogen. Reports suggest that recurrent aeration of VSSF with influent COD/N ratio (from 2.5 to 20) led to nitrogen removal efficiency of 96–99% NH_4-N and 26–94% TN (Fan et al. 2013; Valipour and Ahn 2016). Hu et al. (2014) reported that incorporation of several cycles of short-flooding and long-draining periods can achieve almost ≥94% nitrification, where influent COD/N ratio was >7.

11.5.2 Removal of Phosphorus

Again, in CWs, phosphorus (P) removal may vary considerably (6–99%) as per the environmental factors, bed design and wastewater conditions. Plants usually take up the P directly for their growth, while ligand exchange reactions within the bed also help in P removal (Vohla et al. 2011). Although in normal condition only 40–60% P removal is reported, however, HRT of the system plays an important role which provides extended contact period within the beds (Vymazal 2004). 75–90% decrease of total phosphate in influent COD/N ratios (2.5, 5, and 10) in two-stage hybrid systems is reported by Zhao et al. (2011).

11.6 RBS in Cold Climate Situation

Natural wetland treatment systems are mostly present in tropical warm regions. However, incorporating the same under cold climates is having several challenges. A total of 176 wetlands are listed in northern parts of American continent encompassing the USA and Canada, among which 60% are located in cold areas which mostly treat wastewater from household sources (Mæhlum 1999). Peng et al. (2005) reported that, during cold seasons, a multistage pond-wetland system at Dongying City, China, can remove roughly 85%, 40% and 20% of COD, BOD_5 and NH_4-N, respectively, which is less than the average removal (92%, 73% and 71% for COD, BOD_5 and NH_4-N, respectively) in summer season. However, higher HRT helps to solve the problems for such constructed wetlands in cold regions, which helps to compensate cold-related slower microbial metabolic rate by providing longer contact time. Selections of plants are also an important factor. It is reported that, due to high lignin contents and better adaptability, macrophytes like *Phragmites*, *Typha* and *Scirpus* are good for RBS in cold seasons. Stein and Hook (2005) pointed out that, during the cold condition, due to reduced oxygen utilization and metabolic activities, greater oxygen leakage from plants' roots is evident which helps the sustenance for aerobic microbial respiration in the rhizosphere. Therefore, assortment of suitable plants (including native species) for RBS that can survive in cold climate has many benefits, including reduction of heat loss (insulation) from bed, enhanced transfer of oxygen into the bed and protection of bed water from freezing, wind effects and drifting snow entrapment (due to presence of plant leaves and aerial parts) (Stein and Hook 2005). Li et al. (2011b) reported the creation of an insulating layer using ice and snow through developing ice-air blanket. Effluent levels were raised purposefully to allow it to freeze and then lowering the water level afterwards, which creates an airgap between the top layer of frozen ice and effluent layer (Horváth 2012). Yin and Shen (1995) reported better water quality (SS<20 mgL^{-1}, BOD5<20 mgL^{-1}, TN<15 mgL^{-1} and TP<0.6 mgL^{-1}) under ice layers in a CW of northern China during the winter season. Therefore, a top layer of ice is beneficial for this kind of systems installed at cold areas, which can be done artificially though cold aeration at surface before surrounding temperatures reaches near zero and affecting the bed. However, thickening of ice layer may be prevented to avoid hydraulic failure (Mæhlum 1999). Therefore, greater effective depth of bed (around 1 m) and application of mulch (dead vegetation) or decomposed organic materials are advised for cold conditions (Mæhlum 1999; Wallace and Nivala 2005; Heers 2006). However, care must be taken while designing such RBS to counter freezing cold through uniform insulation and using bed media having low-thermal conductivity (Horváth 2012). Shen et al. (2007) reported 27% for NH_4-N and 31% for total P in a harvested vegetation covered SSF during winter (−3 to 6 °C), which was 5% and 16% higher than that of the control system. Wu et al. (2011) reported, at an ambience of −8 °C, 0.4 m layer of sawdust (as insulator) was sufficient to maintain a bed temperature above 6 °C. The typical removal efficiencies during the winter in this sawdust-insulated system were 88%, 95%, 85% and 96% for total P, BOD_5, NH_4-N and TSS, respectively. Further, artificial aeration at the bed water may be an alternative to keep activated where top layer is completely frozen.

11.7 Molecular Studies on RBS for Deciphering Microorganisms

Microbial activities related to wastewater treatment processes in RBS are an interesting area of study. In recent times, due to development of molecular technologies that avoid cultivation and its characteristic selectivity, deciphering the microbes in RBS can extend our ideas about the key players and their functioning in different climates and conditions. PCR-amplified 16S ribosomal RNA (rRNA) gene fragments, denaturing gradient gel electrophoresis (DGGE) or terminal fragment length polymorphism (TRFLP) were common fingerprinting techniques to assess general outline of the microbial community (Dong and Reddy 2010; He et al. 2016). However, all of the PCR-based technologies are having bias related to primer sets and PCR amplification, which was countered by a specific probe-based technique called fluorescent in situ hybridization (FISH) (Sánchez et al. 2016). Recent developments of high-throughput sequencing (HTS) methods (metagenomics) permit analysis of microbial assemblies in a far greater manner that contributes to an improved impression of the phylogenetic diversity and the genomic potential of the population functioning at the system (Ferrera and Sánchez 2016). Further, gene expression studies through RNA sequencing (metatranscriptomics) and studies on proteins (metaproteomics) for observing metabolism can help to decipher microbial community configuration of the RBSs, in one hand, and functional annotation on the other. Recent studies on SSF revealed bacterial composition of the system linking microbial composition and function (Sánchez et al. 2016). Omics studies help to augment functional features by understanding diverse metabolisms in different environmental conditions. As, for example, Nielsen et al. (2012) suggested a theoretical model ecosystem for relating microbial communities for improved biological removal of phosphorus through developing a core microbiome.

11.8 Conclusive Remarks

Wastewater decontamination using RBSs is an environment-friendly approach, where the relationship between bed media, microbes and reeds plays a crucial role. Operational settings, HRT and design of the beds are mostly important for maintaining sustainable treatment under different climatic conditions. Selection of appropriate reed plants is very important, where species like *Phragmites* sp. are mostly effective for SSF systems. Plant species, COD/N ratio and HRT can influence efficiency for the N and P removal. RBSs can be operated as decentralized engineered secondary effluent treatment system for different regions, including cold areas. Further research on microbial community studies using recent molecular technologies along with advanced engineering designs can help in effective and sustainable application in different climates and conditions.

Acknowledgement S.C. is thankful to Director, DRL (DRDO), Assam, India. The authors apologize for the many colleagues who are not referenced in this work due to space limitations.

References

Allen LH Jr (1997) Mechanisms and rates of O_2 transfer to and through submerged rhizomes and roots via aerenchyma. Annu Proc Soil Crop Sci Soc Fla 56:41–54

Arends JBA, Van Denhouwe S, Verstraete W, Boon N, Rabaey K (2014) Enhanced disinfection of wastewater by combining wetland treatment with bioelectrochemical H_2O_2 production. Bioresour Technol 155:352–358

Austin D, Nivala J (2009) Energy requirements for nitrification and biological nitrogen removal in engineered wetlands. Ecol Eng 35:184–192

Ávila C, Matamoros V, Reyes-Contreras C, Piña B, Casado M, Mita L, Rivetti C, Barata C, García J, Bayona JM (2014) Attenuation of emerging organic contaminants in a hybrid constructed wetland system under different hydraulic loading rates and their associated toxicological effects in wastewater. Sci Total Environ 470–471:1272–1280

Brix H, Schierup H (1989) Use of aquatic macrophyte in water pollution control. Ambio 18:100–107

Brix H, Arias CA, del Bubba M (2001) Media selection for sustainable phosphorus removal in subsurface flow constructed wetlands. Water Sci Technol 44:47–54

Carvalho PN, Arias CA, Brix H (2017) Constructed wetlands for water treatment: new developments. Water 9:397. https://doi.org/10.3390/w9060397

Chen J, Wei XD, Liu YS, Ying GG, Liu SS, He LY, Su HC, Hu LX, Chen FR, Yang YQ (2016) Removal of antibiotics and antibiotic resistance genes from domestic sewage by constructed wetlands: optimization of wetland substrates and hydraulic loading. Sci Total Environ 565:240–248

Ciria MP, Solano ML, Soriano P (2005) Role of macrophyte *Typha latifolia* in a constructed wetland for wastewater treatment and assessment of its potential as a biomass fuel. Biosyst Eng 92:535–544

Dong X, Reddy GB (2010) Soil bacterial communities in constructed wetlands treated with swine wastewater using PCR-DGGE technique. Bioresour Technol 101:1175–1182

Drewe JA, Mwangi D, Donoghue HD, Cromie RL (2009) PCR analysis of the presence and location of *Mycobacterium avium* in a constructed reed bed, with implications for avian tuberculosis control. FEMS Microbiol Ecol 67:320–328

Elsayed OF, Maillard E, Vuilleumier S, Imfeld G (2014) Bacterial communities in batch and continuous-flow wetlands treating the herbicide S-metolachlor. Sci Total Environ 499:327–335

Fan J, Wang W, Zhang B, Guo Y, Ngo HH, Guo W, Zhang J, Wu H (2013) Nitrogen removal in intermittently aerated vertical flow constructed wetlands: impact of influent COD/N ratios. Bioresour Technol 143:461–466

Faulwetter JL, Gagnon V, Sundberg C, Chazarenc F, Burr MD, Brisson J, Camper AK, Stein OR (2009) Microbial processes influencing performance of treatment wetlands: a review. Ecol Eng 35:987–1004

Ferrera I, Sánchez O (2016) Insights into microbial diversity in wastewater treatment systems: how far have we come? Biotechnol Adv 34:790–802

Gontia-Mishra I, Sasidharan S, Tiwari S (2014) Recent developments in use of 1-aminocyclop ropane-1-carboxylate (ACC) deaminase for conferring tolerance to biotic and abiotic stress. Biotechnol Lett 36:889–898

He T, Guan W, Luan Z, Xie S (2016) Spatiotemporal variation of bacterial and archaeal communities in a pilot-scale constructed wetland for surface water treatment. Appl Microbiol Biotechnol 100:1479–1488

Heers M (2006) Constructed wetlands under different geographic conditions: evaluation of the suitability and criteria for the choice of plants including productive species. PhD dissertation, Hamburg University of Applied Sciences

Horváth G (2012) Review of subsurface flow treatment wetland feasibility in Finland. PhD dissertation, Tampere University of Applied Science

Hu Y, Zhao Y, Rymszewicz A (2014) Robust biological nitrogen removal by creating multiple tides in a single bed tidal flow constructed wetland. Sci Total Environ 470–471:1197–1204

Jackson EF, Jackson CR (2008) Viruses in wetland ecosystems. Freshw Biol 53:1214–1227

Jamshidi S, Akbarzadeh A, Woo KS, Valipour A (2014) Wastewater treatment using integrated anaerobic baffled reactor and bio-rack wetland planted with *Phragmites* sp. and *Typha* sp. Environ Health Sci Eng 12:131–143

Kadlec RH (2009) Wastewater treatment at the Houghton Lake wetland: soils and sediments. Ecol Eng 35:1333–1348

Kadlec RH, Knight RL (1996) Treatment wetlands. Lewis Publishers, Boca Raton

Kadlec RH, Wallace S (2009) Treatment wetlands. CRC Press, Boca Raton

Karpiscak MM, Sanchez LR, Freitas RJ, Gerba CP (2001) Removal of bacterial indicators and pathogens from dairy wastewater by a multi-component treatment system. Water Sci Technol 44:183–190

Kinsley C, Kennedy K, Crolla A (2017) A combined reed bed/freezing bed technology for septage treatment and reuse in cold climate regions. Water Sci Technol 76(2):286 293

Kurniadie D (2011) Wastewater treatment using vertical subsurface flow constructed wetland in Indonesia. Am J Environ Sci 7:15–19

Larsen JD (2017) Environmental assessment of sewage sludge management – focusing on sludge treatment reed bed systems. Department of Environmental Engineering, Technical University of Denmark (DTU), Kgs. Lyngby. (Downloaded from: https://orbit.dtu.dk/files/136886674/Thesis_online_version_Julie_Dam_Larsen.pdf)

Li H, Ye ZH, Wei ZJ, Wong MH (2011a) Root porosity and radial oxygen loss related to arsenic tolerance and uptake in wetland plants. Environ Pollut 159:30–37

Li Y, Zhang Y, Zhang X (2011b) Heat preservation of subsurface flow constructed wetland in cold area in winter and its operation effect. Procedia Environ Sci 10:2182–2188

Li Y, Wu B, Zhu G, Liu Y, Ng WJ, Appan A, Tan SK (2016) Highthroughput pyrosequencing analysis of bacteria relevant to cometabolic and metabolic degradation of ibuprofen in horizontal subsurface flow constructed wetlands. Sci Total Environ 562:604–613

Lv T, Zhang Y, Carvalho PN, Zhang L, Button M, Arias CA, Weber KP, Brix H (2017) Microbial community metabolic function in constructed wetland mesocosms treating the pesticides imazalil and tebuconazole. Ecol Eng 98:378–387

Masciandaro G, Iannelli R, Chiarugi M, Peruzzi E (2015) Reed bed systems for sludge treatment: case studies in Italy. 5 Water Sci Technol 72(7):1043–1050

Mæhlum T (1999) Wetlands for treatment of landfill leachates in cold climates. In: Mulamoottil G, McBean EA, Rovers F (eds) Constructed wetlands for the treatment of landfill leachates. CRC Press, Boca Raton, pp 33–46

Mander U, Teiter S, Kuusemets V, Lohmus K, Oovel M, Nurk K, Augustin J (2003) Nitrogen and phosphorous budgets in a subsurface flow wastewater treatment wetland. Water Resour Manag II 61:135–148
Martínez-Viveros O, Jorquera MA, Crowley DE, Gajardo G, Mora ML (2010) Mechanisms and practical considerations involved in plant growth promotion by rhizobacteria. J Soil Sci Plant Nutr 10:293–319
Nielsen PH, Mielczarek AT, Kragelund C, Nielse JL, Saunders AM, Kong Y, Hansen AA, Vollertsen J (2012) A conceptual model system of microbial communities in enhanced biological phosphorus removal plants. Water Res 44:5070–5088
Peng JF, Wang BZ, Wang L (2005) Multi-stage ponds–wetlands ecosystem for effective wastewater treatment. J Zhejiang Univ (Sci) B6:346–352
Rizzo A, Bresciani R, Martinuzzi N, Masi F (2018) French reed bed as a solution to minimize the operational and maintenance costs of wastewater treatment from a small settlement: an Italian example. Water 10:156. https://doi.org/10.3390/w10020156
Sanchez CA, Childers DL, Turnbull L, Upham RF, Weller N (2016) Aridland constructed treatment wetlands II: plant mediation of surface hydrology enhances nitrogen removal. Ecol Eng 97:658–665
Sánchez O (2017) Constructed wetlands revisited: microbial diversity in the –omics era. Microb Ecol 73:722–733
Shen H, Hu HY, Pan YB (2007) Study on enhanced measures for operation of subsurface flow constructed wetlands in winter. China Water Wastewater 23:44–46
Stein OR, Hook PB (2005) Temperature, plants, and oxygen: how does season affect constructed wetland performance? Environ Sci Health 40:1331–1342
Stott R, Jenkins T, Bahgat M, Shalaby I (1999) Capacity of constructed wetlands to remove parasite eggs from wastewaters in Egypt. Water Sci Technol 40:117–123
Taleno VC (2012) Comparison of two constructed wetland with different soil depth in relation to their nitrogen removal. PhD dissertation, Universidad Autónoma De San Luis Potosí
Tee HC, Lim PE, Seng CE, MAM N (2012) Newly developed baffled subsurface-flow constructed wetland for the enhancement of nitrogen removal. Bioresour Technol 104:235–242
Tiwari K, Jha S, Vyas K, Patel K, Patel M (2017) Reed bed technology: a review. Intl J Engg Scien Invention Res Dev III:554–556. e-ISSN: 2349-6185
Valipour A, Ahn YH (2016) Constructed wetlands as sustainable ecotechnologies in decentralization practices: a review. Environ Sci Pollut Res 23:180–197
Valipour A, Hamnabard N, Woo KS, Ahn YH (2014a) Performance of high-rate constructed phytoremediation process with attached growth for domestic wastewater treatment: effect of high TDS and Cu. Environ Manag 145:1–8
Valipour A, Azizi S, Raman VK, Jamshidi S, Hamnabard N (2014b) The comparative evaluation of the performance of two phytoremediation systems for domestic wastewater treatment. Environ Sci Eng 56:319–326
Valipour A, Raman VK, Ahn YH (2015) Effectiveness of domestic wastewater treatment using a bio-hedge water hyacinth wetland system. Water 7:329–347
Vohla C, Koiv M, Bavor HJ, Chazarenc F, Mander U (2011) Filter materials for phosphorus removal from wastewater in treatment wetlands – a review. Ecol Eng 37:70–89
Vymazal J (2004) Removal of phosphorus in constructed wetlands with horizontal subsurface flow in the Czech Republic. Water Air Soil Pollut 4:657–670
Vymazal J (2013) Emergent plants in free water surface constructed wetlands: a review. Ecol Eng 61P:582–592
Vymazal J, Brezinova T (2015) The use of constructed wetlands for removal of pesticides from agricultural runoff and drainage: a review. Environ Int 75C:11–20
Wallace SD, Nivala JA (2005) Thermal response of a horizontal subsurface flow wetland in a cold temperate climate. IWA specialist group on the use of macrophytes in water pollution control. Newsletter 29

Wiessner A, Kuschk P, Kappelmeyer U, Bederski O, Müller RA, Kästner M (2006) Influence of helophytes on redox reactions in their rhizosphere. In: Mackova M, Dowling DN, Macek T (eds) Phytoremediation and rhizoremediation, vol 9A. Springer, Dordrecht, pp 69–82
Wong C, Au E, Truong H, Quan J, Young J, Kathriarachchi S (2009) The reed bed sewage treatment system. http://empslocal.ex.ac.uk
Wu S, Austin D, Liu L, Dong R (2011) Performance of integrated household constructed wetland for domestic wastewater treatment in rural areas. Ecol Eng 37:948–954
Yin H, Shen W (1995) Using reed beds for winter operation of wetland treatment system for wastewater. Water Sci Technol 32:111–117
Zhao YJ, Hui Z, Chao X, Nie E, Li HJ, He J, Zheng Z (2011) Efficiency of two-stage combinations of subsurface vertical down-flow and up-flow constructed wetland systems for treating variation in influent C/N ratios of domestic wastewater. Ecol Eng 37:1546–1554

Inspection of Crop Wild Relative (*Cicer microphyllum*) as Potential Genetic Resource in Transgenic Development

12

Rupesh Kumar Singh, Nitin Bohra, Lav Sharma, Sivalingam Anandhan, Eliel Ruiz-May, and Francisco Roberto Quiroz-Figueroa

Abstract

Crop wild relative (CWR) is a wild ancestor plant/related taxon, closely related to cultivated domesticated plant whose geographic origins can be traced to regions known as Vavilov centers. CWRs have been significantly important to provide efficient traits to crops for improving agricultural production leading toward sustainable agro-ecosystems. Genetic material from CWRs has been utilized since long time as wild maize (*Zea mays* ssp. *parviglumis* H. H. Iltis & Doebley) and is routinely grown alongside maize to promote natural crossing.

R. K. Singh (✉)
Centro de Química de Vila Real (CQ-VR), Universidade de Trás-os-Montes e Alto Douro, Vila Real, Portugal
e-mail: rupesh@utad.pt

N. Bohra
School of Genetics and Biotechnology, University of Trás-os-Montes and Alto Douro, UTAD, Vila Real, Portugal

Department of Biotechnology, National Institute of Technology, Warangal, Telangana, India

L. Sharma
CITAB – Centre for the Research and Technology of Agro-Environmental and Biological Sciences, University of Trás-os-Montes and Alto Douro, UTAD, Vila Real, Portugal

S. Anandhan
ICAR- Directorate of Onion and Garlic Research, Rajgurunagar, Pune, Maharashtra, India

E. Ruiz-May
Red de Estudios Moleculares Avanzados, Instituto de Ecología A. C., ClusterBioMimic®, Xalapa, Veracruz, Mexico
e-mail: eliel.ruiz@inecol.mx

F. R. Quiroz-Figueroa
Instituto Politécnico Nacional, Centro Interdisciplinario de Investigación para el Desarrollo Integral Regional Unidad Sinaloa (CIIDIR-IPN Unidad Sinaloa), Laboratorio de Fitomejoramiento Molecular, Guasave, Sinaloa, México

R. Sathishkumar et al. (eds.), *Advances in Plant Transgenics: Methods and Applications*, https://doi.org/10.1007/978-981-13-9624-3_12

More recently, plant breeders have utilized CWR genes to improve a wide range of crops like maize (*Zea mays* ssp. *mays*), rice (*Oryza sativa*), tomato (*Solanum lycopersicum*), and legumes. Grain legumes have been a major crop group in present time with the objectives to include its tolerance to abiotic stresses, and recently wild pigeon pea (*Cajanus cajan* (L.) Millsp.) was highlighted to improve the CWR germplasm collection with particular emphasis on potential traits for abiotic stress tolerance. CWR may be a potential source for better genes/traits to develop plant breeding or transgenic crops.

Cicer microphyllum is a cold desert plant distributed in dry slopes of the Himalayan mountain range and wild relative of cultivated chickpea (*C. arietenum*). It has been considered to be a repository of genes expressed for extreme cold and drought tolerance and has been reported that wild *Cicer* species possess wealth of genes for biotic and abiotic stresses and also hold promise for enhancing seed yield through introgression of genes in cultivated species. Our previous molecular studies on *Cicer microphyllum* confirm this CWR as a natural repository of potential traits for abiotic stress tolerance. Germplasm availability is challenging to meet the needs of sustainable utilization of this species for routine research. Therefore we report in vitro germplasm conservation of this CWR and physiological responses toward cold and drought stress conditions to ensure long-term availability of germplasm to researchers to explore and utilize it for transgenic crop development.

Keywords

Cicer microphyllum · CWR · Abiotic stress · Micropropagation · Plant tissue culture · Germplasm

12.1 Introduction

Crop wild relative (*CWR*) is a wild progenitor, and their biodiversity centers were proposed by the Russian botanist Nikolai Vavilov, who was the first to realize the importance of crop wild relatives in the early twentieth century (Vavilov 1926). The wild relatives of crops have been significantly important in terms of capability to provide efficient traits to cultivated plants for improving agricultural production leading toward sustainable agro-ecosystems (FAO 1998, 2008; Biodiversity international 2006). With the advent of climate change and greater ecosystem instability, CWRs are likely to prove a critical resource in ensuring food security for the new millennium (Maxted et al. 2008). Genetic material from CWRs has been utilized by humans for thousands of years to improve the quality and yield of crops. Farmers and plant breeders have utilized CWR genes to improve a wide range of crops like rice (*Oryza sativa*), tomato (*Solanum lycopersicum*), and grain legumes (Hajjar and Hodgkin 2007).

Grain legumes have been a major crop group for breeders in present time with the objectives to include its tolerance to abiotic and biotic stresses. Crop wild relative with natural repository of potential traits is a primary need for breeders to ensure the availability of the CWR. Wild genetic resources are underrepresented in ex situ and in situ conservation (Barazani et al. 2017; Warburton et al. 2017), which leads to insufficient availability to plant breeders. Recently Colin et al. (2015) highlighted identification and geographical locations for pigeon pea in order to improve the CWR germplasm collection with particular emphasis on potential traits for abiotic stress tolerance. Abiotic stress management has been a major challenge for the breeders in the past decade as it affects the overall production up to a great extent. Present scenario needs a global transformation toward efficient productive and more sustainable agriculture system. This transformation is possible via enhanced agronomic practices combined with the use of wild crop varieties with reliable yield under more adverse conditions (Foley et al. 2011).

Cicer microphyllum, a wild relative of cultivated chickpea (*C. arietenum*), is considered to be a repository of genes expressed for extreme cold and drought tolerance (Croser et al. 2003). It has been reported that wild *Cicer* species possess wealth of genes for biotic and abiotic stresses and also hold promise for enhancing seed yield through introgression of genes in cultivated species (Singh 1991; Singh and Ocampo 1993). Our previous molecular studies on *Cicer microphyllum* confirm this CWR as a natural repository of potential traits for abiotic stress tolerance (Singh et al. 2011, 2013) and physiologically have a great tolerance under extreme climatic conditions (Singh et al. 2017). Thus, *Cicer microphyllum* may be highly useful for research studies for crossing with cultivated species and for exploring the gene pool for drought and cold adaptation. Although the selection of CWR has greatly been performed through field observation under adverse climatic conditions, inclusive approach with molecular and physiological understanding would be more beneficial (Stoddard et al. 2006). To fulfill these objectives, an in vitro propagation protocol is necessary to maintain the germplasm under laboratory and field conditions. There are many techniques available for conservation of wild, rare, and endemic plant species. These include micropropagation, through direct seed germination, regeneration from callus, embryo rescue, micrografting, and cryopreservation (Nitzsche 1983; Rick 1984; Stanilova et al. 1994). Earlier we have reported an efficient in vitro germination protocol of *Cicer microphyllum* by direct seed treatment and through embryo culture (Singh et al. 2007). However, to date, there has been one report of in vitro regeneration/propagation method via callus induction for *Cicer microphyllum* (Singh et al. 2018). To meet the needs of sustainable utilization of this species for routine research use as well as for germplasm conservation, we have established an efficient and reproducible in vitro regeneration protocol in this study. This effort ensures the conservation of *Cicer microphyllum* and for long-term availability of germplasm to plant breeders.

Cicer microphyllum Benth belongs to family Fabaceae and known by common names as wild gram (E), Seri, or Srad-Kar (L).

12.1.1 Geographical Distribution

Cicer microphyllum is a cold desert plant and distributed in East Asian Himalayas from Afghanistan to Tibet and western Nepal. Cold deserts are usually confined to high altitudes and circumpolar regions. Under the trans-Himalayan zone, the largest cold desert, Ladakh, India, covers more than 70,000 sq. km area of Jammu and Kashmir, and it lies between 31° 15′ N latitude and 75° 15′–80° 15′ E longitude. High mountains and barren landscapes determine the harsh climate particularly temperature during winter months, which fall below −30 °C at Leh and − 70 °C at Dras for about 5–7 months (October to April) every year. The flora of cold desert Ladakh comes under alpine zones which are situated between 2700 and 6000 meters above sea level (Ballabh et al. 2007). *Cicer microphyllum* is specifically distributed at Changthang, Leh, Zanskar, and Lahaul-Spiti valleys of the Indian Himalayan zone between 4230 and 4870 meters above sea level.

12.1.2 Plant Characters

This wild legume (Fig. 12.1) is diffuse, erect, spreading, and shrublet herb in its habitat. Stem up to 60–100 cm, much branched. Leaves are pinnate, ending in a coiled tendril, sessile, cuneate. Many leaflet in widely spaced pairs, ovate-wedge shaped, toothed apex. Flowers solitary or two, in the axils of leaves, stalked, axillary 2–3 cm long. Calyx hairy and lobed, corolla bluish-purple, pods linear-oblong,

Fig. 12.1 *Cicer microphyllum* distributed at Changthang, Leh, Zanskar, and Lahaul-Spiti valleys of Indian Himalayan zone on 4870 meters above sea level

beaked, and silky-hairy. Flowering and fruiting season are from mid-June to September, and seeds can be harvested in end of September to October.

12.1.3 Uses

The whole plant is used as fodder. The green plant is generally grazed by wild animals especially sheep and goats, and the whole plant is harvested in the month of September and dried properly for prolonged winter use. It has good protein content about 15.30% which is very much useful for milch and young animals.

12.2 Regeneration of *Cicer microphyllum* by Tissue Culture

Availability of plant material of this CWR is very difficult due to harsh conditions in natural habitat; therefore a robust and efficient protocol has been developed (Singh et al. 2019). In brief seeds of *Cicer microphyllum* were collected from DIHAR Leh and washed thoroughly under running tap water followed by washing with sterile distill water. Seeds were then treated with 0.1% $HgCL_2$ for 2 min, followed by washing with 70% ethyl alcohol for 2 min and then thoroughly washed with sterile distilled water for three times. Then, embryos were separated and chopped down into small pieces and used as explants for callus induction. Explants were transferred in petri plates with 25 ml MS media added with 0.2 mg/l NAA+ 2.5 mg/l BAP for callus induction, and pH was adjusted to 5.7 prior to autoclaving. Cultures were incubated at 25 °C with 16 h photoperiod. Calli from these primary cultures were transferred to MS medium containing 0.5 mg/l NAA, 2.0 mg/l BAP, and 0.1 mg/l AgNO3 (Fig. 12.2). Data on shoot proliferation efficiency were recorded after 45 days of cultures and presented. Proliferated and elongated shoots were then transferred to rooting media (1.0 mg/l IBA in half strength MS) for root formation. Healthy plantlets with 3–4 cm length with different age (21, 24, 30 days) were individually removed from culture tubes, and their roots were washed carefully with tap water and transferred to pots containing garden soil, vermiculite, and vermicompost (1:1:1).

Callus induction was observed within 18–28 days of incubation of excised embryos as explants. Callus induction was recorded with 95% efficiency and with ideal color and texture. These calli were transferred in MS media containing 0.5 mg/l NAA, 2.0 mg/l BAP, and 0.1 mg/l AgNO3. It was observed that addition of 0.1 mg/l AgNO3 enhances the percentage of shooting as well as the number of shoots per callus. It was also observed that only light green calli produced shoot buds efficiently. The shoot buds first appeared as nodular outgrowth within 20–25 days of cultures, and at the end of 30–35 days of cultures, these nodular outgrowths increased in size followed by production of leaf primordia. Islam and Riazuddin (1993) used BAP (2.0–10.0 mg/l) and IAA (0.1–1.0 mg/l) for shoot proliferation using callus from hypocotyl explants of chickpea. In another study the maximum shoot bud differentiation frequency was observed on MS containing BAP (3.0 mg/l)

Fig. 12.2 In vitro regeneration of plantlets from excised leaves of explants from *Cicer microphyllum*. Induction of callus and multiple shooting on basal MS media supplemented with 0.5 mg/l NAA, 2.0 mg/l BAP, and 0.1 mg/l AgNO3

and NAA (0.5 mg/l) in *Glycine max* by Settu and Ranjithakumari (1999). For adventitious root formation, MS and half strength MS and IBA were tested. It was observed that 1.0 mg/l IBA in half-strength MS was most effective for rooting of in vitro culture shoots of *Cicer microphyllum*. Huda et al. (2003) demonstrated best rooting percentage in half strength of MS supplemented with 1.0 mg/l IBA in previous study on chickpea. Rooted shoots of different ages were transferred to soil to investigate the survivability of transplanted plantlets. Maximum percentage (45) of survivability after 40 days of transplantation was achieved in 30-day-old rooted shoots on mixture of garden soil, vermiculite, and vermicompost (1:1:1).

12.3 Physiological Evaluation for PEG-Mediated Stress

Utilization of CWR in legume breeding has been a significant success to incorporate traits to the crop. *Cicer microphyllum* is a potential CWR for chickpea breeding programs to improve abiotic stress tolerance, which is the current objective for legume breeders of this specie. Chickpea is one of the most common grain legume cultivated worldwide. Abiotic stress (cold/drought stress) is a major source of yield loss and principal limiting abiotic factors of production and has a particularly damaging effect on the plant itself when it occurs during the reproductive phase (Stanley 2003). Plants adopt a combination of physiological and biochemical responses at molecular, cellular, and whole organismal level to cope to drought stress condition. It includes membrane stability, increased antioxidant enzymes to encounter the damaging effects of reactive oxygen species, and enhanced metabolite flux. Genetic assessment for abiotic stress tolerance in legumes is an important area of research in chickpea (Clarke 2001; Clarke et al. 2004). During the past decade, pulse production has significantly fluctuated due to abiotic stress conditions, including cold and drought conditions, and economically put a major pressure in many countries especially Australia, where they are produced on a very large scale (Ahmad et al. 2010). Traditional breeding for cold and drought tolerance has been challenging and limited in pulses due to the lack of reliable screening techniques, potential germplasm, and other effects (Srinivasan et al. 1998; Ali et al. 1999; Clarke et al. 2004; Stoddard et al. 2006). CWR has been successful for introgression of valuable agronomic traits in legumes (Colin et al. 2015). *Cicer microphyllum* can be a great alternative in this program as it survives well in cold deserts. Phenotypic characters like deep root system and smaller leaves to reduce the transpiration loss also make this plant adapted to drought stress. Although CWR selection has generally been performed on basis of field observations, it would benefit from the knowledge of physiological traits (Stoddard et al. 2006). In this context, we assessed a cold desert alpine legume *Cicer microphyllum*, a CWR of cultivated chickpea (*Cicer arietinum*), on the basis of physiological, biochemical, and molecular response toward abiotic stress condition.

At physiological, biochemical, and molecular level, alpine plants go through a complex phenomenon in order to survive in cold alpine climate, but the underlying molecular and physiological mechanisms of this adaptation are still not very clear

(Frei et al. 2012). Cold desert/alpine areas of high mountains and dry landscapes determine the harsh climate for alpine plants including extremely low/freezing temperatures, drought, hypoxia, and high ultraviolet (UV) radiation. Alpine vegetation has developed diversities of defense mechanism against a wide range of adverse climatic conditions, which include both morphological and physiological alterations. For example, higher accumulation of secondary metabolites to counter UV radiation exposure, increased level of antifreeze proteins, and solutes like proline have been reported previously in alpine plants (Korner 2003; Bernal et al. 2013) followed by activation of antioxidant system such as superoxide dismutase (SOD), catalase (CAT), glutathione reductase (GR), and ascorbate peroxidase (APX) to prevent increased accumulation of reactive oxygen species (Foyer and Noctor 2005). Variability in adaptation to cold desert climate is also due to global and differential gene expression pattern as these plants have natural repository of abiotic stress-tolerant genes as proved by our previous studies (Singh et al. 2013). The activation of these genes may be triggered by exposure to stress stimulus, and the alterations in gene expression may end up in various physiological and biochemical changes in a complex phenomenon of tolerance to extreme adverse desert climate (Chinnusamy et al. 2007; Thomashow 2010). Water availability; is limited globally with rapid increase in world's population putting pressure for future food demand and further aggravating drought effects (Somerville and Briscoe 2001). Drought stress is a complex situation for plant survival, and it depends on several factors (Wery et al. 1994), and plants are bound to respond simultaneously to multiple stresses (Zhou et al. 2007).

Cicer microphyllum is the dominant component of the alpine meadow communities of the Indian alpine ecological system and plays an important role in alpine system conservation. Distribution of *Cicer microphyllum* in extremely adverse alpine climate such as the Great Himalayan mountain of Leh and Ladakh, India, along the alpine altitude gradient from 2700 to 6000 meters above sea level (Ballabh et al. 2007) indicates its successful acclimation to the alpine environment. To decipher the mechanisms by which this species adapts to alpine climate, we applied physiological and biochemical approach. Based on these results, we hypothesize that *Cicer microphyllum* uses multiple strategies to acclimate successfully to extreme cold and dry alpine environmental stress. These findings suggest *Cicer microphyllum* a potential genetic resource and unique germplasm for introgression of abiotic stress tolerance traits in cold- and drought-susceptible chickpea varieties through breeding programs and may lead to transformation toward improved agronomic practices. Based on these results, we hypothesized that *Cicer microphyllum* uses multiple strategies to acclimate to alpine environmental stress.

C. microphyllum plantlets were regenerated by in vitro tissue culture as explained earlier and used for further experiments (Singh et al. 2017). Four plants were randomly chosen and subjected to treatment with 50 μM polyethylene glycol (PEG). For proline assay, one set of plants with four replicates was treated with different concentrations of PEG (10, 20, 30, 40, and 50 μM). All the plant samples were frozen on liquid nitrogen, and proline content was assayed accordingly (Bates et al. 1973). For antioxidant enzyme assay, one set of plant was subjected to treatment with 50 μM polyethylene glycol (PEG), and SOD, APX, CAT, and GR activities were analyzed (Beyer and Fridovich 1987; Aebi 1984; Foyer and Halliwell 1976). All assays and measurements were carried out in four replicates. GraphPad program was used for analysis of variance of experiments laid out in a completely randomized design. The percentage data were arcsine (angular) transformed before conducting statistical analyses. The treatment means were compared by using Duncan's new multiple range tests at a significance level of $P \leq 0.05$.

Broad range of compatible organic compounds have been reported to overproduce and accumulate in cells which play significant roles to prevent such adverse stress conditions (Serraj and Sinclair 2002); some of them are proline, glutamate, glycine betane, carnitine, mannitol, sorbitol, fructants, polyols, trehalose, sucrose, and oligosaccharides (Folkert et al. 2001). These solutes tend to accumulate under water-deficit conditions to lower the osmotic potential of corresponding cells followed by water intake into cells to maintain the turgidity. Proline is the most important cytosolute, and its free accumulation is widespread response of higher plants, algae, animals, and bacteria to low water potential (Zhu 2002; Wahid and Close 2007). Increased biosynthesis and slow oxidation in mitochondria enhance its accumulation in leaves during low water potential. Free proline accumulation during different concentrations of PEG (10, 20, 30, 40, and 50 μM) treatment was investigated and recorded significantly increasing with increase in PEG concentration (Fig. 12.3). Increased proline accumulation was reported in pea cultivars under

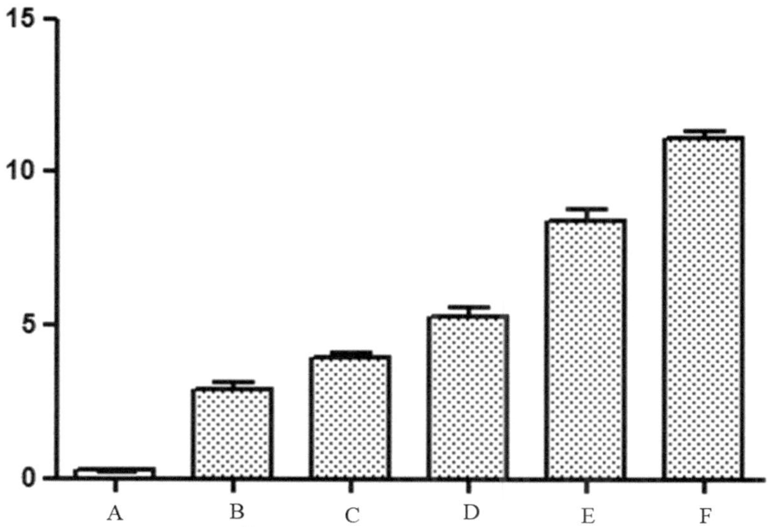

Fig. 12.3 Proline accumulation in *Cicer microphyllum* with10, 20, 30, 40, and 50 μM PEG treatment for 24 h stress condition. **A** Control (plant grown at 24 ± 2 °C), **B–F** with10, 20, 30, 40, and 50 μM PEG treatment, respectively. Four plants were randomly chosen for each treatment. Leaf tissues were harvested from both the treated and control plants, frozen with liquid nitrogen, and used for experiments presented in the figure. The vertical bars represent the mean ± standard error of four replicates

drought stress (Alexieva et al. 2001). Another study on drought-tolerant *Petunia* varieties (*Petunia hybrida*) was reported to accumulate free proline under drought stress that acted as an osmoprotectant and improved drought tolerance (Yamada et al. 2005).

During treatment with 50 μM PEG for 24 h, CAT showed a 2.95-fold increase, SOD showed a 17.09-fold increase, APX showed a 4.56-fold increase, and GR showed 31.43-fold increase. All these antioxidant enzyme activities revealed significant elevation during PEG treatment in comparison to control (Fig. 12.4). These findings indicate that *Cicer microphyllum* has inherent strong antioxidant system to counter the stress conditions. Plant antioxidant defense system constitutes enzymatic (superoxide dismutase, catalase, ascorbate peroxidase, and glutathione reductase) and nonenzymatic (cystein, reduced glutathione, and ascorbic acid) components together (Gong et al. 2005). Antioxidant enzyme system plays most efficient role against oxidative stress to detoxify reactive oxygen species (Hasegawa et al. 2000; Prochazkoya et al. 2001 and Farooq et al. 2008). Ascorbate peroxidase performs key mechanism (Orvar and Ellis 1997), while glutathione reductase plays central role to maintain reduced glutathione during stress (Pastori et al. 2000). Superoxide dismutase performs the first step of reactive oxygen species scavenging system by

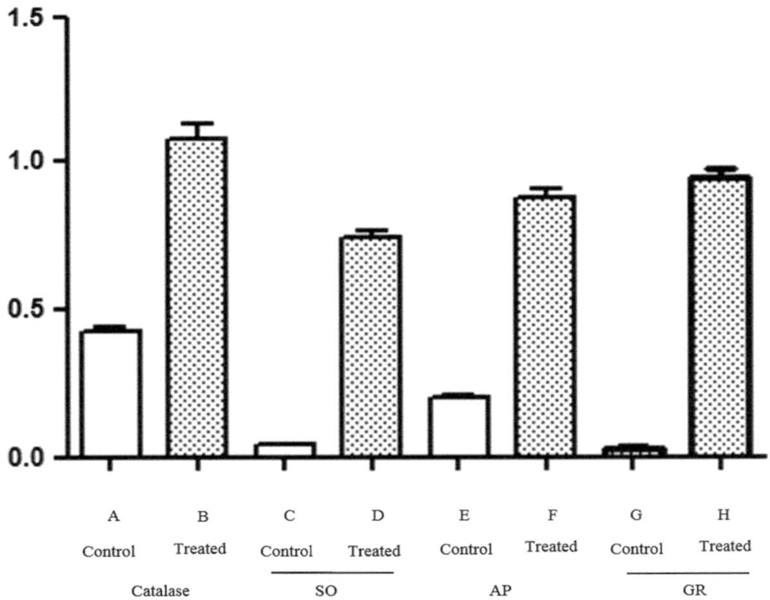

Fig. 12.4 **A** and **B** Spectrophotometric estimation of catalase activity in *Cicer microphyllum* in control (plant grown at 24 ± 2 °C) and treated (after exposure to 50 μM PEG for 24 hr) plant. Both the plants were maintained in culture room at 24 ± 2 °C temperature and 16/8 h of photoperiod. **C** and **D** Spectrophotometric estimation of superoxide dismutase. **E** and **F** Spectrophotometric estimation of APX. **G** and **H** Spectrophotometric estimation of GR. The vertical bars represent the mean ± standard error

catalyzing the dismutation of two molecules of superoxide into O_2 and H_2O_2 (Lima et al. 2002). Superoxide radical has half-life of less than 1 s and rapidly dismutated by superoxide dismutase into H2O2 which is relatively stable and further detoxified by catalase and peroxidae (Apel and Hirt 2004). These metalloenzymes are key components of plant responses against stress, and increased superoxide dismutase activity is well known to confer oxidative stress tolerance (Pan et al. 2006).

Our results showed an increase in antioxidants and proline accumulation and provide a strong evidence of combined role of proline and antioxidant enzymes against major abiotic stresses in alpine region. Another recent report is very significant in the adaptation of plants to the environment and in overcoming stress conditions through modulation of antioxidant enzymes in Tibetan wild barley (Ahmed et al. 2015). Our data suggest that PEG-inducible responses in alpine region could be attributed cumulative to higher activity of antioxidant enzymes, increased proline content.

12.4 *Cicer microphyllum*: Genetic Resource of Candidate Genes

Our previous study focused on cold stress (Singh et al. 2013) revealed that *Cicer microphyllum* has inherent genetic potential to cope with adverse climatic conditions. Forward cDNA subtraction was performed to identify differentially expressed genes. Plants were treated with cold stress, and total RNA was extracted, followed by cDNA synthesis which served as tester population during hybridization. Plants maintained at normal room temperature were treated as control population. Both cDNA were digested and hybridized to screen out the specific transcripts responsive to cold stress, mainly present in tester population. Subtracted product was then cloned in cloning vector for screening where 300 clones were chosen as differentially expressed. These clones were sequenced and deposited in gene database (accession numbers GO241043–GO241326). Sequences were analyzed and grouped in different groups like activators, for degradation, metabolism, cell development, translation/post translation, transporters, stress responsive, signaling, and hypothetical proteins with unknown functions. In that study, many cDNAs have not shown similarity with previous database, and we predict them as novel genes and need further detailed investigation. A wound-induced-like protein/gene (accession number GQ914056) was found upregulated in the subtracted library and was characterized. Expression studies revealed upregulation of this gene during low temperature stress and against salicylic acid. Wound-induced gene was confirmed as in single copy in *C. microphyllum*. Another gene was a metallothionin, which was found responsive to multiple abiotic stresses in our previous report (Singh et al. 2011). Our study demonstrated single-copy presence of this gene in *C. microphyllum* and increasingly expressed toward cold stress, abscisic acid treatment, polyethylene glycol-induced osmotic stress, and zinc sulfate foliar application.

Another very interesting transcriptome study was done between *Cicer microphyllum* and cultivated *desi* cv. ICC4958 recently (Rishi Srivastava et al. 2016). This study demonstrated identification of 12,772 transcripts which included 3242 root- and 1639 shoot-specific *C. microphyllum* genes that possibly participate in adaptation in comparison to cultivated chickpea. In silico characterization revealed polymorphic SSR- and SNP-based markers indicated their potential use in chickpea genome-assisted breeding methods for drought stress management due to the presence of six major QTLs related to drought stress coping mechanism. This recent study proved *C. microphyllum* a novel crop wild donor for drought-tolerant traits in cultivated chickpea to enhance its genetic potential. Specific differential gene identification (Singh et al. 2011, 2013) and recent comprehensive transcriptome, genetic marker, and major QTLs study (Rishi Srivastava et al. 2016) explained how *C. microphyllum* modulates its mechanism at molecular level to cope/adopt to the extreme adverse climatic conditions and suggested this wild species as novel genetic resource for upcoming predicted climatic conditions in order to improve the cultivated chickpea crops for various purposes.

12.5 Importance of CWRs

CWR carry inherent diversified genetic pool in natural habitats with extreme harsh climatic conditions as compared to domestic cultivars. Genetic diversity declines in cultivated crop species as per previous reports in soybean, maize, and rice (Hyten et al. 2006; Zhou et al. 2015; Wright et al. 2005; Castaneda-Alvarez et al. 2016; Larson et al. 2014; Maxted and Kell 2009) in comparison with wild ancestors. Crop wild relatives are distributed across Southern, Eastern, and Western Europe, East Asia, Southeast Asian parts, Mediterranean regions; and South America including many Vavilov centers of diversity and adjacent regions (Castaneda-Alvarez et al. 2016; Larson et al. 2014; Maxted and Kell 2009). Various environmental factors due to climate change in recent decades, increasing human population, and their invasion in natural habitats are major threats of extinction of these crop wild relatives and warrant immediate conservation actions (Ford-Lloyd et al. 2011; Maxted et al. 2010).

Recently China took initiative and revealed 871 endemic valuable plant species which can be good resource for cultivated crops toward socioeconomic interests. It includes major crops like wheat, rice, soybean, millet, potato, and sweet potato and needs immediate strategies to conserve for further exploration. Some endemic species are already at point of extinction including wild relatives of tea (*Camellia fangchengensis* S. and *C. grandibracteata*), apple (e.g., *Malus honanensis*, *M. ombrophila*, and *M. toringoides*), and pear (*Pyrus pseudopashia*) and require a foolproof conservation strategy for China (Shelagh Kell et al. 2015). Recent studies predicted shorter growing periods with lower yield in rice (Hijioka et al. 2014) in China and around the Asian region with most affected Eastern Chinese territory due to increased heat stress in near future (Masutomi et al. 2009) and also reduced total production in maize (Tao et al. 2009). China recognized the utilization of crop wild relatives to cope with extreme heat, drought, and salinity stress as major consequences of climate change in the near future for crop improvement programs by using CWRs as genetic resources toward leading agriculture and horticultural economy under the framework of UK-China Sustainable Agriculture Innovation Network (SAIN) (www.sainonline.org/english.html).

The European region has approximately 3500 endemic plant species and approximately 12,500 native species too; however it is degrading at high rate since last decades (Davis et al. 1994). The conservation strategy in Europe has not been enough to protect various ecosystems; even their contribution is large and fundamental (The Dobris Assessment-http://www.dmu.dk/LibraryNetwork/). At least lately, the European Union agreed to the importance of conservation of crop wild relatives and approved a sixth action plan to estimate and study biodiversity loss by 2002–2012 (http://ec.europa.eu/environment/archives/action-programme/intro.htm) followed by implementation of recommendations (Nigel 2003). A total of 18% of EU's territory has been notified protected to tackle the conservation in sustainable manner under the seventh environment action program as key proposal which is until 2020 and expected to enlarge until 2050 for long-term and sustainable global society (EAP-http://ec.europa.eu/environment/action-programme). For example, *Vitis vinifera* L. ssp. *sylvestris* is a wild ancestor of cultivated *V. vinifera* and is near extinction due to human activities while it played major role in grape industry and economy and presently in very limited population (Arnold et al. 2005). Later *V. riparia* Michx., *V. labrusca* L., and *V. rupestris* Scheele were incorporated as rootstock from America to control phylloxera (Bodor et al. 2010) and became the biggest competitional threat to European wild grape and introduction of new diseases such as black rot (*Guignardia bidwellii*), powdery mildew (*Erysiphe necator*), and downy mildew (*Plasmopara viticola*) in European vineyards. A group of researchers investigated this case and detected that rootstocks from introduced American grape species in European viticulture were reason for the introgression of new diseases while some of the ssp. *sylvestris* accessions were resistant to these diseases and proved to be valuable genetic resources to control the diseases (Stephan Schroder et al. 2015). Remanent *V. vinifera* ssp. *sylvestris* is still present in upper Rhine valley, in Ketsch Peninsula, and in Reiss Peninsula in Germany but at the verge of diminishing (Arnold et al. 2005).

New diseases and pests are always a risk, and CWRs served as important trait donor for breeders to overcome (Hajjar and Hodgkin 2007; Van Treuren et al. 2013) and ensure crop productivity and food security; therefore, conservation of CWRs is highly important but unfortunately is under represented in ex situ condition (Castañeda-Álvarez et al. 2016). Natural habitats of CWRs are predicted to be diminished due to adverse climatic conditions (Jarvis et al. 2008a). Wild relatives of cowpea, potato, and peanuts are supposed to be affected by 16–20%, while some other plant species have a threat of extinction up to 50% (Jarvis et al. 2008b) of their population and warrants an urgent effective road plan to cope it (Maxted et al. 2015). Conservation of natural genetic resources is fundamental in order to ensure food security for future population (Phillips et al. 2017).

Convention for Conservation (Convention on Biological Diversity CBD 1992), Global Strategy for Plant Conservation (CBD 2010a), Strategic Plan for Biodiversity (CBD 2010b), International Seed Treaty (FAO 2001) and European Strategy for

Plant Conservation (Planta Europa 2008) are major initiatives to preserve natural genetic resources globally while tagging in various countries also aside, e.g., Benin (Idohou et al. 2012), United States (Khoury et al. 2013), Finland (Fitzgerald 2013), Spain (Rubio Teso et al. 2013), and Italy (Panella et al. 2014). Recently United Kingdom also agreed and showed interest to conserve the native plant species as priority in situ as well as ex situ across the United Kingdom and territory of Scotland (Fielder et al. 2015a, 2015b). Potatoes, strawberries, rapeseed, wheat, barley, and oats are major crops in Scotland where CWRs for potato and wheat are absent due to different origins (Vavilov 1926), which is another worry to find an alternative global partner for future perspective in climate changes. Scotland urgently needs an inclusive ex situ and in situ conservation approach to maintain its CWRs to meet United Kingdom's and EU's commitments (CBD 1992, 2010a, 2010b; FAO 2001; Planta Europa 2008) toward improved resilience of food security in the near future (Hannah et al. 2016).

Potato (*Solanum tuberosum* L.) is another major food crop worldwide (FAOSTAT 2014) and an excellent model to study the necessity of CWRs. Approximately 200 wild species has been listed for potato in North and South American countries (Hawkes 1990). Recently Carlos F. Marfil et al. (2015) developed an effective method to prove CWRs incorporation at national level in Argentina is also useful for other countries that have land area for potato cultivation but lacking the CWRs, for example, Scotland. Vavilov documented first time the importance of donors in natural environment (Vavilov 1949), but wild crop relatives and natural ancestors were scantily represented; main focus was on cultivated domesticated crop species. The current situation is alarming due to drastic changes in natural environment and expected large growth of human population in coming future; plant scientists are committed to take CWRs as critical factor of conservation of plant genetic resources for food and agriculture (PGRFA) as a first priority, not as an alternative.

12.6 Conclusion

Drought is a major abiotic stress and most critical threat to future food demand for rapidly increasing population and may lead toward great famines in the future. Plants are sessile organisms and bound to adopt and survive under drought stress by induced morphological, biochemical, and physiological responses. Osmoprotectants, osmotic adjustment, antioxidants, and scavenging defense system play a major role to combat drought stress. Our findings in this study have shown that *Cicer microphyllum* has a strong inherent combination of defense responses against drought. Traditionally drought stress effects can be managed by agronomic practices up to a limited extent. But most important strategy is selection of a potential germplasm material using molecular and biotechnological methods. Further the germplasm can be used to incorporate value-added traits by conventional breeding programs or can be exploited at molecular and genetic level to identify the corresponding genes and to transform the candidate genes into drought-susceptible plant varieties irrespective to genetic barriers. The progressive cloning of many stress responsive transcription factors and genes has been proved to represent the molecular basis of stress tolerance (Cattivelli et al. 2002, 2008). In conclusion molecular approaches should be incorporated to elucidate the exact mechanism of plant drought tolerance and improved water-use efficiency. Crop wild varieties exhibiting inherent capability of antioxidant defense system could be highly useful for engineering plants which can tolerate drought stress leading to minimize the economic losses due to drought condition. In this paper, we report in vitro conservation of high altitude cold desert plant *Cicer microphyllum* which ensure availability of germplasm in laboratory conditions, and physiological responses demonstrated in the study also support this plant to exhibit strong antioxidant defense system and may be highly useful for future engineering of plants against drought stress.

References

Ahmad M, Shaista S, Lachlan L (2010) Radiant frost tolerance in pulse crops—a review. Euphytica 172:1–12. https://doi.org/10.1007/s10681-009-0031-4

Ahmed IM, Nadira UA, Bibi N et al (2015) Secondary metabolism and antioxidants are involved in the tolerance to drought and salinity, separately and combined, in Tibetan wild barley. Environ Exp Bot 111:1–12. https://doi.org/10.1016/j.envexpbot.2014.10.003

Alexieva V, Sergiev I, Mapelli S, Karanov E (2001) The effect of drought and ultraviolet radiation on growth and stress markers in pea and wheat. Plant Cell Environ 24:1337–1344. https://doi.org/10.1046/j.1365-3040.2001.00778.x

Ali A, Johnson DL, Stushnoff C (1999) Screening lentil (Lens culinaris) for cold hardiness under controlled conditions. J Agric Sci (Camb) 133:313–319

Apel K, Hirt H (2004) Reactive oxygen species: metabolism, oxidative stress, and signal transduction. Annu Rev Plant Biol 55:373–399. https://doi.org/10.1146/annurev.arplant.55.031903.141701

Arnold C, Schnitzler A, Douard A et al (2005) Is there a future for wild grapevine (Vitisvinifera subsp. silvestris) in the Rhine Valley? Biodivers Conserv 14:1507–1523. https://doi.org/10.1007/s10531-004-9789-9

Ballabh B, Chaurasia OP, Pande PC, Ahmed Z (2007) Raw edible plants of cold desert Ladakh. Indian J Tradit Knowl 6(1):182–184

Barazani O, Mayzlish-Gati E, Lifshitz D et al (2017) Strategies and priorities in field collections for ex situ conservation: the case of the Israel Plant Gene Bank. Genet Resour Crop Evol 64:1–5. https://doi.org/10.1007/s10722-016-0468-y

Bates LS, Waldren RP, Teare ID (1973) Rapid determination of free proline for water-stress studies. Plant Soil 39:205–207. https://doi.org/10.1007/BF00018060

Bernal M, Llorens L, Julkunen-Tiitto R et al (2013) Altitudinal and seasonal changes of phenolic compounds in Buxussempervirens leaves and cuticles. Plant Physiol Biochem 70:471–482. https://doi.org/10.1016/j.plaphy.2013.06.012

Biodiversity International (2006) Crop wild relatives. Bioversity International, Rome

Bodor P, Höhn M, Pedryc A et al (2010) Conservation value of the native Hungarian wild grape (VitissylvestrisGmel.) evaluated by microsatellite markers. Vitis 49:23–27. https://ojs.openagrar.de/index.php/VITIS/article/view/4100

Carlos FM, Verónica H, Ricardo WM (2015) In situ conservation of wild potato germplasm in Argentina: example and possibilities. Global Ecol Conserv 3:461–476. https://doi.org/10.1016/j.gecco.2015.01.009

Castaneda-Alvarez NP, Khoury CK, Achicanoy HA et al (2016) Global conservation priorities for crop wild relatives. Nat Plants 2:16022. https://doi.org/10.1038/nplants.2016.22

Cattivelli L, Baldi P, Crosetti C (2002) Chromosome regions ans stress-related sequences involved in resistance to abiotic stress in triticeae. Plant Mol Biol 48:649–665. https://doi.org/10.1023/A:1014824404623

Cattivelli L, Rizza F, Badeck FW (2008) Drought tolerance improvement in crop plants: an integrative view from breeding to genomics. Field Crop Res 105:1–14. https://doi.org/10.1016/j.fcr.2007.07.004

CBD (2010a) Global strategy for plant conservation. In: Secretariat of the convention onbiological diversity. CBD, Montreal. www.cbd.int/gspc/strategy.shtml

CBD (2010b) Strategic plan for biodiversity 2011–2020. In: Secretariat of the conventionon biological diversity, Montreal, CBD. www.cbd.int/decision/cop/?id=12268

CBD (1992) Convention on biological diversity. Secretariat of the convention onbiological diversity. CBD, Montreal

Chinnusamy V, Zhu J, Zhu JK (2007) Cold stress regulation of gene expression in plants. Trends Plant Sci 12:444–451. https://doi.org/10.1016/j.tplants.2007.07.002

Clarke H (2001) Improving tolerance to low temperature in chickpea. In: AEP (ed) Proceedings of 4th European conference on grain legumes. Cracow, pp 34–35

Clarke HJ, Khan TN, Siddique KHM (2004) Pollen selection for chilling tolerance at hybridisation leads to improved chickpea cultivars. Euphytica 139:65–74. https://doi.org/10.1007/s10681-004-2466-y

Colin K, Nora P, Harold A (2015) Crop wild relatives of pigeon pea [*Cajanus cajan* (L.) Millsp.]: distributions, ex situ conservation status, and potential genetic resources for abiotic stress tolerance. Biol Conserv 184:259–270. https://doi.org/10.1016/j.biocon.2015.01.032

Croser S, Clarke HJ, Siddique KHM, Khan TN (2003) Low temperature stress: implications for chickpea (*Cicer arietenum* L) improvement. Crit Rev Plant Sci 22:185–219. https://doi.org/10.1080/713610855

Davis S, Heywood VH, Hamilton AC (1994–1997) Centres of plant diversity: a guide and strategy for their conservation, vol 3. WWF and IUCN, Gland. https://portals.iucn.org/library/node/8268

Fielder H, Brotherton B, Hosking J (2015a) Enhancing the conservation of crop wild relatives in England. PLoS One 10(6):e0130804. https://doi.org/10.1371/journal.pone.0130804

Beyer WF, Fridovich I (1987) Assaying for superoxidedismutase activity: some large consequences of minorchanges in conditions. Anal Biochem 161:559–566. https://doi.org/10.1016/0003-2697(87)90489-1

Aebi H (1984) Catalase in vitro. Methods Enzymol 105:121–126. https://doi.org/10.1016/S0076-6879(84)05016-3

Foyer CH, Halliwell B (1976) The presence of glutathione and glutathione reductase in chloroplasts: a proposed role in ascorbic acid metabolism. Planta 133:21–25. https://doi.org/10.1007/BF00386001

FAO (1998) The state of the world's plant genetic resources for food and agriculture. FAO, Rome. http://www.fao.org/3/a-w7324e.pdf

FAO (2008) Establishment of a global network for the in situ conservation of crop wild relatives: status and needs. FAO, Rome. http://www.fao.org/docrep/013/i1500e/i1500e18a.pdf

FAO (2001) International treaty on plant genetic resources for food and agriculture. Food and Agriculture Organization of the United Nations. http://www.planttreaty.org/

FAOSTAT (2014) Food and Agriculture Organization of the United Nations. http://faostat3.fao.org/home/E

Farooq M, Aziz T, Basra SMA (2008) Chilling tolerance in hybrid maize induced by seed priming with salicylic acid. J Agron Crop Sci 194:161–168. https://doi.org/10.1111/j.1439-037X.2008.00300.x

Fielder H, Burrows C, Woodman J (2015b) Enhancing the conservation of crop wild relatives in Wales. New J Bot 5:178. https://doi.org/10.1080/20423489.2015.1123965

Fitzgerald H (2013) The national crop wild relative strategy report for Finland. MTT, Report 121, p 95. http://jukuri.mtt.fi/bitstream/handle/10024/481549/mttraportti121.pdf

Foley J, Ramankutty N, Brauman A et al (2011) Solutions for a cultivated planet. Nature 478(7369):337–342. https://doi.org/10.1038/nature10452

Folkert AH, Elena AG, Buitink J (2001) Mechanisms of plant desiccation tolerance. Trends Plant Sci 6:431–438. https://doi.org/10.1016/S1360-1385(01)02052-0

Ford-Lloyd BV, Schmidt M, Armstrong SJ (2011) Crop wild relatives-undervalued, underutilized and under threat? Bioscience 61:559–565. https://doi.org/10.1525/bio.2011.61.7.10

Foyer CH, Noctor G (2005) Redox homeostasis and antioxidant signaling: a metabolic interface between stress perception and physiological responses. Plant Cell 17:1866–1875. https://doi.org/10.1105/tpc.105.033589

Frei ES, Scheepens J, Armbruster GF, Stöcklin J (2012) Phenotypic differentiation in a common garden reflects the phylogeography of a widespread Alpine plant. J Ecol 100:297–308. https://doi.org/10.1111/j.1365-2745.2011.01909.x

Gong H, Zhu X, Chen K et al (2005) Silicon alleviates oxidative damage of wheat plants in pots under drought. Plant Sci 169:313–321. https://doi.org/10.1016/j.plantsci.2005.02.023

Hajjar R, Hodgkin T (2007) The use of wild relatives in crop improvement: a survey of developments over the last 20 years. Euphytica 156:1–13. https://doi.org/10.1007/s10681-007-9363-0

Hannah F, Cécile S, Brian FL, Nigel M (2016) Enhancing the conservation of crop wild relatives in Scotland. J Nat Conserv 29:51–61. https://doi.org/10.1016/j.jnc.2015.11.002

Hasegawa PM, Bressan RA, Zhu JK, Bohnert HJ (2000) Plant cellular and molecular responses to high salinity. Annu Rev Plant Physiol 51:463–499. https://doi.org/10.1146/annurev.arplant.51.1.463

Hawkes JG (1990) The potato: evolution, biodiversity and genetic resource, vol 67. Smithsonian Institution Press, Washington, DC, p 733. https://doi.org/10.1007/BF03044023

Hijioka Y, Lin E, Pereira JJ et al (2014) Impacts, adaptation, and vulnerability. IPCC working group II contribution to AR5. Regional Aspects, vol. II. http://ipccwg2gov/AR5/images/uploads/WGIIAR5-Chap24_FGDall.pdf

Huda S, Islam R, Bari MA, Asaduzzaman M (2003) Shoot differentiation from Cotyledon derived callus of chickpea (*Cicer arietinum* L.). Plant Tissue Cult 13(1):53–59

Hyten DL, Song QJ, Zhu YL et al (2006) Impacts of genetic bottlenecks on soybean genome diversity. Proc Natl Acad Sci U S A 103:16666–16671. https://doi.org/10.1073/pnas.0604379103

Idohou R, EphremAssogbadjo A, Fandohan B et al (2012) National inventory and prioritizationof crop wild relatives: case study for Benin. Genet Resour Crop Evol 60:1337–1352. https://doi.org/10.1007/s10722-012-9923-6

Islam R, Riazuddin S (1993) Shoot organogenesis in chickpea (Cicer arietinum L.) from callus culture of hypocotyl explants. J Biol Sci 1:1–5

Jarvis A, Lane A, Hijmans RJ (2008a) The effect of climate change on crop wild relatives. Agric Ecosyst Environ 126:13–23. https://doi.org/10.1016/j.agee.2008.01.013

Jarvis A, Upadhyaya H, Gowda C et al (2008b) Climate change and its effect on conservation and use of plant genetic resources for food and agriculture and associated biodiversity for food security. http://www.fao.org/docrep/ 013/i1500e/i1500e16.pdf

Khoury CK, Greene S, Wiersema J et al (2013) An inventory of crop wild relatives of the United States. Crop Sci 53:1–13. https://doi.org/10.2135/cropsci2012.10.0585

Korner C (2003) Alpine plant life: functional plant ecology of high mountain ecosystems. Mol Biol 67:589–602. https://doi.org/10.1007/978-3-642-98018-3

Larson G, Piperno DR, Allaby RG et al (2014) Current perspectives and the future of domestication studies. Proc Natl Acad Sci U S A 111:6139–6146. https://doi.org/10.1073/pnas.1323964111

Lima ALS, DaMatta M, Pinheiro HA et al (2002) Photochemical responses and oxidative stress in two clones of *Coffea canephora* under water deficit conditions. Environ Exp Bot 47:239–247. https://doi.org/10.1016/S0098-8472(01)00130-7

Masutomi Y, Takahashi K, Harasawa H, Matsuoka Y (2009) Impact assessment of climate change on rice production in Asia in comprehensive consideration of process/parameter uncertainty in general circulation models. Agric Ecosyst Environ 131(3–4):281–291. https://doi.org/10.1016/j.agee.2009.02.004

Maxted N, Ford-Lloyd BV, Kell SP (2008) Crop wild relatives: establishing the context. In: Crop wild relative conservation and use. CABI Publishing, Wallingford, pp 3–30

Maxted N, Kell SP (2009) Establishment of a global network for the in situ conservation of crop wild relatives: status and needs. Food and Agriculture Organization of the United Nations Commissionon Genetic Resources for Food and Agriculture, Rome. http://www.fao.org/docrep/013/i1500e/i1500e18a.pdf

Maxted N, Kell S, Toledo Á et al (2010) A global approach to crop wild relative conservation: securing the gene pool for food and agriculture. Kew Bull 65:561–576. https://doi.org/10.1007/s12225-011-9253-4

Maxted N, Avagyan A, Frese L et al (2015) ECPGR concept for in situ conservation of crop wildrelatives in Europe. Wild Species Conservation in Genetic Reserves Working Group, European Cooperative Programme for Plant Genetic Resources, Rome

Nigel M (2003) Conserving the genetic resources of crop wild relatives in European protected areas. Biol Conserv 113:411–417

Nitzsche W (1983) Germplasm preservation, handbook of plant cell culture, vol 1. Collier Macmillan Publisher, London, pp 782–805

Orvar BL, Ellis BE (1997) Transgenic tobacco plants expressing antisense RNA for cytosolic ascorbate peroxidase show increased susceptibility to ozone injury. Plant J 11:1297–1305. https://doi.org/10.1046/j.1365-313X.1997.11061297.x

Pan Y, Wu LJ, Yu ZL (2006) Effect of salt and drought stress on antioxidant enzymes activities and SOD isoenzymes of liquorice (*Glycyrrhizauralensis*Fisch). Plant Growth Regul 49:157–165. https://doi.org/10.1007/s10725-006-9101-y

Panella L, Landucci S, Torricelli R et al (2014) The national crop wild relative strategy for Italy: first steps to be taken. PGR Secure. https://www.pgrsecure.bham.ac.uk/sites/default/files/documents/public/National_CWR_Conservation_Strategy_Italy.pdf

Pastori G, Foyer CH, Mullineaux P (2000) Low temperature-induced changes in the distribution of H2O2 and antioxidants between the bundle sheath and mesophyll cells of maize leaves. J Exp Bot 51:107–113. https://doi.org/10.1093/jexbot/51.342.107

Phillips J, Brehm JM, van Oort B et al (2017) Climate change and national crop wild relative conservation planning. Ambio 46:630. https://doi.org/10.1007/s13280-017-0905-y

Planta Europa (2008) A sustainable future for Europe; the European strategy for plant conservation 2008–2014. Plantlife International\Council of Europe, Salisbury\Strasbourg

Prochazkova D, Sairam RK, Srivastava GC, Singh DV (2001) Oxidative stress and antioxidant activity as the basis of senescence in maize leaves. Plant Sci 161:765–771. https://doi.org/10.1016/S0168-9452(01)00462-9

Rick CM (1984) Plant germplasm resources, handbook of plant cell culture, vol 2. Collier Macmillan Publisher, London, pp 9–37

Rishi S, Deepak B, Ayushi M et al (2016) Transcriptome landscape of perennial wild *Cicer microphyllum* uncovers functionally relevant molecular tags regulating agronomic traits in chickpea. Sci Rep **6**:33616. https://doi.org/10.1038/srep33616

Rubio T, Iriondo ML., Parra JM, Torres E (2013) National strategy for the conservation of crop wild relatives of Spain. PGR Secure. http://www.pgrsecure.bham.ac.uk/sites/default/files/documents/public/NationalCWRConservation StrategySpain.pdf

Settu A, Ranjithakumari BD (1999) *In vitro* plant regeneration from leaf explants of soybean *Glycine max* (L.). Plant Tissue Cult 9(1):11–16

Shelagh K, Haining Q, Bin C et al (2015) China's crop wild relatives: diversity for agriculture and food security. Agric Ecosyst Environ 209:138–154. https://doi.org/10.1016/j.agee.2015.02.012

Singh KB, Ocampo B (1993) Interspecific hybridization in annual *Cicer* species. J Genet Breed 47:199–147

Singh KB (1991) Wild species as a source of resistance in cool season food legumes. In: Genetic research and education: current trends in the next fifty years, vol II. Golden Jubilee Symposium, New Delhi, p 297

Singh RK, Singh S, Anandhan S et al (2011) Metallothionin like gene from *Cicer microphyllum* is regulated by multiple abiotic stress. Protoplasma 248(4):839–847. https://doi.org/10.1007/s00709-010-0249-y
Singh RK, Singh S, Pandey P et al (2013) Construction of cold induced subtracted cDNA library from *Cicer microphyllum* and transcript characterization of identified novel wound induced gene. Protoplasma 250-2:459–469. https://doi.org/10.1007/s00709-012-0429-z
Singh RK, Singh S, Anandhan S et al (2017) First insights into the biochemical and molecular response to cold stress in *Cicer microphyllum*, a crop wild relative of chickpea (*Cicer arietinum*). Russian Journal of Plant Physiology 64(5):758–765. https://doi.org/10.1134/S1021443717050120
Singh S, Anandhan S, Arun MB et al (2007) In vitroestablishment of *Cicer microphyllum*, a wild relative of cultivated chickpea from cold desert of Ladakh. J Arid Legumes 4(1):66–68
Singh RK, Singh S, Anandhan S, Quiroz-Figueroa FR, Ruiz-May E (2019) An efficient protocol for in vitro propagation of the wild legume *Cicer microphyllum*, a crop wild relative of chickpea (*Cicer arietinum*). In Vitro Cell Dev Biol Plant 55:9. https://doi.org/10.1007/s11627-018-09958-y
Serraj R, Sinclair TR (2002) Osmolyte accumulation: can it really help increase crop yield under drought conditions? Plant Cell Environ 25:333–341. https://doi.org/10.1046/j.1365-3040.2002.00754.x
Somerville C, Briscoe J (2001) Genetic engineering and water. Science 292:2217. https://doi.org/10.1126/science.292.5525.2217
Srinivasan A, Johansen C, Saxena NP (1998) Cold tolerance during early reproductive growth of chickpea (Cicer arietinum L.): characterisation of stress and genetic variation in pod set. Field Crop Res 57:181–193. https://doi.org/10.1016/S0378-4290(97)00118-4
Stanilova MI, Ilcheva VP, Zagorska NA (1994) Morphogenetic potential and *in vitro*micropropagation of endangered plant species *Leucojumaestivium* L. and *Liliumrhodopaeum*Delip. Plant Cell Rep 13:451–453. https://doi.org/10.1007/BF00231965
Stanley M (2003) Frost damage in crops and pastures. Primary industries and resources SA fact sheet. http://www.pir.sa.gov.au/factsheets
Stephan S, Andreas K, Ernst H et al (2015) Crop wild relatives as genetic resources the case of the European wild grape. Can J Plant Sci 95:905–912. https://doi.org/10.4141/cjps-2015-033
Stoddard FL, Balko C, Erskine W et al (2006) Screening techniques and sources of resistance to abiotic stresses in cool-season food legumes. Euphytica 147:167–186. https://doi.org/10.1007/s10681-006-4723-8
Tao F, Zhang Z, Liu J, Yokozawa M (2009) Modelling the impacts of weather and climate variability on crop productivity over a large area: a new super-ensemblebased probabilistic projection. Agric For Meteorol 149(8):1266–1278. https://doi.org/10.1016/j.agrformet.2009.02.015
Thomashow MF (2010) Molecular basis of plant cold acclimation: insights gained from studying the CBF cold response pathway. Plant Physiol 154(2):571–577. https://doi.org/10.1104/pp.110.161794
Van Treuren R, Van der Arend AJM, Schut JW (2013) Distribution of downy mildew (*Bremia lactucae* Regel) resistances in a genebank collection of lettuce and its wild relatives. Plant Genet Res 11:15–25. https://doi.org/10.1017/S1479262111000761
Vavilov NI (1926) The centres of origin of cultivated plants. Works Appl Bot Plant Breed 1:17–75
Vavilov NI (1949) The origin, variation, immunity and breeding of cultivated plants. Chron Bot 13:1–54
Wahid A, Close TJ (2007) Expression of dehydrins under heat stress and their relationship with water relations of sugarcane leaves. Biol Plant 51:104–109. https://doi.org/10.1007/s10535-007-0021-0
Warburton ML, Rauf S, Marek L et al (2017) The use of crop wild relatives in maize and sunflower breeding. Crop Sci 57:1227–1240. https://doi.org/10.2135/cropsci2016.10.0855
Wery J, Silim SN, Knights EJ et al (1994) Screening techniques and sources and tolerance to extremes of moisture and air temperature in cool season food legumes. Euphytica 73:73. https://doi.org/10.1007/BF00027184

Wright SI, Bi IV, Schroeder SG et al (2005) The effects of artificial selection of the maize genome. Science 308:1310–1314. https://doi.org/10.1126/science.1107891

Yamada M, Morishita H, Urano K et al (2005) Effects of free proline accumulation in petunias under drought stress. J Exp Bot 56:1975–1981. https://doi.org/10.1093/jxb/eri195

Zhou Y, Lam HM, Zhang J (2007) Inhibition of photosynthesis and energy dissipation induced by water and high light stresses in rice. J Exp Bot 58:1207–1217. https://doi.org/10.1093/jxb/erl291

Zhou ZK, Jiang Y, Wang Z et al (2015) Resequencing 302 wild and cultivated accessions identifies genes related to domestication and improvement in soybean. Nat Biotechnol 33:408–414. https://doi.org/10.1038/nbt.3096

Zhu JK (2002) Salt and drought stress signal transduction in plants. Annu Rev Plant Biol 53:247–273. https://doi.org/10.1146/annurev.arplant.53.091401.143329

Genome Modification Approaches to Improve Performance, Quality, and Stress Tolerance of Important Mediterranean Fruit Species (*Olea europaea* L., *Vitis vinifera* L., and *Quercus suber* L.)

13

Hélia Cardoso, Andreia Figueiredo, Susana Serrazina, Rita Pires, and Augusto Peixe

Abstract

In the last decades, the interest on traditional Mediterranean fruits highly increased, not only due to the constant demand of consumers for new crop alternatives but also due to the identification in such species of molecules with important properties for human health (e.g., resveratrol from grapes and oleuropein from olives). Efforts to improve the production capacity and fruit quality, as well as the resistance to biotic and abiotic stress, in such fruit species, were achieved by plant breeders using mainly classical breeding approaches (e.g., selection, hybridization, and mutagenesis); nevertheless, breeding support by plant tissue culture techniques, by marker-assisted selection, as well as by genome modification

All the authors are contributed equally to this chapter.

H. Cardoso (✉)
ICAAM – Instituto de Ciências Agrárias e Ambientais Mediterrânicas, Instituto de Investigação e Formação Avançada, Universidade de Évora – Pólo da Mitra, Évora, Portugal
e-mail: hcardoso@uevora.pt

A. Figueiredo · S. Serrazina
Biosystems & Integrative Sciences Institute (BioISI), Faculdade de Ciências, Universidade de Lisboa, Lisboa, Portugal

R. Pires
Departamento de Fitotecnia, Escola de Ciência e Tecnologia, Universidade de Évora – Pólo da Mitra, Évora, Portugal

A. Peixe
Departamento de Fitotecnia, ICAAM, Escola de Ciência e Tecnologia, Universidade de Évora – Pólo da Mitra, Évora, Portugal

R. Sathishkumar et al. (eds.), *Advances in Plant Transgenics: Methods and Applications*, https://doi.org/10.1007/978-981-13-9624-3_13

was also used. Here we will present the state of the art related with the production of transgenic plants in three Mediterranean fruit species with important impact on the economy, olive, grapevine, and cork oak. The achievements, problems, and future perspectives will be discussed.

Keywords
Mediterranean fruit species; Genome modification; Improved nutrition; Stress tolerance; Transgenic plants

13.1 *Olea europaea* L.

13.1.1 Introduction

Olive (*Olea europaea* L.) belongs to the genus *Olea* that comprises more than 30 species, mainly distributed in 3 different areas: (i) Afro-Mediterranean area (Western Africa, Tanganyika, Canary, and Azores Islands), (ii) Indo-China Malaysia area, and (iii) Natalense-Malgica area (South Africa, Mauritius, and Mascarene Comore) (Ciferri and Breviglieri 1942). Within the species *Olea europaea*, there are two well-known subspecies, the wild *oleaster*, also known as *sylvestris*, and the cultivated *sativa. Olea europaea* subp. *sativa* L. comprises many cultivars. In Italy about 500–600 varieties are known, although many of them being probably only ecotypes (Rugini and Fedeli 1990); 262 cultivars were described in Spain (Barranco and Rallo 2000), and 44 cultivars were registered in the Portuguese Catalogue for National Varieties (DGAV 2016).

The recent interest in olive tree crop, mainly associated with the organoleptic characteristics of oils and the high content in substances useful for human health, led to its diffusion from the traditional area of cultivation, the Mediterranean basin, to countries with different climate conditions. Producers are facing several problems related to olive tree cultivation, like tolerance upon environmental stresses, which are in fact ancient problems that remain unsolved. Additionally, new question appears, mainly related with plant adaptability upon the environmental conditions of the new cultivation areas, the mechanization of harvest and pruning (varieties adapted to intensive and super-intensive orchards), and tolerance upon new diseases that come up (e.g., *Xylella fastidiosa*). Moreover, climate change has raised also several concerns focusing on extreme temperatures and drought tolerance.

Despite the ancient cultivation of olive trees, breeding and genetic improvement is quite recent and remains still far from being comparable to that of other temperate fruit species, like grapevine. Considerable efforts have been made toward the recovery of elite olive cultivars through conventional cross-breeding programs, mainly considering local cultivars as the genotypes well adapted to the environmental conditions, but only few examples are known of new cultivars developed by following these breeding approaches.

The cv. 'Chiquitita' and the cv. 'Oliana' are two of these cultivars. The cv. 'Chiquitita' is the result of a cross between cv. 'Picual' (female) and cv. 'Arbequina' (male) developed in Spain by the cooperative breeding program carried out by the University of Córdoba and the Institute of Agricultural and Fishery Research and Training (IFAPA) (Rallo et al. 2008). The cv. 'Oliana' is a new variety that results from a cross between cv. 'Arbequina' (female) and cv. 'Arbosana' (male), developed in frame of a breeding program carried out by Agromillora Research (Cunill and Duran 2014). In Italy it was recently submitted to registration a new variety named 'Lecciana,' which is the result of a cross between two ancient Italian cultivars, the cv. 'Arbosana' and the cv. 'Leccino,' under a breeding program established between Agromillora Research and the University of Bari (Camposeo 2017).

Notwithstanding the importance of cross-breeding its low efficiency seems to indicate genetic transformation as a promising technique to speed up the development of superior cultivars. The advances on functional genomics studies focused on genotype to phenotype (trait) correlation, candidate genes associated with specific traits identification, and the development of molecular markers open a new window into both olive breeding programs and genetic engineering. Nevertheless, for olive, genetic engineering still faces many challenges, specific knowledge on the genetic control, which means the knowledge of the gene(s) associated with a specific trait and how are they regulated in tolerant/susceptible cultivars, or even associated with other important agronomical traits, is still lacking. However, the data achieved by the whole genome sequencing project of the cv. 'Farga' (Cruz et al. 2016) and also of the wild relative *Olea europaea* subs. *sylvestris* L. (Unver et al. 2017), associated with transcriptomic data, that came up recently, opened new insights into studies that will have a positive impact on olive breeding programs, genetic engineering, and genome editing. Olive plant breeding, following a transgenic approach, requires, in addition to the knowledge of the target gene, availability of an efficient in vitro regeneration protocol, a suitable protocol to transfer the gene into plant genome and an efficient system for selection of the transgenic plant cells.

13.1.2 Somatic Embryogenesis in Olive

Attaining efficient plant regeneration systems from single cells is a crucial step to assist genetic transformation protocols of any plant species. Somatic embryogenesis is up to now the most promising plant regeneration system from single cell and was already achieved in olive, being reported for the first time by Rugini and Tarini (1986), which used as initial explants root segments taken from in vitro germinated seeds. Using the same kind of explants, succeeded protocols were posteriorly described by Rugini (1995), Rugini et al. (1995), and Shibli et al. (2001).

Radicles from mature embryos were also successfully induced to form somatic embryos in wild olive, by Orinos and Mitrakos (1991), as well as in cultivated olive, by Mitrakos et al. (1992), Cerezo et al. (2011), Jafarzadeh-Bajestani et al. (2011), and Pires et al. (2018), with the cvs. 'Koroneiki,' 'Picual,' 'Zard,' and 'Galega vulgar', respectively. Leva et al. (1995), with the cvs. 'Frangivento', 'Picholine' and

'Frantoio'; Brhadda et al. (2003), with the cv. 'Picholine Marocaine'; Trabelsi et al. (2003), with the cvs. 'Chetoui', 'Chemlali' and 'Arbequina'; and Pritsa and Voyiatzis (2004), with the cv. 'Chondrolia Chalkidikis', used cotyledons as initial explants and also succeeded on obtaining the development of somatic embryos.

As stated by Rallo et al. (2018), in all these situations, culture media with high auxin/cytokinin ratio were used for formation of embryogenic calli, followed by its transfer to a new culture media without growth regulators, or with a low concentration of auxins, to allow embryo differentiation and development. Regarding mineral composition of culture media, the OMc (OM basal medium, Rugini 1984), in which macroelements were replaced by those of Bourgin and Nitsch (1967), with addition of 1gl^{-1} casein hydrolysate and omitting glutamine, and the MS (Murashige and Skoog 1962) formulations, were generally used. Nevertheless, the SH basal media (Schenk and Hildebrandt 1972), as well as the ECO (1/4 OM macroelements, 1/4 MS microelements, 1/2 OM vitamins, and 550 mgl^{-1} of glutamine according to Pérez-Barranco et al. 2009), were also successfully tested, respectively by Leva et al. (1995) and Cerezo et al. (2011) (see detailed information on the culture media formulations in Table 13.1). According to Rallo et al. (2018), the mineral requirements for somatic embryogenesis in juvenile origin olive explants seem to depend on explant, more than on the genotype. For radicles, the OMc formulation provided the best results, while on roots taken from seedlings, the MS was preferred. For cotyledons, OMc, MS and 1/2 MS, have shown to be adequate.

Adapting the good results accomplished with the juvenile explants to adult ones, is by now the main bottleneck concerning somatic embryogenesis in olive. In fact, successful reports using adult explants (leaves and petioles) are available only for the cultivars cvs. 'Canino', 'Moraiolo' and 'Dhabia' (Rugini and Caricato 1995; Benelli et al. 2001; Rugini et al. 2005; Mazri et al. 2013). Capelo et al. (2010), working with wild olive, also reported a successful protocol using explants from adult origin (see details in Table 13.1). Rugini and Caricato (1995), were the first to succeed using a two-step protocol: firstly, adventitious shoots were induced over calli formed on leaf petioles from in vitro cultured plants and secondly, petioles extracted from these adventitious shoots were used to induce formation of embryogenic calli.

The petioles from adventitious origin were inoculated on OMc media supplemented with 0.5 μM 2iP (2-isopentenyl adenine), 0.44 μM BAP (6-benzylaminopurine) and 0.25 μM IBA (indole-3-butyric acid). Calli development and formation of embryogenic structures were achieved in the dark, both, in the same media, or in media devoid of growth regulators, supplemented or not with activated charcoal. According to Benelli et al. (2001), somatic embryos proliferation happened directly from epidermal or subepidermal cells, which is of high interest for genetic transformation systems based on particle bombardment. This protocol was later successfully applied by Benelli et al. (2001) and Rugini et al. (2005) to the same cultivars. The results achieved with this two-step procedure proposed by Rugini and Caricato (1995), seam to point for a need of adult explants rejuvenation, in order to be possible cell dedifferentiation and reprograming, leading to the formation of an embryogenic calli. Nevertheless, Capelo et al. (2010) in

Table 13.1 Summary of the work carried on olive somatic embryogenesis

Cultivar	Explant	Calli proliferation	Embryo development	Conversion	References
‘Frantoio’/‘Moraiolo’/‘Dolce Agogia’	Mature embryos	1/2 MS + (0.5-2.5 µM) BA ± 0.5 µM NAA	1/2 MS	OM + (0.5 - 2.5) µM zeatin	Rugini (1988, 1995) and Rugini et al. (2005)
Wild olive	Mature embryo radicles/cotyledons	OMc + 2.5 µM 2iP + 25 µM IBA	OMc + 0.5 µM IBA	1/2 MS	Orinos and Mitrakos (1991)
‘Koroneiki’	Mature embryo radicles	OMc + 2.5 µM 2iP + 25 µM IBA	OMc ± 2.5 µM IBA	–	Mitrakos et al. (1992)
‘San Agostino’	Roots of seedlings	MS + 25 µM NAA + 0.5 µM BA	1/2 MS + 0.5 µM NAA + 0.5 µM BA	1/4 MS + 2.5 µM zeatin	Rugini (1995) and Rugini et al. (1995)
‘Frangivento’/‘Picholine’/‘Frantoio’	Immature cotyledons	SH + 2.5 µM NAA ± 0.5 µM 2iP	SH	SH	Leva et al. (1995)
‘Canino’/‘Moraiolo’	Leaflets/petioles from adult shots of adventitious origin	OMc + 0.2 µM 2iP + 0.44 µM BA + 0.25 µM AIB	OMc liquid OMc solid + 0.1% AC	OMc + 1.3 µM zeatin	Rugini and Caricato (1995), Benelli et al. (2001), and Rugini et al. (2005)
‘Nabali’	Roots of seedlings	MS + 5.0 µm 2,4D+ 0.5 µM Kinetin + 5.0 µM NAA	MS + 10 µM 2iP	MS	Shibli et al. (2001)
‘Picholine Marocaine’	Cotyledons	MS + 2.3 µM zeatin + 10.7 µM NAA	MS + 2.3 µM zeatin	MS + 2.3 µM zeatin	Brhadda et al. (2003)
‘Chetoui’/‘Chemlali’/‘Arbequina’	Cotyledons	OMc + 25 µM IBA + 2.5 µM 2iP	OMc	OM + 4.6 µM zeatin	Trabelsi et al. (2003)
‘Chondrolia Chalkidikis’	Immature and mature cotyledons	1/2 MS + 4.44 µM BA + 2.26 µM 2,4D 1/2 MS + 4.44 µM BA	1/2 MS + 4.44 µM BA + 2.26 µM 2,4 D 1/2 MS + 4.44 µM	–	Pritsa and Voyiatzis (2004)

(continued)

Table 13.1 (continued)

Cultivar	Explant	Calli proliferation	Embryo development	Conversion	References
Wild olive	Adult leaves/ petioles	MS + 12.25 μM IBA + 4.56 μM zeatin	MS	–	Capelo et al. (2010)
'Picual'	Mature embryo radicles	OMc + 2.5 μM 2iP + 25 μM IBA OMc + 0.5 μM IBA ECO + 0.2 μM 2iP + 0.44 μM BA + 0.25 μM AIB	ECO + 0.1% AC	1/3 MS + 10 g/l sucrose	Cerezo et al. (2011)
'Zard'	Mature embryo radicles/cotyledons	OMc + 2.5 μM 2iP + 24 μM IBA OMc + 2.5 μM BA+ 24 μM NAA	OMc	–	Jafarzadeh-Bajestani et al. (2011)
'Dahbia'	Adult leaves	1/2 MS + 30 μM TDZ + 0.5 μM NAA ECO + 0.2 μM 2iP + 0.44 μM BA + 0.25 μM IBA	–	–	Mazri et al. (2013)
'Galega vulgar'	Mature embryo radicles/cotyledons	OMc + 2.5 μM 2iP + 25 μM IBA	OMc	–	Pires et al. (2018)

Adapted from Rallo et al. (2018)

AC (activated charcoal); BA (6-benzylaminopurine); 2,4-D (2-4dichlorophenoxyacetic acid); ECO (1/4 OM macroelements, 1/4 MS microelements, 1/2 OM vitamins, and 550mg/l glutamine); IBA (indole-3-butyric acid); 2iP (6-dimethylaminopurine); MS (Murashige and Skoog 1962); NAA (naphthaleneacetic acid); OM (Olive medium, Rugini 1984); OMc (OM in which macroelements were replaced by those of Bourgin and Nitsch (1967) with addition of 1g/l casein hydrolysate and omitting glutamine); SH (Schenk and Hildebrand 1972); TDZ (thidiazuron) wild olive and Mazri et al. (2013) in the cultivated olive cv. 'Dhabia', achieved the development of embryogenic calli directly over leaf discs and leaf petioles extracted from in vitro cultured adult plants. This probably means that unlike to what happens with juvenile origin material, in the adult ones, the response is highly dependent on the genotype.

Whenever the formation of somatic embryos was attained, significant developmental anomalies were observed, which highly difficulted its conversion into plants. The use of cefotaxime (0.42 mM) on culture media for embryo development, was proposed by Rugini and Caricato (1995), as a way to reduce abnormal embryo formation. Its efficiency is still to prove, but the authors associated the achieved results to a better cell differentiation, as a result of the lower cell multiplication rates due to the antibiotic.

Cerezo et al. (2011), by comparing results with the basal formulations of the OMc and ECO (olive cyclic embryogenesis) media, didn't observe significant differences on the induction and proliferation rates, but verified that the number of abnormal embryos was lower on the ECO medium and higher for the conversion rates of mature embryos into plants. Protocols for cyclic embryogenesis were successfully established for cvs. 'Canino' and 'Maioralo' (Rugini and Caricato 1995), cv. 'Dolce Agogia' (Mencuccini et al. 1997), and cv. 'Picual' (Cerezo et al. 2011), which turned easier to transform and regenerate transgenic plants using embryogenic masses or even somatic embryos at different developmental stages. A recently published protocol for genetic transformation of olive considers the use of callus and globular embryos of yellow-creamed color derived from radicle of mature seeds as the appropriate explant for olive studies involving genetic transformation (Palomo-Ríos et al. 2017).

The olive embryogenic masses proved to growth and provide repetitive embryogenesis when cultured in ECO medium, which is now widely used for long-term maintenance of embryogenic lines in this species. Efficiency of the maturation phase significantly decreased in aged lines, and a decline in embryos quality was observed (Bradaï et al. 2016). Both effects gave rise to a significant decrease in the regeneration potential as duration of the proliferation phase increased. As stated by Rugini and Silvestri (2016), although the application of somatic embryogenesis for large-scale propagation is hampered in olive by the difficulty to use adult explants and also by the low rate of embryo germination, it proves nonetheless to be very useful for genetic improvement tasks based on DNA manipulation techniques.

13.1.3 Olive Breeding by DNA Manipulation

13.1.3.1 Genetic Transformation Mediated by *Agrobacterium* sp.

Genetic transformation mediated by *Agrobacterium* involves transference and integration into the nuclear genome of plant cells of a DNA segment, known as T-DNA from the Ti- or Ri-plasmid, if, respectively, *A. tumefaciens* or *A. rhizogenes* are used. This method is recognized as the most efficient method used till now for studies involving olive genetic transformation.

The first attempts of olive genetic transformation mediated by *Agrobacterium* were reported in the 1980s and beginning of the 1990s by Rugini (1984, 1986), Strobel et al. (1988), and Rugini (1992). Transformation experiments mediated by *Agrobacterium* have been carried out to express *rolABC* genes from *A. rhizogenes* (Mencuccini et al. 1999a, b) and the reporter gene *uidA* from *Escherichia coli*, which encodes β-glucuronidase (GUS) (Torreblanca et al. 2009, 2010; Jafarzadeh-Bajestani et al. 2011). Additional reports are now available concerning the expression of heterologous genes (D'Angeli and Altamura 2007, Jafarzadeh-Bajestani et al. 2011, Behelgardy et al. 2012, Haberman et al. 2017, Narvaez et al. 2018a, b) and overexpression/silencing of endogenous genes in view of functional studies (Cerezo et al. 2015, Hernández et al. 2018).

From those first attempts till now, different procedures have been followed, but only few reports have been focused on the optimization of parameters that influence transformation efficiency, and only few genotypes have been considered on genetic transformation studies (cvs. 'Dolce Agogia,' 'Morailo,' 'Canino,' 'Picual,' and 'Zard'). Efficiency of transformation mediated by *A. tumefaciens* depends on several factors, namely, the explant type, bacterial strain, co-cultivation period, use of substances to promote infection, antibiotic to eliminate the bacteria, promoter driving the gene of interest, and selective pressure that allows the development of only cells integrating the selectable marker gene. The establishment of somatic embryogenesis protocols using different initial explants turned possible the use of those tissues on transformation events. Leaf discs and petioles were directly used in genetic transformation studies of cvs. 'Dolce Agogia' and 'Morailo' (Mencuccini 1994, 1999a, b), but results were not promising. The establishment of cyclic somatic embryogenesis protocols, already described in 13.1.2, allowed the use of embryogenic masses or secondary somatic embryos, at different developmental stages, for successful recovery of transgenic plants of cv. 'Canino' (Rugini et al. 1999), cv. 'Picual' (D'Angeli and Altamura 2007; Torreblanca et al. 2009, 2010; Cerezo et al. 2013; Titouh et al. 2014; Haberman et al. 2017; Narvaez et al. 2018a, b), and cv. 'Zard' (Jafarzadeh-Bajestani et al. 2011).

Concerning the bacterial strain, most of the research experiments based on olive transformation mediated by *A. tumefaciens* have considered the LBA4404 strain (Rugini and Fedeli 1990; Rugini 1995; Rugini et al. 1999; Mencuccini et al. 1999a, b; Rugini et al. 2000; Rugini and Gutiérrez-Pesce 2006) containing the plasmid pBIN19 (developed by Bevan 1984). However, a study conducted to evaluate the influence of the bacterial strain in transformation efficiency of olive, reported the AGL1 as the most efficient strain (Torreblanca et al. 2010). This explains the use of the AGL1 stain in most of the experiments further developed in olive involving genetic transformation (e.g., Haberman et al. 2017; Narvaez et al. 2018a, b). Only one additional strain, the pGV3101, also derived from C58 as the AGL1, has been reported in studies of olive genetic transformation (Jafarzadeh-Bajestani et al. 2011; Behelgardy et al. 2012). Moreover, explants co-cultivation with the *Agrobacterium* suspension culture, known as co-cultivation period, is also a key factor, and different protocols have been described, mainly reporting a short period of inoculation with the bacterial suspension culture, followed by the co-cultivation period of 2 days under dark conditions (Rugini et al. 1999; Mencuccini et al. 1999a, b; Torreblanca et al. 2009, 2010; Jafarzadeh-Bajestani et al. 2011; Cerezo et al. 2013; Titouh et al. 2014; Palomo-Ríos et al. 2017; Haberman et al. 2017).

Selection of the most appropriate antibiotic to control bacterial growth after plant cell transformation, represents an essential procedure that determines the efficiency of a transformation event. In olive, cefotaxime has been reported as the most appropriate antibiotic, since it improves the conversion of the putative transgenic somatic embryos and further plantlets development, without any negative effect during all the process (Rugini and Caricato 1995). Different concentrations have been used involving olive genetic transformation studies; most of them report the use of 250 mgl^{-1} (Rugini et al. 1999; D'Angeli and Altamura 2007; Torreblanca et al. 2009, 2010; Cerezo et al. 2013; Titouh et al. 2014; Haberman et al. 2017; Narvaez et al. 2018a, b). Nevertheless, there are some reports using higher concentrations; the two examples known are the application of 350 mgl^{-1} after co-cultivation of embryos of cv. 'Zard' (Jafarzadeh-Bajestani et al. 2011) and 500 mgl^{-1} after co-cultivation of embryos of cvs. 'Dolce Agogia' and 'Morailo' (Mencuccini et al. 1999a, b). Some studies report the combination of an additional antibiotic, the timentin, to improve effectiveness of bacterial control (Torreblanca et al. 2009, 2010; Haberman et al. 2017; Narvaez et al. 2018a, b). The simultaneous use of cefotaxime and timentin was considered in the protocol recently proposed by Palomo-Ríos et al. (2017), for olive genetic transformation studies.

Selection of the proper antibiotic to allow the regeneration of putative transgenic cells, and simultaneously to inhibit non-transformed cells to proliferate, is the main requirement to efficiently apply the positive selectable marker system. In olive genetic transformation studies, the *Neomycin phosphotransferase II* (*nptII*), which encodes a protein able to inactivate aminoglycoside antibiotics through phosphorylation, has been the unique selectable marker gene used. Among the aminoglycoside antibiotics able to be used as selectable agent associated with the expression of *nptII*, kanamycin appears as the most used. In olive, kanamycin is usually used at concentrations that range from 25 mg l^{-1} to 100 mg l^{-1}, depending on the genotype, the explant type and if applied in solid or liquid culture medium (Jafarzadeh-Bajestani et al. 2011). The 100 mgl^{-1} is the concentration mostly used in solid culture medium to establish selection pressure in callus and embryos (Mencuccini et al. 1999a, b; Rugini et al. 1999; D'Angeli and Altamura 2007) and the 50 mgl^{-1} are used for selection under liquid culture (Mencuccini et al. 1999a, b). Most of the authors using this antibiotic suggest its application 3–4 weeks after co-cultivation with *Agrobacterium* to allow the formation of transgenic cell colonies in large numbers, increasing the probability to differentiate transgenic shoots or embryos (Rugini et al. 2000). When mature tissues were used (leaf petioles e.g.), the selective pressure was applied on calli established on calli proliferation media (Mencuccini et al. 1999a, b).

Considering the variability on the effectiveness on kanamycin selection, Pérez-Barranco et al. (2009) compared the use of that antibiotic with two other aminoglycoside antibiotics, showing that in solid culture medium, paromomycin at 200 mg l^{-1} restricted callus growth earlier than kanamycin, and, after several weeks of culture, the negative effect of kanamycin on the viability of somatic embryos was stronger than that produced by paromomycin. With these results, paromomycin was shown to be a better agent for selection of transgenic olive cells, which explains its use in the reports further published. Torreblanca et al. (2010), proposed a protocol that allows to speed up the regeneration of olive transgenic plants based on the use of paromomycin. Briefly, this protocol consists in inoculation of embryogenic cells 2 days after co-culture, in a solid medium supplied with 200 mg l^{-1} paromomycin, and, 3 months after, the cultured embryogenic cells must be transferred for 3 weeks to liquid medium added with 50 mgl^{-1} of paromomycin. Further, transgenic calli should be filtered and somatic embryos suspensions plated in selection medium containing 200 mgl^{-1} of paromomycin. This procedure was followed by several researchers (Cerezo et al. 2013; Titouh et al. 2014; Haberman et al. 2017; Narvaez et al. 2018a, b), and also Palomo-Ríos and co-authors (2017), proposed this procedure for transformation experiments with olive genotypes. However, besides the suggestion of use paromomycin at 200 mgl^{-1}, those authors highlight that susceptibility/tolerance upon selective pressure is genotype-dependent, and, for that reason, every research involving a different olive cultivar will require a previous study in order to select the most efficient antibiotic concentration to be used in solid and well as in liquid medium (authors suggested 200 mg l^{-1} in solid medium and 25–50 mgl^{-1} in liquid medium).

About the promoter selection, Torreblanca et al. (2010) evaluated the effect of promoters regulating the expression of the reporter gene *uidA*, using two different plasmids: the pBINUbiGUSint (Humara et al. 1999), harboring the *uidA* under the control of the maize polyubiquitin (*UBI1*) promoter, and the pGUSINT (Vancanneyt et al. 1990), harboring the *uidA* under the control of the cauliflower mosaic virus (CaMV35S) promoter. In both cases, *nptII* was used as selectable marker gene, under the control of the nopaline synthase (*NOS*) promoter. Those authors reported that transformation frequencies were higher when using the AGL1 strain harboring the pBINUbiGUSint plasmid, which means that *UBI1* promoter was stronger than *CaMV35S* in olive embryogenic tissues, corroborating the results previously achieved by olive transformation via particle bombardment (Lambardi et al. 1999; Pérez-Barranco et al. 2009).

Although these results suggest the use of *UBI1* as most efficient in terms of *uidA* expression in embryogenic cultures, further data, achieved by same authors, concluded the contrary, and the use of *CaMV35S* promoter instead of *UBI1* was highlighted (Torreblanca et al. 2010). It is important to note that *UBI1* promoter was associated with silencing of *uidA* in acclimated plants, while *CaMV35S* on pGUSINT was expressed at high levels in somatic embryos and also in plants, which indicates that *UBI1* promoter, although useful during the regeneration phase, is not adequate for the expression of genes of interest in olive plants. This way of gene regulation, leading to silencing, could be due to the lack of some important factor that regulates the expression of this monocotyledonous promoter (Torreblanca et al. 2010). In maize, the *UBI1* promoter is expressed constitutively, nevertheless, it could be inducible to higher levels upon stress conditions (Christensen and Quail 1996), which could explain its efficiency under in vitro conditions, as suggested by

Torreblanca et al. (2010). Afterward researchers made use of *CaMV35S* promoter for studies involving expression of genes putatively linked with important agronomical traits in olive (D'Angeli and Altamura 2007, Jafarzadeh-Bajestani et al. 2011; Cerezo et al. 2013; Titouh et al. 2014; Haberman et al. 2017; Narvaez et al. 2018a, b). Palomo-Ríos et al. (2017) also suggested the use of the CaMV35S promoter in the protocol recently published for olive transformation studies.

13.1.3.2 Genetic Transformation by Microparticle Bombardment

Microparticle bombardment as artificial DNA-delivery system, also known as microprojectile bombardment, biolistic, or gene gun, is an efficient technique for transformation of species recalcitrant to *Agrobacterium* infection, allowing studies involving stable and transient gene expression. The development of an efficient transformation protocol by microparticle bombardment requires previous optimization of several parameters as well as physical and biological factors, that can affect the efficiency of transformation (e.g., explant, size, and nature of microparticles used to coat the plasmid DNA, position of the plates from the stopping net, pressure used for bombardments, number of bombardments per plate, DNA concentration, preculture prior to bombardment). The first reports describing genetic transformation by microparticle DNA-delivery in olive were presented by Lambardi and co-authors (1999). These authors used embryogenic cultures of cv. 'Canino' to optimize the conditions for efficient olive genetic transformation by this technique. The size and developmental stage of somatic embryos, the apparatus for DNA delivery, considering the particle inflow gun (PIG) (Vain et al. 1993) and the PDS-1000He (Bio-Rad), the nature of microparticles (tungsten or gold) and the pressure used for bombardments were the parameters that authors considered for optimization.

Results achieved showed that expression of the reporter gene used (*uidA*) was differentially influenced by tested parameters. Somatic embryos size and developmental stage, highly determined the efficiency of the bombardment. Higher efficiency, which means high *uidA* expression detected by glucuronidase histological assay, was associated with the use of large-size somatic embryos. Contrarily, poor efficiency was obtained when smaller embryos (e.g., torpedo) were bombarded. Both tested devices were shown to be suitable for olive somatic embryos transient gene transformation, especially when 1100 psi rupture discs were used. Concerning the bombardment devices, although some authors report an average number of expression units slightly higher using the PIG device, it is the PDS-1000He (Bio-Rad) equipment that is mostly used in olive transformation experiments (Pliego-Alfaro et al. 2005; Pérez-Barranco et al. 2007, 2009).

The effect of bombardment helium pressure was further tested by Pérez-Barranco et al. (2009), using pro-embryogenic masses of small size (1–3 mm diameter) of cv. 'Picual'. Authors reported that higher transformation rates were achieved using a lower pressure (900 psi) than the one previously reported by Lambardi et al. (1999). Better results were achieved in combination with a different target distance. Lambardi et al. (1999) combined high pressure (1100 psi) with larger target distance (12 cm), Pérez-Barranco et al. (2009), achieved the best results by combining a lower pressure (900 psi) with smaller target distance (6 cm). Same authors also reported a lower efficiency obtained at 1500 psi, independently of the target distance. Regarding the nature of the microparticles for plasmid DNA coating, no significant differences were reported (Lambardi et al. 1999; Pérez-Barranco et al. 2009), but a higher level of *uidA* transient expression was observed when large-size somatic embryos were transformed with tungsten microparticles (Lambardi et al. 1999). Tungsten can catalytically degrade DNA and may be toxic to the cells; however, the satisfactory results in transient *uidA* expression could support its selection for genetic transformation studies involving transient gene expression.

Regarding the effect of microparticles size, it was showed that the use of large microparticles (1.6 μm diameter) significantly reduced *uidA* expression in comparison with smaller gold microparticles (1 μm diameter), which could be explained by the smaller size of olive cells (Pérez-Barranco et al. 2009).

The effect of DNA concentration and of a preculture prior to bombardment, was only tested by Pérez-Barranco et al. (2009), on transformation studies carried out in cv. 'Picual.' Authors reported that although the inexistence of significant differences, the increase in DNA concentration to deliver 1.2 to 1.8 μg DNA per bombardment slightly increased the number of *uidA* expression. Concerning the preculture procedure, it was not associated with an increase of transient transformation. Additionally, the promoter associated with the gene of interest could highly influence transformation efficiency in the same way as described in *A. tumefaciens*-mediated system. In order to evaluate the efficiency of promoters, two plasmid vectors were utilized in transformation experiments using the cv. 'Canino' (Lambardi et al. 1999). Both plasmids harbored the *uidA* as reporter gene fused with two different promoters; the pZ085, containing the CaMV35S and the pCGUδ0, harboring the sunflower *UBI* promoter with ubiquitin intron (Binet et al. 1991). High efficiency was detected by glucuronidase histological assay when used the sunflower *UBI* promoter (plasmid pCGUδ0). Similar results were further reported in cv. 'Picual' by comparing the vector pGUSINT, with the CaMV35S promoter (Vancanneyt et al. 1990), the pJGUS5, with the CaMV35S promoter and enhancer (Day et al. 1991) and the pCGUΔ1 (Binet et al. 1991), with the sunflower *UBI* promoter (Pliego-Alfaro et al. 2005; Pérez-Barranco et al. 2007).

13.1.4 Transgenes Introduced in Olive Genome

Facing the new requirements of modern oliviculture, the development of new cultivars, including conventional breeding and genetic transformation strategies, is focused on solving the main agronomic and commercial problems. According to Rugini et al. (1999, 2000), the main desirable traits that should be considered to develop new cultivars are:

1. Alteration of vegetative habit by reducing the size of the plant and the apical dominance in order to increase orchard plant density, shorten the unproductive period and improve radiation of canopy.

2. Develop completely self-fertile plants in order to increase yield or totally self-sterile plants to be used in breeding programs.
3. Increase oil content and quality by producing fruits with balanced oil composition among fatty acids and with a correct modeling of fruit ripening in terms of intensity of the degradation of phenolic and pectic compounds.
4. Increase frost and salt tolerance.
5. Regulation of fruit ripening.
6. Increase disease and pest tolerance.
7. Produce plants with high ability to develop adventitious rooting reducing cost of vegetative propagation and increasing the diversity of cultivars available for orchards plantation.

To go ahead with the development of a superior cultivar by integration of one of those traits, it is required the knowledge of the gene(s) directly correlated with that trait, which consequently determine the phenotype of interest. Many genes have already been isolated and characterized in a diversity of species, which could be of interest to be introduced in olive, singly or associated. Up-to-now few attempts on olive genetic transformation have been reported, and within the available reports, mainly heterologous genes from different plant species have been used aiming to introduce some new characteristics. Below it will be given a detailed description considering the research available in frame of each interesting trait.

13.1.4.1 Changing Vegetative Growth Habitus and Improvement of Rooting Ability

The genes *rolA*, *rolB*, and *rolC*, usually known as *rolABC* genes, containing ORFs (open reading frames) 10, 11, and 12 (Cardarelli et al. 1987), have been described to induce morphological alteration (White et al. 1985); the *rolB*, in particular, was described as stimulating root morphogenesis in recalcitrant woody plants (Capone et al. 1991; Rugini et al. 1991). This gene stimulates the induction and development of roots due to its auxin-like activity, which have made it known as a root-inducing gene, and lead to the establishment of a strong link with the phenotype of "hairy roots" pathogenesis (Cardarelli et al. 1987). The first reports on the expression of *rol* genes, were based on the natural introduction of the T-DNA of *A. rhizogenes*. In vitro grown shoots with three to four nodes, were infected in the middle, or at the base of the shoot, by using a knife or a needle infected with the bacteria. Although the non-detection of the transgenes in roots emerging from inoculated shoots, an increased on rooting ability with more secondary roots was observed, which leaded to the hypothesis that partial integration of T-DNA presents an inductive role on the non-transformed neighbor cells (Strobel et al. 1988, see also review in Rugini and Gutiérrez-Pesce 2006). Afterward, olive genetic transformation studies have been carried out to transfer *rolABC* genes isolated from *A. rhizogenes* but using the system mediated by *A. tumefaciens*.

Mencuccini (1994), reported the use of leaf discs of cvs. 'Moraiolo' and 'Dolce Agogia' for genetic transformation mediated by *A. tumefaciens* strain LBA 4404, carrying the plasmid pBIN19 (Bevan 1984) containing *rolABC* genes isolated from

A. rhizogenes (containing plasmids p1855 and pBR322). However, the authors were unsuccessful to regenerate transgenic plants. Subsequently, Mencuccini et al. (1999a, b), used the same conditions (*Agrobacterium* strain and construction) to transform leaf petioles of cv. 'Dolce Agogia' excised from in vitro plantlets, but no transgenic plantlets were obtained. Further work, developed by Rugini et al. (1999) using embryogenic masses of cv. 'Canino,' and the same construction of Cardarelli et al. (1987), containing *rolABC* genes from *A. rhizogenes*, reported the recovery of putative transgenic plants carrying *rolC* gene. Plants showed the development of many shoots at the embryos epicotyls, which were used to produce cuttings in order to further evaluate their rooting ability. Rugini and Gutiérrez-Pesce (2006), reported as result of field trials with transgenic olive plants for *rolABC*, that plants showed short internodes and increased rooting ability in respect to the nontransgenic controls. Furthermore, they grew slowly and presented many lateral buds giving them a bushy aspect. Regarding rooting ability, an experiment developed with microcuttings from transgenic plants overexpressing the *rolABC* genes has shown that in absence of an auxin, 50% of the explants rooted in comparison with nontransgenic, which showed a maximum rooting rate of 15%. Nevertheless, when used auxin to induce rooting, the rooting efficiency was 100% in *rolABC* transgenic plants, while control clones only rooted 70–80%.

Recently, a research line focused on the expression of the *FTa1* (encoding a protein flowering locus T) gene from *Medicago truncatula* in cv. 'Picual', revealed an association between gene expression and the development of an irregular plant morphology, in addition to unproduction of viable pollen. Transgenic plants showed changes on vegetative growth, such as the loss of apical dominance and continuous growth of lateral meristems. The results achieved highlight the role of *FTa1* in regulating flowering and vegetative *habitus* of olive plants (Cerezo et al. 2013; Haberman et al. 2017).

13.1.4.2 Enhance Tolerance Upon Abiotic Stresses: Cold and Salinity

Olive is a plant species that doesn't require vernalization but still needs a period of low temperatures to induce flower bud differentiation (Hartmann 1953). However, as a frost-sensitive species, olive tree cannot tolerate temperatures below – 6 °C to −10 °C, which could cause irremediable damage to the plants (Fugini and Fedeli 1990). A strategy that may overcome this limitation, to further allow olive cultivation in colder regions and simultaneously support the production of better olive oils, since olive oils can't be provided from olive trees growing under colder climates due to a slowing down of the post-maturation process (Palliotti and Bongi 1996), is of high economic value. The heterologous *osmotin* gene, isolated from tobacco, was expressed in cv. 'Canino' under the control of CaMV35 promoter (Rugini et al. 1999; D'Angeli and Altamura 2007; Bashir et al. 2018), aiming to investigate the involvement of that gene in plant acclimation upon different environmental stress, inclusively cold stress. Osmotin is a protein that belongs to the pathogenesis-related (PR) type-5 protein family (Zhu et al. 1995), which accumulates in defense responses upon biotic and abiotic stresses, like desiccation, salinity, abscisic acid, ethylene, wounding, and low temperatures (Singh et al. 1987; Zhu et al. 1993, 1995). D'Angeli and Altamura (2007) showed that osmotin is positively involved in the acclimation-related programmed cell death, in blocking the cold-induced calcium signaling, and in affecting cytoskeleton in response to cold stimuli, which place the gene as an interesting candidate for further work, regarding the development of superior cultivars with increased tolerance upon low temperatures.

Proline is an osmolyte with osmoprotectant activity, increasing the ability of plant cells to retain water without disturbing normal cellular functions (Kishor et al. 2005). The development of a strategy to increase levels of proline appears as a promising approach to increase the ability of plants to tolerate osmotic stresses caused by drought, salinity, and even cold stresses (Zhu et al. 1998). *P5CS* gene encodes for Δ1-pyrroline-5-carboxylate synthetase, an enzyme that catalyzes two steps of proline biosynthesis in plants (Jafarzadeh-Bajestani et al. 2011 and references therein) and has been followed by different researchers to enhance plant resistance upon environmental stresses. The *P5CS* gene isolated from *Arabidopsis thaliana*, driven by the CaMV35S promoter, was used to transform the cv. 'Zard' aiming to investigate the applicability of that gene on further development of superior cultivars with improved tolerance upon osmotic stress conditions (Jafarzadeh-Bajestani et al. 2011). Results achieved on that research showed that the increase in *P5CS* gene expression was positively correlated with endogenous proline production and accumulation, which corroborates with the results previously achieved in other plant species. Further researcher aiming to investigate performance of transgenic plants, expressing the P5CS, upon osmotic stresses will be of high interest.

13.1.4.3 Pest and Disease Resistance

Osmotin was also expressed aiming to investigate its role on plant tolerance upon pathogen attack, in particular to some fungi (Rugini and Gutiérrez-Pesce 2006). Transgenic plants expressing the *osmotin* gene were obtained from cv. 'Canino' (Rugini et al. 1999). Some trial that run under field conditions showed that transgenic plants were more tolerance to *Spilocaea oleagina* (D'Angeli et al. 2001). Transformation experiments with multiple genes aiming to improve fungal tolerance were developed by Rugini and co-authors, but no further information is given about analysis of plant tolerance (see Rugini et al. 1999 and Rugini and Gutiérrez-Pesce 2006).

Different reports based on the expression of the antifungal protein (AFP) from *Aspergillus giganteus* were also available in cv. 'Picual,' aiming to investigate the applicability of this approach to enhance olive plant tolerance to the pathogens *Verticillium dahliae* and *Rosellinia necatrix* (Titouh e t al. 2014; Narvaez et al. 2018a, b). AFP is a defense-like protein that shows antifungal activity under in vitro conditions, inhibiting the growth of several pathogens, and previous expression in other plant species indicated its usefulness by enhancing protection upon fungal pathogens (Coca et al. 2004). In olive, the achieved results indicate that expression of *AFP* can be useful to enhance resistance against *R. necatrix* but not to *V. dahliae* (Narvaez et al. 2018a, b). A recent research line, also related with enhancing tolerance upon fungal diseases, has been presented by Narváez et al. (2018a, b). This research is based on the expression of the *NPR1* gene isolated from *A. thaliana*. The *NPR1* encodes a key component of SAR signaling mediated by salicylic acid. It is known that after pathogen infection, the accumulation of SA releases NPR1 monomers in the cytosol, that are further translocated to the nucleus to activate the expression of PR proteins. This strategy has conferred tolerance to different pathogens in several plant species. In olive, the involvement of *NPR1* expression in cv. 'Picual' regarding tolerance upon *Verticillium dahliae* and *Rosellinia necatrix* is under analysis.

13.1.4.4 Improvement of Oil Quality

Besides the high interest on this trait, only two reports are available concerning investigation on the genes putatively involved on its regulation (Cerezo et al. 2015, Hernández et al. 2018). Both reports are focused on a single gene, the *HPL* isolated from olive, putatively involved in the production of volatile compounds in fruits. This gene encodes the enzyme 13-liasa hydroperoxide, an enzyme implicated on the biosynthesis pathway of volatile compounds, with an important role on the composition and quality of olive oils. The investigation was based on gene overexpression and silencing and interesting results are expected.

13.2 *Vitis vinifera* L.

13.2.1 Introduction

Grapevine (*Vitis vinifera* L.) is one of the most important fruit crops worldwide, with 7.6 mha of total world area under vines, being Spain, Italy, and France the European countries with the highest vineyard plantation area (OIV report 2018). Outside Europe, China, Turkey and USA lead the vine plantation. Its fruit is mostly processed into wine, but is also destined for fresh consumption as table grapes, dried into raisins, processed into nonalcoholic juice, and distilled into spirits. A global wine production of 250 Mhl was reported for 2017, being Italy the first producer worldwide, followed by France, Spain, and USA (OIV report 2018). Within the Vitaceae family, the *Vitis* genus has a major agronomic importance. Archaeological data suggests that cultivation of the domesticated grape, *V. vinifera* subsp. *vinifera*, began 6,000–8,000 years ago in the Near East from its wild progenitor, *V. vinifera* subsp. *sylvestris* (McGovern 2003). Apart for the use of the fruit, with the development of human culture, fermented beverages gained prominence. Wine making from grapes is documented by artifacts dating back to the Neolithic period (McGovern et al. 1996). Thus, since its domestication *V. vinifera* became the most used in the global wine industry with thousands of cultivars being selected for their quality and adaptation to different climatic conditions and developed by vegetative propagation and crosses. Of the 10000 cultivars known (Robinson et al. 2012), 33 represent almost 50% of the cultivation area where few elite cultivars including 'Cabernet sauvignon,' 'Tempranillo,' 'Chardonnay,' and 'Merlot' are primarily used for wine production (OIV report 2015). Around 56.8% of the known cultivars are used for wine production, 27% for table grape, 7.3% for a mixed utilization for both wine and table grapes, and 0.7% for raisin production (Töpfer et al. 2011).

Vinifera harbors high genetic diversity with polymorphism that dates back tens of millions of years (Lijavetzky et al. 2007; Myles et al. 2010); however, it is highly susceptible to biotic and abiotic stresses, leading to significant economic losses. Wine and table grapes currently receive intense chemical applications to combat severe pathogen pressures. Among the most devastating fungal diseases are the mildews, downy, and powdery mildew caused by *Plasmopara viticola* and *Erysiphe necator* together with gray mold and black rot caused by *Botrytis cinerea* and *Phyllosticta ampelicida* (syn. *Guignardia bidwellii*) (Kiss et al. 2016). Over the past three decades, grapevine trunk diseases also raised several concerns as the pathogens responsible for these diseases attack the long-lasting organs, causing the death of vines on shorter or longer term (Hofstetter et al. 2012). These include canker caused by *Botryosphaeriaceae* or *Diatrypaceae* pathogens (such as *Eutypa dieback*), blackfoot disease (caused by *Dactylonectria* and *Ilyonectria* species), and the *Esca* complex (Fontaine et al. 2016). Other important diseases are Pierce's disease caused by *Xylella fastidiosa*, which can kill the vine in few years (Kiss et al. 2016), and *A. tumefaciens* biovar 3 (*A. vitis*), which causes crown gall disease and induces tumors at wounds on the trunks and canes of the grapevine that can necrotize the plant (Ridé et al. 2000). Viruses cause other diseases that greatly affect grapevines, being fanleaf disease leafroll, rugose wood, and infectious degeneration of the major economic importance (Maliogka et al. 2015).

To reduce the massive use of fungicides to protect the plants, several breeding efforts have been conducted. America *Vitis* species have been employed not only as grapevine rootstocks due to their resistance to pests (*Phylloxera*, nematodes), drought tolerance, salt tolerance, or tolerance of high pH (Keller 2010) but also as resistant traits sources for breeding programs. Breeding goals have been focused not only on disease resistance but also on abiotic stress (water availability, temperature, and light) tolerance and quality improvement (Vivier and Pretorius 2002; Gray et al. 2014; Saporta et al. 2016). Before the advances in genome research, traditional breeding programs were based on the identification of natural and induced mutant alleles for desirable traits and their incorporation into elite genotypes. Aid by genomic information, breeders applied genotype-phenotype associations by linkage mapping (Nakano *et al.* 1994, Franks *et al.* 1998) and more recently both genome-wide association studies (GWAS) and genomic selection (GS), to overcome the problems related to the long generation cycle and high level of heterozygosity (McCarthy et al. 2008). The use of genetic transformation appeared as a promising approach as it allows the introduction of single traits into *V. vinifera* genome (Gray *et al.* 2014). The improvement of genetic editing techniques, such as the use of CRISPR/Cas9 system for genome modification (Wang *et al.* 2016; Malnoy et al. 2016; Ren et al. 2016), harbors new possibilities for the development of new grapevine cultivars.

13.2.2 Somatic Embryogenesis in Grapevine

Grapevine improvement through genetic engineering requires efficient transformation, selection, and regeneration protocols. One of the main drawbacks of grapevine transformation is plant regeneration as it is greatly influenced by the genotype and culture conditions. It may occur via organogenesis or embryogenesis. In *Vitis* it is mainly achieved through somatic embryogenesis. The choice of the best explant is very important, and much research has been conducted on testing different starting explants, such as leaves, anthers, ovaries, whole flowers, mature seeds, stigmas and styles, petioles, tendrils, nodal sections, and protoplasts (reviewed in Laimer 2007; Cardoso et al. 2010, 2019; Saporta et al. 2016).

Regeneration has been obtained for several cultivars, but the appearance of aberrant embryos is a common feature; therefore, embryogenic cultures achieved from different flower organs like anthers, ovaries, or total flowers proved to be most suited for regeneration and gene transfer purposes (Franks et al. 1998; Yamamoto et al. 2000; Perl and Eshdat 1998; Vidal et al. 2003, 2006; Agüero et al. 2005; Gambino et al. 2005; Fan et al. 2008; López-Pérez et al. 2008; Dai et al. 2015) with advantages for *A. tumefaciens*-mediated gene transfer to achieve homogeneously transformed plants (Polito et al. 1989; Perl and Eshdat 1998).

Moreover, the selection of the grapevine genotype greatly influences the regeneration process as it was shown in the study by Locco et al. (2001). Under the same regeneration protocol and transformation condition, the number of plants that regenerated under selective conditions differed greatly with the genotype (Locco et al. 2001).

13.2.3 Grapevine Breeding and DNA Manipulation Strategies

13.2.3.1 Conventional Breeding and DNA Sequencing Advances

Grape breeding began at the end of the eighteenth century, with 'Alexander' being the first cultivar successfully grown (Hedrick 1908). With the introduction of downy and powdery mildew in Europe, breeding activities increased as a consequence of the high susceptibility of *V. vinifera* to these pathogens. The solution was to combine resistant traits from the American species with the fruit quality from *V. vinifera* (Millardet 1885). Several breeding efforts were conducted through the years (reviewed in Töpfer et al. 2011), and not until the end of the twentieth century, good crossing lines combining resistance and quality were achieved. Cultivars like 'Regent' (cross 1967 ('Silvaner' x 'Müller-Thurgau') x 'Chambourcin') gained access to the market becoming one of the most successful cultivars derived from a resistance breeding program developed in Germany (reviewed in Töpfer et al. 2011). Despite the success cases, due to grapevine long generation cycle, heterozygozity, and inbreeding depression which make backcrossing and recurrent selection very difficult (Alleweldt and Possingham 1988), conventional breeding takes a long time, requiring several repetitions due to limited plant material and slow propagation rates. Usually, the development of a new cultivar takes between 25 and 30 years.

With the advances on genome sequencing technologies, *V. vinifera* cv. 'Pinot noir' genome was sequenced in 2007 (Velasco et al. 2007; Jaillon et al. 2007), leading to great advances on the association of genes and traits, how they are inherited, and how they are molecularly organized. The development of molecular markers leads to a breakthrough for grapevine breeding, allowing genetic mapping, genotyping (Williams et al. 1995; Bowers and Meredith 1997), and genome-wide association studies (Korte and Farlow 2013). Several genetic maps were developed, with the particular interest of combining genotypic and phenotypic information-QTL (quantitative trait *locus*) mapping (reviewed in Töpfer et al. 2011). Moreover, marker-assisted selection (MAS) was implemented allowing accelerating grapevine breeding process in 10 years (van Nocker and Gardiner 2014). MAS allows selection during the juvenile phase (Karaagac et al. 2012) and provides advantages for pyramiding disease resistance genes (Eibach et al. 2007; Zini et al. 2015)

With the improvement of next-generation sequencing (NGS) technologies, genotyping-by-sequencing (GBS), with its simultaneous marker discovery and genotyping approach, delivers many benefits including availability of flanking DNA sequence information, high-sample throughput, and high resolution (Elshire et al. 2011). In grapevine, a GBS approach was reported very recently applied to analysis of the recombination rate along the genome in various interspecific contexts represented by three pseudo-backcross populations derived from *V. rotundifolia* and *V. vinifera* at different levels of introgression (Delame et al. 2018). This application opened new insights into the use of *V. rotundifolia* in grape breeding and, more generally, to improve the introgression of gene of interest from wild species related to crops (Delame et al. 2018). Also, with the rapid increase knowledge about grapevine genome sequence and both physiological and molecular role of various grapevine genes, breeding programs were revolutionized.

13.2.3.2 Grapevine Transformation: *Agrobacterium* sp. and Biolistic Methods

Plant genetic transformation appeared as a process through which genes of interest are introduced into the target plant material, allowing the development of new cultivars with high biotic and abiotic stress tolerance, and with improved traits of economic value like color and yield among others. Genetic transformation is a key technology to enhance the potential of existing cultivars by taking advantage of the increasing knowledge availability in grapevine genetics and express exogenous sequences in target plant tissues and to interfere with endogenous genetic expression. It allows the characterization of function and regulation of newly discovered genes if stable transformation is obtained.

On herbaceous model plants, regeneration of stable transformants is straightforward (Goodin et al. 2008; Koornneef and Meinke 2010). Being a woody plant species, grapevine is recalcitrant to the in vitro regeneration protocols, and the first successful transformation events were only reported in 1989 (Baribault et al. 1989) when *nptII* was introduced as selectable marker into cv. 'Cabernet Sauvignon' mediated by *A. tumefaciens* infection. Successful regeneration of transformants was first reported for GUS-positive grapevine rootstock *V. rupestris* 'St. George' (Mullins et al. 1990). *Agrobacterium*-mediated gene transfer has been the methodology most frequently used for grapevine genetic transformation (reviewed in Li et al. 2008; Dhekney et al. 2009). However, the success on the use of biolistic-mediated transformation, first reported in 1993 (Hébert et al. 1993), was already demonstrated in grapevine (e.g., Vidal et al. 2003, 2006).

Through the past three decades, several efforts have been made to targeted traits in the genetic improvement of grapevine (reviewed in Vivier and Pretorius 2000, 2002; Laimer 2007; Saporta et al. 2016), either by T-DNA transfer and its integration into the plant genome by *Agrobacterium* sp. or biolistic methods. Grapevine biotechnology has been focused on disease management and berry ripening. The improvement of these traits rely on an in-depth knowledge of the processes underlying pathogenesis and ripening, cell wall structure, morphogenesis, plant-pathogen interactions, signal transduction, among others. Several target genes have been identified and used for grapevine breeding through genetic transformation. For pathogen resistance, several reports were published (reviewed in Leimer et al. 2007; Saporta et al. 2016). Some successful examples are associated with fungal pathogens resistance, particularly to powdery mildew through the exploitation of rice chitinases (*RCC2* and *Chil1*) that were introduced in cvs. 'Neo Muscat' and 'Pusa Seedless' (Yamamoto et al. 2000; Nirala et al. 2010), of *V. pseudoreticulata* PR gene (*VpPR4-1*) introduced in cv. 'Red Globe' (Dai et al. 2016), and of stilbene synthase (associated with phytoalexin production) transformation into cvs. 'Chardonnay' (Dai et al. 2015) and 'Thompson Seedless' (Fan et al. 2008). Virus resistance was also reported through genetic engineering of cvs. 'Chardonnay' and 'Thompson Seedless' plants through introduction of a gene encoding for the virus coat protein (Mauro et al. 1995; Scorza et al. 1996). Also, several examples were reported for cold tolerance, yield, and berry quality (reviewed in Saporta et al. 2016).

Despite the tremendous progress made in the last decade, it remains difficult to generate stably transformed whole grapevine plants (Vidal et al. 2010). Alternatively, *A. rhizogenes*-transformed roots (hairy roots) provide an interesting system for functional studies (Hu and Du 2006). In grapevine, Gomez et al. (2009) produced hairy roots to localize anthocyanin transporter candidates (*AM1* and *AM3*) to the tonoplast. Using the same system, ectopic expression of *VvMYBPA1* or *VvMYBPA2* provided clues on their roles in the regulation of the proanthocyanidin (PA) pathway (Terrier et al. 2009). Likewise, Höll et al. (2013) recently demonstrated the role of *VvMYB15* in the synthesis of glycosylated stilbenes.

Over the next few years, it is expected that several candidate genes will be identified as being a part of important biochemical pathways or associated with specific traits as biotic stress resistance. Transformation is highly important to elucidate their function but time-consuming (Töpfer et al. 2011). Assessing gene function and activity, before undertaking stable transformation (Sparkes et al. 2006), may become a shortcut for some studies.

13.2.4 Recent Advances in Grapevine Transformation: CRISPR/Cas9

Considerable efforts are being made to improve grapevine traits using both conventional breeding and genetic engineering. However, genetically modified crops are not well accepted in many countries leading to the adoption of tight and stringent regulations concerning their commercial authorization leading to scarce examples of successful commercialization (Nagamangala Kanchiswamy et al. 2015; Sprink et al. 2016). These regulations are not so strict for genetic modified plants with targeted mutagenesis generated by self-repair mechanisms considering that no *Agrobacterium* or transgenic/foreign genetic materials are present, opening new challenges into the use of genome editing approaches for crop improvement (Jones 2015). New breeding techniques that include the most recent approaches for genome editing are attracting plant researchers and breeder's attention. These new techniques employ site-directed nucleases to introduce double-stranded breaks at predetermined sites in DNA. Host cell repair mechanisms will repair those breaks creating insertions or deletions via near homologous end joining or microhomology-mediated end joining or in a modified gene carrying predetermined nucleotide changes copied from a repair matrix via homologous recombination (Mohanta et al. 2017). Nucleases used in genome editing may be meganucleases, zinc finger nucleases, transcription activator-like effector nucleases, and clustered regularly interspaced short palindrome repeats (CRISPR)/CRISPR-associated protein 9 (Cas9). CRISPR/Cas derives from the prokaryotes' immune system is a method of gene edition where Cas nucleases cut the genomic DNA in a specific site, guided by specific RNA sequences, allowing a precise edition of the genome (revised in Fernandez, Marti and Dodd 2018 and Parsons & MacKay 2018). Cas nucleases are programmable RNA-guided endonucleases, which can be customized by changing the RNA unit alone.

CRISPR/Cas9 technology has gain particular attention (Mohanta et al. 2017) for genome editing in eukaryote genomes. There are three types of CRISPR/Cas9 systems (I, II, and III), and the type II system with a single Cas9 protein has emerged as a highly efficient tool for targeted genome editing (Cong et al. 2013; Wu et al. 2014). The Cas9 nuclease is guided by a duplex formed by CRISPR RNA (crRNA) and trans-activating crRNA (tracrRNA) that cleaves the complementary sequence of the crRNA with an NGG protospacer adjacent motif (PAM) (Jinek et al. 2012; Gaj et al. 2013). The PAM is indispensable for target recognition as it helps the crRNA-tracrRNA-Cas9 complex combine with the target sequence (Nishimasu et al. 2014; Sternberg et al. 2014). In addition, the Cas9 nuclease can be converted into a nicking enzyme to facilitate homology-directed repair with mutagenic activity (Cong et al. 2013). Then, the DNA is repaired by mechanisms in the targeted cell, and several types of mutations can be obtained, including knockouts or knockdowns, when the double-strand breaks are repaired by nonhomologous end joining, which may result in insertion and/or deletion (InDel) mutations or knock-ins, obtained by a homology-directed repair system that repairs a template leading to gene replacement or insertion.

The obtained InDel mutations can be previously directed to a gene (or several genes) of interest if the sgRNA includes a sequence to anneal the gene. Design of single guide RNA oligos (sgRNAs) is subject to rules in order to avoid off-target mutations (revised in Osakabe and Osakabe 2017); thus, an annotated genome is a clear advantage for the use of the available in silico tools for sgRNA design. After transformation and plant regeneration, selection of transgenic lines can be achieved by reporter gene localization (Shan et al. 2014; Osakabe et al. 2016; He et al. 2017) when applicable, by Sanger or by deep sequencing of amplicons (Shan et al. 2014; Fan et al. 2015; Woo et al. 2015; Osakabe et al. 2016; Wang et al. 2018).

The application of the CRISPR/Cas9 technology on grapevine is quite recent, with few studies being published so far. In 2016, three studies were published focusing on the identification of protospacers and PAMs present in grapevine genome (Wang et al. 2016), on the edition of grapevine protoplasts with CRISPR/Cas9 ribonucleoproteins (Malnoy et al. 2016), and on the use of CRISPR/Cas9 system to generate targeted genome editing for L-idonate dehydrogenase gene (IdnDH) in cv. 'Chardonnay' using suspension culture of embryogenic cells to regenerate plantlets (Ren et al. 2016). In 2018, a new study was published focusing on biallelic mutations or the extent of off-targeting in the first generation (Wang et al. 2018).

Through the knowledge that NGG (or CCN-complementary strand) sequences are sufficient for targeting (Jiang et al. 2013), Wang and co-workers (2016) have performed a genome-wide bioinformatic analysis and found that protospacers and PAMs were abundantly present in the grape genomes and more or less uniformly distributed among chromosomes and chromosomal regions with predominant abundance in the coding regions of genes. Most genes contain unique protospacers allowing researchers to edit a gene of interest with multiple choices of target sites and great specificity (Wang et al. 2016). The authors have shown that among the five PAM types, TGG and AGG types were the most abundant and created a publicly accessible Grape-CRISPR database allowing users to identify and select multi-protospacers for editing similar sequences in grape genomes simultaneously (Wang et al. 2016).

The first successful grapevine genome editing experiments through CRISPR/Cas9 were performed in cv. 'Chardonay' protoplasts and suspension cell cultures. Protoplasts and direct delivery of purified CRISPR/Cas9 ribonucleoproteins (RNPs) were used to target a MLO-7 susceptibility gene in order to increase resistance to powdery mildew (Malnoy et al. 2016). The Cas9-protein-gDNA ribonucleoproteins (RGEN RNPs) cut gDNA target sites after transfection and then are degraded by endogenous proteases. Targeted deep sequencing was used to analyze the mutation efficiency. The authors have highlighted the efficient targeted mutagenesis in the grapevine MLO-7 locus and the need to optimize plant regeneration from CRISPR RNPs transformed protoplast in order to explore the applications of this technology at field level. Targeting of the L-idonate dehydrogenase gene (IdnDH) that regulates the stable accumulation of tartaric acid in grapevine was also achieved using the CRISPR/Cas9 system by introducing a single plasmid with a specific sgRNA into grape embryogenic cells (suspension cultures) mediated by *A. tumefaciens* transformation (Ren et al. 2016). An average transformation rate of 37.78% was shown, and sequencing analysis revealed a successfully directed target editing in embryogenic cell masses and in the regenerated plantlets. This study also highlighted the importance of the selection of the target sequence for the specificity and efficacy of the CRISPR/Cas9 system (Ren et al. 2016).

More recently, experiments targeting *VvWRKY52* transcription factor gene were published (Wang et al. 2018). The authors have tested four guide and obtained transgenic plants via *Agrobacterium*-mediated transformation, using somatic embryos of the cv. 'Thompson Seedless'. Analysis of the first-generation transgenic plants showed that of the 72 T-DNA-inserted plants, 15 lines carried biallelic mutations. Moreover, functionality of gene editing was shown as the knockout of *VvWRKY52* in grape increased the resistance to *Botrytis cinerea* (Wang et al. 2018).

CRISPR/Cas9 system allows precise genome editing in the first generation of grape and represents a useful tool for gene functional analysis and grape molecular breeding.

13.3 *Quercus suber* L.

13.3.1 Introduction

Cork oak (*Quercus suber* L.) is an evergreen tree, fundamental for the Mediterranean landscape. Cork oak is the main element of the *montado* ecosystem at the Iberian Peninsula. Portugal has 34% of the world's area of *Q. suber* forests, corresponding to 737.000 ha (www.apcor.pt). Besides the production of high-quality cork, used to produce wine bottle-stoppers or thermal and acoustic insulation materials, the acorns are used as food for domestic and wild herbivores, and the wood from prunings as firewood and charcoal source. Even though cork oak is well adapted to dry summers and cool winters, concern exists about its adaptability to the climate change (Spathelf et al. 2014). Cork oak has been suffering from decline since 1980s, when heavy mortality has been noticed. A significant number of reports point to the oomycete *Phytophthora cinnamomi* allied to drought as the cause for oak decline or *la seca* (Brasier et al. 1993; revised in de Sampaio e Paiva Camilo-Alves et al. 2013; Moreira et al. 2006; Serrano et al. 2012). Also related with climate changes, prolongated flooding after a period of severe drought, can lead to opportunistic attack by oomycetes, fungi and insects (Braisier 1996; Scanu et al. 2013).

Cork oak importance to economy and ecology conferred high interest in the development of methodologies for clonal propagation of selected genotypes and improvement. A considerable number of reports have appeared since the last three decades on conventional breeding strategies, mainly focused on molecular markers development and in vitro clonal propagation, especially based on somatic embryogenesis protocols. Presently, constrains related to cork oak propagation, as long life cycle, self-incompatibility, high degree of heterozygosis, and recalcitrance (highly dependent on genotype), have been surpassed for many genotypes (revised in Corredoira et al. 2014), promoting cork oak improvement by genetic transformation. This is a method to precisely express or silence genes related to adaptive traits (as abiotic and biotic stress) or economic traits (as cork quality), adequate for functional genomics approaches to understand molecular and cellular mechanisms.

13.3.2 Somatic Embryogenesis in Cork Oak

Genetic transformation of cork oak has been mainly based in somatic embryogenesis (SE) regeneration protocols (revised in Corredoira et al. 2014 and Gomez-Garay et al. 2014). In the last 6 years, research has focused on long-term maintenance of repetitive embryogenic cultures and SE insights at the molecular level. Jiménez et al. (2013) developed embryogenic cultures in liquid medium for reduction of costs and automation. The authors were able to initiate and maintain embryogenic suspensions for approximately 1 year on 12 cork oak genotypes. Pérez et al. (2013) focused on circumventing the problems associated with liquid cultures (hyperhydricity, asphyxia) and developed temporary immersion systems that permit partial automation of cultures. For successful plant regeneration from genetic transformation through SE, rooting and acclimation must be established. Pintos et al. (2008) and Pintos et al. (2009), after obtaining somatic embryos from immature zygotic embryos, coated them with an encapsulation medium containing alginate, obtaining synthetic seeds. After storage at 4°C, plantlets can be recovered in conversion media or transferred directly to soil. SE from immature male gametophytes (termed microspore or pollen embryogenesis) is known in *Q. suber* since 1997 (revised in Gomez-Garay et al. 2014) to obtain haploid embryos. Chromosome doubling of haploid embryos (spontaneously in culture or after application of colchicine) produces a homozygous plant per generation, which is called doubled-haploid plant. Propagation of selected double-haploid trees after genetic transformation, produced in an expedite way by in vitro microspore embryogenesis, has potential advantages. However, the efficiency of embryogenesis induction by microspores and zygotic embryos is highly variable due to the level of recalcitrance of the genotype. Since 2009 reports have revealed molecular and cellular mechanisms involved in the origin of haploid and diploid embryos, aiming to contribute to *Q. suber* in vitro embryogenesis improvement. Gómez et al. (2009) identified proteins involved in stress and pollen development, and proteins involved in the synthesis of cork components (tannins and phenylpropanoids). Rodríguez-Sanz et al. (2014) performed biochemical and driven cellular analysis on microspore and immature zygotic embryo induction, and revealed modifications in epigenetic marks, cell structural organization, cell wall polymers, and auxin levels. Gómez et al. (2009) and Rodríguez-Sanz et al. (2014) propose molecular markers related to SE induction and cork quality. Pérez et al. (2015) analyzed gene expression during the maturation of cork oak somatic embryos, focusing on epigenetic regulation and abscisic acid-related genes.

13.3.3 Genetic Transformation Protocols Established for Cork Oak

Álvarez et al. (2004) were the first to report a genetic transformation protocol for cork oak from somatic embryos. Several *A. tumefaciens* strains were tested, and the genes transferred were *nptII* for in vitro selection of transformants and *uidA* (*β-glucuronidase*) as reporter gene. *A. tumefaciens* strain AGL1 rendered the best transformation efficiency and embryo maturation; germination and transfer to soil were achieved. Obtained transformation efficiency was up to 30% and 10 months occur from the co-cultivation of embryos with *Agrobacterium* until the obtention of plantlets (Alvarez et al. 2006). The same protocol was improved, and the transformation efficiency raised up to 43% (Álvarez and Ordás 2007). Several variables were tested, and the most relevant were the *Agrobacterium* inoculum density and the co-cultivation period, the time-point of inoculation after embryo subculture, the maturation stage of somatic embryos and the genotype (tested with six different genotypes). Based on the previous method, Álvarez et al. (2009) introduced a gene in *Q. suber* genome for herbicide resistance, *phosphinothricin N-acetyltransferase* (*bar*) gene, in the first attempt to improve cork oak by genetic transformation. Maturation, germination, and transfer to soil of transformed embryos were achieved for several lines.

13.3.4 Candidate Genes and Pathways to Cork Oak Genetic Transformation

Cork oak improvement by genetic transformation depends on the knowledge of genes and metabolic pathways involved in resistance to extreme environmental conditions and to pathogens while maintaining tree yields. Several genetic or physiological studies in *Q. suber* or in closely related species can guide the selection of candidate genes for functional studies or plant improvement and health. Coelho et al. (2011) identified, by cDNA-AFLP, 44 genes differentially expressed in *Q. suber* inoculated with *P. cinnamomi*. Those genes coding for cinnamyl alcohol dehydrogenases, involved in detoxification or lignin and suberin synthesis, transmembrane proteins with LRR domains, transcription factors that interact with transmembrane protein domains and regulate nuclear genes related with pathogenesis, and reactive oxygen species (ROS)-responsive transcription factors, among others. Based on the results, the authors propose a model of *Q. suber* responses to the pathogen. Plants use secondary metabolites, namely, volatile isoprenoids, as a response to the environment, and there are evidences of their involvement in biotic and abiotic stresses (revised in Loreto et al. 2013). Volatile isoprenoids have been implied in pollinator attraction, plant-to-plant communication to signal herbivore attack, and herbivore repellence or as antifeedants, toxins, or antibiotics. On the other hand, volatile isoprenoids may lead to the synthesis of drought stress-defensive products as abscisic acid (ABA), may stabilize thylakoid membranes at high temperatures preventing the destabilization of photosystems, or may have

ROS-scavenging action, preventing oxidative stress. Loreto et al. (2009) show that cork oaks from a humid area in Portugal have a profile of volatile terpene emissions with high levels of limonene, distinct of the profile of cork oaks in xeric areas of central Italy, high in α-pinene, which points out that those compounds are related to plant health in the respective ecosystems subjected to abiotic stress.

Plant hormones of the jasmonate (JA) family seem to regulate volatile isoprenoids in plants, priming and enhancing resistance to fungi, herbivores, and abiotic stresses (revised in Semiz et al. 2012). Semiz et al. (2012) observed the effect of methyl JA elicitation in *Q. ilex*, with the increase of sesquiterpenes, green leaf volatiles, and homoterpene (E)-DMNT emissions and the decrease of monoterpene emissions, predicting a reduction in photosynthesis and stomatal conductance as responses to the simulated stress. These are among the morphological responses that have been observed in plants subjected to drought stress (revised in Kumar et al. 2018). Magalhães et al. (2016) revealed fundamental mechanisms on *Q. suber* roots subjected to long-term drought, with the identification of differentially expressed genes related with the abscisic acid hormone: ABA-dependent signaling network, ABA-related transcription factors, and ABA-responsive genes. During the treatment, physiological symptoms on leaves associated with water privation revealed loss in efficiency of PSII, also indicating photosynthesis decline. Aquaporins are transmembrane proteins that transport water and small solutes across membranes, mostly plasma membrane and tonoplast. Aquaporins may take part of the regulation of water balance, depending on water potential gradient, and are candidates to the response to abiotic stress and environmental adaptation (revised in Kapilan et al. 2018). Rasheed-Depardieu et al. (2012) characterized the expression of nine aquaporins in distinct developmental root zones of *Q. petraea* and *Q. robur*, two closely related species adapted to contrasting ecosystems: *Q. petraea*, a drought-tolerant species adapted to acidic, well-drained soils (as cork oak), and *Q. robur*, adapted to soils where excess of water is a constant. After a comparative expression analysis between the species, the authors selected four aquaporin genes that have differential expression in the different root zones and might be useful to further understand the contrasting drought tolerance levels of those species.

Cork oak has the ability to develop a protective, thick cork tissue that results from the response to an environment with long periods of drought and heat, to damage, or to peeling-off the cork layer. Cork results from the meristematic activity of phellogen which self-regenerates after wounding (Pereira 2015). Teixeira et al. (2018) identified genes and metabolic pathways that are involved in *Q. suber* good-quality cork or bad-quality cork. Besides drought and high temperatures, cork quality may depend on the genotype that ultimately determines how the plant copes with stress (Oliveira et al. 2016; Teixeira et al. 2018). In good-quality cork tissues, there was an upregulation of genes encoding heat-shock proteins (HSPs) and late embryogenesis abundant (LEA) proteins, both important for cell homeostasis and protection. There was also the adequate upregulation of genes encoding regulatory proteins (MYB, bZIP, and NAC). In bad-quality cork, genes involved in the flavonoid pathway were upregulated, favoring free phenolics production and the formation of primary cell walls instead of the main components of cork, suberin, and lignin. On the other hand, in good-quality cork, there is an upregulation of *expansins*, cellulose synthesis (*XTH*), and suberin-associated genes, promoting the elasticity and hydration status of the primary cell wall and secondary cell wall structure. In good-quality cork, more distinct MYC/MYB transcription factors and ABA-related genes were upregulated than in bad-quality cork tissues, being some MYB transcription factors promoters of synthesis, transport, and polymerization of suberin monomers. Also, there was the upregulation of suberin biosynthesis-associated genes in the same tissues.

13.4 Future Perspectives

Genetically transformed plants with enhanced abiotic and biotic resistance constitute alternatives to overcome the limitations of classical breeding and accelerate improvement programs. The area of agricultural land used for production of genetically modified (GM) crops has increased at an unprecedented speed since its introduction almost 20 years ago (Holme et al. 2013). Soybean, maize, cotton, and canola presenting tolerance to herbicides and insect larvae are among the first generation on commercialized GMs. However, genetically modified crops are not well accepted in many countries, and very strict regulations were established for the commercial authorization (Nagamangala Kanchiswamy et al. 2015; Sprink et al. 2016). A general public concern lies on the possibility of genetic material mixing between species (transgenics). To address these concerns, two transformation concepts, cisgenesis and intragenesis, were developed. These concepts rely on the use of genetic transformation methods, but the genetic material is from closely related species (Holme et al. 2013), avoiding the mixing of foreign genetic material within wild relatives. Furthermore, foreign sequences such as selection genes and vector-backbone sequences should be absent. The gene pool exploited by these approaches is the same that is used for conventional breeding; thus, less comprehensive regulatory measures are expected.

More recently, the development of genome editing techniques has created a new paradigm in plant breeding. Classical breeding approaches, based on hybridization and selection, for trait improvement in crops may be complemented by targeted genome editing that exploits knowledge of specific gene sequences. These genome editing techniques exploit natural endonucleases that cut and alter a predetermined gene sequence in the host plant. Genetic modifications are therefore indistinguishable from those generated through natural or induced mutations (Mohanta et al. 2017). GM regulations are not so strict for genetic modified plants with targeted mutagenesis generated by self-repair mechanisms; thus, a higher acceptance may be expected (Jones 2015).

The use of these techniques for crop improvement requires tough genome and transcriptome knowledge. In the last few years, three different projects were initiated for decoding of the whole genome in olive, the Italian OLEA project (Muleo et al. 2012); the IOGC International Consortium, responsible by the genome sequencing of *O. europaea* subp. *sylvestris* (Unver et al. 2017); and the sequencing of the Spanish cv. 'Farga' (Cruz et al. 2016). The first draft of olive genome was achieved by whole genome sequencing of cv. 'Farga,' spanning a length of 1.31 Gb with a median C-value of 1.59pg. The predicted number of protein gene-coding sequences, supported also by RNA sequencing from the different tissues, was estimated to be higher than 56.000 (see review from Sebastini and Busconi 2017). Genomic data achieved by using the new NGS technologies have provide the opportunity for in silico identification and characterization of gene(s) related to important olive traits, as well the mechanism involved in controlling its expression (e.g., transposon elements, miRNA, and functional polymorphisms). Both genes and functional polymorphisms enable new tools for conventional breeding programs by developing new molecular markers and allowing the development of superior cultivars by manipulating endogenous genes following a transgenic approach (silencing or overexpression).

The decoding of olive genome in combination with data arriving from other omics research areas (transcriptomics, proteomics, and metabolomics) represents an important step forward on olive breeding programs, not only related with the development of molecular markers, which could assist conventional breeding programs, but also for the survey of genes involved in important agronomical traits. Regarding the identification of key genes that could be associated with traits of agronomical interest, it is already available some work regarding pollen tube development, pollen allergens, flower and fruit development, oil quality (e.g., genes involved in fatty acids and antioxidants biosynthetic pathways), ability to develop adventitious roots and plant tolerance upon biotic or abiotic stressful conditions (Santos Macedo et al. 2009, Iaria et al. 2016; Alagna et al. 2016; Sebastini and Busconi 2017 and reference therein; Sezer and Taskin 2017; Leyva-Pérez et al. 2018; Mougiou et al. 2018; Roka et al. 2018; Velada et al. 2018). Additionally, the expected release of whole genome sequencing data from a diversity of varieties, will allow a better understanding on the role of functional polymorphisms and to the establishment of a link between genomics and phenomics. All these new data represent a high valuable source of information to be explored on the development of new olive cisgenic cultivars applying the new genome editing tools. However, few protocols for olive propagation under in vitro conditions have been successfully established and only six cultivars have been considered ('Picual', 'Canino', 'Dolce Agogia', 'Mariolo', 'Galega vulgar' and 'Zard'). Since the success of a genetic transformation system depends on the existence of an efficient plant regeneration system, great attention must be given to the establishment of reliable protocols able to regenerate transgenic olive plants. When the aim is to introduce a specific trait in an elite cultivar, it will be required the establishment of protocols using somatic tissues, not considering as explant the seeds. The use of tissues taken from zygotic embryos represents the introduction of a gene (an interesting agronomical trait) in an unknown genome that could be far from the phenotype of a selected elite cultivar.

The first genome draft for *V. vinifera* was obtained in 2007 by two independent projects (Jaillon et al. 2017; Velasco et al. 2007). It contains over 30,400 genes

(Saporta et al. 2016) increasing the possibility to obtaining more cisgenic plants. In 2014, *V. vinifera* gene annotation was reviewed (Vitulo et al. 2014) based on 124 RNAseq SOLiD libraries from leaf, root and berry, from different genotypes under different physiological and stress conditions (http://genomes.cribi.unipd.it/grape). This revision improved the previous version (V1) with 2,258 new coding genes and 3,336 putative long noncoding RNAs. Also several gene models were improved and alternative splicing was described for about 30% of the genes (Vitulo et al. 2014). The availability of the genomic sequence gave the opportunity to conduct several genome-wide studies, focused on different aspects of grape biology, such as berry development, response to different biotic and abiotic stresses, and to identify specific genes associated with desirable traits. However, genome annotation only gives indications of the role of newly discovered genes being insufficient for function and regulation characterization. Gene function may only be validated by knocking out or knocking down its expression. Until recently, grapevine breeding through genetic transformation has been focused on biotic stress resistance, cold tolerance and berry color. Plant regeneration, molecular characterization of the trait of interest and the quantitative character of the trait, were the main drawbacks found. *V. vinifera* genes have been mainly introduced by genetic transformation mediated by *A. tumefaciens*. With improvement on the knowledge regarding gene function and the technical development of gene editing and protocols for plant recovery on different grapevine cultivars and rootstocks (Bortesi and Fischer 2015), it is expected that new breeding programs are established for grapevine.

Since 2014 a significant amount of genetic resources on *Q. suber* became available, opening great opportunities for cisgenics events. At the transcriptome level, 21 cDNA libraries from multiple tissues and organs, developmental stages, or physiological conditions, were pyrosequenced, including embryo development, cork formation and quality and several abiotic stresses or biotic interactions (Pereira-Leal et al. 2014). The first draft version of *Q. suber* genome was produced (Ramos et al. 2018), with the annotation of 79,752 genes and 83,814 transcripts, promoting research that discovers the genes beyond the phenotypes and that are related to cork oak health, or QTLs controlling adaptive traits. An improved version of cork oak genome is underway (Ramos et al. 2018) and cork oak will surely be an adequate candidate for genome edition with CRISPR/Cas9. Together with the genome information, *Q. suber* micropropagation, plant regeneration and gene delivery was well settled for many genotypes, becoming a premium candidate for trait improvement through gene editing or genetic transformation. Current environmental adversities as water insufficiency/excess, higher temperature and/or salinity increase, as well as oomycete-caused diseases, have impact on cork oak endurance and productivity, so, abiotic/biotic stress-related genes and metabolic pathways, are significant targets for gene editing, promoting the ideal scenery for CRISPR/Cas9 utilization to support cork oak research concerning cork oak *montado* health and industry.

Acknowledgments This work was financially supported by national funds through FCT (Foundation for Science and Technology) under the Project UID/AGR/00115/2019 and PTDC/BIA-BQM/28539/2017 and by the Project OLEAVALOR (ALT20-03-0145-FEDER000014) funded by FEDER through the Program Alentejo 2020; Hélia Cardoso and Susana Serrazina were supported by FCT through postdoc fellowship SFRH/BPD/109849/2015 and SFRH/BPD/108653/2015, respectively; Andreia Figueiredo was also supported by FCT, under the investigator FCT program IF/00819/2015.

References

Agüero CB, Uratsu SL, Greve C et al (2005) Evaluation of tolerance to Pierce's disease and Botrytis in transgenic plants of *Vitis vinifera* L. expressing the pear PGIP gene. Mol Plant Pathol 6:43–51. https://doi.org/10.1111/j.1364-3703.2004.00262.x

Alagna F, Cirilli M, Galla G, Carbone F, Daddiego L, Facella P et al (2016) Transcript analysis and regulative events during flower development in olive (Olea europaea L.). PLoSONE 11(4):e0152943. https://doi.org/10.1371/journal.pone.0152943

Alleweldt G, Possingham JV (1988) Progress in grapevine breeding. Theor Appl Genet 75:669–673. https://doi.org/10.1007/BF00265585

Álvarez R, Ordás RJ (2007) Improved genetic transformation protocol for cork oak (Quercus suber L.). Plant Cell Tissue Organ Cult 91:45–52. https://doi.org/10.1007/s11240-007-9276-6

Álvarez R, Alonso P, Cortizo M et al (2004) Genetic transformation of selected mature cork oak (Quercus suber L.) trees. Plant Cell Rep 23:218–223. https://doi.org/10.1007/s00299-004-0810-2

Alvarez M, Millan C, Fernandez O, Ricardo-Javier R,Toribio (2006) Cork Oak Trees (*Quercus suber* L.). Methods Mol Biol (Clifton; NJ) 344:113–123. doi: https://doi.org/10.1385/1-59745-131-2:113

Álvarez R, Álvarez JM, Humara JM et al (2009) Genetic transformation of cork oak (Quercus suber L.) for herbicide resistance. Biotechnol Lett 31:1477–1483. https://doi.org/10.1007/s10529-009-0033-2

Baribault TJ, Skene KGM, Steele Scott N (1989) Genetic transformation of grapevine cells. Plant Cell Rep 8:137–140. https://doi.org/10.1007/BF00716825

Barranco D, Rallo L (2000) Olive cultivars in Spain. HortTechnology 10:107–110

Bashir MA, Silvestri C, Astolfi F, Coppa E, Cristofori V, Rugini E (2018) Transgenic olive (Olea europaea L.) shots overexpressing osmotin gene are less sensitive to in vitro-induced salt stress. OliveBioteq18. Seville, Spain.

Behelgardy MF, Motamed N, Jazii FR (2012) Expression of the P5CS Gene in Transgenic Versus Nontransgenic Olive (Olea europaea) under Salinity Stress. World Applied Sciences Journal 18(4):580–583. https://doi.org/10.5829/idosi.wasj.2012.18.04.1488

Benelli C, Fabbri A, Grassi S, Lambardi M, Rugini E (2001) Histology of somatic embryogenesis in mature tissue of olive. J Horticult Sci Biotechnol 76:112–119

Bevan M (1984) Binary Agrobacterium vectors for plant transformation. Nucleic Acid Res 12:8711–8721

Binet MN, Lepetit M, Weil JH, Tessier LH (1991) Analysis of a sunflower polyubiquitin promoter by transient expression. Plant Sci. 79:87–94

Bortesi L, Fischer R (2015) The CRISPR/Cas9 system for plant genome editing and beyond. Biotechnol Adv 33:41–52. https://doi.org/10.1016/J.BIOTECHADV.2014.12.006

Bourgin JP, Nitsch JP (1967) Production of haploid Nicotiana from excised stamens. Ann Phys Vég 9:377–382

Bowers JE, Meredith CP (1997) The parentage of a classic wine grape, Cabernet Sauvignon. Nat Genet 16:84–87. https://doi.org/10.1038/ng0597-84

Bradaï F, Pliego-Alfaro F, Sánchez-Romero C (2016) Long-term somatic embryogenesis in olive (Olea europaea L.): influence on regeneration capability and quality of regenerated plants. Sci Hortic (Amsterdam) 199:23–31. https://doi.org/10.1016/J.SCIENTA.2015.12.010

Braisier CM (1996) Phytophthora cinnamomi and oak decline in southern Europe. Environmental constraints including climate change. Ann des Sci For 53:347–358. https://doi.org/10.1051/forest:19960217

Brasier CM, Robredo F, Ferraz JFP (1993) Evidence for Phytophthora cinnamomi involvement in Iberian oak decline. Plant Pathol 42:140–145. https://doi.org/10.1111/j.1365-3059.1993.tb01482.x

Brhadda N, Abousalim A, Walali LDE (2003) Effets du milieu de culture et de la lumiére sur lémbriogenése somatique de l'olivier (Olea europaea L.) cv. 'Picholine Marocaine. Fruits 58:167–174

Camposeo S (2017) Nuevas variedades de olivo obtenidas del cruzamiento entre genotipos italianos y españoles adaptadas a los sistemas de cultivo superintensivos. Olint 31:15–16

Capelo AM, Silva S, Brito G, Santos C (2010) Somatic embryogenesis induction in leaves and petioles of a mature wild olive. Plant Cell Tissue Organ Cult 103:237–242. https://doi.org/10.1007/s11240-010-9773-x

Capone I, Cardarelli M, Mariotti D, Pomponi M, De Paolis A, Constantino P (1991) Different promoter regions control level and tissue specificity of expression of Agrobacterium rhizogenes rol B gene in plants. Plant Mol Biol 16:427

Cardarelli M, Mariotti D, Pomponi M, Spano L, Capone L, Costantino P (1987) Agrobacterium rhizogenes T -DNA genes capable of inducing hairy root phenotype. Mol Gen Genet 209:475–480

Cardoso HG, Campos MC, Pais MS, Peixe A (2010) Use of morphometric parameters for tracking ovule and microspore evolution in grapevine (Vitis vinifera L., cv. 'Aragonez') and evaluation of their potential to improve in vitro somatic embryogenesis efficiency from gametophyte tissues. In Vitro Cellular and Developmental Biology - Plant 46:499–508

Cardoso HG, Campos MC, Pais MS, Peixe A (2019) Somatic embryogenesis in Iberian grapevine (Vitis vinifera L.) cultivars using carpels as initial explants – protocol establishment and histological evaluation. J Agric Sci Technol B9:15–30. https://doi.org/10.17265/2161-6264/2019.01.002

Cerezo S, Mercado JA, Pliego-Alfaro F (2011) An efficient regeneration system via somatic embryogenesis in olive. Plant Cell Tissue Organ Cult 106:337–344. https://doi.org/10.1007/s11240-011-9926-6

Cerezo S, Barceló A, Samach A, Mercado JÁ, Pliego-Alfaro F (2013) Transformación genética de olivo (Olea europaea L.) con un gen de Medicago truncatula que codifica para un proteína tipo FT. X Reunión de la Sociedad Española de Cultivo In Vitro de Tejidos Vegetales (SECIVTV), Valencia-Spain

Cerezo S, Hernández ML, Sicardo MD, Mercado JA, Sanz C, Narváez I, Barceló-Muñoz A, Martínez-Rivas JM, Pliego-Alfaro F (2015) Transformación genética de olivo con el gen OeHPL para el análisis funcional del papel de la enzima 13-hidroperóxido liasa (13-HPL) en la producción de compuestos volátiles. XI Reunión de la Sociedad Española de Cultivo In Vitro de Tejidos Vegetales. Valencia-Spain

Christensen AH, Quail PH (1996) Ubiquitin promoter-based vectorsfor high-level expression of selectable and/or screenable markergenes in monocotyledonous plants. Transgenic Res. 5:213–218

Ciferri R, Breviglieri N (1942) Introduzione ad una classificazione morfo-ecologica dell olivo coltivato in Italia. L'Olivicoltore 1:1–2

Coca M, Bortolotti C, Rufat M, Peñas G, Eritja R, Tharreau D et al (2004) Transgenic rice plants expressing the antifungal AFP protein from Aspergillus giganteus show enhanced resistance to the rice blast fungus Magnaporthe grisea. Plant Mol Biol 54:245–259. https://doi.org/10.1023/B:PLAN.0000028791.34706.80

Coelho AC, Ebadzad G, Cravador A (2011) Quercus suber – Phytophthora cinnamomi interaction: a hypothetical molecular mechanism model. New Zeal J For Sci 0134:143–157

Cong L, Ran FA, Cox D et al (2013) Multiplex genome engineering using CRISPR/Cas systems. Science 339:819–823. https://doi.org/10.1126/science.1231143

Corredoira E, Toribio M, Vieitez AM (2014) Clonal propagation via somatic embryogenesis in Quercus spp. Tree Biotechnol:262–302. https://doi.org/10.1201/b16714

Cruz F, Julca I, Gómez-Garrido J, Loska D, Marcet-Houben M, Cano E et al (2016) Genome sequence of the olive tree, Olea europaea. GigaScience 5:1–12

Cunill M, Duran S (2014) "OLIANA®", una nueva variedad de olivo adaptada al sistema super-intensivo. Olint 25:21–29

D'Angeli S, Altamura MM (2007) Osmotin induces cold protection in olive trees by affecting programmed cell death and cytoskeleton organization. Planta 225:1147–1163

D'Angeli S, Gutiérrez-Pesce P, Altamura M, Biasi R, Ruggiero B, Muganu M, Bressan R, Rugini E (2001) Proceedings of the SIGA, Salsomaggiore Terme

Dai L, Zhou Q, Li R et al (2015) Establishment of a picloram-induced somatic embryogenesis system in Vitis vinifera cv. chardonnay and genetic transformation of a stilbene synthase gene from wild-growing Vitis species. Plant Cell. Tissue Organ Cult 121:397–412. https://doi.org/10.1007/s11240-015-0711-9

Dai L, Wang D, Xie X et al (2016) The novel gene VpPR4-1 from Vitis pseudoreticulata increases powdery mildew resistance in transgenic Vitis vinifera L. Front Plant Sci 7:695. https://doi.org/10.3389/fpls.2016.00695

Day AG, Bejarano ER, Burrell M, Buck K, Lichtenstein C (1991) Expression of antisense RNA in transgenic tobacco plants confer resistance to geminivirus infection. Proc Natl Acad Sci USA 88:6721–6725. https://doi.org/10.1073/pnas.88.15.6721

de Sampaio e Paiva Camilo-Alves C, da Clara MIE, de Almeida Ribeiro NMC (2013) Decline of Mediterranean oak trees and its association with Phytophthora cinnamomi: A review. Eur J For Res 132:411–432. https://doi.org/10.1007/s10342-013-0688-z

Delame M, Prado E, Blanc S et al (2018) Introgression reshapes recombination distribution in grapevine interspecific hybrids. Theor Appl Genet:1–15. https://doi.org/10.1007/s00122-018-3260-x

DGAV – Direção Regional de Agriculture e Veterinária (2016) Catálogo Nacional de Variedades, Fruteiras. 23 pg:14–16

Dhekney SA, Li ZT, Compton ME Gray DJ (2009) optimizing initiation and maintenance of Vitis embryogenic cultures. Hortscience 44:1400–1406

Eibach R, Zyprian E, Welter L, Töpfer R (2007) The use of molecular markers for pyramiding resistance genes in grapevine breeding. Vitis 46(2):120–124

Elshire RJ, Glaubitz JC, Sun Q et al (2011) A Robust, Simple Genotyping-by-Sequencing (GBS) Approach for High Diversity Species. PLoS One 6:e19379. https://doi.org/10.1371/journal.pone.0019379

Fan C, Pu N, Wang X et al (2008) Agrobacterium-mediated genetic transformation of grapevine (Vitis vinifera L.) with a novel stilbene synthase gene from Chinese wild Vitis pseudoreticulata. Plant Cell Tissue Organ Cult 92:197–206. https://doi.org/10.1007/s11240-007-9324-2

Fan D, Liu T, Li C et al (2015) Efficient CRISPR/Cas9-mediated Targeted Mutagenesis in Populus in the First Generation. Sci Rep 5:1–7. https://doi.org/10.1038/srep12217

Fernandez i Marti A, Dodd RS (2018) Using CRISPR as a Gene Editing Tool for Validating Adaptive Gene Function in Tree Landscape Genomics. Front Ecol Evol 6:1–7. https://doi.org/10.3389/fevo.2018.00076

Fontaine F, Pinto C, Vallet J et al (2016) The effects of grapevine trunk diseases (GTDs) on vine physiology. Eur J Plant Pathol 144:707–721. https://doi.org/10.1007/s10658-015-0770-0

Franks T, Gang He D, Thomas M (1998) Regeneration of transgenic shape Vitis vinifera L. Sultana plants: genotypic and phenotypic analysis. Mol Breed 4:321–333. https://doi.org/10.1023/A:1009673619456

Fugini E and Fedeli E (1990) Olive (Olea europaea L.) as an oilseed crop. In Bajaj YPS (ed) Biotechnology in agriculture and forestry 10, Legumes and oilseed crops I, vol 682 Springer, pp 594–641

Gaj T, Gersbach CA, Barbas CF (2013) ZFN, TALEN, and CRISPR/Cas-based methods for genome engineering. Trends Biotechnol 31:397–405. https://doi.org/10.1016/j.tibtech.2013.04.004

Gambino G, Gribaudo I, Leopold S et al (2005) Molecular characterization of grapevine plants transformed with GFLV resistance genes: I. Plant Cell Rep 24:655–662. https://doi.org/10.1007/s00299-005-0006-4

Gómez A, López JA, Pintos B et al (2009) Proteomic analysis from haploid and diploid embryos of Quercus suber L. identifies qualitative and quantitative differential expression patterns. Proteomics 9:4355–4367. https://doi.org/10.1002/pmic.200900179

Gomez C, Terrier N, Torregrosa L, Vialet S, Fournier-Level A, Verriès C, Souquet JM, Mazauric JP, Klein M, Cheynier V, Ageorges A (2009) Grapevine MATE-type proteins act as vacuolar H+-dependent acylated anthocyanin transporters. Plant Physiol 150(1):402–415. https://doi.org/10.1104/pp.109.135624

Gomez-Garay A, Manzanera J, Pintos-Lopez B (2014) Embryogenesis in oak species. A review. Forest Systems 23(2):191–198. https://doi.org/10.5424/fs/2014232-05829

Goodin MM, Zaitlin D, Naidu RA, Lommel SA (2008) Nicotiana benthamiana: its history and future as a model for plant–pathogen interactions. Mol Plant-Microbe Interact 21:1015–1026. https://doi.org/10.1094/MPMI-21-8-1015

Gray DJ, Li ZT, Dhekney SA (2014) Precision breeding of grapevine (Vitis vinifera L.) for improved traits. Plant Sci 228:3–10. https://doi.org/10.1016/J.PLANTSCI.2014.03.023

Haberman A, Bakhshian O, Cerezo-Medina S, Paltiel J, Adler C, Ben-Ari G, Mercado JA, Pliego-Alfaro F, Lavee S, Samach A (2017) A possible role for flowering locusT-encoding genes in interpreting environmental and internal cues affecting olive (Olea europaea L.) flower induction. Plant. Cell Environ 40:1263–1280. https://doi.org/10.1111/pce.12922

Hartmann HT (1953) Effect of winter chilling on fruitfulness and vegetative growth in the olive. Am Soc Hort Sci 62:184–190

He Y, Wang R, Dai X, Zhao Y (2017) On improving CRISPR for editing plant genes: ribozyme-mediated guide RNA production and fluorescence-based technology for isolating transgene-free mutants generated by CRISPR. Prog Mol Biol Transl Sci 149:151–166. https://doi.org/10.1016/bs.pmbts.2017.03.012

Hébert D, Kikkert J, Smith F, Reisch B (1993) Optimization of biolistic transformation of embryogenic grape cell suspensions. Plant Cell Rep 12:585–589. https://doi.org/10.1007/BF00233066

Hedrick U (1908) The grapes of New York. J.B. Lyon Co. State Printers, Albany

Hernández ML, Sicardo MD, Arjona PM, Padilla MN, Garcia-Vico L, Velazquez-Palmero D-V, Perez A, Sanz C, Martínez-Rivas JM (2018) Molecular and biochemical characterization of genes and enzymes involved in virgin olive oil quality. OliveBioteq18, Seville-Spain

Hofstetter V, Buyck B, Croll D et al (2012) What if esca disease of grapevine were not a fungal disease? Fungal Divers 54:51–67. https://doi.org/10.1007/s13225-012-0171-z

Höll J, Vannozzi A, Czemmel S et al (2013) The R2R3-MYB transcription factors MYB14 and MYB15 regulate stilbene biosynthesis in Vitis vinifera. Plant Cell 25:4135–4149. https://doi.org/10.1105/tpc.113.117127

Holme IB, Wendt T, Holm PB (2013) Intragenesis and cisgenesis as alternatives to transgenic crop development. Plant Biotechnol J 11:395–407. https://doi.org/10.1111/pbi.12055

Hu ZB, Du M (2006) Hairy root and its application in plant genetic engineering. J Integr Plant Biol 48:121

Humara J, Martin MS, Parra F, Ordas RJ (1999) Improved efficiency of uidA gene transfer in stone pine (Pinus pinea) cotyledonsusing a modified binary vector. Can J For Res 29:1627–1632

Iaria D, Chiappetta A, Muzzalupo I (2016) De Novo Transcriptome Sequencing of Olea europaea L. to Identify Genes Involved in the Development of the Pollen Tube. Scientific World J 2016:Article ID 4305252:1–7

Jafarzadeh-Bajestani M, Khodai-Kalaki M, Motamed N, Noorayin O (2011) Genetic transformation of olive somatic embryos through agrobacterium tumefaciens and regeneration of transgenic plants. Afr J Biotechnol 10(28):5468–5475

Jaillon O, Aury J-M, Noel B et al (2007) The grapevine genome sequence suggests ancestral hexaploidization in major angiosperm phyla. Nature 449:463–467. https://doi.org/10.1038/nature06148

Jaillon O, Aury J-M, Noel B, Policriti A, Clepet C, Casagrande A, Choisne N, Aubourg S, Vitulo N, Jubin C, Vezzi A, Legeai F, Hugueney P, Dasilva C, Horner D, Mica E, Jublot D, Poulain J, Bruyere C, Billault A, Segurens B, Gouyvenoux M, Ugarte E, Cattonaro F, Anthouard V, Vico V, Del Fabbro C, Alaux M, Di Gaspero G, Dumas V, Felice N, Paillard S, Juman I, Moroldo M,

Scalabrin S, Canaguier A, Le Clainche I, Malacrida G, Durand E, Pesole G, Laucou V, Chatelet P, Merdinoglu D, Delledonne M, Pezzotti M, Lecharny A, Scarpelli C, Artiguenave F, Pé E, Valle G, Morgante M, Caboche M, Adam-Blondon A-F, Weissenbach J, Quétier F, Wincker P (2017) The grapevine genome sequence suggests ancestral hexaploidization in major angiosperm phyla. Nature 449:463–468

Jiang W, Bikard D, Cox D et al (2013) RNA-guided editing of bacterial genomes using CRISPR-Cas systems. Nat Biotechnol 31:233–239. https://doi.org/10.1038/nbt.2508

Jiménez J, López-Vela D, Ruiz-Galea M et al (2013) Embryogenic suspensions of adult cork oak: the first step towards mass propagation. Trees Struct Funct 27:13–23. https://doi.org/10.1007/s00468-012-0763-y

Jinek M, Chylinski K, Fonfara I et al (2012) A programmable dual-RNA-guided DNA endonuclease in adaptive bacterial immunity. Science 337:816–821. https://doi.org/10.1126/science.1225829

Jones HD (2015) Regulatory uncertainty over genome editing. Nat Plants 1:14011. https://doi.org/10.1038/nplants.2014.11

Kapilan R, Vaziri M, Zwiazek JJ (2018) Regulation of aquaporins in plants under stress. Biol Res:1–11. https://doi.org/10.1186/s40659-018-0152-0

Karaagac E, Vargas AM, de Andrés MT et al (2012) Marker assisted selection for seedlessness in table grape breeding. Tree Genet Genomes 8:1003–1015. https://doi.org/10.1007/s11295-012-0480-0

Keller M (2010) The science of grapevines: anatomy and physiology, 1st edn. Elsevier Academic Press, Burlington

Kishor PBK, Sangam S, Amrutha RN, Laximi P, Naidu K, Rao KR, Rao S, Reddy KJ, Theriappon P, Sreenivasula N (2005) Regulation of proline biosynthesis, degredation, uptake and transport in higher plants: its implication in plant growth and abiotic stress tolerance. Curr Sci 88:427–438

Kiss L, Holb IJ, Rossi V et al (2016) Foreword: special issue on fungal grapevine diseases. Eur J Plant Pathol 144:693–694. https://doi.org/10.1007/s10658-015-0844-z

Koornneef M, Meinke D (2010) The development of Arabidopsis as a model plant. Plant J 61:909–921. https://doi.org/10.1111/j.1365-313X.2009.04086.x

Korte A, Farlow A (2013) The advantages and limitations of trait analysis with GWAS: a review. Plant Methods 9:29. https://doi.org/10.1186/1746-4811-9-29

Kumar S, Sachdeva S, Bhat KV, Vats S (2018) Plant responses to drought stress: physiological, biochemical and molecular basis In: Biotic and abiotic stress tolerance in plants, pp 1–25

Laimer M (2007) Transgenic Grapevines. Transgenic Plant Journal 1(1):219–227

Lambardi M, Caccavale A, Rugini E, Caricato G (1999) Histological observations on somatic embryos of olive (Olea europaea L.). Acta Hortic 474:67–70

Leva A, Muleo R, Petruccelli R (1995) Long-term somatic embryogenesis from immature olive cotyledons. J Hortic Sci 70:417–421. https://doi.org/10.1080/14620316.1995.11515311

Leyva-Pérez MO, Jiménez-Ruiz J, Cabanás C G-L, Valverde-Corredor A, Barroso JB, Luque F, Mercado-Blanco J (2018) Tolerance of olive (Olea europaea) cv Frantoio to Verticillium dahliae relies on both basal and pathogen-induced differential transcriptomic responses. New Phytologist n.d.217: 671–686 doi: https://doi.org/10.1111/nph.14833

Li ZT, Dhekney SA, Dutt M, Gray DJ (2008) An improved protocol for Agrobacterium-mediated transformation of grapevine (Vitis vinifera L.). Plant Cell Tissue Organ Cult 93:311–321. https://doi.org/10.1007/s11240-008-9378-9

Lijavetzky D, Cabezas J, Ibáñez A et al (2007) High throughput SNP discovery and genotyping in grapevine (Vitis vinifera L.) by combining a re-sequencing approach and SNPlex technology. BMC Genomics 8:424. https://doi.org/10.1186/1471-2164-8-424

Locco P, Franks T, Thomas MR (2001) Genetic Transformation of Major Wine Grape Cultivars of Vitis Vinifera L. Transgenic Res 10(2):105–112

López-Pérez A-J, Velasco L, Pazos-Navarro M, Dabauza M (2008) Development of highly efficient genetic transformation protocols for table grape Sugraone and Crimson Seedless at low Agrobacterium density. Plant Cell Tissue Organ Cult 94:189–199. https://doi.org/10.1007/s11240-008-9404-y

Loreto F, Bagnoli F, Fineschi S (2009) One species, many terpenes: matching chemical and biological diversity. Trends Plant Sci 14:416–420. https://doi.org/10.1016/j.tplants.2009.06.003
Loreto F, Pollastri S, Fineschi S, Velikova V (2013) environmental constraints in the Mediterranean area. Environ Exp Bot. https://doi.org/10.1016/j.envexpbot.2013.09.005
Magalhães AP, Verde N, Reis F et al (2016) RNA-Seq and gene network analysis uncover activation of an ABA-dependent Signalosome during the Cork Oak root response to drought. Front Plant Sci 6:1–17. https://doi.org/10.3389/fpls.2015.01195
Maliogka VI, Martelli GP, Fuchs M (2015) Control of viruses infecting grapevine. Adv Virus Res 91:175–227. https://doi.org/10.1016/BS.AIVIR.2014.11.002
Malnoy M, Viola R, Jung M-H et al (2016) DNA-free genetically edited grapevine and apple protoplast using CRISPR/Cas9 ribonucleoproteins. Front Plant Sci 7:1–9. https://doi.org/10.3389/fpls.2016.01904
Mauro MC, Toutain S, Walter B et al (1995) High efficiency regeneration of grapevine plants transformed with the GFLV coat protein gene. Plant Sci 112:97–106. https://doi.org/10.1016/0168-9452(95)04246-Q
Mazri MA, Belkoura I, Pliego-Alfaro F, Belkoura M (2013) Somatic embryogenesis from leaf and petiole explants of the Moroccan olive cultivar Dahbia. Sci Hortic (Amsterdam) 159:88–95. https://doi.org/10.1016/J.SCIENTA.2013.05.002
McCarthy MI, Abecasis GR, Cardon LR et al (2008) Genome-wide association studies for complex traits: consensus, uncertainty and challenges. Nat Rev Genet 9:356–369. https://doi.org/10.1038/nrg2344
McGovern PE (2003) Ancient wine: the search for the origins of viniculture. Princeton University Press, Princeton
McGovern PE, Glusker DL, Exner LJ, Voigt MM (1996) Neolithic resinated wine. Nature 381:480–481. https://doi.org/10.1038/381480a0
Mencuccini M (1994) Agrobacterium-mediated gene transfer studies for genetic transformation of olive (Olea europaea L.) tissues. In: Proceedings of the VIIIth international congress of plant tissue and cell culture. Florence 12–I 7 June, p 67
Mencuccini M, Micheli M, Angiolillo A, Baldoni L (1997) Trasferimento genico in olivo mediante Agrobacterium tumefaciens. In: Proceedings of the XLI SIGA, p 123
Mencuccini M, Micheli M, Angiolillo A, Baldon L, Frugis G, Mariotti D (1999a) Agrobacterium-mediated DNA transfer in olive callus (Olea europaea L.). Adv Hort Sci 13:25–28
Mencuccini M, Micheli M, Angiolillo A, Baldoni L (1999b) Genetic transformation of olive (Olea europaea L.) using Agrobacterium tumefaciens. Acta Hort. 474:515–519
Millardet A (1885) Historie des principales varieté et espéces de la vigne. Mason G, Paris
Mitrakos K, Alexaki A, Papadimitriou P (1992) Dependence of Olive morphogenesis on callus origin and age. J Plant Physiol 139:269–273. https://doi.org/10.1016/S0176-1617(11)80335-4
Mohanta T, Bashir T, Hashem A et al (2017) Genome editing tools in plants. Genes (Basel) 8:399. https://doi.org/10.3390/genes8120399
Moreira AC, Medeira C, Maia I, Quartin V, Matos MC, Cravador A (2006) Studies on the association of the Quercus suber decline disease with Phytophthora cinnamomi in Portugal. Bol Inf CIDEU 1:31–38
Mougiou N, Trikka F, Trantas E, Ververidis F, Makris A, Argiriou A, Vlachonasiosa KE (2018) Expression of hydroxytyrosol and oleuropein biosynthetic genes are correlated with metabolite accumulation during fruit development in olive, Olea europaea, cv. Koroneiki. Plant Physiol Biochem 128:41–49
Muleo R, Morgante M, Velasco R et al (2012) Olive tree genomics. In: Muzzalupo I (ed) Olive germplasm the olive cultivation table olive and olive oil industry in Italy. InTech, Rijeka, pp 133–148
Mullins MG, Tang FCA, Facciotti D (1990) Agrobacterium-mediated genetic transformation of grapevines: transgenic plants of Vitis rupestris SCHEELE and Buds of Vitis vinifera L. Nat Biotechnol 8:1041–1045. https://doi.org/10.1038/nbt1190-1041
Murashige T, Skoog F (1962) A revised medium for rapid growth and bio assays with tobacco tissue cultures. Physiol Plant 15:473–497. https://doi.org/10.1111/j.1399-3054.1962.tb08052.x

Myles S, Chia J-M, Hurwitz B et al (2010) Rapid genomic characterization of the genus Vitis. PLoS One 5:e8219. https://doi.org/10.1371/journal.pone.0008219
Nagamangala Kanchiswamy C, Sargent DJ, Velasco R et al (2015) Looking forward to genetically edited fruit crops. Trends Biotechnol 33:62–64. https://doi.org/10.1016/j.tibtech.2014.07.003
Nakano M, Hoshino Y, Mii M (1994) Regeneration of transgenic plants of grapevine (Vitis vinifera L.) via Agrobacterium rhizogenesmediated transformation of embryogenic calli. J Exp Bot 45:649–656. https://doi.org/10.1093/jxb/45.5.649
Narvaez I, Khayreddine T, Pliego C, Cerezo S, Jiménez-Díaz RM, Trapero-Casas JL, López-Herrera C, Arjona-Girona I, Martín C, Mercado JA, Pliego-Alfaro F (2018a) Usage of the Heterologous Expression of the Antimicrobial Gene afp From Aspergillus giganteus for Increasing Fungal Resistance in Olive. Front Plant Sci 9:680. https://doi.org/10.3389/fpls.2018.00680
Narvaez I, Pliego C, Fresta L, Jiménez-Díaz RM, Trapero-Casas JL, López-Herrera C, Arjona-López JM, Mercado JA and Pliego-Alfaro F (2018b) Heterologous expression of AtNPR1 gene in olive for increasing fungal tolerance. OliveBioteq18. Seville-Spain
Nirala NK, Das DK, Srivastava PS, et al (2010) Expression of a rice chitinase gene enhances anti-fungal potential in transgenic grapevine (Vitis vinifera L.). Vitis J Grapevine Res 49:181–187
Nishimasu H, Ran FA, Hsu PD et al (2014) Crystal structure of Cas9 in complex with guide RNA and target DNA. Cell 156:935–949. https://doi.org/10.1016/j.cell.2014.02.001
Oliveira V, Lauw A, Pereira H (2016) Sensitivity of cork growth to drought events: insights from a 24-year chronology. Clim Change 137:261–274. https://doi.org/10.1007/s10584-016-1680-7
Orinos T, Mitrakos K (1991) Rhizogenesis and somatic embryogenesis in calli from wild olive (Olea europaea var. sylvestris (Miller) Lehr) mature zygotic embryos. Plant Cell Tissue Organ Cult 27:183–187. https://doi.org/10.1007/BF00041288
Osakabe Y, Osakabe K (2017) Genome editing to improve abiotic stress responses in plants. Prog Mol Biol Transl Sci 149:99–109. https://doi.org/10.1016/bs.pmbts.2017.03.007
Osakabe Y, Watanabe T, Sugano SS et al (2016) Optimization of CRISPR/Cas9 genome editing to modify abiotic stress responses in plants. Sci Rep 6:1–10. https://doi.org/10.1038/srep26685
Palliotti A, Bongi G (1996) Freezing injury in the olive leaf and effects of mefluidide treatment. J Hort Sci Biotech 71(1):57–63
Palomo-Ríos E, Cerezo S, Mercado J, Pliego-Alfaro F (2017) Generation and selection of transgenic olive plants. Bio-Protocol 7. https://doi.org/10.21769/BioProtoc.2611
Parsons J, MacKay J (2018) The theory and applications of CRISPR in plant and tree improvement. CAB Rev Perspect Agric Vet Sci Nutr Nat Resour 13. https://doi.org/10.1079/PAVSNNR201813007
Pereira H (2015) A rationale behind cork properties: a review of structure and chemistry. bioresources.com 10:1–23
Pereira-Leal JB, Abreu IA, Alabaça CS et al (2014) A comprehensive assessment of the transcriptome of Cork oak (Quercus suber) through EST sequencing. BMC Genom 15:1–14. https://doi.org/10.1186/1471-2164-15-371
Pérez M, Bueno MA, Escalona M et al (2013) Temporary immersion systems (RITA®) for the improvement of cork oak somatic embryogenic culture proliferation and somatic embryo production. Trees Struct Funct 27:1277–1284. https://doi.org/10.1007/s00468-013-0876-y
Pérez M, Cañal MJ, Toorop PE (2015) Expression analysis of epigenetic and abscisic acid-related genes during maturation of Quercus suber somatic embryos. Plant Cell Tissue Organ Cult 121:353–366. https://doi.org/10.1007/s11240-014-0706-y
Pérez-Barranco G, Mercado JA, Pliego-Alfaro F, Sánchez-Romero C (2007) Genetic transformation of olive somatic embryos through biolistic. Acta Hortic 738:473–477
Pérez-Barranco G, Torreblanca R, Padilla IMG et al (2009) Studies on genetic transformation of olive (Olea europaea L.) somatic embryos: I. Evaluation of different aminoglycoside antibiotics for nptII selection; II. Transient transformation via particle bombardment. Plant Cell. Tissue Organ Cult 97:243–251. https://doi.org/10.1007/s11240-009-9520-3
Perl A, Eshdat Y (1998) DNA Transfer and Gene Expression in Transgenic Grapes. Biotechnol Genet Eng Rev 15:365–386. https://doi.org/10.1080/02648725.1998.10647962

Pintos B, Bueno MA, Cuenca B, Manzanera JA (2008) Synthetic seed production from encapsulated somatic embryos of cork oak (Quercus suber L.) and automated growth monitoring. Plant Cell Tissue Organ Cult 95:217–225. https://doi.org/10.1007/s11240-008-9435-4

Pintos B, Manzanera JA, Bueno MA, et al (2009) Acclimation and establishment of cork oak (Quercus suber) somatic embryo-derived plantlets and post-acclimation cork quality test. Acta Hortic 812:431–436. https://doi.org/10.17660/ActaHortic.2009.812.61

Pires R, Ribeiro A, Cardoso H, Sobral V, Peixe A (2018). Embriogénese somática e organogénese na variedade de oliveira 'Galega vulgar'. Abstract bokk of the VIII Simpósio Nacional de Olivicultura. Santarém – Portugal, 7–9 junho 2018

Pliego-Alfaro F, Pérez-Barrance G, Sánchez-Romero C, Mercato JA (2005) Genetic transformation of olive somatic embryos through biolistic. Abstracts "International symposium on biotechnology of temperate fruit crops and tropical species, Florida, USA, p 122

Polito VS, McGranahan G, Pinney K, Leslie C (1989) Origin of somatic embryos from repetitively embryogenic cultures of walnut (Juglans regia L.): Implications forAgrobacterium-mediated transformation. Plant Cell Rep 8:219–221. https://doi.org/10.1007/BF00778537

Pritsa TS, Voyiatzis DG (2004) The in vitro morphogenetic capacity of olive embryo explants at different developmental stages, as affected by L-Glutamine, L-Arginine and 2,4-D. J Biol Res 1:55–61

Rallo L, Barranco D, De la Rosa R et al (2008) 'Chiquitita' olive. HortSci 43(2):529–531

Rallo L, Barranco D, Díez CM et al (2018) Strategies for Olive (Olea europaea L.) Breeding: Cultivated Genetic Resources and Crossbreeding. In: Advances in plant breeding strategies: fruits. Springer, Cham, pp 535–600

Ramos AM, Usié A, Barbosa P et al (2018) The draft genome sequence of cork oak. Sci Data 5:1–12. https://doi.org/10.1038/sdata.2018.69

Rasheed-Depardieu C, Parent C, Crèvecoueur M et al (2012) Identification and expression of nine oak aquaporin genes in the primary root axis of two oak species, Quercus petraea and Quercus robur. PLoS One 7. https://doi.org/10.1371/journal.pone.0051838

Ren C, Liu X, Zhang Z et al (2016) CRISPR/Cas9-mediated efficient targeted mutagenesis in Chardonnay (Vitis vinifera L.). Sci Rep 6:32289. https://doi.org/10.1038/srep32289

Ridé M, Ridé S, Petit A et al (2000) Characterization of plasmid-borne and chromosome-encoded traits of Agrobacterium biovar 1, 2, and 3 strains from France. Appl Environ Microbiol 66:1818–1825

Robinson J, Harding J, Vouillamoz J (2012) Wine grapes: a complete guide to 1,368 Vine cultivars, including their origins and flavours. Allen Lane, New York. ISBN 978-0062206367.

Rodríguez-Sanz H, Manzanera JA, Solís MT et al (2014) Early markers are present in both embryogenesis pathways from microspores and immature zygotic embryos in cork oak, Quercus suber L. BMC Plant Biol 14:1–18. https://doi.org/10.1186/s12870-014-0224-4

Roka L, Koudounasa K, Darasa G, Zoidakisb J, Vlahoub A, Kalaitzisc P, Hatzopoulosa P (2018) Proteome of olive non-glandular trichomes reveals protective protein network against (a)biotic challenge. J Plant Physiol 231:210–218

Rugini E (1984) In vitro propagation of some olive (Olea europaea sativa L.) cultivars with different root-ability, and medium development using analytical data from developing shoots and embryos. Sci Hortic (Amsterdam) 24:123–134. https://doi.org/10.1016/0304-4238(84)90143-2

Rugini, E (1986) Olive. In: Bajaj YPS (ed) Biotechnology in agriculture and forestry, vol 10. Springer, Berlin/Heidelberg/New York, pp 253–267

Rugini E (1988) Somatic embryogenesis and plant regeneration in olive (Olea europaea L.). Plant Cell Tissue Organ Cult 14:207–214. https://doi.org/10.1007/BF00043411

Rugini E (1992) Involvement of polyamines in auxin and Agrobacterium rhizogenes-induced rooting of fruit trees in vitro. J Am Soc Hortic Sci 117:532–536

Rugini, E. (1995). Somatic embryogenesis in olive (Olea europaea L.). In: Jain SM, Gupta PK, Newton RJ (eds) Somatic embryogenesis in woody plants, vol II. Kluwer Academic Publishers, Dordrecht, pp 171–189

Rugini E, Caricato G (1995) Somatic embryogenesis and plant recovery from mature tissues of olive cultivars (Olea europaea L.) ‘canino’ and ‘moraiolo’. Plant Cell Rep 14:257–260. https://doi.org/10.1007/BF00233645
Rugini E., Fedeli E (1990) In: Bajaj YPS (ed) Legumes and oilseed crops I. Biotechnology in agriculture and forestry, vol 10. Springer, Berlin/Heidelberg/New York, pp 593–641
Rugini E, Gutiérrez-Pesce P (2006) Genetic improvement of olive. Pomologia Croatica 12:43–74
Rugini E, Silvestri C (2016) Somatic Embryogenesis in Olive (Olea europaea L. subsp. europaea var. sativa and var. sylvestris). Humana Press, New York, pp 341–349
Rugini E, Tarini P (1986) Somatic embryogenesis in olive (Olea europaea L.). In: Moet Hennessy (ed) Proceedings conference fruit tree biotechnology, p.62, Paris-France
Rugini E, Pellegrineschi A, Mencuccini M, Mariotti D (1991) Increase of rooting ability in the woody species kiwi (Actinidia deliciosa A. Chev.) by transformation with Agrobacterium rhizogenes rol genes. Plant Cell Rep 10:291–295
Rugini E, Pezza A, Muganu M et al (1995) Somatic embryogenesis in olive (Olea europaea L.). In: Bajaj YPS (ed) Biotechnology in Agriculture and Forestry, vol 30. Somatic embryogenesis and synthetic seed I. Springer, Berlin, pp 404–414
Rugini E, Gutiérrez-Pesce P, Spampinato PL, Ciarmiello A, D’ambrosio C (1999) New perspective for Biotechnologies in olive breeding: morphogenesis, in vitro selection & gene transformation. Acta Hortic 474:107–110
Rugini E, Biasi R, Muleo R (2000) Olive (Olea europaea var. sativa) Transformation. In: Jain SM, Minocha SC (eds) Molecular biology of woody plants, vol 2. Kluwer Academic Publishers, Dordrecht, pp 245–279. Rugini and Pesce 2006
Rugini E, Mencuccini M, Biasi R, Altamura MM (2005) Olive (Olea europea L.). In Jain SM, Gupta PK (eds) Protocol for Somatic Embryogenesis in Woody Plants. Springer, Dordrecht, pp 345–360
Santos Macedo E, Cardoso HG, Hernández A, Peixe AA, Polidoros A, Ferreira A, Cordeiro A, Arnholdt-Schmitt B (2009) Physiological responses and gene diversity indicate olive alternative oxidase (AOX) as a potential source for markers of efficient adventitious root induction. Physiologia Plantarum 137(4):532–552
Saporta R, San Pedro-Galan T, Gisbert Domenech MC (2016) Attempts at grapevine (Vitis vinifera L.) breeding through genetic transformation: The main limiting factors. VITIS 55:173–186. doi: https://doi.org/10.5073/VITIS.2016.55.173-186
Scanu B, Linaldeddu BT, Franceschini A et al (2013) Occurrence of Phytophthora cinnamomi in cork oak forests in Italy. For Pathol 43:340–343. https://doi.org/10.1111/efp.12039
Schenk RU, Hildebrandt AC (1972) Medium and techniques for induction and growth of monocotyledonous and dicotyledonous plant cell cultures. Can J Bot 50:199–204. https://doi.org/10.1139/b72-026
Scorza R, Cordts JM, Gray DJ, et al (1996) Producing Transgenic “Thompson Seedless” Grape (Vitis vinifera L.) Plants. J Amer Soc Hort Sci 121(4):616–619
Sebastini L, Busconi M (2017) Recent developments in olive (Olea europaea L.) genetics and genomics: applications in taxonomy, varietal identification, traceability and breeding. Plant Cell Rep. https://doi.org/10.1007/s00299-017-2145-9
Semiz G, Blande JD, Heijari J, Isik K, Niinemets U, Holopainen JK (2012) Manipulation of VOC emissions with methyl jasmonate and carrageenan in the evergreen conifer Pinus sylvestris and evergreen broadleaf Quercus ilex. Plant Biol (Stuttg) 14:57–65. https://doi.org/10.1111/j.1438-8677.2011.00485.x
Serrano MS, De Vita P, Fernández-Rebollo P et al (2012) Phytophthora cinnamomi and Pythium spiculum as main agents of Quercus decline in southern Spain and Portugal. IOBC/WPRS Bull 76:97–100
Sezer F, Taskin KM (2017) Molecular characterization of tocopherol biosynthesis genes from Olea europaea (L.) cv. Ayvalık. Turkish J Bot 41:1–10. https://doi.org/10.3906/bot-1701-41
Shan Q, Wang Y, Li J, Gao C (2014) Genome editing in rice and wheat using the CRISPR/Cas system. Nat Protoc 9:2395–2410. https://doi.org/10.1038/nprot.2014.157

Shibli RA, Shatnawi M, Abu-Ein A-JK (2001) Somatic embryogenesis and plant recovery from callus of 'Nabali' Olive (Olea europea L.). Sci Hortic (Amsterdam) 88:243–256. https://doi.org/10.1016/S0304-4238(00)00241-7

Singh NK, Bracker CA, Hasegawa PM, Handa AK, Buckel S, Hermodson MA, Pfankoch ED, Regnier FE, Bressan RA (1987) Characterization of osmotin. Plant Physiol 85:529–536

Sparkes IA, Runions J, Kearns A, Hawes C (2006) Rapid, transient expression of fluorescent fusion proteins in tobacco plants and generation of stably transformed plants. Nature Protoc 1:2019–2025

Spathelf P, Van Der Maaten E, Van Der Maaten-Theunissen M et al (2014) Climate change impacts in European forests: the expert views of local observers. Ann For Sci 71:131–137. https://doi.org/10.1007/s13595-013-0280-1

Sprink T, Eriksson D, Schiemann J, Hartung F (2016) Regulatory hurdles for genome editing: process- vs. product-based approaches in different regulatory contexts. Plant Cell Rep 35:1493–1506. https://doi.org/10.1007/s00299-016-1990-2

Sternberg SH, Redding S, Jinek M et al (2014) DNA interrogation by the CRISPR RNA-guided endonuclease Cas9. Nature 507:62–67. https://doi.org/10.1038/nature13011

Strobel GA, Nachmias A, Hess WM (1988) Improvements in the growth and yield of olive trees by transformation with the Ri plasmid of *Agrobacterium rhizogenes*. Can J Bot 66:2581–2585

Teixeira RT, Fortes AM, Bai H et al (2018) Transcriptional profiling of cork oak phellogenic cells isolated by laser microdissection. Planta 247:317–338. https://doi.org/10.1007/s00425-017-2786-5

Terrier N, Torregrosa L, Ageorges A et al (2009) Ectopic expression of VvMybPA2 promotes proanthocyanidin biosynthesis in grapevine and suggests additional targets in the pathway. Plant Physiol 149:1028–1041. https://doi.org/10.1104/pp.108.131862

Titouh K, Vidoy-Mercado I, Pliego C, Barceló-Muñoz A, San Segundo B, Cerezo S, Mercado JA, Pliego-Alfaro F (2014) Agrobacterium-mediated transformation of olive (Olea europaea L.) with an antifungal protein from Aspergillus giganteus, The third international conference of the IUFRO unit 2.09.02: somatic embryogenesis and other vegetative propagation technologies. Woody Plant Production Integrating Genetic and Vegetative Propagation Technologies. Vitoria-Gasteiz, Spain

Töpfer R, Hausmann L, Harst M, Maul E, Zyprian E, Eibach R (2011) New horizons for grapevine breeding. Methods Temp Fruit Breed 5:79–100

Torreblanca R, Palomo-Ríos E, Cerezo S, Mercado JA, Pliego-Alfaro F (2009) Agrobacterium-mediated transformation of olive (Olea europaea L.) Embryogenic cultures. Acta Hort 839:387–392

Torreblanca R, Cerezo S, Palomo-Rìos E, Marcado JA, Pliego-Alfaro F (2010) Development of a high throughput system for genetic transformation of olive (Olea europaea L.) plants. Plant Cell Tissue Organ Cult 103:61–69

Trabelsi EB, Bouzid S, Bouzid M et al (2003) In-vitro regeneration of olive tree by somatic embryogenesis. J Plant Biol 46:173–180. https://doi.org/10.1007/BF03030446

Unver T, Wu Z, Sterck L et al (2017) Genome of wild olive and the evolution of oil biosynthesis. Proc Nat Acad Sci USA 114:E9413–E9422. https://doi.org/10.1073/pnas.1708621114

Vain P, Keen N, Murillo J, Rathus C, Nemes C, Finer JJ (1993) Development of the Particle Inflow Gun. Plant Cell Tiss Org Cult 33:237–246

van Nocker S, Gardiner SE (2014) Breeding better cultivars, faster: applications of new technologies for the rapid deployment of superior horticultural tree crops. Hortic Res 1:14022. https://doi.org/10.1038/hortres.2014.22

Vancanneyt G, Schmidt R, O'Connor-Sanchez A, Willmitzer L, Rocha-Sosa M (1990) Construction of an intron-containing marker gene: splicing of the intron in transgenic plants and itsuse in monitoring early events in Agrobacterium-mediated plant transformation. Mol Gen Genet 220:245–250

Velada I, Grzebelus D, Lousa D, Soares CM, Santos Macedo E, Peixe A, Arnholdt-Schmitt B, Cardoso HG (2018) AOX1-subfamily gene members in Olea europaea cv. "Galega vulgar" –

gene characterization and expression of transcripts during IBA-induced in vitro adventitious rooting. Intl J Mol Sci 19(2):597
Velasco R, Zharkikh A, Troggio M et al (2007) A high quality draft consensus sequence of the genome of a heterozygous grapevine variety. PLoS One 2:e1326. https://doi.org/10.1371/journal.pone.0001326
Vidal JR, Kikkert JR, Wallace PG, Reisch BI (2003) High-efficiency biolistic co-transformation and regeneration of "Chardonnay" (Vitis vinifera L.) containing npt-II and antimicrobial peptide genes. Plant Cell Rep 22:252–260. https://doi.org/10.1007/s00299-003-0682-x
Vidal JR, Kikkert JR, Donzelli BD et al (2006) Biolistic transformation of grapevine using minimal gene cassette technology. Plant Cell Rep 25:807–814. https://doi.org/10.1007/s00299-006-0132-7
VIDAL JR, Gomez C, Cutanda MC et al (2010) Use of gene transfer technology for functional studies in grapevine. Aust J Grape Wine Res 16:138–151. https://doi.org/10.1111/j.1755-0238.2009.00086.x
Vitulo N, Forcato C, Carpinelli E et al (2014) A deep survey of alternative splicing in grape reveals changes in the splicing machinery related to tissue, stress condition and genotype. BMC Plant Biol 14:99. https://doi.org/10.1186/1471-2229-14-99
Vivier MA, Pretorius IS (2000) Genetic improvement of grapevine: tailoring grape varieties for the third millennium. S Afr J Enol Vitic 21:5–26
Vivier MA, Pretorius IS (2002) Genetically tailored grapevines for the wine industry. Trends Biotechnol 20:472–478. https://doi.org/10.1016/S0167-7799(02)02058-9
Wang Y, Liu X, Ren C et al (2016) Identification of genomic sites for CRISPR/Cas9-based genome editing in the Vitis vinifera genome. BMC Plant Biol 16:96. https://doi.org/10.1186/s12870-016-0787-3
Wang X, Tu M, Wang D et al (2018) CRISPR/Cas9-mediated efficient targeted mutagenesis in grape in the first generation. Plant Biotechnol J 16:844–855. https://doi.org/10.1111/pbi.12832
White FF, Tavlor BH, Huffman GA, Gordon MP, Nester EW (1985) Molecular and genetic analysis of the transfered DNA regions of the root-inducing plasmid of agrobacterium rhizogenes. J Bacteriol 164:33–44
Williams JGK, Hanafey MK, Rafalski JA, Tingey SV (1995) Genetic Analysis Using Random Amplified Polymorphic DNA Markers. Recomb DNA Methodol II:849–884. https://doi.org/10.1016/B978-0-12-765561-1.50061-8
Woo JW, Kim J, Kwon SI et al (2015) DNA-free genome editing in plants with preassembled CRISPR-Cas9 ribonucleoproteins. Nat Biotechnol 33:1162–1164. https://doi.org/10.1038/nbt.3389
Wu X, Scott DA, Kriz AJ et al (2014) Genome-wide binding of the CRISPR endonuclease Cas9 in mammalian cells. Nat Biotechnol 32:670–676. https://doi.org/10.1038/nbt.2889
Yamamoto T, Iketani H, Ieki H et al (2000) Transgenic grapevine plants expressing a rice chitinase with enhanced resistance to fungal pathogens. Plant Cell Rep 19:639–646. https://doi.org/10.1007/s002999900174
Zhu B, Chen TH, Li PH (1993) Expression of ABA-responsive osmotin-like protein genes during the induction of freezingtolerance in *Solanum commersonii*. Plant Mol Biol 21:729–735
Zhu B, Chen TH, Li PH (1995) Expression of three osmotin-like protein genes in response to osmotic stress and fungal infection in potato. Plant Mol Biol 28:17–26
Zhu B, Su J, Chang M, Verma DPS, Fan YL, Wu R (1998) Overexpression of a 1-pyrroline-5-carboxylate synthetase gene and analysis of tolerance to water and salt-stress in transgenic rice. Plant Sci 139:41–48
Zini E, Raffeiner M, Di Gaspero G et al (2015) Applying a defined set of molecular markers to improve selection of resistant grapevine accessions. Acta Hortic:73–78. https://doi.org/10.17660/ActaHortic.2015.1082.9

Part III

Applications in Production of Plant-Made Pharmaceuticals and Other Products

Key Challenges in Developing Products from Transgenic Plants

14

Gauri Nerkar, G. S. Suresha, Bakshi Ram, and C. Appunu

Abstract

The use of plants for the production of plant-made pharmaceuticals (PMPs), plant-made vaccines (PMVs) and plant-made substances of industrial interests through transgenic approach is the sole objective of plant molecular (PM) farming, and it holds great promises in the plant industry. PM farming is advantageous in terms of saving production costs, large-scale production of drugs, animal and human virus-free products, easier storage and transportation of drugs and oral applicability (e.g. edible vaccines). Plant system also offers different platforms, viz. seed-, leaf-, stem-, root- and whole plant-based production in transgenic lines. The other systems such as bacterial, microbial eukaryotes, mammalian cells, insect cells and transgenic animals are expensive. Therefore, transgenic plants have been the subject of considerable attention as new generation bioreactors with respect to their advantages, such as the safety of recombinant proteins (antibodies, enzymes, vaccines, growth factors, etc.) and their potential for the large-scale and low-cost production. However, some of the key challenges that PM farming is facing include technical, economic, safety and regulatory as well as public acceptance. This chapter discusses in detail the challenges in obtaining products from transgenic plants such as expression of the recombinant proteins, downstream processing and purification, glycosylation, regulatory challenges and environmental risks.

Keywords

Plant-made pharmaceuticals · Plant molecular farming · Vaccines · Enzymes · Recombinant proteins · Glycosylation

---

G. Nerkar · G. S. Suresha · B. Ram · C. Appunu (✉)
Division of Crop Improvement, Sugarcane Breeding Institute, Coimbatore, India

R. Sathishkumar et al. (eds.), *Advances in Plant Transgenics: Methods and Applications*, https://doi.org/10.1007/978-981-13-9624-3_14

14.1 Introduction

The production of recombinant proteins through genetically modified (GM) plants for pharmaceutical or industrial applications is referred as plant molecular farming or pharming. During the past two and a half decades, transgenic plants have become potential targets for the cost-effective production of heterologous proteins including therapeutics, vaccines and industrial enzymes. The use of transgenic plants for the production of substances of industrial importance (i.e. plant molecular farming) holds several promises which include saving of the production costs, possibility to realize large-scale production of high-dose drugs, guaranteed animal and human virus-free products, easier storage and transportation of drugs and oral applicability (e.g. edible vaccines; Spök and Karner 2008). Plants are currently used to express antibodies (Hensel et al. 2015), vaccine antigens (Mbewana et al. 2015), bioactive peptides (Bundó et al. 2014), diagnostic proteins (Tinazzi et al. 2015), industrial enzymes (Yoon et al. 2016) and biodegradable plastics (Bohmert-Tatarev et al. 2011). Plant recombinant proteins have been expressed in different platforms such as leafy crops, cereal grains, fruits and vegetables (Nahampun 2015; Dai et al. 2000; Mason et al. 2002; Stoger et al. 2000).Transgenic plants can be used as bioreactors for the production of biopharmaceuticals, and the purified product can be used for oral or parental administration (Yao et al. 2015), or the plant tissue expressing the biopharmaceutical protein may be used for direct oral delivery (Chan and Daniell 2015). The initial research in the field of molecular pharming was focussed on increasing the yields of desirable metabolites in order to use plants as protein factories. However, from a commercial perspective several practical limitations of each platform had to be considered in terms of scale-up and the development of processes that would satisfy the regulatory bodies (Fischer et al. 2012, 2013, 2015). Thus, over the last decade, there has been a renewed focus on the regulatory aspects of plant-based production platforms for recombinant proteins, as researchers have started focussing on a smaller number of platforms based on intact plants, tissues and cell suspension cultures (Paul et al. 2013; Fischer et al. 2015). The prior experience with microbes and animal cells allowed a quick transition from experimental platforms to commercially feasible and optimized manufacturing process as a result of which plants have become increasingly competitive in the market for valuable metabolites and recombinant proteins. Market approval of "Elelyso" in 2012 (Protalix/Pfizer, recombinant glucocerebrosidase produced in carrot cells for treatment of a rare disease) and the recent apparent success in fighting Ebola virus with plant-made antibodies put a spotlight on the enormous potential of next generation plant-made medicines, especially in the name of the guiding principle of reduction of costs. These will help reduce disparities in health rights as well as tools to guarantee adequate health protection in developing countries (Folayan et al. 2016).

14.2 Key Challenges in Developing Products from Transgenic Plants

14.2.1 Selection of the Host for Recombinant Protein Expression

Recent advances in the field of microbial and animal cell platforms for biopharming have allowed us to understand the selection and optimization of production strains which can lead to manifold increase in the productivity of a given process (Gottschalk 2008). Comparison of various systems for recombinant protein production shows that transgenic plants are best suited for recombinant protein production (Table 14.1). The selection of host is a challenging step in plant molecular pharming as it differs according to the underlying goals of the genetic transformation. For instance, the host choice for metabolic engineering tends to be product specific and based on the intrinsic metabolic potential of different host species (Fischer et al. 2015). Thus, commercial platforms currently used for the production of valuable secondary metabolites are based on cultivated plants or, for a few selected compounds, cultivation-optimized wild-type cells, shikonin from *Lithospermum erythrorhizon*, berberine from *Coptis japonica*, sanguinarine from *Papaver somniferum* and paclitaxel from *Taxus* spp., due to their dependence on endogenous metabolism and its complexity (Rischer et al. 2013; Chandra and Chandra 2011; Fischer et al. 2015).

14.2.2 Plant Tissues Used for Expression of Recombinant Proteins

Recombinant proteins can be expressed in edible leaves, seeds, fruits and tubers, depending on the plant species. Many different edible plant species have been investigated, including alfalfa, carrot, lettuce, tomato, potato, maize, wheat, barley, strawberry, soybean, celery, cabbage, spinach, cauliflower, rice and banana (Merlin et al. 2017; Chan and Daniell 2015). Leafy crops (tobacco, alfalfa and lettuce) provide large biomass, fast turnaround production time and multiple harvests per year. However, leaf-based production system is suitable for products that need immediate purification if not stored frozen or dried because the leaf is not a stable environment due to the presence of proteolytic enzymes (Twyman et al. 2003). Furthermore, metabolites such as nicotine or alkaloids in tobacco and oxalic acid in alfalfa are toxic and should be removed from the final products (Fischer et al. 2004).

Vegetable, fruit and seed can be beneficial for the oral delivery of pharmaceuticals and vaccines, allowing direct incorporation in the diet therefore reducing

Table 14.1 Comparison of various systems for recombinant protein production shows that transgenic plants are best suited for recombinant protein production

Costs	Transgenic plants	Plant cell culture	Transgenic animals	Mammalian cultures	Bacteria	Yeast	Plant viruses
Costs	Low	Medium	High	High	Low	Medium	Low
Storage cost	Inexpensive	Moderate	Expensive	Expensive	Moderate	Moderate	Not specified
Productivity	High	Medium	High	Medium	Medium	Medium	Not specified
Timescale	Controversial	Medium	Very long	Medium	Short	Medium	Low
Risks	Low	Low	Viruses, prions	Viruses, prions	Toxins	Low	Not specified
Public perception of risks	High	Not specified	High	Medium	Low	Medium	High
Safety	High	Not specified	High	Medium	Low	Unknown	High
Contamination risks	Low	Low	Viruses, prions and oncogenic DNA	Viruses, prions and oncogenic DNA	Endotoxins	Low	Not specified
Therapeutic risks	Unknown	Not specified	Yes	Yes	Yes	Unknown	Unknown
Ethical concerns	Medium	Low	High	Medium	Low	Low	Not specified
Scale-up capacity	Very high	Medium	Low	Very low	High	High	Not specified
Scale-up costs	Low	High	Medium	High	High	High	Low
Gene size	Not limited	Not specified	Limited	Limited	Unknown	Unknown	Limited
Protein yield	High	Not specified	High	Medium-high	Medium	High	Very high
Propagation	Easy	Easy	Medium	Limited	Easy	Easy	Not specified
Protein folding accuracy	High	Not specified	High	High	Low	Medium	High
Glycosylation	Minor differences	Minor differences	Correct	Correct	None	Incorrect	Minor differences
Product quality	High	High	High	High	Low	Medium	Not specified

Product homogeneity	High	Not specified	High	Medium	Low	Medium	Medium
Distribution	Easy	Not specified	Difficult	Difficult	Feasible	Feasible	Easy
Data monitoring	Difficult	Easy	Difficult	Easy	Easy	Easy	Not specified
GMP conformity	Difficult	Possible	Possible	Possible	Possible	Possible	Not specified
Storage temperature	Room temperature	−20 °C	Liquid nitrogen	Liquid nitrogen	−20 °C	-20 °C	Not specified

Reference: Spök et al. 2008; Appunu et al. 2017

purification cost. Corn, soybean, rice, potato and banana are among popular systems for this purpose (Hudson et al. 2014; Lamphear et al. 2004; Streatfield et al. 2003; Mason et al. 2002; Ou et al. 2014; Tacket et al. 1998). However, fruits and vegetables have short half-life and would not be appropriate for highly valuable products. Moreover, the long development time for transgenic banana plants reduces the feasibility of this species for the oral delivery of biopharmaceuticals (Trivedi and Nath 2004; Chan et al. 2013). On the other hand, a seed-based production system supports a stable environment for long-term storage at ambient temperature without affecting bioactivity of the products (Kusnadi et al. 1997). This feature reflects the simplification of transportation and removes the necessity of cold-chain for vaccines or pharmaceuticals (Ma et al. 2005).

14.3 Expression Systems

14.3.1 Plant Cell Suspension Culture

Plant cell suspension culture method involves gene transfer to the protoplasts obtained by the removal of cell walls of the cells in suspension culture. The purification system and its downstream processing are cheaper and easier (Kim et al. 2008a), it can decrease heterogeneity in proteins and sugar (N-glycans) regarding the uniformity of the type and size of cells (Lienard et al. 2007), and there is no need for the production of transgenic plants; however, the cell lines can be produced after a few months (Aviezer et al. 2009; Shaaltiel et al. 2007). Although plant cell cultures offer a cheaper, safer and easier method for expression of recombinant proteins in comparison to the other methods of production, a major hurdle is due to the enhanced proteolytic activity during the stationary phase which reduces the level of recombinant proteins and lowers its potential for commercial application (Corrado and Karali 2009).

14.3.2 Nuclear Transformation

Stable nuclear transformation refers to the expression of transgenes after integration within the host genomes. Nuclear transformation has been widely used for production of recombinant proteins in transgenic plants. It has a considerable potential for producing crops like cereals that actually grow everywhere. However, a long production cycle for some crops and their potential collisions with natural species or food products have restricted the wide acceptance of this method (Commandeur et al. 2003; Alireza and Nader 2015). So far, cauliflower mosaic virus 35S RNA promoter and maize 1-ubiquitinpromoter, respectively, appear to be the preferred choice of promoters for dicotyledons and monocotyledons (Alireza and Nader 2015). Tissue-specific and organ-specific promoters are used for stimulating the expression of transgenes in some tissues or organs, such as tubers, seeds and fruits (He et al. 2008; Jaeger et al. 2002) which prevents the accumulation of recombinant proteins in vegetative organs and would have negative impact on plant growth (Alireza and Nader 2015).

14.3.3 Chloroplast Transformation

Expression of transgenes from the plastid (chloroplast) genome has recently received considerable attention (Maliga 2004; Daniell et al. 2009; Bock 2007, 2014). In addition to the high level of transgene containment conferred by transgene accommodation in the plastid genome (due to maternal plastid inheritance in most crops leading to exclusion of plastid transgenes from pollen transmission), the chloroplast's enormous capacity to accumulate foreign proteins to very high levels represents a particularly big attraction of the technology. A number of studies have reported examples of foreign proteins that could be produced in transgenic chloroplasts to levels exceeding 10% of the total soluble protein of the plant. However, by far not all attempts to express transgenes to high levels via transformation of the plastid genome have been successful, and it is not always clear why a given transgene works or does not work well in chloroplasts (Bock 2014).

The most important factors that contribute to high levels of protein expression in chloroplasts are the choice of expression elements such as the promoter, the 5′-UTR and the 3'-UTR, albeit to different magnitudes. The codon usage or the coding sequences immediately downstream of the translation initiation codon have been shown to significantly influence expression levels of plastid transgenes. The choice of integration site into the genome is important, e.g. transgene targeting to the inverted repeat region (IR) of the plastid genome doubles the gene dosage, and, at least in some cases, this may result in increased expression levels (Herz et al. 2005). The stability of plastid proteins depends on (1) the action of methionine aminopeptidase, the enzyme that cleaves off the initiator methionine and thereby exposes the penultimate amino acid residue, (2) an N-end rule-like pathway of protein degradation and (3) additional sequence motifs in the N-terminal region (Apel et al. 2010; Bock 2014). Several transplastomic studies have demonstrated that (soluble) recombinant proteins that accumulate in the stroma can be expressed to very high levels, which, in the most extreme cases, can even exceed the accumulation levels of Rubisco in the wild type (Oey et al. 2009) than thylakoid (Hennig et al. 2007) and lumen (Tissot et al. 2008). Other factors that limit the protein production in chloroplasts include post-translational modifications, potential deleterious effects of the recombinant protein on plastid gene expression, photosynthesis or other metabolic processes and genome instability as occasionally induced by unwanted homologous recombination between duplicated expression signals (Bock 2014).

14.3.4 Proteolysis of Endogenous Proteins

The endogenous plant proteolytic activity is a key challenge for optimized accumulation of recombinant proteins in a plant system, which may severely compromise recombinant protein stability (Pillay et al. 2014). Many candidate biopharmaceutical proteins such as plasminogen activators (Schiermeyer et al., 2005), cytokines (Sirko et al., 2011), human serum albumin (Sun et al. 2011) and monoclonal antibodies (Muynck et al. 2009) have been shown to undergo proteolytic processing to different degrees when they are produced in plant cells. Several strategies have been recently applied to improve protein stability (as reviewed by Mandal et al. 2016) by limiting protease action such as recombinant protein production in various subcellular compartments, application of protease inhibitors to limit protease action, addition of protein-stabilizing agents to the culture medium, co-expression of protease inhibitors together with the protein of interest and knockdown of the expression of protease-encoding genes. This was first reported in rice where expression of cysteine protease gene Rep-1 was supressed by RNAi to improve the production of hGM-CSF (Kim et al. 2008b). Recently, with the introduction of different gene targeting strategies based on sequence-specific nucleases, it is now possible to disrupt any protease gene to completely knockout its activity (Fichtner et al. 2014). It is therefore only a matter of time before this technology will be applied to generate plants in which specific protease genes will be disrupted.

14.3.5 Downstream Processing and Purification

The downstream processing and purification constitute up to 80% of the production costs (Gottschalk 2007). Some substances creating problems in downstream processing include lignin, fibres, swelling substances in general, waxes, phenolic compounds, pigments and endogenous proteases. Secondary metabolites are recovered from plant cells or tissues using methods such as adsorption, precipitation and chromatography, often requiring phase portioning and the use of mixtures of organic solvents (Fischer et al. 2015). Contaminating metabolites and host cell proteins cause filter and membrane fouling which represent a significant challenge especially if the product is retained within the cell (Wilson and Roberts 2012). One such example is paclitaxel where it accumulates predominantly in the cell wall and releases by the addition of cellulases (Wilson and Roberts 2012). Similarly, recombinant proteins are easier to purify from plant cells if they are secreted, but higher yields (and specific forms of desirable modification) are often achieved by retaining the protein within the cell, so there is a trade-off between yield and recovery (Twyman et al. 2013). Watery tissues such as tomatoes are easier to extract than dry tissues such as cereals thus reducing the downstream processing cost to a large extent (Schillberg et al. 2005; Yano et al. 2010). Sugarcane is one of the ideal candidates for molecular pharming applications due to well-developed and larger storage vacuole system, high biomass, rapid growth rate, minimal transgene dispersal due to vegetative method of propagation, high quantity of extractable juice with very low protein content (0.01–0.02%) and a well-established downstream processing technology (Nerkar et al. 2018; Appunu et al. 2017). Also, Palaniswamy et al. (2016) have reported a novel vacuolar targeting determinant which could localize recombinant proteins (r-proteins) to the sugarcane vacuoles in high concentrations and such targeted r-proteins can be purified from the sugarcane juice with a few simple steps reducing the costs of downstream processing to a large extent.

Recent innovations implemented to reduce the costs of downstream processing include the replacement of protein A with synthetic affinity ligands such as mercaptoethyl pyridine (MEP HyperCel™) which removes the major cost factor in antibody purification (Arakawa et al. 2014). Processing the initial plant extract using a variety of basic treatments and additives allows more particulates to be retained by inexpensive bag filters before expensive depth filters are challenged by the feed stream (Buyel et al. 2014). For instance, a heat precipitation step removes many soluble host cell proteins (Buyel et al. 2014), flocculants are used to aggregate smaller particulates (Buyel and Fischer 2014a, 2014b), and a cellulose-based filter aids to increase the retentive capacity of the filters (Buyel et al. 2015). Although protein labels facilitate the recovery of proteins, they must be removed after purification so that the structure of purified protein can change into its natural position. For example, the gene sequence of recombinant proteins is fused to the sequence of a special internal oil protein called oleosin in safflower and canola which can be separated after the digestion of internal protease following protein purification (Ahmad 2014).

14.3.6 Glycosylation

Glycosylation is a type of post-translational modification which refers to covalent binding of sugars to proteins in order to increase the close-packing, biological activity, solubility and biological functionality (Lienard et al. 2007) which takes place in plants in the secretory pathway of endoplasmic reticulum and Golgi apparatus. The differences between glycosylation patterns of plants and animals are problematic for humans when medical animal proteins extracted from plants are used (Krupp et al. 2003); consequently, plant engineering is required for a correct human N-glycosylation. The plant N-glycosylation patterns can be changed by using purified enzymes of β-(1,4) galactosyltransferase and sialyltransferase for making glass transitions in the recombinant proteins derived from plants; co-expression of β-(1,4) galactosyltransferase human enzyme with the target transgene intransgenic plants; prohibiting the activity of fucosyltransferase and xylyltransferase enzymes; and targeting pharmaceutical proteins to the endoplasmic reticulum in order to avoid the addition of protective N-glycans (Alireza and Nader 2015). Recently, genome editing based on site-specific nucleases has been used to knockout two α (1,3)-fucosyltransferases and two β(1,2)-xylosyltransferases in *N. benthamiana* to engineer plants that are devoid of plant-specific N-glycosylation patterns (Li et al. 2016).

14.4 Regulatory Challenges and Environmental Risks

Although resolving the technical challenges in plant molecular farming would help in the production of recombinant proteins of a high commercial value through a cost-effective process, the biggest barrier to the commercialization of this technology is the intense scrutiny of these new plant varieties from regulatory agencies and the public as well as the costs associated with regulatory approval. The first plant-derived recombinant pharmaceutical protein approved for human use was produced in plant cells (Aviezer et al. 2009). However, the initial lack of a coherent regulatory framework for pharmaceuticals derived from whole plants (FDA/USDA: Draft Guidance 2002; EMEA (CPMP), 2002; Spök et al. 2008) was later followed by increased efforts to develop a new regulatory pathway for whole plants resulting in updated guidelines [EMEA (CPMP) 2008]. The latest regulatory innovations that have been incorporated in plant-based manufacturing processes include quality by design through the use of statistical experimental models to optimize each production process (Buyel and Fischer 2014c), process analytical technologies to improve product quality and consistency (Holland et al. 2013), automated production platforms to increase process robustness and scalability (Wirz et al. 2012) and single-use technologies to reduce the risk of contamination and product carry-over, including single-use bioreactors (Raven et al. 2015), novel filter materials (Buyel and Fischer 2014d) and chromatography media (Buyel and Fischer 2014e).

While using edible plants for production of biopharmaceuticals, they must be produced, processed and regulated as pharmaceutical products and should therefore be kept separate from food and animal feed supplies (Tacket 2009). Physical separation (growing the pharmaceutical crops in containment or in isolated fields) or biological containment (using species that do not outcross or relying on chloroplast transformation which prevents gene flow because chloroplasts are not found in the pollen of species used for molecular farming) could prevent the pollination of food/feed crops by pharmaceutical crops of the same species. Edible pharmaceutical plants could also be labelled with distinctive features that preserve their identity, facilitate traceability and avoid the contamination of the food supply. As a proof of this concept, transgenic tomato plants expressing neutralizing IgA antibodies against rotavirus were crossed with another transgenic line expressing the *Antirrhinum majus* Roseal and Delila transcription factors in the fruit, thus activating anthocyanin biosynthesis and generating purple fruits (Butelli et al. 2008). The resulting hybrids expressed the recombinant antibodies solely in the purple transgenic fruits making them easy to identify (Juárez et al. 2012). Thus, by making the genetically engineered plants easily traceable, one could account for the risk of admixture.

14.5 Conclusion

Plants have practical and economical advantages over traditional expression systems, making them suitable for the large-scale production of recombinant proteins. Key challenges in obtaining products from transgenic plants and the strategies to overcome are given in Table 14.2. In the efforts to address the key challenges in obtaining products from transgenic plants, the expression platforms have come far away from plant cell cultures to chloroplast transformation and subcellular targeting. Thus, the progress in the development of plant-derived pharmaceuticals is encouraging, and the transgenic plants expressing biopharmaceutical proteins could therefore achieve a significant step toward the goal of developing safe, cost-effective and efficacious biopharmaceuticals. The key factors that are likely to increase the impact of metabolic engineering and molecular farming in the pharmaceutical industry are the advent of precise genome modifications and targeted interventions using the latest genome editing tools based on the CRISPR/Cas9 system, the optimized production processes that can be scaled up to process scale without loss of yield or quality and the implementation of the latest regulatory guidelines to ensure that processes are robust, reproducible, safe and efficient to encourage regulatory approval.

Table 14.2 Key challenges in obtaining products from transgenic plants and the strategies to address them (Compiled by referring to Spök et al. 2008; Alireza and Nader 2015; Pillay et al. 2014; Jouzani and Tohidfar 2013)

Challenges in the production of recombinant proteins in plants	Strategies to address the challenges in production of recombinant proteins
1. Selection of appropriate host	
(a) Economic factors	Selection of host with high total biomass yield, storage characteristics, ease of transport, high value of recombinant proteins, low maintenance costs, less required area, low duration of production cycle, low cost of subsequent products and edibility
(b) Number of transformation events	Selection of host with a high transformation and regeneration efficiency
(c) Despite being a highly economic and transformation friendly host, tobacco cannot be used for recombinant protein production because it contains high amount of toxic combinations like nicotine and other alkaloids that cannot be removed during the purification process	Using alternative forage crops like alfalfa and lettuce for molecular farming
d. Forage plants generally suffer from the problem of instability of expressed proteins, by which drying and freezing of the leaves and immediate processing following the harvest have been inevitable	Expression of proteins in seeds (e.g. rice, corn and legumes) which neither affects the growth of plants nor needs the freezing of leaves or immediate processing after harvest and allows the long-term storage of proteins at a limited temperature without decreasing the level of activity
2. Product shortage	
(a) Low expression of transcripts	Using efficient promoters, (tissue-specific promoters for higher accumulation to particular organs like seeds or fruits), transcription factors and RNA silencing inhibitors for achieving stability of transcripts
(b) Less efficient translation machinery	Using leader sequences, AT rich 3'UTR, codon-optimized gene sequence, position effect can be prohibited by targeting the transgene to plastids
(c) Low stability of protein due to cytosolic gradation and endogenous proteases	Targeting the protein to vacuoles; using protein structure stabilizing agent like cathepsin D inhibitor, rhizo-secretion and proteins that do not need glycosylation for their activity can be targeted to chloroplasts or endoplasmic reticulum
3. Protein quality	

(continued)

Table 14.2 (continued)

Challenges in the production of recombinant proteins in plants	Strategies to address the challenges in production of recombinant proteins
(a) Glycosylation	Using purified enzymes of β-1,4galactosyltransferase and sialyltransferase for making glass transitions in the recombinant proteins derived from plants; co-expression of β-1,4galactosyltransferase human enzyme with the target transgene intransgenic plants; prohibiting the activity of fucosyltransferase and xylyltransferase enzymes; targeting pharmaceutical proteins to the endoplasmic reticulum in order to avoid the addition of protective N-glycans
4. Subcellular targeting of recombinant protein	
(a) Predicting the intracellular localization of the recombinant protein	Employing targeting signals
5. Purification and downstream processing of the recombinant proteins	
(a) Recovery of proteins	Using continuous protein labels (e.g. oleosin fusion)
(b) Downstream processing	Using plant like tomato which has watery texture in their tissues making the extraction procedure easier in comparison with those of dry tissues like grains; targeting of recombinant proteins to the seeds of oilseed as an oleosin fusion, facilitate the extraction of fused proteins from oil bodies and the release of the recombinant proteins from their fusion partner; using edible vaccines which demand no processing costs; targeting proteins to sugarcane vacuoles which occupy a large portion of the stem parenchyma, and their extraction from juice is another strategy to make the extraction procedure easier
6. Biosafety and public concern	
(a) Environmental pollution due to the entrance of transgenes into the food chain	A sound management and supervision required. Targeting the genes which encode r-proteins to the chloroplast genome could provide higher transgene containment owing to the maternal inheritance of chloroplast genome
(b) Possible risks of very high concentration of recombinant proteins on the morphology and physiology of hostplants	Subcellular targeting (e.g. targeting to vacuoles) is recommended where the recombinant protein remains compartmentalized to the vacuole and does not interfere with the normal plant metabolism
(c) Economic risks to farmers and food industry as a result of co-mingling and contamination of MF plants with food/feed chain	Use of non-food/feed crops as source of PMF, production of recombinant proteins by cell suspension cultures in bioreactors, strict physical agronomic confinement and containment strategies for food/feed crops, post-harvest field monitoring and cleaning, use of late maturing or early maturing cultivars or planting at different periods to ensure harvesting at different periods from other crops intended for food/feed and processing

(continued)

Table 14.2 (continued)

Challenges in the production of recombinant proteins in plants	Strategies to address the challenges in production of recombinant proteins
(d) Possible vertical transgene flow and spread by pollen, seed or fruit dispersal	Use of closed isolated physical containment facilities (greenhouses,glasshouses, hydroponics and plant cell suspension cultures), biological containment [self-pollinating species (cleistogamous lines), chloroplast transformation, cytoplasmic male-sterile transgenic plants, sexually incompatible crop with wild relatives, non-germinating seeds or non-sprouting tubers/bulbs, engineered parthenocarpy and apomixis,transgene excision, tissue-specific expression of the transgene and use of inducible promoters]
(e) Horizontal gene transfer by a sexual means	The risk of horizontal gene transfer (especially, antibiotic resistance genes) from plants to microbes is generally believed to be extremely low, as there has been no report of such incidence to date, and in addition, it is important to note that loads of microorganisms, such as symbiotic, pathogenic,endophytic and ectophytic bacteria and fungi with antibiotic resistant genes, are naturally harboured by plants

References

Ahmad K (2014) Molecular farming: Strategies, expression systems and bio-safety considerations. Plant Breed 50:1–10

Alireza T, Nader RE (2015) Molecular farming in plants. In: El-Shemy H (ed) Plants for the future. Intech Open. https://doi.org/10.5772/60757

Apel W, Schulze WX, Bock R (2010) Identification of protein stability determinants in chloroplasts. Plant J 63:636–650

Appunu C, Ram B, Subramonian N (2017) Sugarcane: an efficient platform for molecular farming. In: Mohan (ed) Sugarcane biotechnology: challenges and prospects. Springer Nature, Cham, pp 87–110

Arakawa T, Tsumoto K, Ejima D (2014) Alternative downstream processes for production of antibodies and antibody fragments. Biochim Biophys Acta 1844:2032–2040

Aviezer D, Brill-Almon E, Shaaltiel Y et al (2009) A plant-derived recombinant human glucocerebrosidase enzyme-a preclinical and phase I investigation. PLoS One 4:e4792

Bock R (2007) Plastid biotechnology: prospects for herbicide and insect resistance, metabolic engineering and molecular farming. Curr Opin Biotechnol 18:100–106

Bock R (2014) Engineering chloroplasts for high-level foreign protein expression. In: Maliga P (ed) Chloroplast biotechnology: methods and protocols, methods in molecular biology, vol. 1132. https://doi.org/10.1007/978-1-62703-995-6_5, © Springer Science & Business Media, New York

Bohmert-Tatarev K, McAvoy S, Daughtry S, Peoples OP, Snell KD (2011) High levels of bioplastic are produced in fertile transplastomic tobacco plants engineered with a synthetic operon for the production of polyhydroxybutyrate. Plant Physiol 155:1690–1708

Bundó M, Montesinos L, Izquierdo E, Campo S, Mieulet D, Guiderdoni E et al (2014) Production of cecropin A antimicrobial peptide in rice seed endosperm. BMC Plant Biol 14:102

Butelli E, Titta L, Giorgio M, Mock HP, Matros A, Peterek S et al (2008) Enrichment of tomato fruit with health-promoting anthocyanin by expression of select transcription factors. Nat Biotechnol 26:1301–1308

Buyel JF, Fischer R (2014a) Downstream processing of biopharmaceutical proteins produced in plants: the pros and cons of flocculants. Bioengineered 5:138–142

Buyel JF, Fischer R (2014b) Flocculation increases the efficacy of depth filtration during the downstream processing of recombinant pharmaceutical proteins produced in tobacco. Plant Biotechnol J 12:240–252

Buyel JF, Fischer R (2014c) Characterization of complex systems using the design of experiments approach: transient protein expression in tobacco as a case study. J Vis Exp 83:e51216

Buyel JF, Fischer R (2014d) Scale-down models to optimize a filter train for the downstream purification of recombinant pharmaceutical proteins produced in tobacco leaves. Biotechnol J 9:415–425

Buyel JF, Fischer R (2014e) Generic chromatography-based purification strategies accelerate the development of downstream processes for biopharmaceutical proteins produced in plants. Biotechnol J 9:566–577

Buyel JF, Gruchow HM, Boes A, Fischer R (2014) Rational design of a host cell protein heat precipitation step can simplify the subsequent purification of recombinant proteins fromtobacco. Biochem Engg J 88:162–170

Buyel JF, Opdensteinen P, Fischer R (2015) Cellulose-based filter aids increase the capacity of depth filters during the downstream processing of plant-derived biopharmaceutical proteins. Biotechnol J 10:584–591

Chan HT, Daniell H (2015) Plant-made oral vaccines against human infectious diseases-Are we there yet? Plant Biotechnol J 13:1056–1070

Chan HT, Chia MY, Pang VF, Jeng CR, Do YY, Huang PL (2013) Oral immunogenicity of porcine reproductive and respiratory syndrome virus antigen expressed in transgenic banana. Plant Biotechnol J 11:315–324

Chandra S, Chandra R (2011) Engineering secondary metabolite production in hairy roots. Phytochem Rev 10:371–395

Commandeur U, Twyman R, Fischer R (2003) The biosafety of molecular farming in plants. AgBiotechNet:5

Corrado G, Karali M (2009) Inducible gene expression systems and plant biotechnology. Biotechnol Adv 27:733–743

Dai Z, Hooker B, Anderson D, Thomas S (2000) Expression of *Acidothermus cellulolyticus* endoglucanase E1 in transgenic tobacco: biochemical characteristics and physiological effects. Transgenic Res 9:43–54

Daniell H, Singh ND, Mason H, Streatfield SJ (2009) Plant-made vaccine antigens and biopharmaceuticals. Trends Plant Sci 14:669–679

EMEA Committee for Proprietary Medicinal Products (CPMP) (2002) Points to consider on quality aspects of medicinal products containing active substances produced by stable transgene expression in higher plants. EMA, London (EMEA/CPMP/BWP/764/02)

EMEA Committee for Proprietary Medicinal Products (CPMP) (2008) Guideline on the Quality of Biological Active Substances Produced by Stable Transgene Expression in Higher Plants. London, UK: EMA (EMEA/CHMP/BWP/48316/2006)

FDA/USDA (2002) Draft guidance. Drugs, biologics and medical devices derived from bioengineered plants for use in humans and animals. FDA, Rockville

Fichtner F, UrreaCastellanos R, Ülker B (2014) Precision genetic modifications: a new era in molecular biology and crop improvement. Planta 239:921–939. https://doi.org/10.1007/s00425-014-2029-y

Fischer R, Emans NJ, Twyman RM, Schillberg S (2004) Molecular farming in plants: technology platforms. Encyclopedia Plant Crop Sci:753–756. https://doi.org/10.1081/E-EPCS120024676

Fischer R, Schillberg S, Hellwig S, Twyman RM, Drossard J (2012) GMP issues for recombinant plant-derived pharmaceutical proteins. Biotechnol Adv 30:434–439

Fischer R, Schillberg S, Buyel JF, Twyman RM (2013) Commercial aspects of pharmaceutical protein production in plants. Curr Pharm Des 19:5471–5477

Fischer R, Vasilev N, Twyman RM, Schillberg S (2015) High-value products from plants: the challenges of process optimization. Curr Opin Biotechnol 32:156–162

Folayan MO, Yakubu A, Haire B, Peterson K (2016) Ebola vaccine development plan: ethics, concerns and proposed measures. BMC Med Ethics 17:10. https://doi.org/10.1186/s12910-016-0094-4
Gottschalk U (2007) Large scale manufacturing of monoclonal antibodies: technology update and future trends.. Presentation in the course of conference on Plant-based Vaccines and Antibodies, PBVA 2007, from 18th to 20th June in Verona, Italy
Gottschalk U (2008) Bio separation in antibody manufacturing: the good, the bad and the ugly. Biotechnol Prog 24:496–503
He ZM, Jiang XL, Qi Y, Luo DQ (2008) Assessment of the utility of the tomato fruit-specificE7 promoter for driving vaccine antigen expression. Genetica 133:207–214
Hennig A, Bonfig K, Roitsch T, Warzecha H (2007) Expression of the recombinant bacterial outer surface protein A in tobacco chloroplasts leads to thylakoid localization and loss of photosynthesis. FEBS J 274:5749–5758
Hensel G, Floss DM, Arcalis E, Sack M, Melnik S, Altmann F et al (2015) Transgenic production of an anti HIV antibody in the barley endosperm. PLoS One 10:e0140476
Herz S, Fubl M, Steiger S, Koop H-U (2005) Development of novel types of plastid transformation vectors and evaluation of factors controlling expression. Transgenic Res 14:969–982
Holland T, Blessing D, Hellwig S, Sack M (2013) The in-line measurement of plant cell biomass using radio frequency impedance spectroscopy as a component of process analytical technology. Biotechnol J 8:1231–1240
Hudson LC, Garg R, Bost KL, Piller KJ (2014) Soybean seeds: a practical host for the production of functional subunit vaccines. Biomed Res Int 2014:13
Jaeger G, Scheffer S, Jacobs A et al (2002) Boosting heterologous protein production in transgenic dicotyledonous seeds using *Phaseolus vulgaris* regulatory sequences. Nat Biotechnol 20:1265–1268
Jouzani GS, Tohidfar M (2013) Plant molecular farming: future prospects and biosafety challenges. Biosafety 2:e136
Juárez P, Presa S, Espí J, Pineda B, Antón MT, Moreno V et al (2012) Neutralizing antibodies against rotavirus produced in transgenically labelled purple tomatoes. Plant Biotechnol J 10:341–352
Kim NS, Kim TG, Kim OH, Ko EM, Jang YS, Jung ES et al (2008a) Improvement o f recombinant hGM-CSF production by suppression of cysteine proteinase gene expression using RNA interference in a transgenic rice culture. Plant Mol Biol 68:263–275
Kim TG, Baek MY, Lee EK, Kwon TH, Yang MS (2008b) Expression of human growth hormonein transgenic rice cell suspension culture. Plant Cell Rep 27:885–891
Krupp LB et al (2003) Study and treatment of post Lyme disease (STOP-LD) a randomized double masked clinical trial. Neurology 60:1923–1930
Kusnadi AR, Nikolov ZL, Howard JA (1997) Production of recombinant proteins in transgenic plants: Practical considerations. Biotechnol Bioeng 56:473–484
Lamphear BJ, Jilka JM, Kesl L, Welter M, Howard JA, Streatfield SJ (2004) A corn-based delivery system for animal vaccines: an oral transmissible gastroenteritis virus vaccine boosts lactogenic immunity in swine. Vaccine 22:2420–2424
Li J, Stoddard TJ, Demorest ZL, Lavoie P-O, Luo S, Clasen BM et al (2016) Multiplexed,targeted gene-editing in *Nicotiana benthamiana* for glyco-engineering and monoclonal antibody production. Plant Biotechnol J 14:533–542
Lienard D, Sourrouille C, Gomord V, Faye L (2007) Pharming and transgenic plants. Biotechnol Annu Rev 13:115–147
Ma JKC, Chikwamba R, Sparrow P, Fischer R, Mahoney R, Twyman RM (2005) Plant-derived pharmaceuticals-the road forward. Trends Plant Sci 10:580–585
Maliga P (2004) Plastid transformation in higher plants. Annu Rev Plant Biol 55:289–313
Mandal MK, Ahvari H, Schillberg S, Schiermeyer A (2016) Tackling unwanted proteolysis in plant production hosts used for molecular farming. Front Plant Sci 7:1–6
Mason HS, Warzecha H, Mor T, Arntzen CJ (2002) Edible plant vaccines: applications for prophylactic and therapeutic molecular medicine. Trends Mol Med 8:324–329

Mbewana S, Mortimer E, Pera FF, Hitzeroth II, Rybicki EP (2015) Production of H5N1 influenza virus matrix protein two ectodomain protein bodies in tobacco plants and in insect cells as a candidate universal influenza vaccine. Front Bioengg Biotechnol 3:197

Merlin M, Pezzotti M, Avesani L (2017) Edible plants for oral delivery of biopharmaceuticals. Br J Clin Pharmacol 83:71–81

Muynck B, de Navarre C, Nizet Y, Stadlmann J, Boutry M (2009) Different subcellular localization and glycosylation for a functional antibody expressed in Nicotiana tabacum plants and suspension cells. Transgenic Res 18:467–482

Nahampun HN (2015) Plant molecular farming for the production of industrial enzymes and vaccines. Graduate theses and dissertations. 14962. http://lib.dr.iastate.edu/etd/14962

Nerkar G, Thorat A, Sheelavantmath S, Kassa HB, Devarumath R (2018) Genetic transformation of sugarcane and field performance of transgenic sugarcane. In: Gosal S, Wani S (eds) Biotechnologies of crop improvement, vol 2. Springer, Cham

Oey M, Lohse M, Kreikemeyer B, Bock R (2009) Exhaustion of the chloroplast protein synthesis capacity by massive expression of a highly stable protein antibiotic. Plant J 57:436–445

Ou J et al (2014) Transgenic rice endosperm as a bioreactor for molecular pharming. Plant Cell Rep 33:585–594

Palaniswamy H, Syamaladevi DP, Mohan C, Philip A, Petchiyappan A, Narayanan S (2016) Vacuolar targeting of r-proteins in sugarcane leads to higher levels of purifiable commercially equivalent recombinant proteins in cane juice. Plant Biotechnol J 14:791–807

Paul MJ, Teh AY, Twyman RM, Ma JK (2013) Target product selection-where can molecular pharming make the difference? Curr Pharm Des 19:5478–5485

Pillay P, Schlüter U, van Wyk S, Kunert K, Vorster BJ (2014) Proteolysis of recombinant proteinsin bioengineered plant cells. Bioengineered 5:15–20

Raven N, Rasche S, Kuehn C, Anderlei T, Klöckner W, Schuster F, Henquet M, Bosch D, Büchs J, Fischer R, Schillberg S (2015) Scaled-up manufacturing of recombinant antibodies produced by plant cells in a 200-L orbitally-shaken disposable bioreactor. Biotechnol Bioeng 112:308–321

Rischer H, Häkkinen ST, Ritala A, Seppänen-Laakso T, Miralpeix B, Capell T, Christou P, Oksman-Caldentey KM (2013) Plant cells as pharmaceutical factories. Curr Pharm Des 19:5640–5660

Schiermeyer A, Schinkel H, Apel S, Fischer R, Schillberg S (2005) Production of *Desmodus rotundus* salivary plasminogen activator a1(DSPAa1) into bacco is hampered by proteolysis. Biotechnol Bioeng 89:848–858

Schillberg S, Twyman RM, Fischer R (2005) Opportunities for recombinant antigen and antibody expression in transgenic plants-technology assessment. Vaccine 23:1764–1769

Shaaltiel Y, Bartfeld D, Hashmueli S et al (2007) Production of glucocerebrosidase with terminal mannose glycans for enzyme replacement therapy of Gaucher's disease using a plant cell system. Plant Biotechnol J 5:579–590

Sirko A, Vanek T, Gora-Sochacka A, Redkiewicz P (2011) Recombinant cytokines from plants. Int J Mol Sci 12:3536–3552

Spök A, Karner S (2008) Plant molecular farming- Opportunities and challenges. JRC Scientific and technical reports. In Stein A-J, Rodríguez-Cerezo E (eds) EUR-Scientific and Technical Research Series. https://doi.org/10.2791/30861

Spök A, Twyman RM, Fischer R, Ma JKC, Sparrow PAC (2008) Evolution of a regulatory framework for plant-made pharmaceuticals. Trends Biotechnol 26:506–517

Stoger E et al (2000) Cereal crops as viable production and storage systems for pharmaceutical scFv antibodies. Plant Mol Biol 42:583–590

Streatfield SJ, Lane JR, Brooks CA, Barker DK, Poage ML, Mayor JM et al (2003) Corn as a production system for human and animal vaccines. Vaccine 21:812–815

Sun QY, Ding LW, Lomonossoff GP, Sun YB, Luo M, Li CQ et al (2011) Improved expression and purification of recombinant human serum albumin from transgenic tobacco suspension culture. J Biotechnol 155:164–172

Tacket CO (2009) Plant-based oral vaccines: results of human trials. Curr Top MicrobiolImmunol 332:103–117

Tacket CO, Mason HS, Losonsky G, Clements JD, Levine MM, Arntzen CJ (1998) Immunogenicity in humans of a recombinant bacterial antigen delivered in a transgenic potato. Nat Med 4:607–609

Tinazzi E, Merlin M, Bason C, Beri R, Zampieri R, Lico C, et al. (2015) Plant-derived chimeric virus particles for the diagnosis of primary Sjogren syndrome. Front Plant Sci 6:1080

Tissot G, Canard H, Nadai M, Martone A, Botterman J, Dubald M (2008) Translocation of a protinin, a therapeutic protease inhibitor, into the thylakoid lumen of genetically engineered tobacco chloroplasts. Plant Biotechnol J 6:309–320

Trivedi PK, Nath P (2004) *MaExpl*, an ethylene-induced expansin from ripening banana fruit. Plant Sci 167:1351–1358

Twyman RM, Stoger E, Schillberg S, Christou P, Fischer R (2003) Molecular farming in plants: host systems and expression technology. Trends Biotechnol 21:570–578

Twyman RM, Schillberg S, Fischer R (2013) Optimizing the yield of recombinant pharmaceutical proteins in plants. Curr Pharm Des 19:5486–5494

Wilson SA, Roberts SC (2012) Recent advances towards development and commercialization of plant cell culture processes for the synthesis of biomolecules. Plant Biotechnol J 10:249–268

Wirz H, Sauer-Budge AF, Briggs J, Sharpe A, Sudong S, Sharon A (2012) Automated production of plant-based vaccines and pharmaceuticals. J Lab Automat 17:449–457

Yano M, Hirai T, Kato K, Hiwasa-Tanase K, Fukuda N, Ezura H (2010) Tomato is a suitable material for producing recombinant miraculin protein in genetically stable manner. Plant Sci 178:469–473. https://doi.org/10.1016/j.plantsci.2010.02.016

Yao J, Weng Y, Dickey A, Wang KY (2015) Plants as factories for human pharmaceuticals: applications and challenges. Int J Mol Sci 16:28549–28565

Yoon S, Devaiah SP, Choi SE, Bray J, Love R, Lane J et al (2016) Overexpression of the cucumber expansin gene (Cs-EXPA1) in transgenic maize seed for cellulose deconstruction. Transgenic Res 25:173–186

Enhanced Production of Therapeutic Proteins in Plants: Novel Expression Strategies

15

Gowtham Iyappan, Rebecca Oziohu Omosimua, and Ramalingam Sathishkumar

Abstract

Transient gene expression is a rapid and reproducible approach for the production of therapeutics in large scale. *Agrobacterium tumefaciens* (diarmed tumorigenicity) is employed by introducing recombinant plasmids into plant intracellular spaces in plant cell through syringe or vacuum infiltration that results in ectopic integration into the plant genome. Gene expression usually occurs 3–4 days post-infiltration. Cytoplasmic proteolytic cleavage and posttranscriptional gene silencing (PTGS) are the limiting factors negatively affecting transgene expression. PTGS may be the limiting factor for heterologous protein expression in plant host system. In order to overcome the issues associated with PTGS and other factors that limit protein expression, several strategies are followed. Transient expression studies for high-level expression or modification in viral vectors by incorporating strong promoter or deconstructed virus and viral regulatory elements/co-expression of PTGS suppressor protein are among a few. The advent of modified yield-giving strategies is also regrettably inadequate to meet

G. Iyappan
Plant Molecular Farming Laboratory, DRDO-BU Centre for Life Sciences, Bharathiar University, Coimbatore, India

R. O. Omosimua
Biotechnology Advanced Research Centre, Sheda Science and Technology complex, FCT-Abuja, Nigeria

Plant Genetic Engineering Laboratory, Department of Biotechnology, Bharathiar University, Coimbatore, Tamil Nadu, India

R. Sathishkumar (✉)
Plant molecular farming Laboratory, DRDO-BU Centre for Life Sciences, Bharathiar University, Coimbatore, India

Plant Genetic Engineering Laboratory, Department of Biotechnology, Bharathiar University, Coimbatore, Tamil Nadu, India
e-mail: rsathish@buc.edu.in

R. Sathishkumar et al. (eds.), *Advances in Plant Transgenics: Methods and Applications*, https://doi.org/10.1007/978-981-13-9624-3_15

the expectations; this may be due to cytoplasmic proteolysis of the heterologous protein. Cytoplasmic persistence of transgenic protein may be improved by intracellular compartmentalization. There are many latent shortcomings addressed in this book chapter on strategies to improve overall yield of high-value protein therapeutics in plants.

Keywords
Transient expression · Vacuum infiltration · Posttranscriptional gene silencing · Gene expression · Viral vectors

15.1 Introduction

The expression constructs, their transcript and proteins levels, determine the level of transcription and translation of the foreign genes in different species. An optimized factor for the transcription, translation, posttranslational events, and gene regulatory elements incorporated into the vectors achieves redress expression of the desired protein. Since many strategies are followed and more effort has been taken in getting a good yield, astonishingly many of the experiments often result in very low protein yield. This could be an overwhelming challenge in plant molecular pharming approach. One of the main factors responsible for this is epigenetic silencing of a transgene by the host plant. Epigenetic silencing is a native defense mechanism in the plants for shielding themselves from viral infections. In this scenario, plant will regulate the expression-specific gene(s) by means of either gene silencing by homology-dependent gene silencing (HDGS), transcriptional gene silencing (TGS) or posttranscriptional gene silencing (PTGS). While enhancing the expression of proteins in plants, many reports support the use of gene silencing suppressor proteins to overwhelm the PTGS; one such suppressor is a P19 gene from tomato bushy stunt virus (TBSV). Gene silencing in plants is generally mediated either by small interfering RNA (siRNA) or microRNA (miRNA) pathways (Garabagi et al. 2012). Both structural and chemical miRNA resemble siRNA and mediate the RNA interference, PTGS, and TGS (Jones-Rhoades et al. 2006). The non-translated miRNAs are processed by Dicer-like ribonucleases from stem-loop regions of longer RNA transcripts. On the other hand, siRNAs are generated from long double-stranded RNA precursors as like the same machinery, but many *Arabidopsis* species require RNA-dependent RNA polymerase for siRNA biogenesis. MicroRNAs are highly conserved for its targets, when the target transcript available in the cellular environment favors the silencing most likely occur. In contrast, siRNA targets the gene from which it is originally generated or to the closely related genes (Fig. 15.1). Both miRNA and siRNA recognize the transgene/viral genes through RNA-induced silencing complex (RISC). For the less stable duplex siRNA (less stable 5′end pairing) loaded into the RISC, these two strands of siRNA are called guide and passenger strands. Guide strand in RISC complex identifies the target and mediates cleavage of the target. The viral infection triggers the accumulation of virus-derived

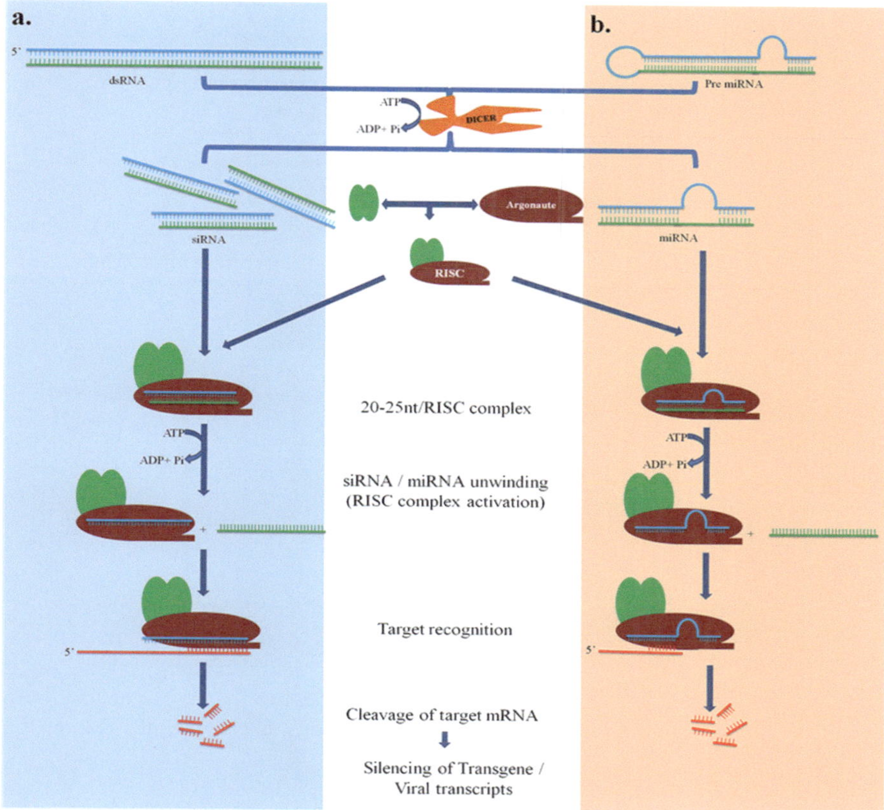

Fig. 15.1 Model of gene silencing at RNA level. Both siRNA and miRNA share much resemblance and are processed by Dicer-like ribonucleases. (**a**) siRNA generated from long dsRNAs and targets the gene from which it is originally derived. The less stable duplex siRNA (less stable 5′end pairing) loaded into the RISC, now these two strands of siRNA are called guide and passenger strands. Guide strand in RISC complex identifies the target and mediates cleavage. (**b**) MicroRNA generated from single RNA molecule with an intermittent stem-loop secondary structure. Conserved nature and imperfect stem-loop prevent the silencing of gene locus that actually encodes the miRNA. Mechanism of miRNA-mediated cleavage is not completely reported yet; hence it is modeled from siRNA mechanism. Duplex miRNA loaded into RISC based on 5′end pairing stability, followed by guide strand-directed cleavage of the target

siRNA levels in plants. P19/suppressor protein present in both cytoplasm and nucleus will act during such situation as a counter defense to reduce the accumulation of siRNA by mediating the non-specific binding and prevent the loading of siRNA into the RISC complex (Fig. 15.2).

A transient expression study involves the delivery of *Agrobacterium* suspension to the leaf intracellular spaces, leading to the accumulation of multiples copies of transgene mRNA transcripts; this will increase the risk of duplex formation/complementary base pairing from which siRNA/miRNA biogenesis will occur. In the

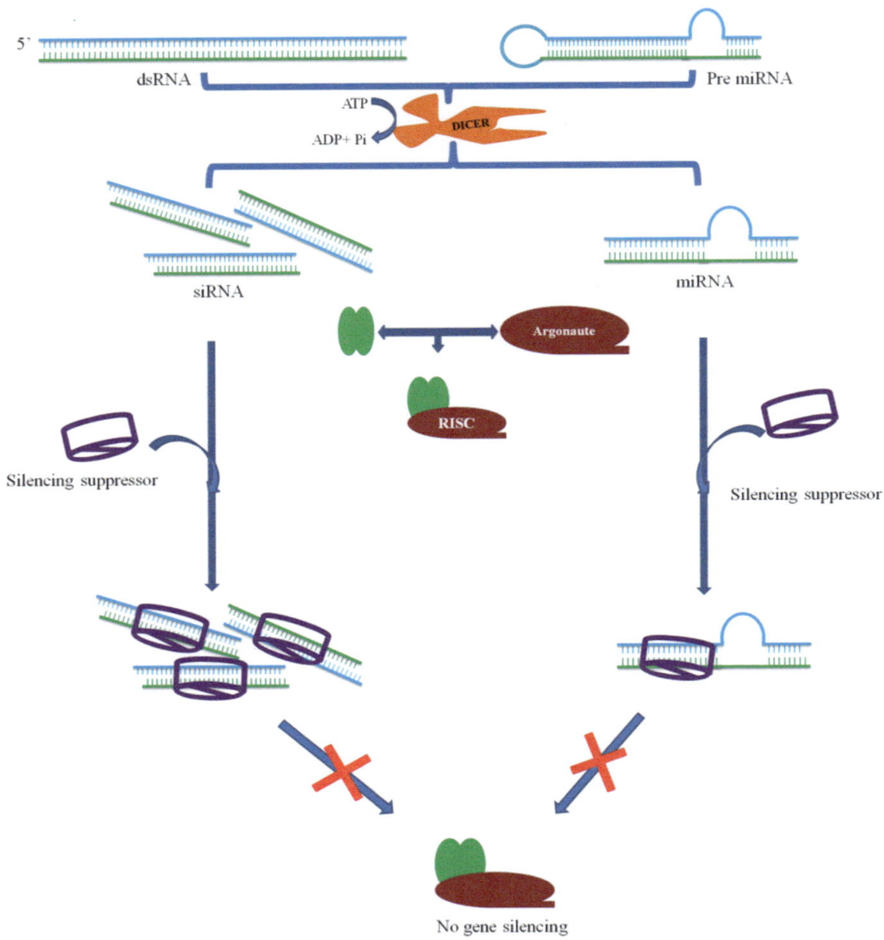

Fig. 15.2 Overview of gene silencing suppressor protein functions in PTGS. A Gene suppressor protein nonspecifically binds with siRNA/miRNA and prevents RISC complex loading. The high titer of suppressor proteins accumulated in the transient expression analysis ensures high probability of binding with siRNA/miRNA and allows target gene expression

presence of high titer gene silencing suppressor proteins, it prevents the guide strand (siRNA/miRNA) from loading into RISC complex, and hence the PTGS may be suppressed; ultimately transgene expression will be more. Many reports suggest that P19 may increase the protein yield 10–30-fold only in transient expression studies, not in stable transgenic plants. However, P19 does not give the same pattern of expression for different transgenes and in transgenic hosts. The literal mode of action of P19 in many aspects is hitherto underexplored. This review focuses on novel approaches to overcome latent shortcomings in the plant system for improving the yield through various strategies incorporating the use of regulatory elements, viral vectors, and cellular compartmentalizations.

15.2 Regulation Accessories for Improved Expression

Irrespective of tissues, constitutive promoters used transcribe the genes effectively; on the other hand, this may be good enough to improve the expression in both monocotyledon and dicotyledon plants. Largely CaMV 35S is still being used as a strong constitutive promoter for protein therapeutic production in dicots; for monocots ubiquitin-1 promoter is used (Ko and Koprowski 2005). CaMV 35S, subterranean clover stunt virus (SCSV), segment 4 (S4) promoter element, and duplication of S4 (S4:S4) efficiency were compared in three different plant expression systems; among these three, CaMV 35S increased the expression effectively than S4, and S4:S4 duplication has given 153 μg/g of reagent per gram seed weight (Schunmann et al. 2002). Since the CaMV 35S is derived from the viral origin, it may show the far-reaching effect on the plant genome after transformation and change the deeds of other genes, even on adjacent chromosomes. Such unwanted events lead to silencing of transgene/epigenetic silencing. Many reports suggest that co-expression/co-infiltration of helper constructs will improve the expression thereby yield. Multigene stacks/increase in a number of transgene under the control of the same promoter may lead to homology-dependent gene silencing (Mette et al. 2000; Peremarti et al. 2010). For plants, we don't have many options for improving expression with various strong promoters like CaMV 35S. This creates a major gap and scope to the promoter research studies. A multigene stacking stratagem involves, each gene under the control of different promoters may be the better solution for homology-dependent gene silencing (Dutt et al. 2016). Four different full-length promoters, cauliflower mosaic virus (D35S), the cassava vein mosaic virus (DCsVMV), the figwort mosaic virus (DFMV), the mirabilis mosaic virus (DMMV), and the peanut chlorotic streak virus (DPCLSV), were derived from *Caulimovirus*; GUS fusion expression pattern has been tested in transgenic citrus. Among all, DMMV promoter served as a strong promoter as like D35S (Dutt et al. 2016). A gene from bacterial origin also simply got modified to shuttle in the plant system by the addition of legumin signal peptide (Bäumlein et al. 1991). Moreover, the expression may be further enhanced by using 5′UTR, regulatory elements and promoters from the plant origin (Bao et al. 2015). Chloroplast has its own 5′UTR for elevating expression levels. Endogenous 5′UTR of photosynthetic proteins serves as good regulatory sequences for achieving more expression (Gimpel and Mayfield 2013; Rosales-Mendoza et al. 2012). Transcription factors are the protein which binds to *cis*-regulatory elements of the gene and thereof take over the entire control of transcription based on timely requirement and rate. The noncoding intervening sequences elevate the accumulation of mRNA transcripts and proteins by unknown mechanisms while using intron-containing transgenes as compared to that of intronless gene constructs. The ability of introns to stimulate transcription and exon junction complex-enhanced translation are observed in the study where intron-containing GUS gene expression in *Arabidopsis* is found to be higher (Rose 2004). Enormous modifications and developments in the expression constructs and various strategies (listed in Table 15.1) have stricken the importance and scope of understanding the gene regulatory processes and the noncoding regions of the genome.

Table 15.1 Expression strategies for enhancing protein yield in plant systems

S.No	Modification in the construct	Strategy and host	Protein	Expression levels	Reference
1.	Oleosin promoter used for targeted expression	Seed-specific expression for post-harvest storage	Oleosin-hirudin fusion protein	1% of total seed protein	Parmenter et al. (1995)
2.	CaMV 35S with double enhancer and C terminal KDEL motif	Enhancer duplication and ER targeting	Single-chain antibody scFv	0.01–1% TSP	Schouten et al. (1996)
3.	CaMV 35S promoter and seed-specific legumin B4 LeB4 and the unknown seed protein USP promoters from *Vicia faba* for organ-specific storage.	Targeted expression to ER and seed	scFv	4–6.8% TSP in ER 3–4% TSP in seed	Fiedler et al. (1997)
4.	Legumin B4 signal peptide – scFv and KDEL motif	Targeted expression to ER and tuber, potato	scFv	2% TSP	Fiedler et al. (1997)
5.	TMV 30B vector	In vitro transcription and viral vector used, *N. benthamiana*	mAb CO17-1A	Not reported	Verch et al. (1998)
6.	Endogenous *Prrn* promoter for the expression	Chloroplast transformation, *N. benthamiana*	Human somatotropin	>7% TSP	Staub et al. 2000
7.	Signal sequence of the sweet potato sporamin storage protein with or without KDEL	Targeted expression, tobacco	HBsAg-specific scFv	Apoplastic fluid – 0.031% of TSP Vacuole – 0.032% of TSP ER – 0.22% of TSP	Ramirez et al. (2002)
8.	Different promoters	Different promoters were used Barley, potato, tobacco	HIV diagnostic reagent	35S promoter – 0.77% TSP S4 promoter – 0.31% TSP S4:S4 promoter – 0.63% TSP	Schunmann et al. (2002)

9.	Signal peptide of the 2S2 storage protein of *Arabidopsis thaliana* and KDEL addition	Site-specific expression and ER and apoplast targeting for improving post-harvest storage, potato	Light, heavy, and derived Fd chains of the MAK33 antibody	0.5% of TSP in tuber	Wilde et al. (2002)
10.	Used sweet potato sporamin signal peptide and KDEL	Transgenic approach, ER retention signal	Anti-hepatitis B surface antigen	5% TSP in F1	Marlene et al. (2003)
11.	P19 of TBSV	P19 co-expression	GFP	Not reported	Voinnet (2003)
12.	rRNA operon prrn constitutive promoter. Individual ribosome binding sites for each gene. Transgene introduced immediate downstream of 5′UTR of PsbA gene	Chloroplast transformation, *Nicotiana tabacum*	CTB-2L21	31.1% TSP	Molina et al. (2004)
13.	Pathogenesis-related protein signal peptide added to 5′ terminus, ER retention signal at 3′ of gene.	Site-specific expression and tomato plant	HBV large surface antigen gene PRS-S1S2S	0.02% TSP	Lou et al. (2007)
14.	AlcA:AlcR switch and GAL4/ VP16-UAS	AlcA:AlcR switch and GAL4/ VP16-UAS enhancer trap transgenic lines of *Arabidopsis*	GUS/GFP	Not reported	Jia et al. (2007)
15.	CaMV 35S with duplicated enhancer, chalcone synthase 5′UTRRb7 scaffold attachment region SAR, chloroplast transit peptide of potato *rbcS1* gene	Transient expression chloroplast localization Stable expression chloroplast localization	HPV-16 L1 capsid protein	>17% TSP in transient 11% TSP in transgenic	Maclean et al. (2007)
16.	RNA virus-derived replicons	Viral vectors	dsReD, HBcAg, GFP, and anti-HIV IgG	~10–20% of TSP	Sainsbury and Lomonossoff (2008)

(continued)

Table 15.1 (continued)

S.No	Modification in the construct	Strategy and host	Protein	Expression levels	Reference
17.	P25 of potato virus *X*	Co-expression of viral silencing suppressor proteins.	HIV-1 Nef	0.71% TSP	Lombardi et al. (2009)
	P19 of artichoke mottled crinkle virus			1.33% TSP	
	P19 tomato bushy stunt virus			0.93% TSP	
18.	Codon-optimized genes	Nuclear and chloroplast transformation in *Chlamydomonas reinhardtii*	VEGF HMGB1	3% 2.5%	Rasala et al. (2011)
19.	AlcA:AlcR switch	Viral replicon – transgenic plant under strong ethanol inducible switch, *Nicotiana benthamiana* plants	GFP	4.3 mg/g fresh biomass	Werner et al. (2011)
20.	Chloroplast psbA promoter and codon optimization	Chloroplast transformation in *Chlamydomonas reinhardtii*	αCD22 αCD22PE40 αCD22HCH23PE40	~0.3–0.4% ~0.2–0.3%	Tran et al. (2013)
21.	AlcA:AlcR switch	In plant activation of viral replication system, tobaccum	GUS	50% TSP	Dugdale et al. (2013)
22.	Mutation in HC-Pro	Co-expression of viral silencing suppressor proteins	GFP/GUS	Not reported	Haikonen et al. (2013)
23.	Cloning of P19 and transgene in the single construct	Comparison of different vectors with P19-containing vector	GFP/GUS	Not reported	Shah and Almaghrabi (2013)
24.	Chloroplast psbA promoter	Growth supplements (selenocystamine, cystamine, diselenide, and disulfide molecules) to grow *Chlamydomonas reinhardtii*	GFP/luciferase (Gluc)	Not significant increase in GFP 0.5–5% in Gluc	Ferreira-Camargo et al. (2015)

25.	Expression cassette cloned in vector containing *psbH-trnE2* intergenic region	Chloroplast transformation in *Chlamydomonas reinhardtii*	CYP79A1 (model cytochrome P450)	0.4% of total chloroplast protein	Gangl et al. (2015)
26.	AlcA:AlcR switch	Chloroplast transformation in tobacco	Dengue envelop protein domain III (EDIII)	0.8 and 1.6%	Gottschamel et al. (2016)
27.	Expression cassette cloned in vector containing *psbH-trnE2* intergenic region	Chloroplast transformation	Human growth hormone	0.5 mg/L of *C. reinhardtii* culture	Wannathong et al. (2016)
28.	MagnICON transient expression system	Deconstruction of viral vectors *N. benthamiana*	Human glutamic acid decarboxylase (GAD65)	Maximum 226.9 ± 42.1 μg/fresh leaf weight	Merlin et al. (2016)
29.	High-[CO2]-inducible promoter *Os.hb2* promoter	Inducible expression Rice	Phytotoxic antimicrobial peptideBP100	Up to 1% accumulation	Ruiz et al. (2016)
30.	Tobacco mosaic virus (TMV) RNA-based overexpression vector (TRBO)	Targeted expression to ER and apoplast *N. benthamiana*	Butyrylcholinesterase	~20 μg/g fresh leaf weight	Alkanaimsh et al. (2016)
31.	Vacuolar signal peptides KISIA and NIFRGF and ER targeting	Vacuolar and ER targeting *N. benthamiana*	Monoclonal antibody 14D9	Accumulation of ER- and vac-Abs was 10- to 15-fold higher than sec-Ab	Ocampo et al. (2016)
32.	N terminal domain of the maize prolamin 27 kD γ-zein fusion with gene of interest	Transgenic plants targeting ER *N. benthamiana*	Human bone morphogenetic protein-2 (hBMP2)	1% TSP	Ceresoli et al. (2016)
33.	Signal peptide and transmembrane domain in N and C terminal domain of the gene, respectively, and the insert cloned in pCambia1300-based vector	Co-infiltration of P19, P0, and P1 RNA silencing suppressors *N. benthamiana*	Promastigote surface antigen (PSA) from the parasitic protozoan *Leishmania infantum*	–	Lacombe et al. (2018)

15.3 Viral Vector System

Viral vector system is an efficacious system for therapeutic protein production in plants except the fact that vector stability issues are of major concern. The conventional approach for the enhanced expression using viral vectors is by opting for suitable promoter-regulated expression of target gene. Alternately plant viral vector system has been developed and used for achieving high expression levels. Tobacco mosaic virus (TMV) and potato virus X (PVX) are the widely studied system for such purposes. The highest level of 2% total soluble protein (TSP) was achieved by expressing the α-trichosanthin gene under the transcriptional control of *Tobamovirus* subgenomic promoter in the plant viral RNA vector (Kumagai et al. 1993). Hence, plant viral vectors serve as a good carrier for the expression of multi-subunit proteins like antibodies. In addition, it is more rapid as compared to the non-viral vectors, where regeneration after transformation, hardening, and sexual crossing of plants to achieve multiple-subunit assemblies is a highly time-consuming process (Verch et al. 1998). Viral vectors confer systemic infection, which is an added beneficial feature. However, there is a high probability of transgene loss during systemic infection. Insertion of transgene into small viral genomes augments genetic overload. This may negatively affect the systemic invasion in plant vascular system, and meanwhile, reduction of transgene also occurs. This unstable nature of viral vectors has been improved for achieving the high-level expression in the plant molecular farming stream. Alterations can be made in the viral genome for compensating the genetic load, such as mutations in viral movement protein for improving systemic invasion. DNA shuffling of movement protein has also improved its function; this may be helpful for achieving desirable features in the viral vectors and genetic stability and good expression of the transgene (Toth et al. 2002). One such approach involves engineering of viral coat proteins to function as a carrier molecule to express the transgene. Coat protein of alfalfa mosaic virus was used as a carrier protein to express rabies and HIV antigenic peptides; this has demonstrated the stable nature of recombinant virus during systemic movement in the plant tissues (Yusibov et al. 1997). Viral assembly failure is most often observed if larger sequences are incorporated with the coat protein (Dietzschold et al. 1990; Yusibov et al. 1997). The stable transgenic approach has been employed to overcome this issue by integrating the viral vector cassette into the genome of plants in every cell. Even though the level of mRNA transcripts in the transgenic plant and the virus particle inoculated plant were found to be similar and the final yield also low (Mori et al. 2001). This may be due to PTGS of plants against viral particles. Bean yellow dwarf virus (BeYDV) *cis*-acting elements were flanked with the transgene and BeYDV Rep protein in a separate vector with strong inducible promoter incorporated into the NT1 cell suspension, tomato and potato plants. BeYDV Rep system revealed significant yields in NT1 cells and transgenic tomato plants. Even though stringent strategies were followed while designing the system, the expression was ceased probably by PTGS after 7/12 days of induction. HC-Pro viral suppressor incorporation to the BeYDV Rep system also failed to overcome PTGS (Zhang and Mason 2006).

Initially, the viral vectors were constructed as a ditto of wild-type viruses carrying transgene with slight modifications; as unfortunate, they all behaved like a stable infectious virus. An alternative approach called "deconstructed viral vector" strategy combines the advantages of viral and non-viral vectors (Yuri Gleba et al. 2004). An attempt employing the *Agrobacterium*-mediated delivery of two subgenomic components has been successfully demonstrated in *Nicotiana benthamiana* and cowpea for its compatibility of CPMV-based vector derivatives in molecular farming (Yuri Gleba et al. 2004; Liu and Lomonossoff 2002). In vivo/in planta assembly of functional viral vectors using separate pro-vector module is another milestone in the deconstructed viral vector approach (Marillonnet et al. 2004). Magnifection the recently developed transfection technology eliminates the shortcomings in viral vector usage and adapts *Agrobacterium* as an infective systemic mediator that transform viral particles in the plants. This involves the construction T-DNA constructs with viral replicons and its agroinfiltration which leads to in vivo assembly of viral particles that resulted in the high expression level (Gleba et al. 2005). The magnifection technology incorporates the benefits of conventional systems, i.e., use of *Agrobacterium* for T-DNA transfer, replication and expression machinery of RNA viral vector, posttranslational modifications, and economic production strategies in plants. In the interim, posttranslational modification may not necessarily be important in many of the vaccine candidates and antibodies and limits the use of plantibodies for passive immunization applications; production of antibodies through magnifection system requires extensive vector manipulations; also the high expression levels of recombinant proteins is detrimental to the plants (Gleba et al. 2005). The advent of magnICON (trade name of manifestation technology) is better than their earlier versions and successfully produced many protein therapeutics (Chen and Davis 2016; Hefferon 2012a, b, 2014). To be short, the new system must have the features for transgene replication, containment, PTM, and suppression system for PTGS in a plant host to be developed to crossover the bottlenecks in the use of viral vector systems for plant molecular farming.

15.3.1 Suppressor-Enhanced Expression

Transgene expression in plant system is limited by various host defense mechanisms. Posttranscriptional gene silencing (PTGS) is one such defensive approach of the plants against viral and bacterial invasions. As a counter defense, many plant viruses have evolved proteins that specifically suppress plant adaptive defenses like PTGS. Many gene silencing suppressor proteins reported are entirely different from each other and not even share structural similarity/the mode by which it acts (Roth et al. 2004). Suppressor proteins either reduce siRNA's accumulation by acting on Dicer which process dsRNA and generate siRNA or bind with siRNAs through active RISC complex (Fig. 15.2). P19 is one such suppressor protein that binds to siRNAs and inhibits the function of siRNA (Silhavy et al. 2002). Helper component proteinase (HC-Pro) does the same; in addition it binds to endogenous microRNAs (miRNAs). Fascinatingly, HC-Pro reduces accumulation of siRNA and in other ways enhances miRNA accumulation but blocks miRNA assembly with RISC complex (Mallory et al. 2002; Kasschau et al. 2003; Roth et al. 2004). The highest level of transgene expression been achieved even in non-viral vectors by co-infiltrating with PTGS silencing suppressors (Shah et al. 2013). Co-infiltration of P19 gene into the plant system elevates transgene expression up to 50-folds in *N. benthamiana* plants; also the effect was not saturated in the infiltrated leaves, till senescence and trace presence of P19 (Voinnet et al. 2003). Gene silencing suppressor proteins from various sources (P25 of potato virus X, P19 gene from artichoke mottled crinkle virus and tomato bushy stunt virus) were studied for its support in the transgene expression; the highest level 1.3% TSP was achieved with co-infiltrating AMCV-P19 gene (Lombardi et al. 2009). HC-proof *Potyvirus* is considered as an efficient viral suppressor protein. Alterations introduced to the HC-Pro improves silencing properties (Torres-Barcelo et al. 2008). Mutant HC-Pro improves heterologous protein sustainability in *N. benthamiana* when it is co-infiltrated. Also, mutant PVA HC-Pro co-infiltration doesn't show any stress symptoms in the plant leaves as compared to the wild PVA HC-Pro (Haikonen et al. 2013). Transgenic plants expressing BMV replicase factors (i.e., replication and transcription factors 1a and 2a which is encoded by the BMV RNA1 and BMV RNA2, respectively) under inducible promoter and derivative of RNA3-transgene under a constitutive promoter showed high expression levels when the integrated replicase system was induced in transgenic plants. BMV replicase transcripts increased close to 300-fold (Mori et al. 2001). Cowpea mosaic virus-based transient expression system CPMV RNA-2 and CPMV-HT developed by Sainsbury and Lomonossoff (2008) with further modifications in 5′ and 3′UTRs (i.e., expression cassette cloned in between the 5′ and 3′UTR) improved translation levels. This is relatively time-consuming, and co-infiltration of separate P19 construct may not ensure all the infiltrated cells received necessary expression cassettes. Considering all these limitations, CPMV-HT and P19 expression cassettes were incorporated into same T-DNA region that resulted in high expression construct named pEAQ and its series (Sainsbury et al. 2009). The preponderance of pEAQ-HT compared with TMV RNA-based viral vectors showed similar expression and also suggests P19 incorporation in the expression cassette imperative for achieving high expression (Shah et al. 2013). As P19 has built into the pEAQ-HT series under CaMV 35S, one could achieve high expression levels in transient expressions. Constitutive expression of gene silencing suppressor has a negative impact on plant routines; hence, this has to be introduced under inducible promoters. INPACT (in plant activation) system based on tobacco yellow dwarf mastrevirus (TYDV) collates advantages of stable and transient means of expression. INPACT cassette has TYDV LIR region (Large Intervening Region) next to the LB/RB, the gene of interest/GUS cloned under CaMV 35S promoter. Viral Rep/RepA expression is under control of the AlcA:AlcR gene switch, which is receptive to trace levels of ethanol. In vivo assembly of replisome and high expression results in the ethanol application (Dugdale et al. 2013, 2014). The deconstructed viral vectors undoubtedly contributed the protein production in plant system.

15.4 Persistence of Transgene Protein in Cytoplasm

Transient expression of protein peaks 60–72 h of post-infiltration and then declines thereafter due to PTGS. Intracellular proteolysis also another major constraint leads to protein degradation. This could be overcome by compartmentalizing the protein into different subcellular organelles. Targeted protein expression has shown a decrease in the risk of intracellular proteolytic cleavage and post-harvest storage of functional proteins even at room temperature (Fiedler et al. 1997; Schunmann et al. 2002). Various strategies have been followed by using different sets of constitutive and tissue-specific promoter's with duplications, double enhancers, and endoplasmic reticulum (ER) retention signal/chloroplast/apoplast transit fusion to functional proteins for achieving the highest yield of functional proteins. Modification with ER signal peptide (KDEL) reserves the topmost yield-improving strategy (Fiedler et al. 1997). Chloroplast localization (chloroplast transformation) is another compartment that possesses more advantages in yield, stability, and posttranslational modifications.

15.4.1 ER Compartment

To improve the intracellular stability and yield percentage, KDEL motif is incorporated into the gene sequence, which translocates the protein into ER (Munro and Pelham 1987; Schouten et al. 1996; Schunmann et al. 2002). KDEL is the sequence which distinguishes the newly synthesized secretory proteins and the proteins that should permanently reside at the endoplasmic reticulum (Munro and Pelham 1987); 2S2 seed storage protein of *Arabidopsis* marks the protein to translocate into the ER (Wilde et al. 1996). Upstream sequences of legumin gene *LeB4* from *Vicia faba* and *Arabidopsis* oleosin promoter showed seed-specific heterologous expression in tobacco transgenic plants and oil seed plants, respectively (Bäumlein et al. 1991; Parmenter et al. 1995; Wilde et al. 1996). Plant-based antibody fusion protein targeted to ER was effectively used for the HIV agglutination assay and was the first report of using plant extract directly without purification for the diagnostic purpose (Schunmann et al. 2002). HIVDR protein targeted to potato tuber and barely seeds were analyzed after a couple of months, 43% and 78% of pre-stored HIVDR activity were retained, respectively (Artsaenko et al. 1998; Schunmann et al. 2002). Two different double transgenic plants expressing both heavy and light chain of murine IgG1 (mIgG1) as functional antibody rhizosecreted and another one expressed with transmembrane sequence retained the protein in plasma membrane in the hydroponic system. Surprisingly these plants transported the antigen, and antigen-mIgG1 complex transported to the uppermost leaf of plants where it is sequestered as an immune complex. Hence, the dual application of antibody expressing plants could also be used in phytoremediation (Drake et al. 2002). The recombinant human butyrylcholinesterase intended to accumulate either in ER or apoplastic compartment; the FLAG-tagged recombinant protein resided in ER was recovered with >95% activity with different extraction buffers (Alkanaimsh et al. 2016). Transgene protein accumulated in ER was found to be more stable even at room temperature; proteins confined in the membranous organelle are protected from other toxic substances present in the host plant cells. Hence, therapeutic targeted to the ER can be directly used for the application and reveals purification is not necessary.

15.4.2 Chloroplast Compartment

Chloroplast transformation is an alternative approach for achieving a maximum yield of protein up to 47% of total soluble protein. This may be due to the site-specific integration of the transgene into the chloroplast genome, and the high copy numbers are also important factor for such amazing increase in yield (Almaraz-Delgado et al. 2014; Murphy 2007; Rasala et al. 2011; Molina et al. 2004; Staub et al. 2000). The lack of position effect, protease degradation, and gene silencing in chloroplast makes chloroplast transformation unique in all the aspect and ultimate choice for the production of both prokaryotic and eukaryotic proteins (Daniell et al. 2001). Chloroplast transformation is a potent strategy for achieving the good yield; however, the process is time-consuming similar to nuclear transformation. Unfortunately, this may not be an appropriate technique for producing protein therapeutics in epidemic/pandemic emergencies. Alternatively, incorporation of chloroplast signal peptide to transgene eliminates the cellular proteolytic degradation of proteins (Gleba et al. 2005) and confines the proteins into the chloroplast. The chloroplast is very pertinent for the expression of specialized protein and also it lacks glycosylation machinery. If posttranslational modification is the concern, chloroplast is the appropriate site where correct folding and formation of disulfide bonds occur with high accumulation of peptides (Ruf et al. 2001), meanwhile there are no position effect, no gene silencing, and less ethical issues since it is following maternal inheritance (Ko and Koprowski 2005). Chloroplast targeting by means of adding a chloroplast transit peptide yields a maximum of 11% TSP (Maclean et al. 2007). Chloroplast targeting resulted in high accumulation of soluble and biologically active protein in *Chlamydomonas reinhardtii* as compared to other traditional production platforms. Various human proteins were expressed, and their expression levels were compared in the algal host system; 2–3% of TSP results in chloroplast compartment (Rasala et al. 2011). Compared to other plant host systems, expression is still low. However, this new strategy needs to be studied further for increasing protein yield in the chloroplast.

15.5 Final Remarks and Conclusions

The cumulative effect of different approaches and modifications in the expression constructs, construct designing strategies, use of various tags, and signal peptide fusion with genes shows highly appreciable efforts of researchers. All the above efforts led to increase in final yield and stability of active proteins in subcellular organelles and cytoplasm. Unfortunately, only a few strategies resulted in higher yield for particular genes and cannot be considered as common in every case. This emerged as the major lacuna why the proven yield-improving strategy is not giving the same pattern of expression for other genes. A deep understanding is obligatory in the following aspects of protein expression: (1) the role of intervening sequences and noncoding sequences in the context of expression, (2) posttranscriptional regulation of gene expression, and (3) role played by enhancer, activator, 5′ and 3′UTRs, and other modulators of regulatory machinery.

Acknowledgment The authors thank financial support to Department of Biotechnology, Bharathiar University, under DST- PURSE scheme.

References

Alkanaimsh S, Karuppanan K, Guerrero A, Tu A, Hashimoto B, Hwang MS et al (2016) Transient expression of tetrameric recombinant human butyrylcholinesterase in *Nicotiana benthamiana*. Front Plant Sci 7(June):743. https://doi.org/10.3389/fpls.2016.00743

Almaraz-Delgado AL, Flores-Uribe J, Pérez-España VH, Salgado-Manjarrez E, Badillo-Corona JA (2014) Production of therapeutic proteins in the chloroplast of Chlamydomonas reinhardtii. AMB Express 4(1):57. https://doi.org/10.1186/s13568-014-0057-4

Artsaenko O, Kettig B, Fiedler U, Conrad U, Düring K (1998) Potato tubers as a biofactory for recombinant antibodies. Mol Breed 4(4):313–319. https://doi.org/10.1023/A:1009676832273

Bao SUN, Guo-qing SUN, Zhi-gang M, Rui Z, San-dui GUO (2015) A novel constitutive promoter and its downstream 5 ′ UTR derived from cotton (Gossypium spp.) drive high-level gene expression in stem and leaf tissues. J Integr Agric 3119(15):1–11. https://doi.org/10.1016/S2095-3119(15)61054-1

Bäumlein H, Boerjan W, Nagy I, Panitz R, Inzé D, Wobus U (1991) Upstream sequences regulating legumin gene expression in heterologous transgenic plants. Mol Gen Genet MGG 225(1):121–128. https://doi.org/10.1007/BF00282650

Ceresoli V, Mainieri D, Del Fabbro M, Weinstein R, Pedrazzini E (2016) A Fusion between domains of the human bone morphogenetic protein-2 and maize 27 kD γ-zein accumulates to high levels in the endoplasmic reticulum without forming protein bodies in transgenic tobacco. Front Plant Sci 7(March):358. https://doi.org/10.3389/fpls.2016.00358

Chen Q, Davis KR (2016) The potential of plants as a system for the development and production of human biologics. F1000Research 5, F1000 Faculty Rev-912. https://doi.org/10.12688/f1000research.8010.1

Daniell H, Streatfield SJ, Wycoff K (2001) Medical molecular farming: production of antibodies, biopharmaceuticals and edible vaccines in plants. Trends Plant Sci 6(5):219. https://doi.org/10.1016/S1360-1385(01)01922-7

De Wilde C, De Neve M, De Ryckea R, Bruynsa A (1996) Intact antigen-binding MAK33 antibody and F ab fragment accumulate in intercellular spaces of Arabidopsis thaliana. Plant Sci 114:233–241. https://doi.org/10.1016/0168-9452(96)04331-2

De Wilde C, Peeters K, Jacobs A, Peck I, Depicker A (2002) Expression of antibodies and Fab fragments in transgenic potato plants: a case study for bulk production in crop plants. Mol Breed 9:271–282

Dietzschold B, Gore M, Marchadier D, Niu HS, Bunschoten HM, Otvos L et al (1990) Structural and immunological characterization of a linear virus-neutralizing epitope of the rabies virus glycoprotein and its possible use in a synthetic vaccine. J Virol 64(8):3804–3809. http://www.pubmedcentral.nih.gov/articlerender.fcgi?artid=249675&tool=pmcentrez&rendertype=abstract

Drake PMW, Chargelegue D, Vine ND, Van Dolleweerd CJ, Obregon P, Ma JK (2002) Transgenic plants expressing antibodies: a model for phytoremediation. FASEB J 16:1855–1860

Dugdale B, Mortimer CL, Kato M, James TA, Harding RM, Dale JL (2013) In plant activation: an inducible, hyperexpression platform for recombinant protein production in plants. Plant Cell 25(7):2429–2443. https://doi.org/10.1105/tpc.113.113944

Dugdale B, Mortimer CL, Kato M, James TA, Harding RM, Dale JL (2014) Design and construction of an in-plant activation cassette for transgene expression and recombinant protein production in plants. Nat Protoc 9(5):1010–1027. https://doi.org/10.1038/nprot.2014.068

Dutt M, Erpen L, Ananthakrishnan G, Barthe GA, Brlansky RH, Maiti IB, Grosser JW (2016) Comparative expression analysis of five caulimovirus promoters in citrus. Plant Cell Tissue Org Cult (PCTOC). https://doi.org/10.1007/s11240-016-0993-6

Ferreira-Camargo LS, Tran M, Beld J, Burkart MD, Mayfield SP (2015) Selenocystamine improves protein accumulation in chloroplasts of eukaryotic green algae. AMB Express 5(1):126. https://doi.org/10.1186/s13568-015-0126-3

Fiedler U, Phillips J, Artsaenko O, Conrad U (1997) Optimization of scFv antibody production in transgenic plants. Immunotechnology 3:205–216

Gangl D, Zedler JAZ, Włodarczyk A, Jensen PE, Purton S, Robinson C (2015) Expression and membrane-targeting of an active plant cytochrome P450 in the chloroplast of the green alga Chlamydomonas reinhardtii. Phytochemistry 110:22–28. https://doi.org/10.1016/j.phytochem.2014.12.006

Garabagi F, Gilbert E, Loos A, Mclean MD, Hall JC (2012) Utility of the P19 suppressor of gene-silencing protein for production of therapeutic antibodies in Nicotiana expression hosts. Plant Biotechnol J 10(9):1118–1128. https://doi.org/10.1111/j.1467-7652.2012.00742.x

Gimpel JA, Mayfield SP (2013) Analysis of heterologous regulatory and coding regions in algal chloroplasts. Appl Microbiol Biotechnol 97(10):4499–4510. https://doi.org/10.1007/s00253-012-4580-4

Gleba Y, Marillonnet S, Klimyuk V (2004) Engineering viral expression vectors for plants: the "full virus" and the "deconstructed virus" strategies. Curr Opin Plant Biol 7(2):182–188. https://doi.org/10.1016/j.pbi.2004.01.003

Gleba Y, Klimyuk V, Marillonnet S (2005) Magnifection – a new platform for expressing recombinant vaccines in plants. Vaccine 23:2042–2048. https://doi.org/10.1016/j.vaccine.2005.01.006

Gottschamel J, Lössl A, S, Wang Y, Skaugen M, Bock R, Clarke JL (2016) Production of dengue virus envelope protein domain III-based antigens in tobacco chloroplasts using inducible and constitutive expression systems. Plant Mol Biol:497–512. https://doi.org/10.1007/s11103-016-0484-5

Haikonen T, Rajamäki ML, Valkonen JPT (2013) Improved silencing suppression and enhanced heterologous protein expression are achieved using an engineered viral helper component proteinase. J Virol Methods 193(2):687–692. https://doi.org/10.1016/j.jviromet.2013.07.054

Hefferon KL (2012a) Plant virus expression vectors set the stage as production platforms for biopharmaceutical proteins. Virology 433(1):1–6. https://doi.org/10.1016/j.virol.2012.06.012

Hefferon KL (2012b) Recent advances in virus expression vector strategies for vaccine production in plants. Virol Mycol 01(S1):1–5. https://doi.org/10.4172/scientificreports.174

Hefferon K (2014) Plant virus expression vector development: new perspectives. Biomed Res Int 2014:785382. https://doi.org/10.1155/2014/785382

Jia H, Van Loock B, Liao M, Verbelen JP, Vissenberg K (2007) Combination of the ALCR/alcA ethanol switch and GAL4/VP16-UAS enhancer trap system enables spatial and temporal control of transgene expression in Arabidopsis. Plant Biotechnol J 5(4):477–482. https://doi.org/10.1111/j.1467-7652.2007.00255.x

Jones-Rhoades MW, Bartel DP, Bartel B (2006) MicroRNAS and their regulatory roles in plants. Annu Rev Plant Biol 57:19–53. https://doi.org/10.1146/annurev.arplant.57.032905.105218

Kasschau KD, Xie Z, Allen E, Llave C, Chapman EJ, Krizan KA, Carrington JC (2003) P1/HC-Pro, a viral suppressor of RNA silencing, interferes with Arabidopsis development and miRNA unction. Dev Cell 4(2):205–217

Ko K, Koprowski H (2005) Plant biopharming of monoclonal antibodies. Virus Res 111(1):93–100. https://doi.org/10.1016/j.virusres.2005.03.016

Kumagai, M. H., et al. (1993) "Rapid, High-Level expression of biologically active α-trichosanthin in transfected plants by an RNA viral vector." Proceedings of the national academy of sciences of the United States of America, vol. 90, no. 2, pp. 427–430. JSTOR, http://www.jstor.org/stable/2361067

Lacombe S, Bangratz M, Brizard J-P, Petitdidier E, Pagniez J, Sérémé D, Lemesre J-L, Brugidou C (2018) Optimized transitory ectopic expression of promastigote surface antigen protein in Nicotiana benthamiana, a potential anti-leishmaniasis vaccine candidate. J Biosci Bioeng 125(1):116–123

Liu L, Lomonossoff GP (2002) Agroinfection as a rapid method for propagating Cowpea mosaic virus-based constructs. J Virol Methods 105(2):343–348. https://doi.org/10.1016/S0166-0934(02)00121-0

Lombardi R, Circelli P, Villani ME, Buriani G, Nardi L, Coppola V et al (2009) Mottled crinckle virus, 1–11. https://doi.org/10.1186/1472-6750-9-96

Lou X, Yao Q, Zhang Z, Peng R, Xiong A, Wang H (2007) Expression of the human hepatitis B virus large surface antigen gene in transgenic tomato plants. Clin Vaccine Immunol 14(4):464–469. https://doi.org/10.1128/CVI.00321-06

Maclean J, Koekemoer M, Olivier AJ, Stewart D, Hitzeroth II, Rademacher T et al (2007) Optimization of human papillomavirus type 16 (HPV-16) L1 expression in plants: comparison of the suitability of different HPV-16 L1 gene variants and different cell-compartment localization. J Gen Virol 88(5):1460–1469. https://doi.org/10.1099/vir.0.82718-0

Mallory AC, Reinhart BJ, Bartel D, Vance VB, Bowman LH (2002) A viral suppressor of RNA silencing differentially regulates the accumulation of short interfering RNAs and micro-RNAs in tobacco. Proc Natl Acad Sci USA 99(23):15228–15233. https://doi.org/10.1073/pnas.232434999

Marillonnet S, Giritch A, Gils M, Kandzia R, Klimyuk V, Gleba Y (2004) In planta engineering of viral RNA replicons: efficient assembly by recombination of DNA modules delivered by agrobacterium. Proc Natl Acad Sci U S A 101(18):6852–6857. https://doi.org/10.1073/pnas.0400149101

Marlene P, Ram N, Rodr M, Ayala M, Mart A, Linares M et al (2003) Expression and characterization of an anti- (hepatitis B surface antigen) glycosylated mouse antibody in transgenic tobacco (Nicotiana tabacum) plants and its use in the immunopurification of its target antigen. Biotechnol Appl Biochem 230:223–230

Merlin M, Gecchele E, Arcalis E, Remelli S, Brozzetti A, Pezzotti M, Avesani L (2016) Enhanced GAD65 production in plants using the MagnICON transient expression system: optimization of upstream production and downstream processing. Biotechnol J 11(4):542–553. https://doi.org/10.1002/biot.201500187

Mette MF, Aufsatz W, van der Winden J, Matzke MA, Matzke AJ (2000) Transcriptional silencing and promoter methylation triggered by double-stranded RNA. EMBO J 19(19):5194–5201. https://doi.org/10.1093/emboj/19.19.5194

Molina A, Hervás-Stubbs S, Daniell H, Mingo-castel AM, Veramendi J (2004) High-yield expression of a viral peptide animal vaccine in transgenic tobacco chloroplasts. Plant Biotechnol J 2:141–153. https://doi.org/10.1111/j.1467-7652.2004.00057.x

Mori M, Fujihara N, Mise K, Furusawa I (2001) Inducible high-level mRNA amplification system by viral replicase in transgenic plants. Plant J 27(1):79–86. https://doi.org/10.1046/j.1365-313x.2001.01079.x

Munro S, Pelham HR (1987) A C-terminal signal prevents secretion of luminal ER proteins. Cell 48(5):899–907. https://doi.org/10.1016/0092-8674(87)90086-9

Murphy DJ (2007) Improving containment strategies in biopharming. Plant Biotechnol J 5:555–569. https://doi.org/10.1111/j.1467-7652.2007.00278.x

Ocampo CG, Lareu JF, Marin Viegas VS, Mangano S, Loos A, Steinkellner H, Petruccelli S (2016) Vacuolar targeting of recombinant antibodies in Nicotiana benthamiana. Plant Biotechnol J:1–11. https://doi.org/10.1111/pbi.12580

Parmenter DL, Boothe JG, Van Rooijen GJH, Yeung EC, Moloney MM (1995) Production of biologically active hirudin in plant seeds using oleosin partitioning. Plant Mol Biol 29:1167–1180

Peremarti A, Twyman RM, Gómez-Galera S, Naqvi S, Farré G, Sabalza M et al (2010) Promoter diversity in multigene transformation. Plant Mol Biol 73(4):363–378. https://doi.org/10.1007/s11103-010-9628-1

Ramirez N, Ayala M, Lorenzo D, Palenzuela D, Herrera L, Doreste V et al (2002) Expression of a single-chain Fv antibody fragment specific for the hepatitis B surface antigen in transgenic tobacco plants. Transgenic Res 11:61–64

Rasala BA, Muto M, Lee PA, Jager M, Cardoso RMF, Behnke A et al (2011) Production of therapeutic proteins in algea, analysis of expression of seven human proteins in the chloroplast of Chlamydomonas reinhardtii. Plant Biotechnol J 8(6):719–733. https://doi.org/10.1111/j.1467-7652.2010.00503.x.Production

Rosales-Mendoza S, Paz-Maldonado LMT, Soria-Guerra RE (2012) Chlamydomonas reinhardtii as a viable platform for the production of recombinant proteins: current status and perspectives. Plant Cell Rep 31(3):479–494. https://doi.org/10.1007/s00299-011-1186-8

Rose AB (2004) The effect of intron location on intron-mediated enhancement of gene expression in Arabidopsis. Plant J 40(5):744–751. https://doi.org/10.1111/j.1365-313X.2004.02247.x

Roth BM, Pruss GJ, Vance VB (2004) Review-Plant viral suppressors of RNA silencing. Virus Res 102(1):97–108. https://doi.org/10.1016/j.virusres.2004.01.020

Ruf S, Hermann M, Berger IJ, Carrer H, Bock R (2001) Stable genetic transformation of tomato plastids and expression of a foreign protein in fruit. Nat Biotechnol 19(9):870–875. https://doi.org/10.1038/nbt0901-870

Ruiz C, Pla M, Company N, Riudavets J, Nadal A (2016) High CO2 concentration as an inductor agent to drive production of recombinant phytotoxic antimicrobial peptides in plant biofactories. Plant Mol Biol 90(4–5):329–343. https://doi.org/10.1007/s11103-015-0419-6

Sainsbury F, Lomonossoff GP (2008) Extremely high-level and rapid transient protein production in plants without the use of viral replication. Plant Physiol 148(3):1212–1218. https://doi.org/10.1104/pp.108.126284

Sainsbury F, Thuenemann EC, Lomonossoff GP (2009) PEAQ: Versatile expression vectors for easy and quick transient expression of heterologous proteins in plants. Plant Biotechnol J 7(7):682–693. https://doi.org/10.1111/j.1467-7652.2009.00434.x

Schouten A, Van Engelen FA, De Jong GAMI, Borst-vrenssen AWMT, Zilverentant JF, Bosch D et al (1996) The C-terminal KDEL sequence increases the expression level of a single-chain antibody designed to be targeted to both the cytosol and the secretory pathway in transgenic tobacco. Plant Mol Biol 30:781–793

Schunmann PHD, Coia G, Waterhouse PM (2002) Biopharming the SimpliRED??? HIV diagnostic reagent in barley, potato and tobacco. Mol Breed 9(2):113–121. https://doi.org/10.1023/A:1026752805494

Shah KH, Almaghrabi B, Bohlmann H (2013) Comparison of expression vectors for transient expression of recombinant proteins in plants. Plant Mol Biol Report 31:1529–1538. https://doi.org/10.1007/s11105-013-0614-z

Silhavy D, Molnár A, Lucioli A, Szittya G, Hornyik C, Tavazza M, Burgyán J (2002) A viral protein suppresses RNA silencing and binds silencing-generated, 21- to 25-nucleotide double-stranded RNAs. EMBO J 21(12):3070–3080. https://doi.org/10.1093/emboj/cdf312

Staub JM, Garcia B, Graves J, Hajdukiewicz PT, Hunter P, Nehra N et al (2000) High-yield production of a human therapeutic protein in tobacco chloroplasts. Nat Biotechnol 18(3):333–338. https://doi.org/10.1038/73796

Torres-Barceló C, Martín S, Daròs JA, Elena SF (2008) From hypo- to hypersuppression: effect of amino acid substitutions on the RNA-silencing suppressor activity of the Tobacco etch potyvirus HC-pro. Genetics 180(2):1039–1049. https://doi.org/10.1534/genetics.108.091363

Toth RL, Pogue GP, Chapman S (2002) Improvement of the movement and host range properties of a plant virus vector through DNA shuffling. Plant J 30(5):593–600

Tran M, Van C, Barrera DJ, Pettersson PL, Peinado CD, Bui J, Mayfield SP (2013) Production of unique immunotoxin cancer therapeutics in algal chloroplasts. Proc Natl Acad Sci USA 110(1):E15–E22. https://doi.org/10.1073/pnas.1214638110

Verch T, Yusibov V, Koprowski H (1998) Expression and assembly of a full-length monoclonal antibody in plants using a plant virus vector. J Immunol Methods 220(1–2):69–75. http://www.ncbi.nlm.nih.gov/pubmed/9839927

Voinnet O, Rivas S, Mestre P, Baulcombe D (2003) An enhanced transient expression system in plants based on suppression of gene silencing by the p19 protein of tomato bushy stunt virus. Plant J 33(5):949–956. https://doi.org/10.1046/j.1365-313X.2003.01676.x

Wannathong T, Waterhouse JC, Young REB, Economou CK, Purton S (2016) New tools for chloroplast genetic engineering allow the synthesis of human growth hormone in the green alga Chlamydomonas reinhardtii. Appl Microbiol Biotechnol:1–11. https://doi.org/10.1007/s00253-016-7354-6

Werner S, Breus O, Symonenko Y, Marillonnet S, Gleba Y (2011) High-level recombinant protein expression in transgenic plants by using a double-inducible viral vector. Proc Natl Acad Sci USA 108(34):14061–14066. https://doi.org/10.1073/pnas.1102928108

Yusibov V, Modelska A, Steplewski K, Agadjanyan M, Weiner D, Hooper DC, Koprowski H (1997) Antigens produced in plants by infection with chimeric plant viruses immunize against rabies virus and HIV-1. Proc Natl Acad Sci USA 94(11):5784–5788. https://doi.org/10.1073/pnas.94.11.5784

Zhang X, Mason H (2006) Bean yellow dwarf virus replicons for high-level transgene expression in transgenic plants and cell cultures. Biotechnol Bioeng 93(2):271–279. https://doi.org/10.1002/bit.20695

Transcriptional Engineering for Enhancing Valuable Components in Photosynthetic Microalgae

16

Srinivasan Balamurugan, Da-Wei Li, Xiang Wang, Wei-Dong Yang, Jie-Sheng Liu, and Hong-Ye Li

Abstract

Photosynthetic microalgae can accumulate a wide array of valuable components. However, higher titer of desired product usually occurs under sub-optimal conditions with the compromise of cellular biomass, which seriously hindered their commercial applications. Conventional metabolic engineering has been employed by targeted perturbation of selected genes in the metabolic pathway without compromising cellular growth. Nevertheless, previous studies have shown mixed and inconsistent success owing to the intricate nature of the target metabolic pathways. Transcriptional engineering represents a promising strategy to govern multiple metabolic pathways by regulation of critical transcription factors, thereby controlling the expression of target gene(s). It has exhibited potential significance and advancements in synthetic biology for microalgal strain improvement. In this chapter, we focus on the significance and status of transcriptional engineering strategies and also speculate on future development to enhance production of microalgal valuable components.

Keywords

Transcription factors · Transcriptional engineering · Biofuel · Polysaccharides · Transgenic microalgae

S. Balamurugan · D.-W. Li · X. Wang · W.-D. Yang · J.-S. Liu · H.-Y. Li (✉)
Key Laboratory of Eutrophication and Red Tide Prevention of Guangdong Higher Education Institutes, College of Life Science and Technology, Jinan University, Guangzhou, China
e-mail: thyli@jnu.edu.cn

R. Sathishkumar et al. (eds.), *Advances in Plant Transgenics: Methods and Applications*, https://doi.org/10.1007/978-981-13-9624-3_16

16.1 Introduction

Increasing population, urbanization, industrialization, and individual energy demand together with the depletion of fossil fuel reserves aggravate global energy and environmental and economic stability (Hajjari et al. 2017). Given the detrimental impact of fossil fuels on global climate and economic stability, there exists a pressing need to explore the sustainable, carbon neutral, and alternative energy reserves to potentially replace the fossil fuels (Li et al. 2016). Biofuels are referred to as the fuels derived from the biological feedstocks including vegetable fats, animal oils, and triglycerides derived from other biological feedstocks (Niu et al. 2016). Given the salient characteristics of biofuels over fossil fuels such as carbon neutral, preferable cetane number, cold filter plugging point, fatty acid composition, and feasibility to use as petroleum-based fuel additives, biofuels have emerged as the promising alternatives to conventional fossil fuels (Bergthorson and Thomson 2015; Li et al. 2016).

Based on the source of the biological feedstocks, biofuels are classified into three generations such as first-, second-, and third-generation biofuels (Saladini et al. 2016). Biofuels that are derived from edible crops such as maize, barley, potato, sugarcane, soybean, corn, coconut, sunflower, rapeseed, and palm are considered as first-generation feedstock. However, unsustainability in biofuel production, limited yield, and enormous requirement of water and food versus fuel debate significantly impede their commercial applications. Subsequently, nonedible oil crops such as *Jatropha curcas*, *Miscanthus* sp., and cassava have been employed to generate biofuels, and they are considered as second-generation biofuels. Even though it obviates the food versus fuel conflicts, massive requirement of arable land and water, lower yield, and higher feedstock input (Aro 2016) remain as the major drawbacks. Consequently, these conclusions have provided the impetus to explore the sustainable and promising bioenergy feedstocks to obviate the aforementioned techno-biological drawbacks associated with first- and second-generation biofuel feedstocks. Third-generation biofuels have been considered as the sustainable biological feedstocks that are derived from microalgae, cyanobacteria, seaweeds, etc. Among them, photosynthetic microalgae have emerged as the promising sustainable biological feedstocks when compared to other biological feedstocks because of their inherent salient characteristics such as higher growth rate, lipid content, photosynthetic efficiency, non-encroachment of agricultural land, capability to grow in a wide range of waters, cheaper cultivation media, capability to adapt to a range of harsh and fluctuating environmental conditions, higher lipid yield, noncompetitive with edible crops, and potential to accumulate various valuable by-components (Ma et al. 2018; Niu et al. 2016). Interestingly, oleaginous microalgae could accumulate lipids more than 60% of their cell dry weight when cultivated under sub-optimal conditions (Chen et al. 2017). However, subjecting the cells under the stress conditions significantly hinders their biomass and eventually reduces the overall yield and productivity, thereby hindering their commercial applications. Collectively, these data propelled the researchers to explore the engineering strategies to overproduce lipids without hindering their growth and biomass accumulation. In this chapter, we described the biochemical strategies employed to enhance lipid accumulation and outlined the progress in genetic engineering strategies to improve microalgal lipid content. Particularly, we narrated the importance of transcriptional engineering to obviate the existing hurdles and the future directions for empowering microalgal trait by using transcriptional engineering strategy.

16.2 Current Strategies to Enhance Microalgal Lipids

Previous studies have demonstrated that subjecting the microalgal cells to adverse conditions could significantly increase lipid content, which in turn hinders the routine cellular metabolic activities and induces the cells to synthesize energy-rich molecules. There are various physical and chemical treatments such as light (He et al. 2015), pH (Bartley et al. 2014), chemicals (Ma et al. 2018), temperature (Wang et al. 2016), and nutrient deprivation (Yang et al. 2013) that were employed to increase the microalgal lipid accumulation. Among various stressors, nutrient deprivation has been established as the effective stress condition to hyperaccumulate lipid content during stationary phase (Xue et al. 2017). Among these, nitrogen and phosphorous deprivation have been shown to effectively increase lipid content during late growth phase. Nitrogen is an essential nutrient component for algal growth and metabolism. Besides, it is an indispensable constituent of important macromolecules such as amino acids and proteins and thus governs a crucial role in cellular growth, biomass, and metabolism. Nitrogen deficiency significantly hinders the cellular physiological activities, photosynthesis, and protein synthesis, and consequently metabolic fluxes could be reallocated toward lipogenesis to accumulate energy-rich molecules. Previous studies have shown that nitrogen deprivation remarkably increased the lipid accumulation during stationary phase. Jiang et al. (2014) subjected the free-living *Symbiodinium* cells to nitrogen deprivation and found that nitrogen deprivation remarkably increased the lipid droplet formation in the treated cells (Jiang et al. 2014). Similarly, lipid accumulation was found to be increased by 30% under nitrogen limited conditions in *Scenedesmus* (Xin et al. 2010). Various previous studies have demonstrated the impact of nitrogen deprivation on lipid accumulation (Negi et al. 2016; Rios et al. 2015; Zhu et al. 2014); however, these stress treatments resulted in impaired cellular biomass and consequently reduced overall yield and productivity, which in turn hinders the application of such stress treatments for commercial applications (Chen et al. 2017). Representative biochemical strategies and their impact on microalgal lipid accumulation have been provided in Table 16.1.

Table 16.1 Biochemical strategies that have been reported to enhance microalgal lipid content

Stressors	Microalgae	Observations	References
High light (400 μmol photon $m^{-2} s^{-1}$)	*Chlorella* sp.	Lipid accumulation and photosynthesis was increased and decreased, respectively	He et al. (2015)
Light-emitting diode wave length	*Picochlorum atomus*	Green LED increased the lipid accumulation by 50.3% (w/w)	Ra et al. (2016)
pH	*Nannochloropsissalina*	Growth, lipid accumulation was found to be higher at pH 8; however, adjusting pH during late growth phase did not yield significant lipid enhancement	Bartley et al. (2014)
Dark and low temperature	*Rhodotorula glutinis*	Both biomass and lipids were increased	Zhang et al. (2014a, b)
Combined nitrogen-, phosphorous-, and iron-deficiency stress	*Ankistrodesmus falcatus KJ671624*	Lipid content was increased, and nitrogen was found to be the influential stressor for lipid enhancement	Singh et al. (2015)
Phosphate stress	*Chlorella* sp.	Low phosphate condition increased lipid content and productivity; however, no change was observed in terms of carbohydrate content	Liang et al. (2013)
Nitrogen deprivation	Various algal strains	Nitrogen deficiency enhanced lipid content Among the studied algae, *Scenedesmus obliquus* and *Chlorella zofingiensis* were found to be the promising for lipid production	Breuer et al. (2012)
Nitrogen limitation during late growth phase	*Chlorella vulgaris*	Cells accumulated lipids up to 53% of dry cell weight after 24 h of treatment	Mujtaba et al. (2012)
Nutrient limitation	*Parachlorella kessleri*	Nutrient depletion increased lipid content up to 29% of dry cell weight Chlorophyll and starch content were decreased	Fernandes et al. (2013)

16.3 Genetic Improvement Strategies for Microalgal Lipid Accumulation

Genetic engineering strategy has emerged as a potential tool to increase the titer of desired metabolites without impairing cellular physiological properties by manipulating the specific metabolic nodes in the host cells. Identification of a key metabolic target is considered a crucial factor for successful trait improvement by genetic engineering means. Particularly, lipid metabolism is being controlled by various layers of regulations such as availability of increased carbon metabolic precursors (Wang et al. 2018a, b), reducing equivalents (Xue et al. 2018), overexpression of rate-limiting enzymes involved in lipogenesis, etc., and hence, genetic engineering strategies to increase algal trait resulted in inconsistent success (Bajhaiya et al. 2017). Among these, identification of potential metabolic target is considered a critical parameter that impacts genetic improvement. Previous studies have elucidated the key metabolic nodes and genes and their molecular mechanisms underpinning the lipid accumulation in oleaginous microalgae. Interestingly, Yang et al. (2013) elucidated the molecular mechanisms underlying the nitrogen-deprivation-mediated lipid hyperaccumulation in *P. tricornutum* by investigating the transcriptional analysis of the key metabolic networks under nitrogen deprivation. These data potentially uncovered membrane remodeling and other intricate metabolic nodes and provided valuable insights into the identification and characterization of signature lipogenic genes (Yang et al. 2013).

Consequently, enormous research efforts have been devoted to improve the microalgal lipid accumulation capability by genetically manipulating the key genes associated with lipogenesis. Particularly, genes involved in TAG biosynthetic pathway were studied intensively, and various reports have demonstrated their crucial role in lipid overproduction. For instance, overexpression of glycerol-sn-3-phosphate acyltransferase (GPAT) resulted in significant enhancement of lipids in transgenic *P. tricornutum* than that of wild-type (WT) cells (Niu et al. 2016). Congruently, heterologous expression of *Lobosphaera* GPAT resulted in 50% lipid increment in transgenic *Chlamydomonas reinhardtii* than WT cells (Iskandarov et al. 2016). Overexpression of 1-acyl-sn-glycerol-3-phosphate acyltransferase (AGPAT/LPAAT) resulted in 20% lipid increment in transgenic *C. reinhardtii* (Yamaoka et al. 2016). Congruently, overexpression of plastidial acyl AGPAT/LPAAT significantly increased the lipid content by 1.81-fold in transgenic diatom (Balamurugan et al. 2017). Similarly, overexpression of diacylglycerol acyltransferase, which catalyzes the final and committed TAG synthesis resulted in significant elevation of lipid content in transgenic strains (Ahmad et al. 2015; Li et al. 2016; Wei et al. 2017). Meanwhile, attempts have been made to increase the lipid content by increasing the supply of NADPH, the reducing equivalent required for lipid biosynthesis. Malic enzyme (ME) which catalyzes the decarboxylation of malate to pyruvate and yields NADPH has been targeted for adequate supply of NADPH (Xue et al. 2015). Consequently, various studies have overexpressed ME in microalgae which resulted in significant lipid elevation in engineered strains (Xue et al. 2016). The role of oxidative pentose phosphate pathway (OPPP) in providing lipogenic NADPH has also been demonstrated (Wasylenko et al. 2015). Strikingly, overexpression of glucose-6-phosphate dehydrogenase (G6PD) resulted in remarkable elevation of NADPH and lipid content in microalgae and elucidated the crucial source of lipogenic NAPDH in microalgae (Xue et al. 2017; Xue et al. 2018). On the other hand, attempts have been made to overexpress more than one key gene in order to overproduce lipids. Wang et al. (2018a, b) overexpressed both GPAT1 and AGPAT1 in *P. tricornutum* and found that lipid content was increased in the dual expressed cells than that of individual expression of the respective genes (Wang

Table 16.2 Genetic engineering strategies to improve lipids by enhancing carbon precursors and lipid-associated proteins and redirecting carbon flux

Strategy	Microalgae	Observations	References
Plastidial expression of acetyl-CoA carboxylase	*Phaeodactylum tricornutum*	Transplastomic microalgae was generated Lipid content was increased by 1.77-fold	Li et al. (2018a, b)
Nucellar expression of acetyl-CoA carboxylase (ACCase)	*Cyclotella cryptica, Navicula saprophila*	Increased transcript of ACCase gene was observed Lipid enhancement was not achieved due to the feedback inhibition increased ACCase activity	Dunahay et al. (1995)
Overexpression of malonyl CoA-acyl carrier protein transacylase	*N. oceanica*	Growth and photosynthetic rate were improved Lipid content was increased up to 31%	Chen et al. (2017)
Diacylglycerol acyltransferase overexpression	*P. tricornutum*	Lipid content was increased by 35% PUFA content was remarkably increased	Niu et al. (2013)
RNA-interference-mediated silencing of phosphoenolpyruvate carboxykinase	*P. tricornutum*	Neutral lipids were increased Photosynthetic activity was decreased	Yang et al. (2016)
RNA-interference-mediated UDP-glucose pyrophosphorylase (UGPase) silencing	*P. tricornutum*	UGPase silencing significantly reduced the accumulation of carbohydrate polymer Lipid content was increased remarkably	Zhu et al. (2016)
Antisense knockdown of pyruvate dehydrogenase kinase	*P. tricornutum*	Lipid content was increased up to 82% as determined by fluorometric analysis Cell growth rate was decreased	Ma et al. (2014)
Heterologous expression of yeast DGAT1 and plant oleosin	*P. tricornutum*	Number and volume of lipid droplets were increased Fatty acid profile was unaltered in the transgenic strains	Zulu et al. (2017)
Overexpression of lipid-droplet-associated protein	*P. tricornutum*	Size of lipid droplets and content of TAG were increased	Wang et al. (2017)

(continued)

Table 16.2 (continued)

Strategy	Microalgae	Observations	References
Overexpression of patatin-like phospholipase domain-containing protein 3 (PNPLA3)	*P. tricornutum*	Lipid content was increased by 70%	Wang et al. (2015)
Heterogenous expression of synthetic human PNPLA3	*P. tricornutum*	HsPNPLA3 with site mutation was expressed TAG content was increased by 1.55-fold	Wang et al. (2018a, b)
Overexpression of seipin	*P. tricornutum*	Volume of lipid droplets was increased Neutral lipid content was increased by 57%	Lu et al. (2017)

et al. 2018a, b). Besides, various studies have implied the significance of various crucial lipogenesis associated proteins on lipid overproduction in oleaginous microalgae (Lu et al. 2017; Wang et al. 2017, 2018a, b). Representative genetic engineering strategies and their impact on microalgal lipid accumulation have been provided in Table 16.2. Collectively, these studies demonstrated the complexities exist in algal lipogenesis and the crucial role of recruiting the potential candidate gene for genetic engineering and also exemplified the necessity of targeting multiple metabolic nodes to achieve the successful lipid engineering in host cells. However, expressing multiple metabolic targets is considered cumbersome and warrants the potential target to simultaneously govern multiple metabolic nodes.

16.4 Transcription Factors and Their Role

Even though various studies have reported the impact of genetic engineering strategy on lipid accumulation, traditional genetic engineering approaches have yielded inconsistent success (Bajhaiya et al. 2017). Due to the crucial regulatory role of various key points, lipid overproduction systematically warrants the coordinated expression of multiple metabolic nodes. Transcription factors (TFs) are the functional proteins that play a predominant role in regulating the transcriptional activation of various genes under adverse conditions, thereby controlling growth and developmental processes (Nakashima et al. 2014; Li et al. 2019). Particularly, TFs such as APETALA2/ethylene response factor (AP2/ERF); NAM, ATAF, and CUC (NAC); v-myb myeloblastosis viral oncogene homolog (MYB); basic leucine zipper (bZIP); and WRKY families are known as the important class of transcription factors that respond to the environmental signals and activate the transcriptional initiation of various genes to cope with the particular adverse conditions. In vascular plants, TFs have been extensively studied and their role on improving the phenotypic trait has been reported in a wide range of plants for various trait improvements. For instance, overexpression of a MYB-related TF referred to as *MYB48-1* resulted in increased tolerance to drought stress possibly by regulating the expression of genes involved in abscisic acid biosynthesis and signaling pathways in rice (Xiong et al. 2014). *NAC10* TF was expressed specifically in the roots of rice which conferred drought tolerance and also increased grain yield under field conditions (Jeong et al. 2010). Overexpression of nuclear factor Y (NFY) TF improved the tolerance to drought and saline stress in transgenic rice (Chen et al. 2015). On the other hand, studies have examined the impact of TFs on regulating fatty acid synthetic pathways. Overexpression of glycoalkaloid metabolism 9 (GAME9) regulated the expression of genes involved in cholesterol synthesis in transgenic tomato and tobacco (Cárdenas et al. 2016). Overexpression of sterol regulatory element binding protein 1 (SREBP1), the crucial factor that governs the mammalian lipid homeostasis, increased the *de novo* syntheissi of fatty acid and TAG in goat mammary epithelial cells (Xu et al. 2016). These results have emphasized the potential of TFs on simultaneously governing various key genes of multiple metabolic pathways. However, the reports on the significance of TFs on microalgae are scarce; particularly impact of TE on enhancing algal growth and lipid content and the impact of TE on the target gene expression are yet to be explored.

16.5 Microalgal Transcriptional Engineering

Recently, transcriptional engineering (TE) has garnered significant research focus to simultaneously initiate the transcriptional activation of various key genes in microalgae. TE is referred to the controlled expression of target TFs to govern the expression of multiple metabolic targets. With the advancement of bioinformatic tools and the availability of completely sequenced genome, prediction and identification of TFs have been significantly increased. Recently, Hu and colleagues reported the genome-wide identification of transcription factors and their binding sites that are involved in lipid biosynthesis by computational approaches, thereby sketching the transcriptional network of lipogenesis in oleaginous heterokont *Nannochloropsis* (Hu et al. 2014).

Recently, transcriptional engineering has attracted huge research attention to regulate the transcriptional activation of various key metabolic nodes in microalgae. Various attempts have been made to employ TE to improve the biofuel potential of microalgae. Overexpression of basic helix-loop-helix (bHLH) resulted in increased biomass and fatty acid content in transgenic *N. salina* (Kang et al. 2015). Interestingly, transgenic *C. reinhardtii* expressing Dof-type transcription factor resulted in significant enhancement of fatty acid content without altering the carbon reallocation mechanisms in the cells. Besides, the report showed that overexpression of Dof-type TF increased the expression of genes such as lecithin cholesterol acyltransferase, glycerol-3-phosphatase acyltransferase, and phosphatidylcholine-sterol O-acyltransferase, thereby engineering the fatty acid profile of lipids (Salas-Montantes et al. 2018). Overexpression of Dof TF remarkably increased total lipids by twofold without altering fatty acid composition, and Dof overexpression increased the expression of key genes involved in glycerolipid synthesis (Ibáñez-Salazar et al. 2014). Similarly, heterologous expression of TF GmDof4 from soybean enhanced the lipid content without impairing cellular physiological parameters under mixotrophic conditions and also increased the expression of acetyl-coenzyme A carboxylase in *Chlorella ellipsoidea* (Zhang et al. 2014a, b). Heterologous expression of WRI TF from *Arabidopsis thaliana* resulted in lipid increment and also differentially regulated the expression of various key genes (Kang et al. 2017). Overexpression of bHLH TF resulted in increased biomass and lipid accumulation particularly during early growth phase, and bHLH overexpression significantly increased the expression of various key genes involved in growth and lipid metabolism in heterokont *N. salina* (Kang et al. 2018). A bZIP transcription factor was overexpressed in *N. salina* which increased the expression of key lipogenic genes, and the transgenic cells were shown to exhibit higher growth and lipid content. Besides, the transgenic cells were found to have tolerance to nitrogen limitation and salt stress conditions (Kwon et al. 2018). Previous studies demonstrated the role of transcription factors in regulating the transcription of key genes in microalgae; however, identification and characterization of the target genes regulated by the corresponding TFs have not been corroborated (Kang et al. 2015).

Interestingly, a *NobZIP1* transcription factor was identified and overexpressed, which resulted in concurrent lipid overproduction and secretion in oleaginous heterokont *N. oceanica* (Li et al. 2019). These findings instigated to uncover the novel mechanistic role underlying the intricate lipid overproduction and secretion phenomenon in *N. oceanica*. Chromatin immunoprecipitation-qPCR (ChIP-qPCR) analysis showed that NobZIP1 up- and downregulated key genes involved in lipid and carbohydrate metabolism. The NobZIP1-driven carbohydrate metabolic genes were individually overexpressed and silenced in the cells, which uncovered that NobZIP1-regulated UDP-glucose dehydrogenase (UGDH) was the key gene responsible for alteration of cell wall structure, thereby facilitating the lipid secretion across weakened cells. Besides, this report extends the prior findings by identifying the target genes and elucidating their mechanistic role underlying the intricate phenomenon which provides a novel mechanistic insight into the role of bZIP transcription factor.

16.6 Possible Approaches to Empower Transcriptional Engineering for Microalgal Biotechnological Applications

Various previous studies have demonstrated the potential of TE in regulating the transcription of key lipid metabolic genes in oleaginous microalgae, thereby simultaneously enhancing the lipid content and cellular biomass; however the information on the target genes and the regions regulated by the corresponding TFs have not been characterized in microalgae. Hence, understanding the functional role and identifying the target genes and the binding region of TF in oleaginous microalgae are warranted. In vascular plants, various studies have precisely elucidated the mechanistic role of TFs and their binding sites by using chromatin immunoprecipitation followed by high-throughput sequencing (ChIP-seq) analysis (Mach 2018).

Zhan et al. (2018) employed the RNA sequencing and ChIP-seq analyses and dissected that bZIP transcription factor could directly target a range of key genes in maize endosperm (Zhan et al. 2018). Interestingly, Li et al. (2018a, b) employed the probe affinity purification and mass spectrometry analyses and uncovered that maize bZIP22 TF bound directly to the ACAGCTCA box in the 27-kD γ-zein promoter and initiated its expression (Li et al. 2018a, b). These studies highlighted the necessity of employing several molecular tools such as transcriptome analysis, RNA sequencing, ChIP-seq, and ChIP-qPCR analyses to draw potential information on the intricate mechanistic role of TFs to exploit its potential. However, such analyses on microalgal TFs are lacking so far in microalgae, thereby hindering the extensive application of TE despite their functional consequences. These data in maize have provided the impetus to uncover the precise mechanistic role of microalgal TFs in order to unravel the potential of TE.

16.7 Conclusions and Perspectives

Over the past few years, significant research efforts have been devoted to potentiate microalgae for commercially viable biofuel production; particularly remarkable progress on metabolic engineering has been made. However, complex nature of lipogenic circuit and lack of potential metabolic target to govern the regulation of various key metabolic branches have impaired its commercial exploitation. Amidst these, transcriptional engineering has emerged as the robust and versatile approach to regulate the complex transcriptional regulatory network such as lipogenic and carotenogenic pathway. The controlled expression of transcription factors has been shown to be effective on targeting multiple genes of key metabolic pathways in various plant systems and also in some microalgal species. However, in microalgae the mechanistic role of TFs including their target genes, binding region, and mode of regulation has not been thoroughly elucidated.

Given the functional significance of TFs on simultaneously regulating various genes, it is of paramount importance to exploit systems and molecular biology tools to precisely uncover the putative target genes and functional regions, in order to uncover its mechanistic role that underpins the complex metabolic regulatory networks. Employing transcriptional engineering for improving the cellular content of various by-components originated from multiple key metabolic nodes is feasible and can enhance the overall process economics. Further systematic studies are warranted to exemplify the potential TFs and their mechanistic role, which will provide a biotechnological breakthrough to improve energy and valuable components from microalgae.

References

Ahmad I, Sharma AK, Daniell H, Kumar S (2015) Altered lipid composition and enhanced lipid production in green microalga by introduction of brassica diacylglycerol acyltransferase 2. Plant Biotechnol J 13:540–550

Aro E (2016) From first generation biofuels to advanced solar biofuels. Ambio 45:24–31

Bajhaiya AK, Ziehe Moreira J, Pittman JK (2017) Transcriptional engineering of microalgae: prospects for high-value chemicals. Trends Biotechnol 35:95–99

Balamurugan S, Wang X, Wang H-L, An C-J, Li H, Li D-W, Yang W-D, Liu J-S, Li H-Y (2017) Occurrence of plastidial triacylglycerol synthesis and the potential regulatory role of AGPAT in the model diatom *Phaeodactylum tricornutum*. Bitechnol Biofuels 10:97

Bartley ML, Boeing WJ, Dungan BN, Holguin FO, Schaub TJJ (2014) pH effects on growth and lipid accumulation of the biofuel microalgae *Nannochloropsis salina* and invading organisms. J Appl Phycol 26:1431–1437

Bergthorson JM, Thomson MJ (2015) A review of the combustion and emissions properties of advanced transportation biofuels and their impact on existing and future engines. Renew Sust Energ Rev 42:1393–1417

Breuer G, Lamers PP, Martens DE, Draaisma RB, Wijffels RH (2012) The impact of nitrogen starvation on the dynamics of triacylglycerol accumulation in nine microalgae strains. Bioresour Technol 124:217–226

Cárdenas PD, Sonawane PD, Pollier J, Vanden Bossche R, Dewangan V, Weithorn E, Tal L, Meir S, Rogachev I, Malitsky S, Giri AP, Goossens A, Burdman S, Aharoni A (2016) GAME9 regulates the biosynthesis of steroidal alkaloids and upstream isoprenoids in the plant mevalonate pathway. Nat Commun 7:10654

Chen M, Zhao Y, Zhuo C, Lu S, Guo Z (2015) Overexpression of a NF-YC transcription factor from bermudagrass confers tolerance to drought and salinity in transgenic rice. Plant Biotechnol J 13:482–491

Chen J-W, Liu W-J, Hu D-X, Wang X, Balamurugan S, Alimujiang A, Yang W-D, Liu J-S, Li H-Y (2017) Identification of a malonyl CoA-acyl carrier protein transacylase and its regulatory role in fatty acid biosynthesis in oleaginous microalga *Nannochloropsis oceanica*. Biotechnol Appl Biochem 64:620–626

Dunahay TG, Jarvis EE, Roessler PG (1995) Genetic transformation of the diatoms *Cyclotella cryptica* and *Navicula saprophila*. J Phycol 31:1004–1012

Fernandes B, Teixeira J, Dragone G, Vicente AA, Kawano S, Bišová K, Přibyl P, Zachleder V, Vítová M (2013) Relationship between starch and lipid accumulation induced by nutrient depletion and replenishment in the microalga *Parachlorella kessleri*. Bioresour Technol 144:268–274

Hajjari M, Tabatabaei M, Aghbashlo M, Ghanavati H (2017) A review on the prospects of sustainable biodiesel production: a global scenario with an emphasis on waste-oil biodiesel utilization. Renew Sust Energ Rev 72:445–464

He Q, Yang H, Wu L, Hu C (2015) Effect of light intensity on physiological changes, carbon allocation and neutral lipid accumulation in oleaginous microalgae. Bioresour Technol 191:219–228

Hu J, Wang D, Li J, Jing G, Ning K, Xu J (2014) Genome-wide identification of transcription factors and transcription-factor binding sites in oleaginous microalgae *Nannochloropsis*. Sci Rep 4:5454

Ibáñez-Salazar A, Rosales-Mendoza S, Rocha-Uribe A, Ramírez-Alonso JI, Lara-Hernández I, Hernández-Torres A, Paz-Maldonado LMT, Silva-Ramírez AS, Bañuelos-Hernández B, Martínez-Salgado JL, Soria-Guerra RE (2014) Over-expression of Dof-type transcription factor increases lipid production in *Chlamydomonas reinhardtii*. J Biotechnol 184:27–38

Iskandarov U, Sitnik S, Shtaida N, Didi-Cohen S, Leu S, Khozin-Goldberg I, Cohen Z, Boussiba S (2016) Cloning and characterization of a GPAT-like gene from the microalga *Lobosphaera incisa* (Trebouxiophyceae): overexpression in *Chlamydomonas reinhardtii* enhances TAG production. J Appl Phycol 28:907–919

Jeong JS, Kim YS, Baek KH, Jung H, Ha S-H, Do Choi Y, Kim M, Reuzeau C, Kim J-K (2010) Root-specific expression of *Osnac10* improves drought tolerance and grain yield in rice under field drought conditions. Plant Physiol 153:185–197

Jiang P-L, Pasaribu B, Chen C-S, Witt SN (2014) Nitrogen-deprivation elevates lipid levels in *Symbiodinium* spp. by lipid droplet accumulation: morphological and compositional analyses. PLoS One 9:e87416

Kang NK, Jeon S, Kwon S, Koh HG, Shin S-E, Lee B, Choi G-G, Yang J-W, Jeong B-r, Chang YKJ (2015) Effects of overexpression of a bHLH transcription factor on biomass and lipid production in *Nannochloropsis salina*. Biotechnol Biofuels 8:200

Kang NK, Kim EK, Kim YU, Lee B, Jeong W-J, Jeong B-r, Chang YKJ (2017) Increased lipid production by heterologous expression of AtWRI1 transcription factor in *Nannochloropsis salina*. Biotechnol Biofuels 10:231

Kang NK, Kim EK, Sung MG, Kim YU, Jeong BR, Chang YKJB (2018) Increased biomass and lipid production by continuous cultivation of *Nannochloropsis salina* transformant overexpressing a bHLH transcription factor. Biotechnol Bioeng 116:555–568

Kwon S, Kang NK, Koh HG, Shin S-E, Lee B, Jeong B-r, Chang YK (2018) Enhancement of biomass and lipid productivity by overexpression of a bZIP transcription factor in *Nannochloropsis salina*. Biotechnol Bioeng 115:331–340

Li D-W, Cen S-Y, Liu Y-H, Balamurugan S, Zheng X-Y, Alimujiang A, Yang W-D, Liu J-S, Li H-Y (2016) A type 2 diacylglycerol acyltransferase accelerates the triacylglycerol biosynthesis in heterokont oleaginous microalga *Nannochloropsis oceanica*. J Biotechnol 229:65–71

Li D-W, Xie W-H, Hao T-B, Cai J-X, Zhou T-B, Balamurugan S, Yang W-D, Liu J-S, Li H-Y (2018a) Constitutive and chloroplast targeted expression of acetyl-CoA carboxylase in oleaginous microalgae elevates fatty acid biosynthesis. Mar Biotechnol 20:566–572

Li C, Yue Y, Chen H, Qi W, Song R (2018b) The ZmbZIP22 transcription factor regulates 27-kD γ-Zein gene transcription during maize endosperm development. Plant Cell 30:2402–2424

Li DW, Balamurugan S, Yang YF, Zheng JW, Huang D, Zou LG, Yang WD, Liu JS, Guang Y, Li HY (2019) Transcriptional regulation of microalgae for concurrent lipid overproduction and secretion. Sci Adv 5:eaau3795

Liang K, Zhang Q, Gu M, Cong WJ (2013) Effect of phosphorus on lipid accumulation in freshwater microalga *Chlorella* sp. J Appl Phycol 25:311–318

Lu Y, Wang X, Balamurugan S, Yang W-D, Liu J-S, Dong H-P, Li H-Y (2017) Identification of a putative seipin ortholog involved in lipid accumulation in marine microalga *Phaeodactylum tricornutum*. J Appl Phycol 29:2821–2829

Ma Y-H, Wang X, Niu Y-F, Yang Z-K, Zhang M-H, Wang Z-M, Yang W-D, Liu J-S, Li H-Y (2014) Antisense knockdown of pyruvate dehydrogenase kinase promotes the neutral lipid accumulation in the diatom *Phaeodactylum tricornutum*. Microb Cell Factories 13:100

Ma Y, Balamurugan S, Yuan W, Yang F, Tang C, Hu H, Zhang H, Shu X, Li M, Huang S, Li H, Wu L (2018) Quercetin potentiates the concurrent hyper-accumulation of cellular biomass and lipids in *Chlorella vulgaris*. Bioresour Technol 269:434–442

Mach J (2018) Corn ChIPs and RNA-seq: researchers dip into advanced tools and resources to examine bZIP transcription factor function in the maize endosperm. Plant Cell 30:2641–2642

Mujtaba G, Choi W, Lee C-G, Lee K (2012) Lipid production by *Chlorella vulgaris* after a shift from nutrient-rich to nitrogen starvation conditions. Bioresour Technol 123:279–283

Nakashima K, Yamaguchi-Shinozaki K, Shinozaki K (2014) The transcriptional regulatory network in the drought response and its crosstalk in abiotic stress responses including drought, cold, and heat. Front Plant Sci 5:170

Negi S, Barry AN, Friedland N, Sudasinghe N, Subramanian S, Pieris S, Holguin FO, Dungan B, Schaub T, Sayre RJ (2016) Impact of nitrogen limitation on biomass, photosynthesis, and lipid accumulation in *Chlorella sorokiniana*. J Appl Phycol 28:803–812

Niu Y-F, Zhang M-H, Li D-W, Yang W-D, Liu J-S, Bai W-B, Li H-Y (2013) Improvement of neutral lipid and polyunsaturated fatty acid biosynthesis by overexpressing a type 2 diacylglycerol acyltransferase in marine diatom *Phaeodactylum tricornutum*. Mar Drugs 11:4558

Niu Y-F, Wang X, Hu D-X, Balamurugan S, Li D-W, Yang W-D, Liu J-S, Li H-Y (2016) Molecular characterization of a glycerol-3-phosphate acyltransferase reveals key features essential for triacylglycerol production in *Phaeodactylum tricornutum*. Biotechnol Biofuels 9:60

Ra CH, Kang C-H, Jung J-H, Jeong G-T, Kim S-K (2016) Enhanced biomass production and lipid accumulation of *Picochlorum atomus* using light-emitting diodes (LEDs). Bioresour Technol 218:1279–1283

Rios LF, Klein BC, Luz LF, Maciel Filho R, Wolf Maciel MR (2015) Nitrogen starvation for lipid accumulation in the microalga species *Desmodesmus* sp. Appl Biochem Biotechnol 175:469–476

Saladini F, Patrizi N, Pulselli FM, Marchettini N, Bastianoni S (2016) Guidelines for emergy evaluation of first, second and third generation biofuels. Renew Sust Energ Rev 66:221–227

Salas-Montantes CJ, González-Ortega O, Ochoa-Alfaro AE, Camarena-Rangel R, Paz-Maldonado LMT, Rosales-Mendoza S, Rocha-Uribe A, Soria-Guerra RE (2018) Lipid accumulation during nitrogen and sulfur starvation in *Chlamydomonas reinhardtii* overexpressing a transcription factor. J Appl Phycol 30:1–13

Singh P, Guldhe A, Kumari S, Rawat I, Bux F (2015) Investigation of combined effect of nitrogen, phosphorus and iron on lipid productivity of microalgae *Ankistrodesmus falcatus* KJ671624 using response surface methodology. Biochem Eng J 94:22–29

Wang X, Liu Y-H, Hu D-X, Balamurugan S, Lu Y, Yang W-D, Liu J-S, Li H-Y (2015) Identification of a putative patatin-like phospholipase domain-containing protein 3 (PNPLA3) ortholog involved in lipid metabolism in microalga *Phaeodactylum tricornutum*. Algal Res 12:274–279

Wang Y, He B, Sun Z, Chen Y-F (2016) Chemically enhanced lipid production from microalgae under low sub-optimal temperature. Algal Res 16:20–27

Wang X, Hao T-B, Balamurugan S, Yang W-D, Liu J-S, Dong H-P, Li H-Y (2017) A lipid droplet-associated protein involved in lipid droplet biogenesis and triacylglycerol accumulation in the oleaginous microalga *Phaeodactylum tricornutum*. Algal Res 26:215–224

Wang X, Dong H-P, Wei W, Balamurugan S, Yang W-D, Liu J-S, Li H-Y (2018a) Dual expression of plastidial GPAT1 and LPAT1 regulates triacylglycerol production and the fatty acid profile in *Phaeodactylum tricornutum*. Biotechnol Biofuels 11:318

Wang X, Wei W, Li N-J, Yuan W, Ding Y, Yang W-D, Liu J-S, Balamurugan S, Li H-Y (2018b) Heterogeneous expression of human PNPLA3 triggers algal lipid accumulation and lipid droplet enlargement. Algal Res 31:276–281

Wasylenko TM, Ahn WS, Stephanopoulos G (2015) The oxidative pentose phosphate pathway is the primary source of NADPH for lipid overproduction from glucose in *Yarrowia lipolytica*. Metab Eng 30:27–39

Wei H, Shi Y, Ma X, Pan Y, Hu H, Li Y, Luo M, Gerken H, Liu J (2017) A type-I diacylglycerol acyltransferase modulates triacylglycerol biosynthesis and fatty acid composition in the oleaginous microalga, *Nannochloropsis oceanica*. Biotechnol Biofuels 10:174

Xin L, Hong-ying H, Ke G, Ying-xue S (2010) Effects of different nitrogen and phosphorus concentrations on the growth, nutrient uptake, and lipid accumulation of a freshwater microalga *Scenedesmus* sp. Bioresour Technol 101:5494–5500

Xiong H, Li J, Liu P, Duan J, Zhao Y, Guo X, Li Y, Zhang H, Ali J, Li Z (2014) Overexpression of OsMYB48-1, a novel MYB-related transcription factor, enhances drought and salinity tolerance in rice. PLoS One 9:e92913

Xu HF, Luo J, Zhao WS, Yang YC, Tian HB, Shi HB, Bionaz M (2016) Overexpression of SREBP1 (sterol regulatory element binding protein 1) promotes de novo fatty acid synthesis and triacylglycerol accumulation in goat mammary epithelial cells. J Dairy Sci 99:783–795

Xue J, Niu Y-F, Huang T, Yang W-D, Liu J-S, Li H-Y (2015) Genetic improvement of the microalga *Phaeodactylum tricornutum* for boosting neutral lipid accumulation. Metab Eng 27:1–9

Xue J, Wang L, Zhang L, Balamurugan S, Li D-W, Zeng H, Yang W-D, Liu J-S, Li H-Y (2016) The pivotal role of malic enzyme in enhancing oil accumulation in green microalga *Chlorella pyrenoidosa*. Microb Cell Factories 15:120

Xue J, Balamurugan S, Li D-W, Liu Y-H, Zeng H, Wang L, Yang W-D, Liu J-S, Li H-Y (2017) Glucose-6-phosphate dehydrogenase as a target for highly efficient fatty acid biosynthesis in microalgae by enhancing NADPH supply. Metab Eng 41:212–221

Xue J, Chen T-T, Zheng J-W, Balamurugan S, Cai J-X, Liu Y-H, Yang W-D, Liu J-S, Li H-Y (2018) The role of diatom glucose-6-phosphate dehydrogenase on lipogenic NADPH supply in green microalgae through plastidial oxidative pentose phosphate pathway. Appl Microbiol Biotechnol 102:10803–10815

Yamaoka Y, Achard D, Jang S, Legéret B, Kamisuki S, Ko D, Schulz-Raffelt M, Kim Y, Song W-Y, Nishida I, Li-Beisson Y, Lee Y (2016) Identification of a *Chlamydomonas* plastidial 2-lysophosphatidic acid acyltransferase and its use to engineer microalgae with increased oil content. Plant Biotechnol J 14:2158–2167

Yang Z-K, Niu Y-F, Ma Y-H, Xue J, Zhang M-H, Yang W-D, Liu J-S, Lu S-H, Guan Y, Li H-Y (2013) Molecular and cellular mechanisms of neutral lipid accumulation in diatom following nitrogen deprivation. Biotechnol Biofuels 6:67

Yang J, Pan Y, Bowler C, Zhang L, Hu H (2016) Knockdown of phosphoenolpyruvate carboxykinase increases carbon flux to lipid synthesis in *Phaeodactylum tricornutum*. Algal Res 15:50–58

Zhan J, Li G, Ryu C-H, Ma C, Zhang S, Lloyd A, Hunter BG, Larkins BA, Drews GN, Wang X, Yadegari R (2018) Opaque-2 regulates a complex gene network associated with cell differentiation and storage functions of maize endosperm. Plant Cell 30:2425–2446

Zhang J, Hao Q, Bai L, Xu J, Yin W, Song L, Xu L, Guo X, Fan C, Chen Y, Ruan J, Hao S, Li Y, Wang RR-C, Hu Z (2014a) Overexpression of the soybean transcription factor GmDof4 significantly enhances the lipid content of *Chlorella ellipsoidea*. Biotechnol Biofuels 7:128

Zhang Z, Zhang X, Tan T (2014b) Lipid and carotenoid production by Rhodotorula glutinis under irradiation/high-temperature and dark/low-temperature cultivation. Bioresour Technol 157:149–153

Zhu S, Huang W, Xu J, Wang Z, Xu J, Yuan Z (2014) Metabolic changes of starch and lipid triggered by nitrogen starvation in the microalga *Chlorella zofingiensis*. Bioresour Technol 152:292–298

Zhu B-H, Shi H-P, Yang G-P, Lv N-N, Yang M, Pan K-H (2016) Silencing UDP-glucose pyrophosphorylase gene in *Phaeodactylum tricornutum* affects carbon allocation. New Biotechnol 33:237–244

Zulu NN, Popko J, Zienkiewicz K, Tarazona P, Herrfurth C, Feussner I (2017) Heterologous co-expression of a yeast diacylglycerol acyltransferase (ScDGA1) and a plant oleosin (AtOLEO3) as an efficient tool for enhancing triacylglycerol accumulation in the marine diatom *Phaeodactylum tricornutum*. Biotechnol Biofuels 10:187

MIX
Papier aus verantwortungsvollen Quellen
Paper from responsible sources
FSC® C105338

If you have any concerns about our products,
you can contact us on
ProductSafety@springernature.com

In case Publisher is established outside the EU,
the EU authorized representative is:
Springer Nature Customer Service Center GmbH
Europaplatz 3, 69115 Heidelberg, Germany

Printed by Libri Plureos GmbH
in Hamburg, Germany